Lecture Notes in Mathematics

Edited by A. Dold, Heidelberg and B. Eckmann, Zürich

343

Hermitian K-Theory and Geometric Applications

Proceedings of the Conference held at the Seattle Research Center of the Battelle Memorial Institute, from August 28 to September 8, 1972

Edited by H. Bass, Columbia University, New York, NY/USA

Springer-Verlag
Berlin · Heidelberg · New York 1973

AMS Subject Classifications (1970): 13D15, 14F15, 16A54, 18F25

ISBN 3-540-06436-2 Springer-Verlag Berlin · Heidelberg · New York
ISBN 0-387-06436-2 Springer- Verlag New York · Heidelberg · Berlin

Offsetdruck: Julius Beltz, Hemsbach/Bergstr.

Introduction

A conference on algebraic K-theory was held at the Battelle Seattle Research Center from August 28 to September 8, 1972, with the joint support of the National Science Foundation and the Battelle Memorial Institute. The present volume consists mainly of papers presented at, or stimulated by, that conference, plus some closely related papers by mathematicians who did not attend the conference but who have kindly consented to publish their work here. In addition there are several papers devoted to surveys of subjects treated at the conference, and to the formulation of open research problems. It was our intention thus to present a reasonably comprehensive documentation of the current research in algebraic K-theory, and, if possible, to give this research a greater coherence than it has heretofore enjoyed. It was particularly gratifying to see the latter aim largely achieved already in the course of preparing these Proceedings.

Algebraic K-theory has two quite different historical roots both in geometry. The first is concerned with certain topological obstruction groups, like the Whitehead groups, and the L-groups of surgery theory. Their computation, which is in principle an algebraic problem about group rings, is one of the original missions of algebraic K-theory. It remains a rich source of new problems and ideas, and an excellent proving ground for new techniques.

The second historical source of algebraic K-theory, from which the subject draws its name, is Grothendieck's proof of the Riemann-Roch theorem, and the topological K-theory of Atiyah-Hirzebruch, which has the same point of departure. Starting from the analogy between projective modules and vector bundles one is led to seek a K-theory for rings analogous to that of Atiyah-Hirzebruch for spaces. This enterprise made, at first, only very limited progress. In the few years preceding this conference, however, several interesting definitions of higher K-groups were proposed; the relations between them were far from clear.

Meanwhile the detailed study of K_1 and K_2 had revealed some beautiful arithmetic phenomena within the classical groups. This contact with algebraic number theory had become a major impulse in the subject as well as a theme for conjectures about the significance of the higher K-groups.

More recently there have appeared definitions and potential applications of higher K-theory in the framework of algebraic geometry.

As this brief account suggests, a large number of mathematicians, with quite different motivations and technical backgrounds, had become interested in aspects of algebraic K-theory. It was not altogether apparent whether the assembling of these efforts under one rubric was litte more than an accident of nomenclature. In any case it seemed desireable to gather these mathematicians, some of whom had no other occasion for serious technical contact, in a congenial and relaxed setting, and to leave much of what would ensu.e to mathematical and human chemistry. A consensus of those who were present is that the experiment was enormously successful. Testimony to this is the fact that many of the important new results in these volumes were proved in the few months following the conference, growing out of collaborative efforts and discussions begun there.

One major conclusion of this research is that all of the higher K-theories which give the "classical" K_n's for $n \leq 2$ coincide. Thus, in some sense, the subject of higher algebraic K-theory "exists", an assertion some had begun to depair of making. Moreover one now has, thanks largely to the extraordinary work of Quillen, some very effective tools for calculating higher K-groups in interesting cases.

The papers that follow are somewhat loosely organized under the headings: I. Higher K-theories; II. "Classical" algebraic K-theory, and connections with arithmetic; and III. Hermitian K-theories and geometric applications. Certain papers, as their titles indicate, contain collections of research problems. The reader should be warned, however, that because of the vigorous activity ensuring the conference, some of the research problems posed below are in fact resolved elsewhere in these volumes. The editional effort necessary to eliminate such instances would have cost an excessive delay in publication.

I am extremely grateful to the following participants who contributed

to the preparation of the survey and research problem articles:
S. Bloch, J. Coates, Keith Dennis, S. Gersten, M. Karoubi, M.P. Murthy, Ted Petrie, L. Roberts, J. Shaneson, M. Stein, and R. Swan.

On behalf of the participants I express our thanks to the National Science Foundation and the Battelle Memorial Institute for their generous financial support. For the splendid facilities and setting of the Battelle Seattle Research Center, and for the efficient and considerate services of its staff, the conference participants were uniformly enthusiastic in their praise and gratitude.

Finally, I wish to thank Kate March of Columbia University for her invaluable secretarial and administrative assistance in organizing the conference, and Robert Martin of Columbia University for his aid in editing these Proceedings.

H. Bass

Paris, April, 1973

LIST OF PARTICIPANTS AND AUTHORS

Dr. Neil Paul Aboff
Department of Mathematics
Harvard University
2 Divinity Avenue
Cambridge, MA 02138

Dr. Yilmaz Akyildiz
Department of Mathematics
University of California
at Berkeley
Berkeley, CA 94704

Dr. Roger Alperin
Department of Mathematics
Rice University
6100 Main Street
Houston, TX 77001

Dr. Donald W. Anderson
Department of Mathematics
University of California
at San Diego
San Diego, CA 92037

Dr. David M. Arnold
Department of Mathematics
New Mexico State University
Las Cruces, NM 88001

Dr. Anthony Bak
Departement des Mathématiques
2-4 rue du Lièvre
Genève, Switzerland

Dr. Hyman Bass
Department of Mathematics
Columbia University
Broadway and West 116th Street
New York, NY 10027

Dr. Israel Berstein
Department of Mathematics
White Hall
Cornell University
Ithaca, NY 14850

Dr. Spencer J. Bloch
Department of Mathematics
Fine Hall
Princeton University
Princeton, NJ 08540

Dr. Armand Borel
Institute for Advanced Study
Princeton University
Princeton, NJ 08540

Dr. Kenneth S. Brown
Department of Mathematics
Cornell University
Ithaca, NY 04850

Dr. Sylvain Cappell
Department of Theoretical
Mathematics
The Weizman Institute of Science
Rehovoth, Israel

Mr. Joe Carroll
Department of Mathematics
Harvard University
2 Divinity Avenue
Cambridge, MA 02138

Dr. A. J. Casson
Department of Mathematics
Trinity College
Cambridge, London, England

Dr. Stephen U. Chase
Department of Mathematics
Cornell University
White Hall
Ithaca, NY 14850

Dr. K. G. Choo
Department of Mathematics
University of British Columbia
Vancouver, British Columbia
Canada

Dr. John Henry Coates
Department of Mathematics
Stanford University
Stanford, California 94305

Dr. Edwin H. Connell
Department of Mathematics
University of Miami
Coral Gables, FL 33124

Dr. Francis X. Connolly
Department of Mathematics
University of Notre Dame
Notre Dame, IN 46556

Dr. R. Keith Dennis
Department of Mathematics
Cornell University
White Hall
Ithaca, NY 14850

Dr. Andreas W. M. Dress
Fakultat fur Mathematik
Universitat Bielefeld, FRG
48 Bielefeld
Postfach 8640, Germany

Dr. Richard Elman
Department of Mathematics
Rice University
6100 Main Street
Houston, TX 77001

Dr. E. Graham Evans, Jr.
Department of Mathematics
University of Illinois
Urbana, Illinois 61801

Dr. Howard Garland
Department of Mathematics
State University of New York
at Stony Brook
Stony Brook, NY 11790

Dr. Steve M. Gersten
Department of Mathematics
Rice University
6100 Main Street
Houston TX 77001

Dr. Charles H. Giffen
Department of Mathematics
University of Virginia
Charlottesville, VA 22904

Mr. Jimmie N. Graham
Department of Mathematics
McGill University
P. O. Box 6070
Montreal 101, Quebec Canada

Dr. Bruno Harris
Department of Mathematics
Brown University
Providence, RI 02912

Dr. Allen E. Hatcher
Department of Mathematics
Fine Hall
Princeton University
Princeton, NJ 08540

Dr. Alex Heller
Department of Mathematics
Institute for Advanced Study
Princeton University
Princeton, NJ 08540

Dr. Wu-chung Hsiang
Department of Mathematics
Fine Hall
Princeton University
Princeton, NJ 08540

Dr. James E. Humphreys
Courant Institute
New York University
251 Mercer Street
New York, NY 10012

Dr. Dale Husemoller
Department of Mathematics
Haverford College
Haverford, PA 19041

Dr. J. P. Jouanolou
Université de Strasbourg
Département de Mathématique
7 Rue René Descartes
67-Strasbourg, France

Dr. Max Karoubi
Faculté des Sciences
Département de Mathématiques
Univesité de Paris VII
Quai St.Bernard
Parix 5, France

Dr. Stan Klasa
Department of Mathematics
Carleton University
Ottawa 1, Ontario, Canada

Dr. Mark I. Krusemeyer
Institute for Advanced Study
Princeton, NJ 08540

Dr. Kee Y. Lam
Department of Mathematics
University of British Columbia
Vancouver, British Columbia
Canada

Dr. T. Y. Lam
Department of Mathematics
University of California
at Berkeley
Berkeley, CA 94720

Dr. Ronnie Lee
Department of Mathematics
Yale University
New Haven, CT 06520

Dr. Stephen Lichtenbaum
Department of Mathematics
Cornell University
Ithaca, NY 14850

Dr. Jean Louis Loday
Université de Strasbourg
Département de Mathematique
Rue René Descartes
67-Strasbourg, France

Dr. Erhard Luft
Department of Mathematics
University of British Columbia
Vancouver, British Columbia
Canada

Mr. Robert D. Martin
Room 207, Mathematics Building
Columbia University
Broadway and West 116th Street
New York, NY 10027

Dr. Serge Maumary
Faculte des Sciences
Departement des Mathematiques
Universite de Lausanue
Lausanue, Switzerland

Dr. Dusa McDuff
Department of Pure Mathematics
Cambridge University
16 Mill Lane
Cambridge, England

Dr. M. Pavaman Murthy
Department of Mathematics
University of Chicago
Chicago, IL 60637

Dr. Richard R. Patterson
Department of Mathematics
University of California
at San Diego
La Jolla, CA 92037

Dr. Claudio Pedrini
Instituto di Matematica
Via L. B. Alberti 4
16132-Genova, Italy

Dr. Ted Petrie
Department of Mathematics
Rutgers University
New Brunswick, NJ 08903

Dr. Irwin Pressman
Department of Mathematics
Carleton University
Ottawa 1, Ontario, Canada

Dr. Stewart B. Priddy
Department of Mathematics
Northwestern University
633 Clark Street
Evanston, IL 60201

Dr. Daniel Quillen
Department of Mathematics
Massachusetts Institute of Technology
Cambridge, MA 02139

Dr. Andrew A. Ranicki
Department of Pure Mathematics
Cambridge University
16 Mill Lane
Cambridge, England

Dr. Leslie G. Roberts
Department of Mathematics
Queen's University
Kingston, Ontario, Canada

Dr. Graeme Segal
Department of Mathematics
Massachusetts Institute of Technology
Cambridge, MA 02139

Dr. Julius L. Shaneson
Department of Theoretical Mathematics
The Weizman Institute of Science
Rehovoth, Israel

Dr. Rick W. Sharpe
Department of Mathematics
Columbia University
Broadway and West 116th Street
New York, NY 10027

Dr. Man Keung Siu
Department of Mathematics
University of Miami
Coral Gables, FL 33124

Dr. James D. Stasheff
Department of Mathematics
Temple University
Philadelphia, PA 19122

Dr. Michael R. Stein
Department of Mathematics
Northwestern University
Evanston, IL 60201

Dr. Jan R. Strooker
Mathematische Instituut der Rijksuniversiteit
Budapestlaan, De Vithof
Utrecht, The Netherlands

Dr. Richard G. Swan
Department of Mathematics
University of Chicago
Chicago, IL 60637

Dr. John T. Tate
Department of Mathematics
Harvard University
2 Divinity Avenue
Cambridge, MA 02138

Dr. Lawrence Taylor
Department of Mathematics
University of Chicago,
Chicago, IL 60637

Mr. Neil Vance
Department of Mathematics
University of Virginia
Charlottesville, VA 22904

Dr. Orlando E. Villamayor
Department of Mathematics
Northwestern University
Evanston, IL 60201

Dr. John B. Wagoner
Department of Mathematics
University of California at Berkeley
Berkeley, CA 94720

Dr. Friedhelm Waldhausen
Fakultat fur Mathematik
Universitat Bielefeld
4800 Bielefeld
West Germany

Dr. C. T. C. Wall
University of Liverpool
Department of Pure Mathematics
Liverpool L69 3BX
England

Table of Contents

ALGEBRAIC K-THEORY III

Lecture Notes in Mathematics, Vol. 343: "Hermitian K-theory and geometric applications"

HERMITIAN K - THEORY IN TOPOLOGY

by

JULIUS L. SHANESON

This report will survey some points of connection between Hermitian[1] algebraic K-theory and geometric topology. The earlier and important links between topology and algebraic K-theory (see [M1]) and the more recent links with higher algebraic K-groups (see the papers of Hatcher, Wagoner, and Hsiang-Sharpe in these proceedings) will not be discussed systematically. Indications of and/or references to known (to me) or announced calculations of Wall groups (up to December 1972) will be given, together with occasional and brief indications of important topological applications, where these exist. Of course, the works referred to often give fuller descriptions than is possible here.

[1] Karoubi convinced me that this name is better than "unitary K-theory".

1. Classifying Manifolds and Surgery.

1.1. Let X be a space. A fundamental problem may be stated as follows: Find all smooth, piecewise linear (P.L.), or topological manifolds M of the homotopy type of X, up to diffeomorphism, P.L. homeomorphism, or just homeomorphism, respectively. For example, if $X = S^n$ is the n-sphere, $n \geq 5$, the generalized Poincare conjecture of Smale [Sm 1](also Stallings [St]) implies that a closed P.L. manifold homotopy equivalent to X is P.L. homeomorphic to it. The same holds for a topological "homotopy sphere", using [KS].

On the other hand, Milnor [M2] constructed many non-diffeomorphic smooth manifolds homeomorphic to S^7, even before the generalized Poincare conjecture was settled. Kervaire and Milnor [KM] introduced Hermitian K-theory into the problem of classifying manifolds for the case $X = S^n$, $n \geq 5$, in order to classify smooth "homotopy spheres". Browder and Novikov [B] [N1]extended this to the case of simply-connected manifolds by introducing the notion of surgery on a "normal map", to be discussed below. See [Br] for a complete description of the simply-connected case.

1.2. To simplify the discussion, let us begin by assuming that X is a closed, connected H-manifold, H = O (for smooth), PL (for P.L.), or TOP (for topological [1]), and let us restrict attention to closed H-manifolds of the same homotopy type (and hence same dimension) as X. The problem of whether a given compact

---

[1] In the discussion below, we really have the smooth and PL case more in mind. The results are transferred to the topological case by using work of Lees [L] and Kirby-Siebenmann [KS] [S1].

space has the homotopy type of a manifold can be approached in a spirit similar to what will be described shortly, and was first so approached by Browder [B].

Let $\mathcal{S}_H(X)$ denote the equivalence classes of simple[1] homotopy equivalences $h: M \to X$, M a connected, closed H-manifold; h and h' are said to be <u>equivalent</u> iff there is an H-isomorphism $\phi: M \to M'$ with $h'\phi$ homotopic to h. This definition is due mainly to Sullivan, who reformulated the problem as follows: Study $\mathcal{S}_H(X)$. In general it seems quite difficult to solve the original problem of 1.1 from the type of analysis of $\mathcal{S}_H(X)$ to be given below. For $X = S^n$, the two problems are obviously equivalent, and for some other fairly symmetric or homotopically simple manifolds, a solution of the original problem can still be deduced or obtained directly, e.g. $X = CP^n$ (see [Sv]). $X = RP^n$ (see[L], [W3]), X = lens space (see [BPW], [TW]), X = certain 5-manifolds [Sh2], $X = T^n = S^1 x \ldots x S^1$ [HS].

Given a simple homotopy equivalence $h: M \to X$, as above, it is a standard fact that there is a stable (i.e. fibre dimension >dim X) H-bundle ξ over X and a stable bundle map $c: \nu_M \to \xi$ covering h, ν_M the stable normal bundle of M. This bundle map is unique up to isotopy and composition with bundle maps of ξ over the identity. So we have:

$$\begin{array}{ccc} \nu_M & \xrightarrow{c} & \xi \\ \downarrow & & \downarrow \\ M & \xrightarrow{h} & X \end{array} \qquad (1.2.1)$$

[1] Of course, this condition can be expressed as the vanishing of an invariant in the Whitehead group of X, $Wh(\pi_1 X)$.

Suppose that M and X are oriented, in the gneralized sense of twisted integer coefficients if M and X are not orientable in the usual sense. A diagram like (1.2.1), but with h required only to be of <u>degree one</u>, is called a <u>normal map.</u> A <u>cobordism</u> of normal maps is a normal map into $X \times I$, $I = [0,1]$, that respects the boundary, and the normal maps into $X \times i$, $i=0,1$, obtained by restriction, are called <u>normally cobordant</u>. Two normal maps are <u>equivalent</u> if they differ by composition with a bundle map, over the identity of X, of bundles over X. Let NM(X) denote the cobordism classes of equivalence classes of normal maps into X. It is clear that we have a map

$$\eta : \mathcal{S}_H(X) \to NM(X) \ .$$

called the <u>normal invariant</u>.

Let G/H be the classifying space for stable fiber homotopy trivializations of stable H-bundles. Then, by transversality, one may show $NM(X) \cong [X ; G/H]$, the set of homotopy classes of maps of X into G/H ; this is due to Sullivan [Sv], who used it to reformulate in a more global way the theory of Browder and Novikov in the simply connected case, as below. For H = PL [Sv] or TOP, [Sv1] [KS], the space G/H is well known and [X;G/H] can often be computed, though for H = O much less is known. Here we will confine ourselves to the question of the "kernel" and image of η.

1.3. Consider the problem of finding the image of η; i.e. when is a normal map normally cobordant to a simply homotopy equivalence. Surgery theory produces a Hermitian K-theoretic invariant that solves this problem; this is the surgery obstruction $\sigma(f,b)$ that lies in the Wall group $L_n^s(\pi,w)$, $n = \dim X$, $\pi = \pi_1 X$, $w:\pi \to \{\pm 1\}$ the orientation character of π. For $\pi = \{e\}$ this invariant was given by Kervaire and Milnor for $X = S^n$ or D^n, the n-disk, and in general by Browder and Novikov. For arbitrary π, it is due to Wall.

To begin, we need an algebraic criterion for recognizing a simple homotopy equivalence. To tell whether a homotopy equivalence is simple, we have the torsion invariant in the Whitehead group. To recognize a homotopy equivalence $h:(M,\partial M) \to (X,\partial X)$ of manifolds with possibly non-empty boundary, assumed to induce a simple homotopy equivalence of boundaries[1], we use the following criterion, which is a combination of Poincare duality and the theorem of Hurewicz:

(1.3.1) h _is a homotopy equivalence iff it induces an isomorphism of fundamental groups and isomorphisms_ $H_i(M;Z\pi) \to H_i(X,Z\pi)$ $\pi = \pi_1 X$, _of homology groups with local coefficients in the group ring_ $Z\pi$, $i \leq [n/2]$, $n = \dim X$. (_For example,_ $H_i(X,Z\pi)$ _is the integral homology_ $H_i(\tilde{X})$ _of the universal cover_).

A theorem of Smale [Sm1] (see also [M3]) tells us that any normal cobordism, _relative_ to the boundary if there is any, can be obtained by a sequence of handle attachments along embedded copies of $S^i \times D^{n-i}$ in the interior that represent elements in the kernel of the map on i-th homotopy groups induced by the normal

---

[1] Actually, only a simple homology equivalence with coefficients in $Z\pi_1 X$ is required throughout surgery theory. This improvement is contained in the "homology surgery theory" developed in the paper [CS1] on the theory of codimension two submanifolds.

map. Thus, all normal maps normally cobordant to a given one, (h,c) say, are obtained from it by "surgery" [M4] [B] [N1] [Br] [TW]. On the other hand, it is easy to see that by surgery one may at least obtain that h induces an isomorphism of fundamental groups and of the homology groups $H_i(M;Z\pi)$ with $H_i(X;Z\pi)$ for $i < [n/2]$; i.e. to obtain h that is [n/2]-connected.

Suppose $n = 2k \geq 6$ is even. Let K be the kernel of $h_*:H_k(M;Z\pi) \to H_k(X;Z\pi)$. By Poincare duality , K is a stably true, stably based $Z\pi$-module (denoted $K_k(M)$ in [TW]). By the Hurewicz theorem, every element of K is represented by a sphere $\alpha: S^k \to M$. Using transversality ("general position") one may define a non-singular $(-1)^k$-Hermitian form

$$\phi : K \times K \to Z\pi ,$$

and a self intersection form to which ϕ is associated,

$$\mu : K \to Z\pi/\{x-(-1)^k \bar{x}\} ,$$

as done in [TW,§5] . (Actually ϕ is purely homological, but μ requires some extra care. Here if $x = \Sigma \alpha_g g$, $\bar{x} = \Sigma w(g)\alpha_g g^{-1}$). By $(-1)^k$- Hermitian, we mean that the form is sesquilinear and satisfies $\phi(x,y) = (-1)^k\overline{\phi(y,x)}$. Also, $\mu(\alpha x) = \alpha\mu(x)\bar{\alpha}$, $\alpha \in Z\pi$, $\phi(x,x) = \mu(x) + (-1)^k \mu(\bar{x})$, and $\mu(x+y) - \mu(x) - \mu(y) \equiv \phi(x,y)$. Finally, non-singularity in this context means that

$$A\phi : K \to \mathrm{Hom}_{Z\pi}(K,Z\pi) = K^* ,$$

given by $A\phi(x)(y) = \phi(x,y)$, is an isomorphism of s-based modules, where the dual module ($\alpha\lambda(x) = \lambda(\bar{\alpha}x)$ for $\lambda \in K^*$) has the dual class of stable bases.

Now, suppose (K,ϕ,μ) happens to be hyperbolic; i.e. $K = L \oplus L'$, L and L' stably based modules of 1/2 the rank of K whose stable bases "add up" to an s-base of K, with $\phi|L \times L = 0$ and $\mu|H = 0$. The module L is called a subkernel. After stabilizing, which may be geometrically realized by taking the connected sum with $S^k \times S^k$ sufficiently many times, we may assume s-based modules are based and choose representative bases. (Also, we could choose L' with ϕ and μ trivial on L'; see [TW, §5]). The Whitney trick for removing intersections and self-intersections imply the basis for L can be represented by r disjointly embedded framed spheres $S_i^k \times D^k$ in M. Now an argument of Kervaire and Milnor [KM, §7] for $\pi = \{e\}$ and Wall [TW, §5] in the general case shows that performing surgery on these classes "kills" the kernel of h_* so as to produce (see (1.3.1)) a simple homotopy equivalence.

Thus we define the Wall group $L_{2k}^s(\pi,w)$ as the Grothendieck group of forms (K,ϕ,μ) as above, modulo the subgroup generated by hyperbolic forms. The preceeding discussion can be formalized to give the following result of Wall [TW, §5], at least for n even.

1.3.2 Theorem. Let (h,c), $h:(M^n,\partial M) \longrightarrow (X^n,\partial X)$ be a normal map that induces a homotopy equivalence of boundaries. Assume $n \geq 5$. Then the invariant $\sigma(h,c) \in L_n^s(\pi,w)$ vanishes if and only if (h,c) is normally cobordant relative to the boundary to a simple homotopy equivalence.

(For the best available general result when $n=4$, see [CS2]).

Thus one has the sequence

(.3.3) $\mathcal{S}_H(x) \overset{\eta}{\to} [X;G/H] \overset{\sigma}{\to} L_n^s(\pi,w)$ $n = \dim X \geq 5$,

X a closed H-manifold.

<u>Note:</u> The definition, as the definition of odd Wall groups to be indicated below, is given in its most primitive form. See [W2] and [R] for more sophisticated and general definitions.

For $n=2k+1 \geq 5$, 1.3.2 and 1.3.3 are also valid. In this case Kervaire and Milnor [KM] showed directly that, in the simply-connected case the studied, one could always perform surgery to obtain a homotopy equivalence. A satisfactory Hermitian K-theoretic invariant was first introduced by Wall [TW]; of course it follows from [KM] (see also [B]) that $L_{2k+1}(e) = 0$. Recently, Connolly [C1,2] has rehabilitated somewhat the approach of [W4].

To describe the odd dimensional case, let (h,c) be a normal map, $h:(M,\partial M) \to (X^{2k+1},\partial X)$, assumed to induce a simple homotopy equivalence of boundaries. (Again, only a simple homology equivalence with coefficients in $Z\pi_1 X$ is really needed; see [CS1]). Again we suppose h induces isomorphisms of fundamental groups and homology over $Z\pi$, $\pi = \pi_1 X$, through dimension $[n/2]-1$. Let H be the kernel of $h_* : H_k(M;Z\pi) \to H_k(X,Z\pi)$, and choose a set of generators of H . These may be represented (by a Whitney theorem) by disjointly embedded spheres with trivial normal bundle, connected to the (pre-assigned) base point. Let U be their union. One can arrange $h^{-1}D = U$, D an n-disk in X^n .

The normal map $(h|\partial U, b|\partial U)$ is k-connected, and (K,ϕ,μ), as in the discussion preceeding (1.3.2), is defined for this normal map. In fact, (K,ϕ,μ) is hyperbolic in two ways. In this case $K = H_k(\partial U; Z\pi)$ and the kernels[1] L and L', respectively of the maps $H_k(\partial U;Z\pi) \to H_k(U;Z\pi)$ and $H_k(\partial U;Z\pi) \to H_k(M_o,Z\pi)$, M_o = closure of $M - U$, are subkernels [TW,5.7]. For Poincare duality shows that they are stably based summands of a suitable type, and a general position and counting (over $Z\pi$) argument shows that ϕ and μ vanish on bounding classes, i.e. on L and L'. Further, it is easy to check that if L and L' are dual, i.e. $K = L \oplus L'$ as stably based modules, then h is a simple homotopy equivalence(see [TW, p56]). Also, it is shown in [TW] that there is an automorphism of (K,ϕ,μ) carrying L to L'. at least after stabilizing so that L and L' are actually based.

Hence one may assemble the information as follows: Let κ_r be the "standard $(-1)^k$ hyperbolic form of dimmension $2r$ (= "standard kernel" of [TW]), with "standard subkernel" L_r and with compatible natural identifications $\kappa_{r+1} = \kappa_r \oplus \kappa_1$. Thus, for some r, we have an isomorphism ρ,

$$(K,\phi,\mu,L) \cong (\kappa_r, L_r) .$$

Let $SU_r^\epsilon(Z\pi)$ be the automorphisms of κ_r (preserving preferred basis class, of course), $\epsilon = (-1)^k$. Then $SU_r^\epsilon(Z\pi) \subset SU_{r+1}^\epsilon(Z\pi)$

---

[1] To avoid technical stabilization difficulties it is actually better to use the obvious dual subkernel to L, generated by non-bounding k-spheres in $U = S^k \times S^k \# \ldots \# S^k \times S^k$. This was pointed out by Sharpe.

by adding id_{κ_1} . Let $SU^{\varepsilon}(Z\pi)$ be the limit (i.e. the stabilization). The construction of the preceeding paragraph together with ρ leads to an element

$$\alpha \in SU_r^{\varepsilon}(Z\pi) \subset SU^{\varepsilon}(Z\pi) .$$

The choices made, e.g. generators of H , possible automorphisms α throwing L to L' , dictate dividing out $SU^{\varepsilon}(Z\pi)$ by the subgroup $TU^{\varepsilon}(Z\pi)$ of stabilizations of elements that preserve L_r (as an s-based module). Further, if we choose the generators of H so that the first one, say, is represented by an embedded framed sphere on which it is proposed to perform surgery , then one sees that the effect of this surgery is to replace α by $\sigma\alpha$. Here σ is the element of $SU^{\varepsilon}(Z\pi)$ that is the stabilization of the element of $SU_1^{\varepsilon}(Z\pi)$ with matrix

$$\begin{pmatrix} 0 & 1 \\ \varepsilon & 0 \end{pmatrix}$$

with respect to the "standard base" of κ_1 ; i.e. σ interchanges L_1 and its dual.

Wall [TW, §6] (see also [SP2]) proves that the subgroup generated by $TU^{\varepsilon}(Z\pi)$ and σ is normal, with abelian quotient. We define this quotient to be $L_n^s(\pi, w)$. Then a rigorous version of the above discussion yields (1.3.2) for n odd also.

To analyze the "kernel" of η , one uses a geometric realization thoerem for elements of the Wall groups; for $\pi = \{e\}$ this is the "plumbing" of Kervaire and Milnor (see [Br]), and in general, a result of [TW,§5,6].

(1.3.4) Theorem. Let $h: M^n \to X$ be a simple homotopy equivalence of (say) compact closed n-manifolds, $n \geq 5$. Let $\gamma \in L_{n+1}(\pi,w)$, $w = \pi_1 X$ with w the orientation character of X (first Stiefel-Whitney class). Then $\exists$ a normal cobordism (f,b) of h, to a homotopy equivalence $h': M' \to X$, with $\sigma(f,b) = \gamma$.

For $n = 4$, see [CS2] for the best general result currently known.

Note that if $\gamma = 0$, then (1.3.2) and the s-cobordism theorem ([Sm2]see also [K], [H], [Sl]) imply that for any (f,b) as in 1.3.4, with $\sigma(f,b) = 0$, we have that h and h' represent the same element in $\mathcal{S}_H(X)$. So if in the notation of (1.3.4), we put $\gamma.[h] = [h']$, this gives a well-defined action of the group $L^s_{n+1}(\pi,w)$ on $\mathcal{S}_H(X)$.

(1.3.5) Corollary. The sets of $\eta^{-1}(y)$, for $y \in [X,G/H]$ are precisely the orbits of the action of $L^s_{n+1}(\pi,w)$.

We may sum up this section with the "exact sequence"

(1.3.6) $$L^s_{n+1}(\pi,w) \to \mathcal{S}_H(X) \overset{\eta}{\to} [X;G/H] \overset{\sigma}{\to} L^s_n(\pi,w) \quad (n \geq 5).$$

This sequence extends further to the left. Also, σ is not in general a homomorphism.

1.4. An analogous theory exists for describing the obstruction to making a normal map, by surgery, into a (not necessarily simple) homotopy equivalence. The obstruction groups are usually denoted $L^h_n(\pi,w)$; they were given in [W4] for n even; for n odd, they were introduced in [Sh]. They are constructed in a way similar to the groups $L_n^s(\pi,w)$, but without the introduction of stable bases and the demand, at certain points, that certain automorphisms preserve a class of bases. The groups $L_n^h(\pi,w)$ are related to $L_n^s(\pi,w)$ by the exact sequence of [Sh, 4.1] ("Rothenberg's exact sequence"). Note that $A_n(\pi,w)$ in [Sh] is just $H^n(Z_2, Wh(\pi))$, where the Z_2-action on $Wh(\pi)$ is given by $*$, induced by $-$. In particular, these groups are isomorphic modulo elements of order two.

As a further refinement, one can consider the problem of performing surgery on a normal map (f,b) , $f : M \to X$, to get a homotopy equivalence with torsion in a subgroup $A \subset Wh(\pi_1 X)$ with $A = A^*$. The groups $L_n^A(\pi,w)$ can be related to each other ($L^{Wh(\pi)} = L^h$ and $L^{\{0\}} = L^s$) by the natural generalization of [Sh, 4.1] . This is due to Cappell. (The third term is of the form $H^n(Z_2,B/A)$, when comparing L^A and L^B , $A \subset B$) .

Finally, there is an analogue of Wall groups based on projective modules, the "Novikov surgery groups" $L_n^p(\pi,w)$. Maumary [M] discusses these groups and uses them to study proper surgery on open manifolds. They are related to $L_n^h(\pi,w)$ by an exact sequence like [Sh, 4.1] involving $H^k(Z_2, \tilde{K}_o(\pi))$; the action of Z_2 is induced by $-$. Also, there are intermediate groups based on subgroups of $\tilde{K}_o(\pi)$, invariant under the action of Z_2 , and appropriate exact sequences relating them.

For more details and generalizations on the material of 1.4, see [Sh] [R] [W2][M].

1.5. A theory exists in which it is not required that the boundary be held fixed. This requires a definition of relative Wall groups $L_n(f)$, for $f : (\pi,w) \to (\pi',w')$ a morphism. These can be defined as in [TW,§§7,9],[R1], or [Sp1]. The obvious long exact sequences exist. From an algebraic point of view, the most interesting definition (for the more difficult case of even dimensional manifolds with odd dimensional boundaries) is that of Sharpe, since it is closely connected to Hermitian K_2 and its relation to Hermitian K_o , i.e. L_{2k} . After all, from an algebraic point of view one might naively expect or hope that Wall groups would be a sequence of higher Hermitian K-functions rather than a constant repetition of Hermitian K_o and K_1 functions. See also [Kr].

There is also a theory for groupoids, but this will not be discussed here.

1.6. All types of surgery groups are periodic, i.e. $L_n^x = L_{n+4}^x$, for all possible x . Further, if (f,b) is a normal map and CP^2 is complex projective 2-space (a four-dimensional real manifold) then

$$\sigma(f,b) = \sigma((f,b) \times CP^2) .$$

1.7. We have not discussed any results about classifying manifolds M^{n+k} of the homotopy type of an n-dimension complex, $k > 0$. For $k \geq 3$, M compact. There are results of Browder, Haefliger, Wall and Sullivan that use the theory of surgery. (See [TW, § 11] and references in [TW]) . However, these results rely more upon the existence of a "surgery theory" then upon its actual character as a Hermitian algebraic K-theory. In fact, the existence of the theory can be established [TW,§9] in a purely geometric fashion, without reference to Hermitian K-theory. (However, the periodicity, (1.6), which is very important in applications (e.g [HS]), has never been proven without relying on the algebraic descriptions).

For $k=1$, one has [S2] at one's disposal, for example. This involves ordinary algebraic K-theoretic invariants.

For $k=2$, one has to confront complexities connected with possible fundamental groups of the complements of submanifolds of codimension two, e.g. knot groups. To study codimension two problems, the author and S. Cappell [CS 1] have introduced new Hermitian K-functions for the problem of performing surgery on a normal map (f,b), $f : M^n \to X$, to get a (simple) homology equivalence with local coefficients in, for example, a quotient ring (with involution and 1) of $Z\pi_1 X$. For n odd, these are essentially Wall groups again, of the quotient ring. However, for n even, let $\mathfrak{F} : Z\pi_1 X \to \Lambda$ be an epimorphism of rings with involution. We consider $(-1)^k$-forms (K,ϕ,μ) as in defining the Wall group $L^h_{2k}(Z\pi_1 X)$, for example, except that K is any, finitely generated $Z\pi$-module, $\pi = \pi_1 X$, and there is no non-singularity requirement on ϕ. However, $K \otimes_{Z\pi} \Lambda$ is required to be stably free, and the form induced over Λ by ϕ is required to be non-singular. These forms may be added by orthogonal direct sum. One forms the Grothendieck group of such forms, modulo the subgroup generated by forms that are "hyperbolic" in the following sense: There is a submodule ("presub-kernel") $L \subset K$, with $\phi | K \times K = 0$ and $\mu | K = 0$, so that $(K,\phi,\mu) \otimes_{Z\pi} \Lambda$ is hyperbolic with subkernel the image of $L \otimes \Lambda$ in $K \otimes \Lambda$. This group is called $\Gamma^h_{2k}(\mathfrak{F})$. Similarly, one can define $\Gamma^s_{2k}(\mathfrak{F})$, etc. Note that $\Gamma^x_{2k}(\mathrm{id}) = L^x_{2k}(\pi, w)$. However, the Γ-groups are very far from the relative L-groups; often they are not even finitely generated. Further results on Γ-groups would have many important geometric applications.

1.7. Recently, Hsiang and Sharpe have elucidated a geometric interpretation of Hermitian K_2, in terms of surgery on pseudo-isotopies. See their paper in these proceedings.

2. Wall Groups of Finite Groups.

2.1 For $\pi = \{e\}$ Kervaire and Milnor [KM] (see also [Bt]) showed that

$$L_n(e) = \begin{cases} Z \\ 0 \\ Z_2 \\ 0 \end{cases} \quad \text{if } n = \begin{cases} 0 \\ 1 \\ 2 \\ 3 \end{cases} \pmod 4.$$

For $n=0$ and $n=2 \pmod 4$ they gave the index and Arf invariant, respectively, as complete invariants.

2.2 Improving results of Petrie-Passmann [PP], Connolly [C] has shown that for π finite and $w:\pi \to \{\pm 1\}$ an arbitrary homomorphism, the following holds:

(2.2.1) $L_3^h(\pi,w)$ <u>has exponent</u> 2 <u>and</u> $L_1^h(\pi,w)$ <u>has exponent</u> 4 .

(Actually, Connolly also shows that $L_1^h(\pi)$ has exponent 2 if $|\pi|$ is odd).

In this connection Bass [Ba 4] has observed that as a consequence of unitary stability theorems [Ba 1,2] and a recent theorem of Kajdan one has:

(2.2.2) <u>For</u> π <u>finite</u>, $L_n^x(\pi,w)$ <u>is finite for</u> n <u>odd, for all possible</u> x .

When w is trivial, we write just $L_n^x(\pi)$. Here is another general result, for <u>abelian</u> groups.

(2.2.3) *If* $f : \pi \to \pi'$ *induces an isomorphism of the 2-primary components of the abelian groups* π *and* π' , *then the induced map* $f_* : L_3^x(\pi) \to L_3^x(\pi')$ *is an isomorphism*, $x=s,h$.

In particular, $L_3^x(\pi) = 0$ *for* π *abelian, of odd order.*

This result appears as (1.2) of [Ba 4]. For π cyclic, the vanishing of $L_3^x(\pi)$ has also been announced by Bak [Bk]. For π cyclic of odd order, it is due to Lee [Le 1]. Wall's announced results on dihedral groups [W1] indicate that no such result holds in the non-abelian case. Bak has also announced the next result.

(2.2.4) *For* π *abelian of odd order*, $L_1^x(\pi) = 0$, $x = s,h$.

Let $A(\pi)$ be the kernel of the natural map $Wh(\pi) \to K_1(Q\pi)/\pm\pi$.

(2.2.5) (Wall [W1]) . *For* π *finite of odd order*, $L_n^{A(\pi)}(\pi) = 0$.

Note that this implies that $L_h^x(\pi)$ is a finite-dimensional vector space over Z_2 , π of odd order, x any of the possibilities of §1 . Earlier, R. Lee [Le 2] showed that for π metacyclic pq , p and q odd primes, $L_3^h(\pi) = 0$. These results have applications to the spherical space form problem, e.g. [Le 2].

To complete the discussion on odd Wall groups of finite groups, we must try to describe the situation of groups of even order.

(2.2.6) (Bass [Ba 4]). *If* π *is finite abelian of exponent two, then* $L_1^s(\pi) \to L_1^h(\pi)$ *is an isomorphism and* $L_1^s(\pi) \cong {}_2Pic(\pi)$, *the kernel of* $Pic(\pi) \xrightarrow{\times 2} Pic(\pi)$.

(In fact, $Wh(\pi) = 0$ [BMS, 4.13]) .

(Bass also shows that the 2-rank of ${}_2Pic(\pi)$ is *at least* $2^r - r - 1 - \binom{r}{2}$, r the 2-rank of π) .

On the other hand, Wall [W4] announced the following:

(2.2.7) *Let π be finite abelian, of 2-rank r and with s summands of order two. Assume $w: \pi \to \{\pm 1\}$ is trivial on elements of order two. Then $L_1^{A(\pi)}(\pi, w)$ has exponent two and 2-rank* $2^r - r - 1 - \binom{s}{2}$.

Note that (2.2.7) implies that $L_1^x(\pi, w)$ has exponent four, and that for a cyclic group $L_1^{A(\pi)}(\pi, w) = 0$. However, for a cyclic group, $K_1(Z\pi) \to K_1(Q\pi)$ is injective. [Ba 4, X 3.6 and XI 7.3]. Also, $L_1^s(Z_2, id) = 0$ [TW, 13, A.1]. Hence we have

(2.2.8) *For π cyclic and any $w: \pi \to \{\pm 1\}$* , $L_1^s(\pi, w) = 0$.

Notice that the results claimed in [W4] imply that, in general $L_1^h(\pi)$ and $L_1^h(\pi/\pi^2)$ are not isomorphic.

The results of Bass and Wall also coincide for L_3 of abelian groups of exponent two.

(2.2.9) *For π finite abelian of exponent 2 , then $L_3^h(\pi)$ is a group of exponent 2 and rank $2^r - 1$, r the 2-rank of π . (Recall* $Wh(\pi) = 0$) .

Wall [W1] also announces:

(2.2.10) <u>For</u> π <u>abelian of 2-rank</u> r, $L_3^{A(\pi)}(\pi)$ <u>has exponent two and rank</u> 2^r-1 .

If π is cyclic, $A(\pi) = \{0\}$, as noted above, e.g., in [W1]. So we have part of the following result of [Ba 4]:

(2.2.11) <u>For</u> π <u>cyclic</u>, $x = s$ <u>or</u> h, $L_3^x(\pi) \to L_3^x(\pi/\pi^2)$ <u>is an isomorphism, and</u> $L_3^x(\pi)$ <u>has the same order as</u> π/π^2.

In 1.4 of [Ba 4] , Bass gives a generator explicitly. In fact, he gives explicit generators for $L_3^x(\pi)$, $x = s,h$ π finite abelian. For $\pi = Z_4$, (2.2.11) was proved independently by Berstein (private communication).

Bass' general results on L_3 can be described as follows: Let $B(\pi) \subset Z\pi$ be the fixed ring of the involution on $Z\pi$ given by $\bar{g} = g^{-1}$ for $g \in \pi$. Let $q: B(\pi) \to Z_2[{}_2\pi]$ be the restriction of the natural map. Assume π abelian, finite. Then there is a subgroup $G_o(\pi)$ of the additive group $Z_2[\pi/\pi^2]$, a subgroup $G_1(\pi)$ of the multiplicative group (of units) $Z_2[{}_2\pi]^{\cdot}$, both functorial in π , and an exact functorial commutative diagram

$$\begin{array}{ccccccccc} 0 & \to & \dfrac{Z_2[\pi/\pi^2]}{Z_2\cdot 1+G_o(\pi)} & \to & L_3^s(\pi) & \to & \dfrac{Z_2[{}_2\pi]^{\cdot}}{{}_2\pi\cdot G_1(\pi)} & \to & 0 \\ & & \downarrow & & \downarrow & & \downarrow & & \\ 0 & \to & \dfrac{Z_2[\pi/\pi^2]}{Z_2\cdot 1+G_o(\pi)} & \to & L_3^h(\pi) & \to & \dfrac{Z_2[{}_2\pi]^{\cdot}}{q(B(\pi)^{\cdot})\cdot G_1(\pi)} & \to & 0 \end{array} \qquad (2.2.12)$$

with the middle vertical map surjective. Some information about $G_o(\pi)$ is given, leading to the upper bound

$$2^{(2^{n+m+1} - 2^n - m - 1)} \quad \text{for the order of } L_3^s(\pi) .$$

3. For the non-orientable case, Wall [W1] announcec the next two results.

(2.3.1) For π abelian, $w:\pi \to \{\pm 1\}$ non-trivial, $L_3^{A(\pi)}(\pi,w)$ has exponent four or two and order $2^{(2^r+2^{r-1}-1)}$. Actually, according to Taylor's paper in these proceedings, the exponent is two. (Again, r denotes the 2-rank).

(2.3.2) If there is x in the finite abelian group π with $x^2 = 1$ and $w(x) = -1$, then $L_n^{A(\pi)}(\pi) \cong L_n^h(Z_2,id) \oplus E$, E an elementary 2-group of rank $N/2 - N/2^r - r+1$, where N is the order of π and r the 2-rank. (See [TW] for Wall groups of Z_2).

If w is non-trival but vanishes on elements of order two, then $L_o^{A(\pi)}(\pi,w)$ has exponent two and rank $2^r-1-r-\binom{s}{2}$, and $L_2^{A(\pi)}(\pi,w)$ is isomorphic with Z_2.

The reader is also referred to the results of the paper of Taylor in these proceedings and to the results of [W1] on dihedral groups of order 2p or 8 and the quaternion group.

4. Let π be a finite group with complex representation ring $R(\pi)$, with conjugation denoted $-$. Then there is a signature homomorphism

$$\Sigma : L_n^h(\pi) \to R(\pi) \quad ,$$

for n even. For example, Petrie (page 176 of [P]) defines it as a type of G-signature, in the style of Atiyah-Singer. In [TW, §13 A] and [W4], Wall defines it as the "multisignature". We also denote by Σ the obvious map $L_n^x(\pi) \to R(\pi)$ determined by Σ on $L_n^h(\pi)$; recall that all these groups are isomorphic modulo elements of order two. Note that for $n \equiv 0 \pmod 4$ the image of Σ is contained in the real representations and, for $n \equiv 2 \pmod 4$, in the purely imaginary representations.

(2.4.1) [W1]. _For_ π _finite of odd order,_

$$\Sigma : L_{2k}^{A(\pi)}(\pi) \to R(\pi)$$

has image $\{4(x + (-1)^k \bar{x}) \mid x \in R(\pi)\}$, _kernel_ Z_2 _if_ k _is odd and_ 0 _if_ k _is even._

As noted above, for π cyclic $L_{2k}^{A(\pi)}(\pi) = L_{2k}^s(\pi)$. In this case, the rank of the image of Σ was given by Petrie [P]. One can further conclude that if π cyclic of odd prime order, the image of Σ on $L_{2k}^h(\pi)$ is $\{2(x + (-1)^k\bar{x} \mid x \in R(\pi)$, and the kernel 0 , k even, or Z_2 , k odd. (Combine [Sh, 4.1] , [TW, 13.A.5], [M1,6.4], and [Ba 5, _XI_, 7.3], for example).

For cyclic groups, these calculation were one of the central steps in the classification of homotopy lens spaces [BPW] and, together with results of 2. and 3. for Z_2, in the classification of homotopy projective spaces [Md] [W3]. They are also important for studying semi-free actions of groups on spheres; e.g. see [BP].

5. We have indicated in the course of this discussion some of the known results relating the various types of Wall groups. It is clear from this discussion, and will be even clearer in Section 3, that a precise knowledge is needed in order to complete calculations. In this connection, we mention the new result of Dennis and Stein (these proceedings) that $KSL_1(Z\pi) \neq 0$ if π is an elementary p group of order p^n , $n \geq 3$.

We also mention the result of [D] on induction as a possible tool in the further study of (odd) Wall groups of finite groups.

Added December 20, 1972. I have just received a letter from A. Bak, announcing some new results. The results on even Wall groups will appear in a joint paper with W. Scharlau in Inventions Math. For odd Wall groups he claims at least the following: If π is finite, nilpotent, of odd order then $L_i^s(\pi) = 0$ for $i=1,3$.

3. Infinite Groups.

3.1. Let π be a finitely presented group and $w : \pi \to \{\pm 1\}$ is a homomorphism, let $G = \pi \times Z$ (Z = infinite cyclic group) and let $w_1 : G \to \{\pm 1\}$ be the composition of w with the natural projection

(3.1.1) Theorem [Sh, 5.1]

$$L_n^s(G,w_1) \cong L_n^s(\pi,w) \oplus L_{n-1}^h(\pi,w)$$

A related result appears in [TW, §12]. In [CS1, $\underline{V}$], the analogous result is proven for Γ-groups.

Using this theorem and 2.1, the Wall group of $Z\oplus \ldots \oplus Z$ can be calculated[1]; note that $L_n^s(Z\oplus \ldots \oplus Z) = L_n^h(Z\oplus \ldots \oplus Z)$, as $Wh(Z\oplus \ldots \oplus Z) = 0$ [Ba 5, $\overline{\underline{XII}}$]. This calculation, together with 1.3.6 and 1.6, leads to a complete classification of PL (and topological) manifolds homotopy equivalent to the n-torus $T^n = S^1 \times \ldots \times S^1$, $n \geq 5$ [HS] [TW, §15A]. As is well-known, classification of such PL manifolds was used by Kirby in solving the annulus conjecture and in [KS], and also [LR], where Kirby's ideas were extended to solve the Hauptvermutung and triangulation problems.

We shall indicate the ideas of the geometric proof in (3.1.1) given in [Sh]. Recently, Novikov [N2] developed ideas for a purely algebraic proof of (3.1.1). For an algebraic proof, see

[1] For Z Browder used the Browder-Levine fibering theorem to study surgery directly [B1] in some cases, without using Wall groups.

the paper of Ranicki [R] [1] . Karoubi [Kr], using different methods, has also obtained many results in this direction, and generalizations to his higher Hermitian K-functions. Naturally, the algebraic proof also extends[2] the result to rings more general then the group ring $Z\pi$. On the other hand, the geometric proof is susceptible to generalizations, to be described below, to other constructions besides taking the product with Z (i.e. forming the Laurent series of $Z\pi$) . So far (Dec. 1972), the algebraic methods have not been extended beyond $Z \times_\alpha \pi$.

To prove (3.1.1), we actually construct a homomorphism

$$\alpha\colon L_n^s(G) \to L_{n-1}^k(\pi)$$

and then show that the sequence

$$(3.1.2) \qquad 0 \to L_n^s(\pi) \overset{j_*}{\to} L_n^s(G) \overset{\alpha}{\to} L_{n-1}^h(\pi) \to 0$$

is split exact. (For this discussion, we suppress the orientation character).

To define α , one would like to proceed as follows: Let (f,b) , $f : M^r \to N \times S^1$, $\pi_1 N = \pi$, be a normal map with $\sigma(f,b)$ a given element γ in $L_n^s(G)$. Then, assuming (without) loss of generality) that f is transverse regular to

---

[1] The twisted product $Z \times_\alpha \pi$ (see below) is also covered by Ranicki.

[2] Using essentially the observation that certain Γ-groups of [CS1] can be interpreted as Wall groups, (3.1.1) can actually be established geometrically for group rings $R\pi$, $Z\subset R\subset Q$. The same remark applies to the rest of the results in this section.

$N = N \times pt \quad N \times S^1$, so that $f^{-1}N \quad M$ is a submanifold, we would like to define

$$\alpha(\gamma) = \sigma(f\,|f^{-1}N\ ,\ b\ |\ f^{-1}N)\ .$$

If M is a closed manifold the right side at least makes sense. However, not all elements of $L_n^s(G)$ can be realized by normal maps into closed manifolds, in general. But for $n \geq 6$, (1.3.4) allows us to realize γ as $\sigma(f,b)$ with $f|\partial M : \partial M \to \partial N \times S^1$ a simple homotopy equivalence. If $n \geq 7$, we may then apply the fibering theorem of Farrell [F] or the splitting principle of Farrell-Hsiang [FH], in dimension $n-1$, so that, after a homotopy, we may assume also that $f|f^{-1}(\partial N) : f^{-1}(\partial N) \to \partial N$ is a homotopy equivalence. Then the right side does make sense. A simple homotopy equivalence cannot be hoped for, in general, because crossing with S^1 always annihilates Whitehead torsion.

In fact, one can show by geometric arguments that this gives a well-defined homomorphism, $n \geq 7$. (This requires, among other things, applications of the splitting principle in dimension n also). It actually depends only on how $\pi_1 N$ is identified with π (see [Sh1]). The map from $L_{n-1}^h(\pi)$ to $L_n^s(G)$ given by crossing with S^1 defines a splitting of α. Of course, by periodicity (1.6), the requirement $n \geq 7$ entails no loss of generality.

Suppose $\alpha(\gamma) = 0$. By the "cobordism extension theorem", it may be assumed, after a normal cobordism, that $f\ |\ f^{-1}N : f^{-1}N \to N$ is a homotopy equivalence. From this it follows that the image of γ in $L_n^h(G)$ is the image of a (unique) element $\xi \in L_n^h(\pi)$ under the natural map. Using the (obvious) fact that the natural map of $H^n(Z_2, Wh(\pi))$

into $H^n(Z_2, Wh(G))$ is a monomorphism and the diagram on page 323 [Sh] (but with different notation, e.g. $A_n = H^n$, etc.), it follows from [Sh, 4.1] that ξ is in the image of the natural map $L_n^s(\pi) \to L_n^h(\pi)$. So changing γ by an element in the image of the natural map j_*, we may assume that $\xi = 0$.

From this it follows immediately that (the new) γ is in the image of $H^{n+1}(Z_2, Wh(G)) \to L_n^s(G)$. Further, by further surgery, it can be arranged that

$$f \,|(M-f^{-1}N) : M-f^{-1}N \to (N \times S^1-N)$$

is <u>also</u> a <u>homotopy equivalence.</u> This fact, the geometric interpretation of the above map, and an elementary result about torsion of the map f now obtained (already proven by Farrell and Hsiang) imply that γ is the image of an element in $H^{n+1}(Z_2, Wh(G))$ that actually comes from $H^{n+1}(Z_2, Wh(\pi))$, and so (again p.323 of [Sh]) is in the image of j_*.

3.2. Clearly, a similar approach can be proposed to calculate Wall groups in terms of component groups for any construction on groups that can be realized geometrically by constructions on manifolds. To carry this out, one needs an appropriate analogue of the Farrell-Hsiang splitting principal. For example, Ronnie Lee proposed to calculate the Wall groups of a free product (which corresponds to the geometric operation of connected sum) in this way, However, he could establish the required splitting principal, at least for groups with no 2-torsion, in only even dimensions. (see [Le2]).

However, Cappell [Ca] subsequently established a very general geometric splitting principle. Let X^n , $n \geq 6$, be a compact connected (smooth, PL, TOP), manifold (or even a suitable Poincare complex), with two-sided connected codimension 1 submanifold or sub-Poincare complex $Y \subseteq Y \times (-1,1) \subseteq X$. Assume $\partial Y \cap (-1,1) = (Y \times (-1,1)) \cap \partial X$. Let $h : (M, \partial M) \to (X, \partial X)$ be a homotopy equivalence, transverse regular to $(Y, \partial Y)$, with

$$h | \partial h^{-1} Y : \partial h^{-1} Y \to \partial Y \quad (\text{note } \partial h^{-1} Y = h^{-1}(\partial Y))$$

a homotopy equivalence. When $\pi_1 Y \to \pi_1 X$ is a monomorphism, then there is a homomorphism ϕ: $Wh(\pi_1 X) \to K_o(\pi_1 X)$ defined by Waldhausen [Wh] (and [FH] for $\pi_1 X = Z x_\alpha \pi_1 Y$).

(3.2.1) (<u>One formulation of Cappell's splitting principle</u>)

<u>Suppose</u> $\pi_1 Y \to \pi_1 X$ <u>is a monomorphism whose image is closed under square roots</u> (i.e. <u>contains</u> x <u>if it contains</u> x^2) . <u>Then</u> h <u>is h-cobordant relative</u> ∂M <u>to</u> g , <u>transverse regular to</u> Y , <u>with</u>

$$g \mid g^{-1} Y : g^{-1} Y \to Y$$

<u>a homotopy equivalence, if and only if</u> $\phi(\tau(f)) = 0$, $\tau(f)$ <u>the torsion of</u> f . <u>Further, the h-cobordism can be taken to have torsion in the kernel of</u> ϕ , <u>also</u>.

For $\pi_1 X = X \, x_\alpha \pi_1 Y$, the square root closed condition is always satisfied. Using the decomposition of $Wh(\pi_1 X)$ given in [FH] and an argument involving h-cobordisms, one can derive the Farrell-Hsiang splitting principle from (3.2.1).

Theorem (3.2.1) is certainly false without the hypothesis that $\pi_1 Y \to \pi_1 X$ be a monomorphism. It is also widely believed that the square-root closed condition cannot be eliminated. For the best available result if $n = 5$, see [CS2].

To apply this to Wall groups, let $G = \pi_1 *_{\pi} \pi_2$ be an amalgamated free product of the finitely presented groups π_1 and π_2 along the finitely presented subgroup π. Let $w : G \to \{\pm 1\}$. Let $B \in Wh(G)$ be the kernel of the Waldhausen map $\phi : Wh(G) \to \tilde{K}_o(\pi)$.

(3.2.2) [Ca]. <u>If π is square root closed in π_1 and π_2 (which</u> **<u>implies</u>** <u>a square root closed in G), then there is an exact sequence</u>

$$\ldots \to L_n^h(\pi,w|\pi) \xrightarrow{j} L_n^h(\pi_1,w|\pi_1) \oplus L_n^h(\pi_2,w|\pi_2) \to L_n^B(G,w) \xrightarrow{\alpha} L_{n-1}^h(\pi,w|\pi) \to \ldots$$

In (3.2.2), the unlabelled map is a natural map and $j(\xi) = (j_1(\xi), -j_2(\xi))$, j_1 and j_2 natural maps.

Note that, as a special case of (3.2.2), if π_1 and π_2 have no 2-torsion, then $\tilde{L}_n(\pi_1 * \pi_2) = \tilde{L}_n(\pi_1) \oplus \tilde{L}_n(\pi_2)$, where $\tilde{L}_n(\pi)$ denote the kernel of the natural map $L_n(\pi) \to L_n(e)$. For n even, this result is due to R. Lee [Le 2].

The image of ϕ is the kernel L of the natural map $\tilde{K}_o(\pi) \to \tilde{K}_o(\pi_1) \oplus \tilde{K}_o(\pi_2)$. Hence $L_n^B(G,w)$ can be related to $L_n^h(G,w)$ by a long exact sequence involving $H^n(Z_2, L)$ as the third term.

Again, the first point in proving (3.2.2) is to construct α. Assume $n \geq 7$. Let U^{n-1} V^{n-2} be a closed manifold with closed connected two sided codimension one submanifold V with $U-V=U$, U_2, U_1 and U_2 connected and disjoint, so that the diagrams

$$\begin{array}{ccccc} & & \pi_1(U_1) & & \\ & \nearrow & & \searrow & \\ \pi_1(V) & & & & \pi_1(U) \\ & \searrow & & \nearrow & \\ & & \pi_1(U_2) & & \end{array}$$

and

$$\begin{array}{ccccc} & & \pi_1 & & \\ & \nearrow & & \searrow & \\ \pi & & & & G \\ & \searrow & & \nearrow & \\ & & \pi_2 & & \end{array}$$

are isomorphic via an isomorphism carrying w to the orientation character of U. Let $(X,Y) = (U,V) \times I$. Given $\gamma \in L_n^B(G)$, we may realize γ as $\sigma(f,b)$, by a refinement of 1.3.4 $f:(M,\partial M) \to (X,Y)$ with f a homotopy equivalence with torsion in B on both boundary components. Hence, by (3.2.1), we may assume after adding h-cobordisms that don't affect γ, that f is transverse to Y and that

$$f \mid f^{-1}Y : f^{-1}Y \to Y$$

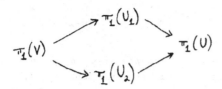

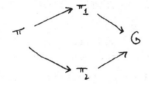

induces a homotopy equivalence of $\partial(f^{-1}Y)$ with ∂Y. So if we write

$$\alpha(\gamma) = \gamma(f \mid f^{-1}Y, b \mid f^{-1}Y) \quad ,$$

the right side again makes sense as an element of $L^h_{n-1}(\pi)$.

Again, one can show by geometric arguments that α is a well-defined homomorphism and depends only on the identification of fundamental groups involved. Then, by geometric arguments, one shows the exactness at $L^B_n(G)$ and $L^h_{n-1}(\pi)$ using general principles about Wall groups and surgery theory, and the definition of α. At the direct sum, one also uses the fact ker ϕ contains the image of $Wh(\pi_1) \oplus Wh(\pi_2)$ under the natural map and (3.2.1) to split up a normal cobordism to a homotopy equivalence with torsion in B of the union along Y of normal maps into Y_1 and Y_2.

Corresponding to the case $X - Y$ connected, one also obtains a result. Let $v_1, v_2: \pi \to \pi_1$ be monomorphisms of finitely presented groups. Let G be the quotient of the free product $Z * \pi_1$ by the smallest normal subgroup containing the elements $tv_1(x)t^{-1}v_2(x)$, $x \in \pi$, t a given generator of Z. Let $w : G \to Z_2$ with $w \circ v_1 = w \circ v_2 = w|\pi$. Let B be the kernel of the Waldhausen map $Wh(G) \to \tilde{K}_o(\pi)$.

(3.2.3) [Ca] <u>Assume</u> $v_1(\pi)$ <u>and</u> $v_2(\pi)$ <u>are square-root closed. Then there is an exact sequence</u>

$$\cdots \to L^h_n(\pi_1, w(\pi) \xrightarrow{v_1 - v_2} L^h_n(\pi_1, w|\pi_1) \to L^B_n(G,w) \xrightarrow{\alpha} L^h_{n-1}(\pi, w|\pi) \to \cdots$$

Here v_1 and v_2 also denote the maps these morphisms induce on Wall groups. Again, $L^B_n(G,w)$ is related to $L^h_n(G,w)$ by a sequence involving $H^n(Z_2,L)$, L the kernel of $K_o(\pi) \xrightarrow{v_1-v_2} K_o(\pi_1)$.

For the case $G = Z \times \pi$, $v_1 = v_2$ and one obtains a short exact sequence. Using the analogues of [Sh,4.1] for comparing various theories and the decomposition of $Wh(Z \times \pi)$ [Ba 5], (3.1.1) can be deduced from this short exact sequence. For the case $G = Z \times_\alpha \pi$, a version of the result was first given by Farrell-Hsiang.

Of course, these results can be combined, especially in cases where the Whitehead groups vanish. Thus Cappell calculated the Wall groups of the fundamental groups of orientable surfaces and knot complements of fibered knots of S^1 in S^3. More generally let $\{e\}$ be accessible or square root-closed accessible of order zero, and say π is (square root-closed) accessible of order n if it can be obtained as G is obtained from π,π_1 and π_2 in (3.2.2) or from π,π_1,v_1 and v_2 in (3.2.3), where $\pi \subset \pi_1$ and $\pi \subset \pi_2$ (as square root-closed subgroups) or v_1 and v_2 are monomorphisms (with square root closed images), and π, π_1 and π_2 are (square root closed) accessible of order $(n-1)$. For square root closed accessible groups, (3.2.2) and (3.2.3) can be thought of as Mayer-Vietoris sequences, at least up to 2-torsion. On the other hand, the Wall groups of $\{e\}$ are the homotopy groups of G/TOP, in dimensions at least one. So let $\mathbb{L}$ denote the Ω-spectrum with 0-th term $G/TOP \times Z$: i.e. if M_i is the i-th term, $M_i = \Omega M_{i+1}$. Then homotopy of the Eilenberg-Maclane space $K(\pi,1)$ with coefficients in $\mathbb{L}$ also has Mayer-Vietoris sequences. So we have a result of Quinn [Q].

(3.2.4) <u>If</u> π <u>is square-root closed accessible of some order then, modulo elements of order a power of two</u>,

$$L_n^h(\pi) \cong H_n(K(\pi,1) : \mathbb{L}) .$$

Actually, Quinn constructs a natural map $A_\pi : H_n(K(\pi,1) ; \mathbb{L}) \to L_n^h(\pi)$ that gives the isomorphism in this case. Further, Quinn gives the following sharpening in a special case (see [Q]) :

(3.2.5) <u>If</u> π <u>is square-root closed accessible of order</u> ≤ 3 , <u>then</u> A_π <u>is an isomorphism.</u>

This uses a result of Waldhausen [Wh] that $Wh(\pi) = 0$ if π is accessible of order ≤ 3 . For example, if π is the fundamental group of a fibered knot-complement, then $Wh(\pi) = 0$ and the map $L_n^h(\pi) \to L_n^h(Z)$ induced by abelianization is an epimorphism.

3.3. Let $\mathfrak{F} : Z\pi \to Z\pi'$ be a homomorphism induced by a homomorphism of finitely presented groups $\pi \to \pi'$, preserving orientation characters. By $\mathfrak{F}[T]$ denote the extension of to rings of finite Laurent series. $\mathfrak{F}[T] : (Z\pi)[T] \to (Z\pi')[T]$, that carries T to T then the following result about the Γ-groups (1:7) is proven in [CS1,$\underline{V}$] :

(3.3.1) $\Gamma_n^s(\mathfrak{F}[T]) \cong \Gamma_n^s(\mathfrak{F}) \oplus L_{n-1}^h(\pi',w')$.

This result is analogous to (3.1.1) and is in fact the same result for n odd (so $\Gamma_n^s(\mathcal{F}) = L_n^s(\pi,w)$, for example) or for $\mathcal{F}$ = identity. The proof is also similar to that of (3.1.1) outlined above, but uses a homology splitting principal in even dimensions [CS1,15.1] analogous to the Farrell-Hsiang theorem for homotopy equivalences. (No such principle exists in odd dimensions, in general).

For $\mathcal{F}_o : Z[Z] \to Z$ the augmentation, $\Gamma_n(\mathcal{F}_o) = 0$ for n odd and $\tilde{\Gamma}_{2k}(\mathcal{F}_o)$ is isomorphic to the group of cobordism classes of P.L. locally flat embeddings of S^{2k-3} in S^{2k-1}, $k \geq 3$, with vanishing index (k odd) or Arf invariant (k even) [CS1]. By Levine [Lv], this group is isomorphic to the direct sum of a countably infinite sum of copies of Z, a countably infinite sum of copies of Z_2, and a countably infinite sum of copies of Z_4. (However, an explicit isomorphism is not given in [Lv], only a complete set of (rational) invariants). Thus (3.3.1) allows a calculation of $\Gamma_n(\mathcal{F}_o[T_1,\ldots,T_m])$, at least as an abstract group.

REFERENCES

This list is not quite a complete bibliography, but is complete at the present time (Dec. 1972), when combined with the bibliography of [TW] and of [Ba 1,2,3,4].

[Bk] A. Bak, to appear. (See also a preprint with W. Scharlan, Witt groups of orders and finite groups).

[Ba 1,2] H. Bass. papers in these proceedings.

[Ba 3] H. Bass. American Journal of Math., to appear.

[Ba 4] H. Bass. L_3 of finite abelian groups, to appear.

[Ba 5] H. Bass. "Algebraic K-theory", Benjamin, 1968.

[Ba 6] H. Bass. The stable structure of quite general linear groups, Bull. Amer. Math. Soc. 70 (1964), 429-433.

[BMS] H. Bass. J.W. Milnor, and J.-P. Serre. Solutions of the congruence subgroup problem for $SL_n (n \geq 3)$ and $Sp_{2n} (n \geq 2)$, Publ. IHES No.33 (1967), 59-137.

[Bt] I. Berstein. A proof of the vanishing of the simply-connected surgery obstruction in the odd-dimensional case, preprint, Cornell University, 1969.

[Bt 1] I. Berstein. Some algebraic calculations of Wall groups for Z_2 (unpublished).

[B] W. Browder. Homotopy type of differentiable manifolds, In "Coloq. on Alg. Top"., Aarhus notes, 1962, 42-46.

[Bl] W. Browder. Manifolds with $\pi_1 = Z$, Bull. Amer. Math. Soc. 71 (1966), 238-244.

[Br] W. Browder. "Surgery on simply-connected manifolds". Springer-Verlag, 1972.

[BP] W. Browder and T. Petrie. Diffeomorphisms of manifolds and semi free actions on homotopy spheres. Bull. Amer. Math. Soc. 77 (1971), 160-3.

[BPW] W. Browder, T. Petrie, and C.T.C. Wall. The classification of free actions of cyclic groups of odd order on homotopy spheres, Bull. Amer. Math. Soc. 77 (1971), 455-9.

[Ca] S. Cappell. Lecture notes on the splitting theorem, mimeographed, Princeton University. [See also Bull. Amer. Math. Soc. 77(1971), 281-6).

[Ca 1] S. Cappell. Papers in these proceedings.

[CS 1] S. Cappell and J.L. Shaneson. The codimension two placement problem and homology equivalent manifolds, to appear. (See also: Submanifolds group actions and Knots I, II, and Non-locally flat embeddings, smoothings, and group actions. to appear in Bull. A.M.S.).

[CS2] S. Cappell and J.L. Shaneson. Surgery on 4-manifolds and applications. Comm. Math. Helv. 46(1971), 500-528.

[C] F.X. Connolly, a paper in these proceedings.

[D] A. Dress. Induction and Structure theorems for Grothendieck and Witt rings of orthogonal representations of finite groups, Bull. A.M.S., to appear. (See also a paper in these proceedings).

[F] F.T. Farrell. Fibering manifolds over a circle. Indiana Jour. of Math. 21 (1971-2)

[FH] F.T. Farrell and W.C. Hsiang. Manifolds with $\pi_1 = Z \times_\alpha G$, Amer. J. of Math., to appear (See also B.A.M.S. 74(1968), 548-553).

[HS] W.C. Hsiang and J.L. Shaneson. Fake tori, in "Topology of manifolds" (Proceedings of the 1969 Georgia Conference on Topology of Manifolds), Markham Press, 1970, 19-50.

[HSP] W.C. Hsiang and R. Sharpe. A geometric interpretation of KU_2, these proceedings.

[Kr] M. Karoubi. Developpements recents en K-theorie algebrique to appear.

[K] M. Kervaire. Le theoreme de Barden-Mazur-Stallings. Comm. Math. Helv. 40 (1965), 31-42.

[KM] M. Kervaire and J.W. Milnor. Groups of homotopy spheres, I, Annals of Math. 77(1963), 5-4-537.

[KS] R. Kirby and L.C. Siebenmann. On the triangulation of manifolds and the Hauptvermutung. Bull. Amer. Math. Soc. 75 (1969), 742-749. (See also Notices of the Amer. Math. Soc. 16 (1969), 848).

[LR] R. Lashof and M. Rothenberg. Triangulation of manifolds I, II, Bull. A.M.S. 75 (1969), 750-757.

[Le 1] R. Lee. Computation of Wall groups, Topology 10(1971), Yale University, 1971.

[Le 2] R. Lee. The spherical space form problem, preprint, Yale University, 1971.

[L] J. Lees. Immersions and surgeries of topological manifolds, Bull. A.M.S. 75 (1969), 529-534.

[Lv] J. Levine. Invariants of Knot cobordism, Inventiones Math. 8 (1969), 98-110.

[M] S. Maumary. Paper in these proceddings.

[Md] S. Lopez de Medrano. "Involutions on manifolds", Springer Verlag, 1971.

[M 1] J.W. Milnor. Whitehead torsion, Bull. A.M.S. 72 (1966), 358-426.

[M 2] J.W. Milnor. On manifolds homeomorphic to the 7-sphere, Ann. of Math. 64 (1956), 399-405.

[M 3] J.W. Milnor. "Lectures on the h-cobordism theorem", Princeton University, 1965.

[M 4] J. W. Milnor. A procedure for killing the homotopy groups of differentiable manifolds, In "Proc. Symp. in Pure Math. 3 (Differential Geometry), Amer. Math. Soc., 1961, 39-55.

[N 1] S.P. Novikov. Homotopy equivalent smooth manifolds I, Translations of the Amer. Math. Soc. 48 (1965), 271-396. (Russian: Izv. Akad. Nauk. SSSR, Ser. Mat. 28 (2) (1964), 365-474).

[N 2] S. P. Novikov. Hermitian Analogs of K-theory I, II, Mathematics of the USSR - Izvestia 4(1970), 257-292 and 479-505. (Russian: Tom 34 (1970)).

[N 3] S.P. Novikov. Pontrjagin classes, the fundamental group, and some problems in stable algebra, In "Essays on Topology and Related Topics. Memoires dedies a Georges de Rham. "Springer 1970, 147-155.

[PP] D. S. Passman and T. Petrie. Surgery with coefficients in a field, Ann. of Math. 95 (1972) 385-405.

[P] T. Petrie. The Atiyah-Singer invariant, the Wall groups, and the function te^x+1/te^x-1, Ann. of Math. 92(1970), 174-187.

[Q] F. Quinn. $B_{(TOP_n)}\sim$ and the surgery obstruction, Bull. A.M.S. 77 (1971), 596-600.

[R] A.A. Ranicki, Algebraic L-theory I., II. Proc. Lond. Math. Soc., (to appear) III. these proceedings.

[R 1] A. A. Ranicki, Geometric L-theory (to appear).

[S 1] L.C. Siebenmann. Proceedings of the International Congress of Mathematicians, Nice, 1970, Gauthier-Villiars, 1971.

[S 2] L.C. Siebenmann. "The obstruction to finding a boundary for an open manifold of dimension greater than five", Ph.D. Thesis, Princeton, 1965.

[S i] M.K. Siu. "Computation of Unitary Whitehead groups of cyclic groups", Ph.D. Thesis, Columbia University, 1971.

[Sh] J.L. Shaneson. Wall's surgery obstruction groups for $Z \times G$, Annals of Math. 90 (1969), 296-334.

[Sh 1] J.L. Shaneson. Product formulas for $L_n(\pi)$, Bull. A.M.S. 76 (1970), 787-791.

[Sh 2] J.L. Shaneson. Non-simply-connected surgery and some results in low dimensional topology, Comm. Math. Helv. 45 (1970), 333-352.

[Sp 1] R. Sharpe. Surgery on compact manifolds: The bounded even dimensional case, to appear.

[Sp 2] R. Sharpe. On the structure of the unitary Steinberg group, to appear.

[Sp 3] R. Sharpe. These proceedings.

[Sm 1] S. Smale. Generalized Poincare's conjecture in dimensions greater than form. Ann. of Math. 74 (1961), 391-406.

[Sm 2] S. Smale. On the structure of manifolds. Amer. J. of Math. 84 (1962), 387-399.

[St] J. Stallings. The piecewise linear structure of euclidean space, Proc. Camb. Phil. Soc. 58 (1962), 481-488.

[Sv] D. Sullivan. "Geometric Topology seminar notes", Princeton University, 1967.

[Sv 1] D. Sullivan. Geometric periodicity and the invariants of manifolds, in "Manifolds-Amsterdam 1970", Springer-Verlag 1971.

[T] L.R. Taylor. These proceedings.

[TW] C.T.C. Wall. "Surgery on compact manifolds", Academic Press, 1970.

[W 1] C.T.C. Wall. _Some L groups of finite groups._ Bull. A.M.S., to appear.

[W 2] C.T.C. Wall. _Foundations of algebraic L theory_, these proceedings.

[W 3] C.T.C. Wall. _Free piecewise linear involutions on spheres_ Bull. A.M.S. 74 (1968), 554-558.

[W 4] C.T.C. Wall. _Surgery of non-simply-connected manifolds._ Ann. of Math. 84(1966), 212-276.

[W 5] C.T.C. Wall. _Classification of Hermitian forms IV. Global rings_, to appear.

[Wh] F. Waldhausen. _Whitehead groups of generalized free products_, Preliminary report, mimeographed.

SOME PROBLEMS IN HERMITIAN K - THEORY

by

Julius L. Shaneson

References are to the bibliography of my paper "Hermitian K-theory and topology" (denoted [HKT]) in these proceedings.

1. Calculate the Wall groups of Bieberbach groups.

An n-dimensional Biberbach group is a torsion free group π which satisfies an exact sequence

$$0 \to N \to \pi \to G \to 1 ,$$

where N is maximal abelian in π and free abelian on n generators, and G is finite. These are the fundamental groups of compact flat Riemannian manifolds, which they classify. (See L.S.Charlap, Compact Flat Riemannian Manifolds, I, Ann. of Math. 81 (1965), 15-30). These groups may be one of the simplest classes of torsion free (beyond free abelian groups) which do not appear to fall within the scope of the results of 3.1 and 3.2 in [HKT]. This calculation would be a crucial first step in classifying "homotopy flat manifolds".

2. (Cappell's conjecture). Let π be a 3-dimensional knot group. Then abelianization induces an isomorphism $L_n^h(\pi) \to L_n^h(Z)$.

A 3-dimensional knot group is the fundamental group of the complement of a smooth (or tame P.L. or topological) embedding of S^1 in S^3. (See L. Neuwirth, "Knot Groups", Annals of Mathematics studies 56, Princeton University Press, for example). By Alexander duality, $\pi/[\pi,\pi]$ is infinite cyclic. According to Waldhausen [Wh], $Wh(\pi) = 0$, so the same problem for $L_n^s(\pi)$ is equivalent.

The truth of this conjecture was asserted in [Q]. However, as far as I know, it is at least not known if a 3-dimensional knot group is square-root-closed [HKT, 3.2] accessible. (Actually, there is even a strong rumor that this is not the case).

3. Give algebraic proofs of (3.2.2) and (3.2.3) of [HKT].

Algebraic proofs would of course lead to extensions of these results to rings other than integral group rings. As noted in [HKT], for R a ring between Z and Q, the results can be extended to group rings over R, by geometric methods. Furthermore, as pointed out by Cappell, if 2 is invertible in R, then the square root closed condition can be omitted. In fact, the proof of the relevant geometric splitting principle is somewhat simplified. So perhaps a good first step would be to attack this problem algebraically when 2 is invertible in the ring, as Karoubi did in studying (3.1.1) from an algebraic point of view.

4. Find explicit invariants for elements of $L_n^x(\pi)$.

This is a vague problem, of course, and in some of the calculations the invariants are given as explicitly as one could wish. To a topologist, this problem means the following: given a normal map

$$\begin{array}{ccc} \nu_M & \xrightarrow{b} & \xi \\ \downarrow & & \downarrow \\ M & \xrightarrow{f} & X \end{array}$$

one wants to be able to compute (f,b), or its invariants in some complete system of invariants, in terms of "a priori" information about M, X, f, ξ, and b. This is made a little more precise in the next problem.

5. Find a definition of Wall groups so that the surgery obstruction a normal map can be taken instantly (i.e. without any preliminary surgeries to make it highly connected).

An idea to such a candidate has been in the air for some type: try to define Wall groups in terms of chain complexes with Poincare duality. See for example, §17 of [TW] and A. Miscenko, Homotopy invariants of manifolds, I, Mathematics of the USSR-Izvestia, Amer. Math. Soc. vol. 4(1970), 506-519. The chain complex of M with local coefficients in $Z\pi$, $\pi = \pi_1 M$, would then represent an element in the Wall group, and one would wish naively for the surgery obstruction of a normal map to be a difference of the elements represented by the two manifolds; see Theorem 6.2 of Miscenko's paper, just cited. Actually, Miscenko works over $Z[1/2]$, unfortunately denoted Q in his paper. In fact, it seems likely that some extra relative structure will be needed, corresponding to some information carried by the bundle map, to catch the Kervaire invariant, for example. (Warning: The result on page 263 of [TW] is not a solution to this problem).

To state the next problem, let us go back to 3.2 of [H K T.] in particular to the universal homomorphism of Quinn, $A_\pi: H_n(K(\pi,1):\mathbb{L}) \to L_n(\pi)$ mentioned in connection with (3.2.4). It is well known that, rationally, G/TOP is a product of the Eilenberg-MacLane spaces $K(Q,4i)$. Hence A leads[1] to a homomorphism $H_n(\pi,Q) \to L_n(\pi) \otimes Q$, which we call ℓ_π, following the notation of [TW,17H], where an equivalent definition is given.

6. Prove (or disprove) that ℓ_π is injective.

Wall [TW, §17H] conjectures that this is equivalent to the assertion that a certain higher signature $\sigma_\pi(M,f) \in H_n(\pi,Q)$, defined in [TW] for M a closed manifold and $f:M \to K(\pi,1)$, is a homotopy invariant. Perhaps this has been proven, though I am not aware of it. In any case, in his lecture in the Arbeitstagung, June 1972, Miscenko showed that $\sigma_\pi(M,f)$ is a homotopy invariant for π the fundamental group of a manifold of negative curvature. Of course, the results in [HKT] give many (other) examples for which 6 is true.

7. Give an algebraic description of ℓ_π.

The solution to this problem might be a bridge between 5. and 6.

---

[1] By a construction similar to Sullivan's for his $\mathcal{L}$-class.

8. <u>Which elements of Wall groups are the surgery obstructions of normal maps of closed manifolds.</u>

For example, in [P] it is shown that for p odd only the image of $L_{2k}(e)$ in $L^h_{2k}(Z_p)$ is so realized.

Cappell suggests the following approach. Given a simple, homotopy equivalence $f_i:(X,\partial X) \to (Y,\partial Y)$ of manifolds, $\exists$ an underlying sequence of elementary operations ("collapses", geometrically) from the chain complex $C_*(X;\Lambda)$ to $C_*(Y;\Lambda)$, $\Lambda = Z\pi_1 Y$. Poincare duality, in a strong form, gives a similar passage from $C_*(Y;\Lambda)$ to $C^*(Y,\partial Y;\Lambda)$. Then using f^* back to $C^*(X,\partial X;\Lambda)$ and Poincare duality again, we obtain a sequence of "elementary chain homotopies" of $C_*(X;\Lambda)$ with itself. This should determine an element of $H^n(Z_2, Wh_2(\pi))$ whose vanishing can be used to define the notion of " super simple" homotopy equivalence. (Problem for a topologist: Give a geometric definition of "super simple" homotopy equivalence).

Using such a notion, Wall groups $L^{s^2}_n(\pi)$, for surgery to obtain "super simple" homotopy equivalences, should be defined. More generally, define $L^{s^r}_n(\pi)$, based on some notion of simple equivalence involving K_1 , $K_2,\ldots,K_r$, with natural maps $L^{s^r}_n(\pi) \overset{\beta_r}{\to} L^{s^{r-1}}_n(\pi)$, with $s^o = h$. Clearly if $\gamma = \gamma(f,b)$ for (f,b) a normal map of closed manifolds, $\gamma \in \mathrm{Im}(\beta_r \circ \ldots \circ \beta_1)$ for all r . As a wild guess, one might conjecture the converse. In any case we have the problem:

9. <u>Define the groups $L^{s^r}_n(\pi)$ rigorously $r \geq 2$ and relate them</u>

<u>to each other and</u> $L_n^s(\pi)$ <u>and</u> $L_n^h(\pi)$

In 9. one expects exact sequences involving cohomology of Z_2 with coefficients in suitable higher K-groups.

(I can't resist mentioning a related topological problem, even through it's really outside the present context. One has notions of Poincare complex, simple Poincare complex (called finite Poincare complex in [TW]). One could define super-simple Poincare complexes by requiring the duality map on the chain level to be super-simple in the sense of a suitable K_2-group. Similarly define r-super-simple Poincare complexes for $r > 2$ using higher K-groups, and let ∞-super-simple mean (consistantly) r-super-simple for all r. Closed manifolds will be infinite super-simple. As a wild conjecture, one might formulate the following: Every ∞-super-simple Poincare complex is ∞-super-simple homotopy equivalent to a closed manifold.)

The next few problems concern the Γ-groups of [CS1]; see also [HKT, 1.7] and Cappell's paper in these proceedings on Γ-groups.

10. <u>Find an algebraic proof of</u> (3.3.1) <u>of</u> [HKT] (i.e. [CS 1, 14.2])

11. <u>Prove analogues of</u> (3.2.2) <u>and</u> (3.2.3) <u>for</u> <u>Γ-groups.</u>

There is some reason to believe that, in some cases, the results may be easier to obtain than for Wall groups. For example, by a simple geometric argument, M. Gutierez has proven that any link $S_1^n \cup \dots \cup S_k^n \subset S^{n+2}$, $n \geq 3$, is link cobordant to a <u>split</u> link (i.e. the component S_i^n are contained in disjoint disks).

This implies that if $\mathfrak{J}_r : Z[F_{r+1}] \to Z$ is the augmentation, F_r the free groups of rank r, then $\Gamma_n^h(\mathfrak{J}_k)$ is the direct sum, (k+1)-times, $\Gamma_n^h(\mathfrak{J}_o) \oplus \ldots \oplus \Gamma_n^h(\mathfrak{J}_o)$ (Actually, since $Wh(e) = 0$, $\Gamma_n^h(\mathfrak{J}_k) = \Gamma_n^s(\mathfrak{J}_k) = \ldots$ etc.).

12. Let $\mathfrak{J}^p : Z[Z] \to Z[Z_p]$ be the natural map, $p \geq 0$. Calculate $\Gamma_{2k}^\varepsilon(\mathfrak{J}^p)$, $\varepsilon = s,h$.

(Note $\Gamma_{2k+1}^\varepsilon(\mathfrak{J}^p) = L_{2k+1}^\varepsilon(Z_p)$).

For $k \geq 3$, $\Gamma_{2k}(\mathfrak{J}^o)$ is the subgroup of the group of cobordism classes of knots $S^{2k-3} \subset S^{2k-1}$ with vanishing index (k odd) or Arf invariant (k even). As an abstract group, this has been computed by Levine [Lv]. The solution of this problem would lead, by [CS 1], to a complete classification of the equivariant cobordism classes of knots invariant under free P.L. (or topological) actions of cyclic groups on spheres, and would also give much information in the smooth case.

13. <u>Show that the natural map</u>

$$\Gamma_{2k}^s(\mathfrak{J}^p) \to \Gamma_{2k}(\mathfrak{J}^o)$$

<u>is a monomorphism.</u>

Using the fact that the rational group ring, $Q[Z]$ is a principal ideal ring, it is not difficult to show that

$$\Gamma_{2k}^{s}(\mathfrak{F}_Q^{p}) \to \Gamma_{2k}^{s}(\mathfrak{F}_Q)$$

is a monomorphism; $\mathfrak{F}_Q^p$, for example, denotes the homomorphism $Q[Z] \to Z[Z_p]$ induced by $\mathfrak{F}^p$, and $\Gamma_{2k}^s(\mathfrak{F}_Q^p)$ is defined analogously to $\Gamma_{2n}^s(\mathfrak{F}^p)$. Hence 12. would be implied by the next problem, with $\pi = Z$, $\pi' = Z_p$.

14. <u>Let π' be finite of odd order. Let $\Phi : Z[\pi] \to Z[\pi']$ be induced by an epimorphism $\pi \to \pi'$ of groups. Show that the natural map $\Gamma_{2k}^x(\Phi) \to \Gamma_{2k}^x(\Phi_Q)$ is a monomorphism,</u> $x = s,h$.

Perhaps it is too much to expect the result in this generality In any case, such a general statement must exclude the possibility of elements of even order, in view of the results mentioned in [HKT, §2] and the fact that if $\Phi = id$, $\Gamma_{2k}^x(\Phi) = L_{2k}^x(\pi)$.

Solution of 13 and 14 also have important consequences for the theory of invariant knots of group actions.

15. <u>Can the square-root closed condition be eliminated in</u> (3.2.2) <u>and</u> (3.2.3)?

<u>As a special case, we have</u>

16. <u>Compute</u> $L_n^h(Z_2 * Z_2)$.

This is no doubt the simplest case in which the square-root closed condition in (3.2.2) and (3.2.3) fails. Cappell has shown that $\tilde{L}_4^h(Z_2 * Z_2) = \tilde{L}_4(Z_2) \oplus \tilde{L}_4(Z_2)$, by extending the splitting theorem to this case.

An (unexpected) positive solution would also solve 2. In Cappell's paper in the splitting principle in these proceedings, he gives a unitary nil-group as the obstruction to the general splitting principal. So another version of 14 would be the question of whether or not this group is trivial. As noted above, (3.2.2) and (3.2.3) can be established geometrically over a ring R, $Z \subset R \subset Q$, with $1/2 \varepsilon R$, without the square-root closed condition.

17. We conclude with a problem of Petrie on orders over $Q[t,t^{-1}]$ and bilinear forms.

(1) Let $R = Q[t,t^{-1}]$, F its field of fractions, $\Theta = \prod_{r=o}^{n} R$, (n+1) copies of R. Then Θ is an R-order in $\Theta \otimes_R F$. Fix an order $\Lambda \subset \Theta$ and clasisfy all orders Γ such that

(a) $\Lambda \subset \Gamma$

(b) All orders Λ, Γ, Θ are closed under the Adams operations on Θ defined by

$$\psi^k(\phi_o(t), \phi_1(t), \ldots, \phi_n(t)) = (\phi_o(t^k), \phi_1(t^k), \ldots, \phi_n(t^k)) .$$

and

(c) $\Lambda_\zeta = \Gamma_\zeta$, $\zeta = (\Phi_{p^r}(t))$ the prime ideal in R defined by the cyclictomic polynomial $\Phi_{p^r}(t)$, p prime.

(2) In addition to the above assumptions, assume Λ , Γ , Θ all have non-singular R-valued bilinear forms

$$< , >_\Lambda \; < , >_\Gamma \; , \; < , >_\Theta \; ,$$ so that there are elements $\alpha_\Lambda \in \Lambda$, $\alpha_\Gamma \in \Gamma$, with

$$< x,y >_\Lambda = \mathrm{Id}\left(\frac{x \cdot y}{\alpha_\Lambda}\right)$$

$$< z,w >_\Gamma = \mathrm{Id}\left(\frac{z \cdot w}{\alpha_\Gamma}\right)$$

$$< u,v > = \mathrm{Id}\,(u \cdot v)$$

where $\mathrm{Id} : \Theta \to R$ is defined by

$$\mathrm{Id}\,(\phi_o,\ldots,\phi_n) = \phi_o + \ldots + \phi_n \; .$$

One can assume α_Λ and α_Γ have the form

$t^{\lambda_i} \prod_{j=1}^{n} (1-t^{x_{ij}})$ in Θ , x_{ij} integers, $j=1,2,\ldots,n$, $i=o,\ldots,n$

(3) Petrie points out that an important starting point is the order

$$\Lambda = \frac{R[x]}{\left(\prod_{i=o}^{n}\right)(x-t^{a_i}))} \; , \quad a_i \text{ distinct integers.}$$

Then $\Lambda \subset \vdash\!\Theta\!\dashv$ by sending x to $(t^{a_0},\ldots,t^{a_n})$. Here $\alpha_\Lambda = (\prod_{j\neq i}(1-t^{a_j-a_i}))$; i.e. the i-th co-ordinate of α_Λ is $\prod_{j\neq i}(1-t^{a_j-a_i})$.

(4) Describe the ideal in Λ that annihilates Γ/Λ .

(5) Petrie remarks that he has some interesting examples of $\Lambda \subset \Gamma \subset \vdash\!\Theta\!\dashv$ with $\Lambda \neq \Gamma$ and with the annihilator of Γ/Λ the principal ideal in Λ generated by $\Phi_{pq}(t)\cdot 1$, p and q prime.

Some problems and conjectures in algebraic K-theory

by Max Karoubi

I. DEFINITIONS. Let A be a ring with unit. We denote by $K_n(A)$ the group K_n of the ring A as defined by Quillen. If A is provided by an involution $a \longmapsto \bar{a}$, we denote by ${}_{\varepsilon}O_{n,n}(A)$ the group of "ε- orthogonal matrices", i.e. the multiplicative group of $2n \times 2n$ matrices of the form

$$M = \begin{pmatrix} a & b \\ c & d \end{pmatrix}$$

such that $M.M^* = M^*.M = I$ where

$$M^* = \begin{pmatrix} {}^t\bar{d} & \varepsilon\,{}^t\bar{b} \\ \varepsilon\,{}^t\bar{c} & {}^t\bar{a} \end{pmatrix}$$

and where ε is an element of the center of A such that $\varepsilon.\bar{\varepsilon} = 1$. Let ${}_{\varepsilon}O(A) = \varinjlim {}_{\varepsilon}O_{n,n}(A)$ and let $B_{{}_{\varepsilon}O(A)}$ be the classifying space of ${}_{\varepsilon}O(A)$ (For simplicity we assume from now that 2 is invertible in A). Following Quillen, one may add 2 and 3 celle to $B_{{}_{\varepsilon}O(A)}$ in order to make $\pi_1(B_{{}_{\varepsilon}O(A)})$ abelian without changing the homology of $B_{{}_{\varepsilon}O(A)}$. In that way one obtains a space $B^+_{O(A)}$ which homotopy groups are called ${}_{\varepsilon}L_n(A)$. One has natural homomorphisms $L_n(A) \longrightarrow K_n(A)$ and $K_n(A) \longrightarrow {}_{\varepsilon}L_n(A)$ induced by the forgetful functor and the hyperbolic functor respectively. The kernel (resp. the cokernel) of this homomorphism is called ${}_{\varepsilon}L'_n(A)$ (resp. ${}_{\varepsilon}W_n(A)$). On the other hand the maps $B^+_{GL(A)} \longrightarrow B^+_{{}_{\varepsilon}O(A)}$ and $B^+_{{}_{\varepsilon}O(A)} \longrightarrow B^+_{GL(A)}$ have homotopic fibers called ${}_{\varepsilon}\mathcal{U}(A)$ and ${}_{\varepsilon}\mathcal{V}(A)$. Let

${}_\varepsilon V_n(A) = \pi_n(\mathcal{V}(A))$ and ${}_\varepsilon U_n(A) = \pi_n(\mathcal{U}(A))$, $n > 0$. For $n = 0$ the definitions have to be modified in an obvious way.

II. PROBLEMS AND CONJECTURES.

1. Let Z_2 acts on GL(A) by the formula

$$\alpha \longmapsto ({}^t\bar{\alpha})^{-1}$$

where ${}^t\bar{\alpha}$ is the conjugate of the transpose of A. Then $\mathbb{Z}_2$ acts also on $B_{GL(A)}$, $B^+_{GL(A)}$ and $K_n(A)$. Let $k_n^{odd}(A) = H^{odd}(\mathbb{Z}_2; K_n(A))$ and $k_n^{ev}(A) = H^{ev}(\mathbb{Z}_2; K_n(A))$. Then I conjecture that $k_n(A) \approx k_n(A[x])$ with $k_n = k_n^{ev}$ or k_n^{odd}. By the fundamental theorem in hermitian K-theory (see the "clock exact sequence" below) this problem is equivalent to ${}_\varepsilon L'_n(A) \approx {}_\varepsilon L'_n(A[x])$ and ${}_\varepsilon W_n(A) \approx {}_\varepsilon W_n(A[x])$ (this is true for $n = 0$). More generally let A be a ring with an antiautomorphism σ such that $\sigma^{2k} = 0$. Then $\mathbb{Z}_{2k}$ acts on $K_n(A)$ and one may ask about the homotopy invariance of the group $H^i(\mathbb{Z}_{2k}; K_n(A))$, at least if $(2k)!$ is invertible in A.

2. Shaneson, Wall, Novikov, Ranicki and others have proved that ${}_\varepsilon L'_n(A_z) \approx {}_\varepsilon L'_n(A) \oplus {}_\varepsilon L'_{n-1}(A)$ and ${}_\varepsilon W_n(A_z) \approx {}_\varepsilon W_n(A) \oplus {}_\varepsilon W_{n-1}(A)$ for $n = 0,1,2$ where A_z is the ring of Laurent polynomials $A[z, z^{-1}]$ with the involution $a z \longmapsto \bar{a} z^{-1}$. The natural conjecture is that these statements must be true for arbitrary n.

3. The conjecture 2 is related to the following: one conjectures an exact sequence

$$0 \longrightarrow K_n(A) \longrightarrow K_n(A[z]) + K_n(A[z^{-1}]) \longrightarrow K_n(A_z) \longrightarrow K_{n-1}(A) \longrightarrow 0$$

where the last map (the "index map") is induced by the canonical homomorphism from A_z to SA, the suspension of the ring A (one has then to use the isomorphism $K_n(SA) \approx K_{n-1}(A)$ proved by Wagoner and Gersten). More precisely this exact sequence implies the isomorphisms

$$k_n^{ev}(A_z) \approx k_n^{ev}(A) \oplus k_{n-1}^{odd}(A)$$

$$k_n^{odd}(A_z) \approx k_n^{odd}(A) \oplus k_{n-1}^{ev}(A)$$

(a weaker form of conjecture 3). Then conjecture 2 follows from the five lemma applied to the "clock exact sequence" which is analogous to the Rothenberg exact sequence:

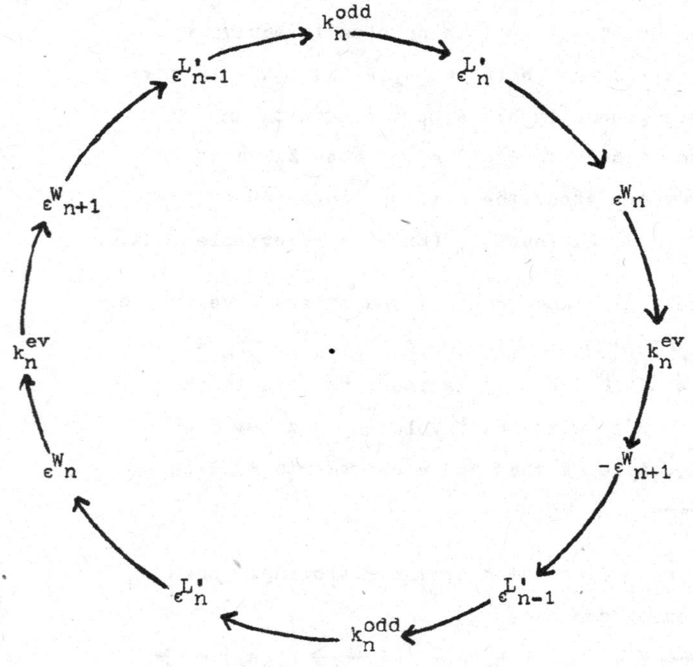

4. The conjecture 3 can be generalized in many directions. One of them is to conjecture a Mayer-Vietoris type of exact

sequences

$$ {}_{\epsilon}L'_n(C) \longrightarrow {}_{\epsilon}L'_n(A) \oplus {}_{\epsilon}L'_n(B) \longrightarrow {}_{\epsilon}L'_n(A \underset{C}{*} B) \longrightarrow $$

$$ {}_{\epsilon}L'_{n-1}(C) \longrightarrow $$

$$ {}_{\epsilon}W_n(C) \longrightarrow {}_{\epsilon}W_n(A) \oplus {}_{\epsilon}W_n(B) \longrightarrow {}_{\epsilon}W_n(A \underset{C}{*} B) \longrightarrow $$

$$ W_{n-1}(C) $$

(for $n = 1,2$ this seems to have been proved by Cappell).

5. Using hermitian K-theory and Quillen's recent results one can prove that $K_3(Z)$ is finite and $\#K_3(Z) \geq 48$ (this contradicts a recent conjecture of Lichtenbaum). The problem now is to compute exactly $K_3(\mathbf{Z})$ and ${}_{\epsilon}L_3(\mathbf{Z}')$ where where $\mathbf{Z}' = \mathbf{Z}[\frac{1}{2}]$. Is the homomorphism $\pi_n^s \longrightarrow K_n(\mathbf{Z})$ injective?

6. If 2 is invertible in A one can prove an isomorphism between the theories ${}_{\epsilon}V_n(A)$ and ${}_{-\epsilon}U_{n+1}(A)$. If A is arbitrary the problem is to find reasonable definitions of ${}_{\epsilon}L_n$, ${}_{\epsilon}U_n$, ${}_{\epsilon}V_n$,... such that we have this theorem again.

7. I offer the following problem: Find a theory $S_n(A)$ for any commutative ring A such that

1) $S_n(A) \otimes Z[\frac{1}{2}] \approx W_n(A) \otimes Z[\frac{1}{2}]$

2) For any exact sequence of rings

$$ 0 \longrightarrow A' \longrightarrow A \longrightarrow A'' \longrightarrow 0 $$

one has a long exact sequence

$$ S_{n+1}(A) \longrightarrow S_{n+1}(A'') \longrightarrow S_n(A') \longrightarrow S_n(A) \longrightarrow $$

$$ S_n(A'') $$

3) If A is the ring of real continuous functions functions on X, then $S_n(A) \approx KO^{-n}(X)$.

8. Find a K-theoretical proof of Quillen's results about the K-theory of a finite field and of Borel's results about the K-theory of number fields.

9. Clifford algebras play an important role in topographical K-theory. On the other hand the periodicity theorems in algebraic K-theory don't use Clifford algebras. What is the exact role of Clifford algebras in both theories? Is it possible to find a unified approach for the periodicity theorems in the topological and algebraic contexts using Clifford algebras?

10. Volodin has defined new K-groups of a ring A (called $\overset{V}{K}_n(A)$ here). It remains to prove that $\overset{V}{K}_n(A) \approx K_n(A)$ (this seems to have been done by Wasserstein and Wagoner). Anyway, it looks reasonable to define in the same manner groups ${}_\varepsilon\overset{V}{L}_n(A)$, ${}_\varepsilon\overset{V}{U}_n(A)$, ${}_\varepsilon\overset{V}{V}_n(A)$. Is it true then that ${}_\varepsilon L_n(A) \approx {}_\varepsilon\overset{V}{L}_n(A)$, ${}_\varepsilon V_n(A) \approx {}_\varepsilon V^V_n(A)$, ${}_\varepsilon U_n(A) \approx {}_\varepsilon U^V_n(A)$?

Unitary algegraic K-theory

by Hyman Bass

This paper contains an exposition of foundational material on the theory of non-singular hermitian forms over rings with involution. A great deal of work in this area has recently appeared. Much of it, including that presented here, has been influenced by surgery theory, in particular the problem of computing surgery obstruction groups. Applications of this work to the computation of certain odd dimensional surgery groups will be given in a third paper.[1] The second paper[2] will give a detailed treatment of the orthogonal group and spinor norm over a commutative ring; this theory is also needed for the above computations.

It will be clear to an informed reader that much of the material presented here can already be found in some form in the literature [1,4,5,7,8,10,12,13,14,15] (though such was less the case when this was first written). My main aim, which is to furnish a fairly comprehensive reference for the algebraic tools (exact sequences, stability theorems,...) needed to compute "unitary Whitehead groups," does not seem to be met by any of the existing published material. Moreover the deepest results, which are due to Bak [1] and Wasserstein [15] are susceptible

[1] L_3 of finite abelian groups (to appear).

[2] Clifford algebras and spinor norms over a commutative ring (to appear).

to important technical improvements of which I have need in making surgery group calculations.

Finally this paper is intended to make propaganda for the notions of "unitary ring" (Ch. I, §4) and "unitary ideal" (Ch. I, §6) due to Bak [1]. These concepts permit a refinement of the notion of quadratic hermitian form, say as presented in Wall [13]. Their inevitability, once one begins a serious study of unitary groups over rings in which 2 is not invertible, is very strikingly revealed in the results of Ch. II, §7, on the normal subgroups of stable unitary groups. The essential role of the notion of unitary ring in the stability theorems of Ch. IV will also be apparent. Further, while the general notion of unitary ring is not needed to define the surgery groups, it will be seen in the third paper of this series 1) how general unitary rings intervene naturally in the computation of surgery groups.

These papers are what survives a larger project undertaken initially in collaboration with Amit Roy. That collaboration concerned mainly the computation of even dimensional surgery groups, but our results were superceded by other work since published. Roy furnished also a major stimulus to the research presented here, for which I thank him warmly. I would also like to thank the Tata Institute of Fundamental Research for their generous hospitality during a portion of this research.

§1. Rings with involution

By ring with involution we understand a ring A equipped with an antiautomorphism $a \mapsto \bar{a}$ of order ≤ 2. Thus $\overline{a+b} = \bar{a} + \bar{b}$, $\overline{ab} = \bar{b}\bar{a}$, and $\bar{\bar{a}} = a$ for all $a, b \in A$.

Examples and remarks:

(1.1) If the involution is trivial, i.e. if $\bar{a} = a$ for all $a \in A$, then A must be commutative. Conversely any commutative ring may be considered a ring with (trivial) involution.

(1.2) If A is a ring with involution we equip the n by n matrix ring $M_n(A)$ with the "conjugate transpose" involution,

$$\overline{(a_{ij})} = (\overline{a_{jk}})$$

(1.3) Let A be a ring with involution and let π be a group. The group ring $A[\pi]$ then admits the involution

$$\overline{\sum_{x \in \pi} a_x x} = \sum_{x \in \pi} \overline{a_x}\, x^{-1}.$$

More generally suppose π is a monoid equipped with an antiautomorphism $x \mapsto \bar{x}$ of order ≤ 2. Let $\chi: \pi \to$ Center (A) be a multiplicative map such that $\chi(x)\chi(\bar{x}) = 1$ for all $x \in \pi$. (If π is a group and $\bar{x} = x^{-1}$ the last condition implies $\overline{\chi(x)} = \chi(x)$.) Then the monoid algebra $A[\pi]$ admits the involution $\sum_x a_x x \mapsto \sum_x \overline{a_x}\chi(x)\bar{x}$.

(1.4) For any ring A we denote

$$H(A) = A \times A^{op}$$

equipped with the involution $\overline{(a,b)} = (b,a)$, and call this the <u>hyperbolic</u> <u>ring</u> on A.

(1.5) For any ring with involution A the Jacobson radical rad A is involution invariant, so B = A/rad A inherits an involution from A. Suppose B is semi-simple, say $B = B_1 \times \ldots \times B_n$ where each B_i is a simple ring. Then the involution induces a permutation of order ≤ 2 of the set of simple factors B_i. Collecting the B_i s into orbits we see that B is a product of rings with involution C_j each of which is isomorphic to some B_i or to $H(B_i)$ for some i.

In particular a simple (Artin) ring with involution is either simple as a ring or else hyperbolic on a simple ring.

(1.6) Let F be a field and let A be a finite dimensional simple F-algebra. An involution on A induces one on C = Center (A), say with fixed field C_0. One says the involution on A is of first or second kind according as $[C:C_0] = 1$ or 2.

§2. Sesquilinear forms

(2.1) Conventions. We fix a ring with involution $A^{(*)}$. By an "A-module" we shall understand a right A-module. A left A-module N may be viewed as a right A-module via the convention

$$na = \bar{a}n \qquad \text{for } n \in N,\ a \in A,$$

and similarly right A-modules may be converted to left A-modules.

For example if M is an A-module then $M^* = \mathrm{Hom}_A(M,A)$ is naturally a left A-module. We write $\bar{M}$ for M^* viewed as a right A-module. Putting

$$\langle f,m\rangle_M = f(m)$$

for $f \in \bar{M}$ and $m \in M$ we then have $\langle fa,mb\rangle_M = \bar{a}\langle f,m\rangle_M b$ for all $a,b \in A$. A homomorphism $g\colon M \to N$ of A-modules induces one $\bar{g}\colon \bar{N} \to \bar{M}$ by

$$\langle \bar{g}f,m\rangle_M = \langle f,gm\rangle_N.$$

(2.2) Sesquilinear forms. Let M and N be A-modules. A function $h\colon N \times M \to A$ is called a sesquilinear form if it is biadditive and satisfies $h(na,mb) = \bar{a}h(n,m)b$ for all $a,b \in A$, $n \in N$, $m \in M$. Such a form induces maps

(*) With slight modifications the material of this section can be developed without the condition $\bar{\bar{a}} = a$ (cf. Bourbaki, [4], §1).

$$ {}_h d\colon N \longrightarrow \bar{M}, \qquad \langle {}_h dn, m\rangle_M = h(n,m) $$

$$ d_h\colon M \longrightarrow \bar{N}, \qquad \langle d_h m, n\rangle_N = \overline{h(n,m)} $$

both of which are A-linear. We call h <u>non singular</u> if ${}_h d$ and d_h are isomorphisms.

Associated with h is the sesquilinear form $\bar{h}\colon M \times N \to A$ defined by

$$ \bar{h}(m,n) = \overline{h(n,m)}. $$

It is immediate from the definition that ${}_{\bar{h}}d = d_h$ and $d_{\bar{h}} = {}_h d$; hence $\bar{h}$ is non singular if and only if h is. Further $\bar{\bar{h}} = h$.

From the sesquilinear form $\langle\ ,\ \rangle_M\colon \bar{M} \times M \to A$ we obtain a homomorphism $d_M = d_{\langle\ ,\ \rangle_M}\colon M \to \bar{\bar{M}}$,

$$ \langle d_M m, f\rangle_{\bar{M}} = \overline{\langle f, m\rangle_M}. $$

We call M <u>reflexive</u> if d_M is an isomorphism.

(2.3) PROPOSITION. <u>Let</u> $h\colon N \times M \to A$ <u>be a sesquilinear form. Then</u> ${}_h d = \overline{d_h} \circ d_N$ <u>and</u> $d_h = \overline{{}_h d} \circ d_M$.

If $n \in N$ and $m \in M$ then

$$ \langle \overline{d_h}\, d_N n, m\rangle_M = \langle d_N n, d_h m\rangle_{\bar{N}} $$
$$ = \overline{\langle d_h m, n\rangle_N} = \overline{\overline{h(n,m)}} = \langle {}_h dn, m\rangle_M. $$

The other formula follows similarly.

(2.4) COROLLARY. *The following conditions are equivalent*

a) h *is non singular*

b) N *is reflexive and* d_h *is an isomorphism*

c) M *is reflexive and* ${}_h d$ *is an isomorphism*.

(2.5) *Orthogonality*. Let $h: N \times M \to A$ be a sesquilinear form. If N_0 is a submodule of N we put $N_0^{\perp} = \{m \in M | h(n,m) = 0$ for all $n \in N_0\}$. It is the kernel of the homomorphism $M \to \overline{N_0}$ induced by h. Similarly a submodule M_0 of M leads to an exact sequence $0 \to {}^{\perp}M_0 \to N \to \overline{M_0}$.

Suppose $N = N_0 \oplus N_1$ and that h is non singular. Then it is clear that $M = N_0^{\perp} \oplus N_1^{\perp}$.

(2.6) *Direct sum*. If $h_i: N_i \times M_i \to A$ $(i = 0,1)$ are sesquilinear forms we define $h = h_0 \oplus h_1$ on $(N_0 \oplus N_1) \times (M_0 \oplus M_1)$ by $h((n_0,n_1),(m_0,m_1)) = h_0(n_0,m_0) + h_1(m_0,m_1)$. Then ${}_h d$ can be identified with ${}_{h_0}d \oplus {}_{h_1}d$, and similarly for d_h. Hence h is non singular if and only if h_0 and h_1 are non singular.

(2.7) *Matrices*. If M is a free A-module with basis $e_1,\ldots,e_m$ the dual basis $\bar{e}_1,\ldots,\bar{e}_m$ of $\bar{M}$ is defined by $\langle \bar{e}_i, e_j \rangle_M = \delta_{ij}$. Suppose N is free with basis $d_1,\ldots,d_n$ and that $g: N \to M$ is A-linear. Then we associate to g the $m \times n$ matrix (g_{ij}) defined by $g(d_j) = \sum_i e_i g_{ij}$. Thus

$$g_{ij} = \langle \overline{e_i}, gd_j \rangle_M$$

$$= \langle \bar{g}\, \bar{e}_i, d_j \rangle_N$$

$$= \overline{\langle d_j, \bar{g}\, \bar{e}_i \rangle_{\bar{N}}} = \bar{g}_{ji}$$

so $(\bar{g})_{ji} = \overline{g_{ij}}$, i.e. the matrix of $\bar{g}$ is the conjugate transpose of that of g.

Let $h: N \times M \to A$ be a sesquilinear form. Giving h is equivalent to giving the $n \times m$ matrix (h_{ij}) defined by $h_{ij} = h(d_i, e_j)$. Note that $(\bar{h})_{ji} = \overline{h_{ij}}$, so that $h \mapsto \bar{h}$ corresponds to conjugate transpose of matrices. We have $(d_h)_{ij} = \langle d_i, d_h e_j \rangle_{\bar{N}} = \overline{\langle d_h e_j, d_i \rangle_N} = \overline{\overline{h(d_i, e_j)}} = h_{ij}$. Thus h and d_h have the same associated matrix.

§3. λ-hermitian forms and modules

(3.1) Sesquilinear modules. They are pairs (M,h) consisting of an A-module M and a sesquilinear form $h: M \times M \to A$. Adjectives applying to M or to h will be applied to the pair (M,h) also. Thus we may speak of (M,h) as being finitely generated, projective, reflexive, non singular . . . We shall sometimes refer to the sesquilinear module M without explicit reference to h.

Let (M,h) and (M',h') be sesquilinear modules. Their direct sum $(M \oplus M', h \oplus h')$ will be denoted $(M,h) \perp (M',h')$, or simply $M \perp M'$ if no confusion results.

A morphism $f:(M,h) \to (M',h')$ is an A-linear map $f: M \to M'$ such that $h'(fx,fy) = h(x,y)$ for all $x,y \in M$.

(3.2) PROPOSITION. *Let $f:(M,h) \to (M',h')$ be a morphism of sesquilinear modules. Suppose that h is non singular. Then f is injective and $M' = (fM) \oplus {}^{\perp}(fM)$.*

If $fx = 0$ then for all $y \in M$ we have $h(x,y) = h'(fx,fy) = 0$ so $x = 0$ because ${}_{h}d$ is injective. Thus f is injective, so we may identify M with a submodule of M' so that $h = h'|M \times M$. If $x \in M'$ the linear form $y \mapsto h'(x,y)$ on M is of the form $y \mapsto h(x_M,y)$ for a unique $x_M \in M$, because h is non singular.

We thus have x expressed uniquely as the sum of $x_M \in M$ and $x - x_M \in {}^{\perp}M$.

(3.3) Sesq (M) <u>and</u> λ-<u>hermitian</u> <u>forms</u>. Let C be the center of A, and let M be an A-module. The set Sesq(M) of all sesquilinear forms on M is a (left) C-module: $(ch)(m,n) = ch(m,n)$. It is equipped with the conjugate C-linear involution $h \mapsto \bar{h}$.

Similarly $\mathrm{Hom}_A(M,\bar{M})$ is a C-module: $(cd)(m) = d(mc) = d(m)c$. The natural isomorphism $h \mapsto d_h$ from Sesq(M) to $\mathrm{Hom}_A(M,\bar{M})$ is C-linear. Moreover $d_{\bar{h}} = {}_hd = \overline{d_h} \circ d_M$ (see (2.2) and (2.3)). Hence if M is reflexive and we identify M with $\bar{\bar{M}}$ via d_M then $h \mapsto d_h$ preserves involutions (using $d \mapsto \bar{d}$ in $\mathrm{Hom}_A(M,\bar{M})$).

Similarly the isomorphism $h \mapsto {}_hd$ from Sesq(M) to $\mathrm{Hom}_A(M,\bar{M})$ is <u>conjugate</u> C-linear, and involution preserving when M is reflexive.

Let $\lambda \in C$ satisfy $\lambda\bar{\lambda} = 1$. A form $h \in$ Sesq(M) is called λ-<u>hermitian</u> if $h = \lambda\bar{h}$. For such forms the relation of orthogonality is symmetric: $h(x,y) = 0 \Leftrightarrow h(y,x) = 0$. Further we have $d_h = d_{\lambda\bar{h}} = \lambda d_{\bar{h}} = \lambda\,{}_hd$ (see (2.3)) so that non-singularity of h need be checked only on one of d_h and ${}_hd$.

The map $h \mapsto \lambda\bar{h}$ is a conjugate C-linear automorphism of order ≤ 2 of Sesq(M). We define $S_\lambda(h) = h + \lambda h$ and put

$$\mathrm{Sesq}_\lambda(M) = \mathrm{Im}(S_\lambda)$$

and

$$\mathrm{Sesq}^{\lambda}(M) = \mathrm{Ker}(S_{-\lambda})$$

The latter is just the module of λ-hermitian forms on M. It contains $\mathrm{Sesq}_{\lambda}(M)$, whose members we call *even* λ-hermitian forms.

If $a \in A$ we put $S_{\lambda}(a) = a + \lambda\bar{a}$ and thus define $S_{\lambda}(A) = \mathrm{Im}(S_{\lambda})$ and $S^{\lambda}(A) = \mathrm{Ker}(S_{-\lambda})$.

(3.4) PROPOSITION. *Let* (M,h) *be a* λ-*hermitian module with* M *projective. Then* h *is even if and only if* $h(x,x) \in S_{\lambda}(A)$ *for all* $x \in M$.

If $h = S_{\lambda}(g)$ then $h(x,x) = S_{\lambda}(g(x,x))$. For the converse choose M' so that $L = M \oplus M'$ is free and consider the form $(L,k) = (M,h) \perp (M',0)$. If we show that $k = S_{\lambda}(g)$ for some g then $h = S_{\lambda}(g|M \times M)$. Moreover $k(x,x) \in S_{\lambda}(A)$ for all $x \in L$, clearly, so it suffices to prove the proposition for (L,k). Let (e_i) be an ordered basis for L. Write $k(e_i,e_i) = S_{\lambda}(a_i)$ and define $g(e_i,e_j)$ to be $k(e_i,e_j)$ if $i < j$, a_i if $i = j$, and 0 if $i > j$. Then it is clear that $k = S_{\lambda}(g)$.

(3.5) *Metabolic modules.* Let (M,g) be a λ-hermitian module. The associated *metabolic module* is

$$\mathbb{H}(M,g) = (M \oplus \bar{M}, h_{(M,g)})$$

where $h = h_{(M,g)}$ is defined by

$$h((m,f),(m',f')) = \langle f,m'\rangle_M + \lambda\overline{\langle f',m\rangle_M} + g(m,m').$$

If we put

$$q_M((m,f),(m',f')) = \langle f,m'\rangle_M \quad .$$

then we see that $h_{(M,g)} = S_\lambda(q_M) + (g \oplus 0)$. Hence $h_{(M,g)}$ *is even if and only if* g *is even*. The homomorphism

$${}_h d\colon M \oplus \bar{M} \longrightarrow \overline{(M \oplus \bar{M})} = \bar{M} \oplus \bar{\bar{M}}$$

is represented by the matrix

$$\begin{pmatrix} g^d & 1_{\bar{M}} \\ \lambda d_M & 0 \end{pmatrix}$$

This is invertible if and only if d_M is, whence assertion a) of the next proposition.

(3.6) PROPOSITION. *Let* (M,g) *and* (M',g') *be* λ-*hermitian modules*.

a) $H(M,g)$ *is non singular if and only if* M *is reflexive*.

b) *Let* $j \in \mathrm{Sesq}(M)$. *Then* $\beta = \begin{pmatrix} 1 & 0 \\ j^d & 1 \end{pmatrix} \in GL(M \oplus \bar{M})$ *is an isomorphism from* $H(M,g)$ *to* $H(M,g - S_\lambda(j))$.

c) *If* g *is even then* $H(M,g) \cong H(M,0)$.

d) *If* $\alpha\colon (M,g) \to (M',g')$ *is an isomorphism then* $H(\alpha) = \alpha \oplus \bar{\alpha}^{-1}$ *is an isomorphism* $H(M,g) \to H(M',g')$, *thus making* H *a functor*.

e) <u>**The** map</u> $((m,f),(m',f')) \mapsto ((m,m'),(f,f'))$ <u>is an isomorphism</u> $H(M,g) \perp H(M',g') \to H((M,g) \perp (M',g'))$ <u>of</u> λ-<u>hermitian modules</u>.

To prove b) put $g' = g - S_\lambda(j)$ and $h' = h_{(M,g')} = h + S_\lambda(j)$. Then

$h'(\beta(m,f),\beta(m',f'))$

$$= h'((m,f + {}_j dm),(m',f' + {}_j dm'))$$

$$= h'((m,f),(m',f')) + h'((0,{}_j dm),(m',f')) + h'((m,f),(0,{}_j dm'))$$

$$= h'((m,f),(m',f')) + \langle {}_j dm,m'\rangle_M + \lambda \overline{\langle {}_j dm',m\rangle}$$

$$= h'((m,f),(m',f')) + S_\lambda(j)(m,m')$$

$$= h((m,f),(m',f')).$$

If g is even we can write $g = S_\lambda(j)$ so c) follows from b).

To prove d) put $h' = h_{(M',g')}$. Then

$$h'(H(\alpha)(m,f),H(\alpha)(m_1,f_1))$$

$$= h'((\alpha m,\bar{\alpha}^{-1}f),(\alpha m_1,\bar{\alpha}^{-1}f_1))$$

$$= \langle \bar{\alpha}^{-1}f,\alpha m_1\rangle_M + \lambda \overline{\langle \bar{\alpha}^{-1}f_1,\alpha m\rangle_{M'}} + g'(\alpha m,\alpha m_1)$$

$$= \langle f,m_1\rangle_M + \lambda \langle f_1,m\rangle_M + g(m,m')$$

$$= h((m,f),(m_1,f_1)),$$

as required. Functoriality of H is clear.

Finally assertion e) follows from a trivial calculation.

(3.7) PROPOSITION. _Let_ (M,g) _be a_ λ-_hermitian module._ _Then_

$$\sigma : (M,g) \perp (M,-g) \longrightarrow H(M,g)$$

$$\sigma(m,n) = (m - n, {}_g dn)$$

is a morphism inducing the identity: $M \oplus 0 \to M \oplus 0$. _It is an isomorphism if and only if_ g _is non-singular_.

If $\Delta : M \to M \oplus M$ is the diagonal then $\mathrm{Ker}(\sigma) = \Delta\,\mathrm{Ker}({}_g d)$ and $\mathrm{Im}(\sigma) = M \oplus \mathrm{Im}({}_g d)$, whence the last assertion.

With $h = h_{(M,g)}$ we have

$$h(\sigma(m,n),\sigma(m',n')) = h((m-n,{}_g dn),(m'-n',{}_g dn'))$$

$$= \langle {}_g dn, m'-n' \rangle_M + \lambda\,\overline{\langle {}_g dn', m-n \rangle_M} + g(m-n,m'-n')$$

$$= g(n,m'-n') + \lambda\,\overline{g(n',m-n)} + g(m-n,m'-n').$$

Combining the first and last terms, and using the fact that $g = \lambda\bar{g}$ we then obtain

$$g(m-n,n') + g(m,m'-n')$$

$$= g(m,m') - g(n,n')$$

$$= (g \oplus (-g))\ ((m,n),(m',n')),$$

whence the proposition.

From Propositions (3.6) and (3.7) we obtain:

(3.8) COROLLARY. *A λ-hermitian module (M,g) is non singular if and only if it is an orthogonal direct summand of a reflexive metabolic module, $H(L,k)$, in which case we can choose $L = M$. If g is even we can choose $k = 0$.*

(3.9) *Totally isotropic direct summands.* Let (M,h) be a λ-hermitian module. If N is a submodule of M we denote $h|N \times N$ for short by $h|N$. We call N a *non-singular submodule* if $h|N$ is non-singular. We call N a *totally isotropic submodule* if $h|N = 0$, i.e. if $N \subset N^{\perp}$.

(3.10) PROPOSITION. *Let (M,h) be a non-singular λ-hermitian module. Let M_0 be a totally isotropic direct summand of M.*

(i) *$M_0^{\perp}$ is a direct summand of M, say $M = M_0^{\perp} \oplus M_1$.*

(ii) *The homomorphism $d\colon M_0 \to \bar{M}_1$, $\langle dm_0, m_1\rangle_{M_1} = h(m_0, m_1)$, is an isomorphism.*

(iii) *The submodule $N = M_1 \oplus M_0$ of M is non-singular, and $M = N \oplus N^{\perp}$.*

(iv) *$1_{M_1} \oplus d\colon (N, h|N) \to H(M_1, h|M_1)$ is an isomorphism of λ-hermitian modules.*

Assertions (i) and (ii) follow easily from the non-singularity of h and the fact that M_0 is a direct summand of M (see (2.5)). Since $M_0 \subset M_0{}^{\perp}$ it follows that the sum $N = M_1 + M_0$ is direct and that the matrix of $_{(h|N)}d$ has the form $\begin{pmatrix} * & d \\ \lambda\bar{d} & 0 \end{pmatrix}$. This is invertible since, by (ii), d is invertible. Assertion (iii) follows from this and Prop. (3.2). Finally, to verify (iv), we must show that, if $h' = h_{(M_1,h|M_1)}$, we have $h'((m_1,dm_0),(m_1',dm_0'))$ $= h((m_1,m_0),(m_1',m_0'))$. The first member equals $\langle dm_0,m_1'\rangle_{M_1}$ $+ \lambda\,\overline{\langle dm_0',m_1\rangle_{M_1}} + (h|M_1)(m_1,m_1') = h(m_0,m_1') + \lambda\,\overline{h(m_0',m_1)} + h(m_1,m_1')$ $= h(m_0,m_1') + h(m_1,m_0') + h(m_1,m_1') + h(m_0,m_0')$ (the last term being zero) $= h(m_1 + m_0, m_1' + m_0')$, which is the second member above. This proves Prop. (3.10).

§4 Unitary rings (A,λ,Λ) and (λ,Λ)-quadratic modules. The groups $KU_i^\lambda(A,\Lambda)$.

(4.1) A unitary ring is a triple (A,λ,Λ) where A is a ring with involution, λ is an element of the center of A such that $\lambda\bar{\lambda} = 1$, and where Λ is an additive subgroup of A satisfying

(4.1.1)
$$\begin{array}{c} S^{-\lambda}(A) = \{a \in A \mid a = -\lambda\bar{a}\} \\ \cup \\ \Lambda \\ \cup \\ S_{-\lambda}(A) = \{a - \lambda\bar{a} \mid a \in A\} \end{array}$$

and

(4.1.2) $\quad a\, r\, \bar{a} \in \Lambda \quad$ for all $a \in A,\ r \in \Lambda$.

A morphism $f:(A,\lambda,\Lambda) \to (A',\lambda',\Lambda')$ of unitary rings is a morphism $f:A \to A'$ of rings with involution such that $f\lambda = \lambda'$ and $f\Lambda \subset \Lambda'$. We call f a unitary surjection if $fA = A'$ and $f\Lambda = \Lambda'$.

(4.2) REMARKS. (4.2.1) If 2 is invertible in A then $S_{-\lambda}(A) = S^{\lambda}(A)$ so λ determines Λ.

(4.2.2) Conversely, Λ almost determines λ. For if (A,μ,Λ) is also a unitary ring then we have $(S_{-\lambda}(A) + S_{-\mu}(A)) \subset \Lambda \subset (S^{-\lambda}(A) \cap S^{-\mu}(A))$. Hence $(\bar{\lambda} - \bar{\mu})\Lambda = 0$ so $\lambda = \mu$ provided that Λ has zero annihilator in the center of A, e.g. if $S_{-\lambda}(A)$ or if $1 - \lambda$ has this property.

(4.2.3) If f is a morphism as above and if $fA = A'$ then $f\Lambda \supset fS_{-\lambda}(A) = S_{-\lambda}(A')$. Hence f is automatically a unitary surjection in case $\Lambda' = S_{-\lambda}(A')$.

(4.2.4) Suppose A <u>is commutative</u>. Let A_0 denote the subring of A generated by all "norms" $a\bar{a}$ $(a \in A)$. This is a subring of the fixed ring $A_1 = S^1(A)$ of the involution, and it contains the set $S_1(A) = \{a + \bar{a} | a \in A\}$ of all "traces", as we see from the equation $(1 + a)\overline{(1 + a)} = 1 + (a + \bar{a}) + a\bar{a}$. Conditions (4.1.1) and (4.1.2) are equivalent to: Λ is an A_0-module between $S_{-\lambda}(A)$ and $S^{-\lambda}(A)$.

Suppose further that <u>the involution on A is trivial.</u> Then $2A \subset A_0$ and $A_0/2A$ is the image of Frobenius $(a \mapsto a^2)$ in $A/2A$. We have $\lambda^2 = 1$. The case <u>$\lambda = -1$</u> is called the <u>restricted symplectic case</u>. Then $S_{-\lambda}(A) = 2A \subset \Lambda \subset S^{-\lambda}(A) = A$; the <u>symplectic case</u> is that when $\Lambda = A$.

If <u>$\lambda = 1$</u> then $S_{-\lambda}(A) = 0 \subset \Lambda \subset S^{-\lambda}(A) = \mathrm{Ker}(A \xrightarrow{2} A)$. When $\Lambda = 0$ we call this the <u>orthogonal case</u>.

(4.2.5) Suppose the ring A is a <u>product</u> $A_1 \times A_2$ of rings A_i with involution. Then $\lambda = (\lambda_1, \lambda_2)$ with $\lambda_i \in \text{Center}(A_i)$ and $\lambda_i\bar{\lambda}_i = 1$. Let $e_1 = (1,0)$ and $e_2 = (0,1)$. If $r = (r_1, r_2) \in \Lambda$ then $e_i r \bar{e}_i = e_i r \in \Lambda$ so $(r_1, 0), (0, r_2) \in \Lambda$, whence $\Lambda = \Lambda_1 \times \Lambda_2$ where Λ_i is the projection of Λ in A_i. It follows that

$(A,\lambda,\Lambda) = (A_1,\lambda_1,\Lambda_1) \times (A_2,\lambda_2,\Lambda_2)$, a _product of unitary rings_.

For the remainder of this section we fix a unitary ring (A,λ,Λ).

(4.3) _The modules_ $\mathrm{Sesq}^{\lambda}_{\Lambda}(M)$ _and_ $\Lambda(M)$. Let M be an A-module. Recall that for $h \in \mathrm{Sesq}(M)$ we have $S_\lambda(h) = h + \lambda\bar{h}$, $\mathrm{Sesq}^{-\lambda}(M) = \mathrm{Ker}(S_\lambda)$, and $\mathrm{Sesq}_{-\lambda}(M) = \mathrm{Im}(S_{-\lambda}) \subset \mathrm{Sesq}^{-\lambda}(M)$ (see (3.3)). We define

$$\mathrm{Sesq}^{-\lambda}_{\Lambda}(M) = \{h \in \mathrm{Sesq}^{-\lambda}(M) \,|\, h(m,m) \in \Lambda \text{ for all } m \in M\}$$

It is immediate that

$$\mathrm{Sesq}_{-\lambda}(M) \subset \mathrm{Sesq}^{-\lambda}_{\Lambda}(M) \subset \mathrm{Sesq}^{-\lambda}(M).$$

Moreover, these are all modules over C_0, the fixed ring of the involution on C = Center(A). If $\Lambda = S^{-\lambda}(A)$ the second inclusion is an equality. If $\Lambda = S_{-\lambda}(A)$ and if M is a projective A-module the first inclusion is an equality (see (3.4)).

We now assume, for convenience, that M is reflexive, and we identify M with $\bar{\bar{M}}$ via d_M.

The C-linear isomorphism $h \mapsto d_h$ from Sesq(M) to $\mathrm{Hom}_A(M,\bar{M})$ carries $\mathrm{Sesq}^{-\lambda}_{\Lambda}(M)$ to

$$\Lambda(M) = \{d \,|\, d = -\lambda\bar{d},\ \langle m,dm\rangle_{\bar{M}} \in \Lambda \text{ for all } m \in M\}.$$

The last condition can be restated as: $\langle dm,m\rangle_M \in \bar{\Lambda}$ for all $m \in M$.

Similarly the conjugate C-linear isomorphism $h \mapsto {}_h d$ from $\mathrm{Sesq}(M)$ to $\mathrm{Hom}_A(M,\bar{M})$ carries $\mathrm{Sesq}_{\Lambda}^{-\lambda}(M)$ to

$$\bar{\Lambda}(M) = \{d \mid d = -\bar{\lambda}\bar{d},\ \langle dm,m\rangle_M \in \Lambda \text{ for all } m \in M\}.$$

The condition $\langle dm,m\rangle_M \in \Lambda$ is equivalent to $\langle m,dm\rangle_{\bar{M}} \in \bar{\Lambda}$. Note therefore that

$$\bar{\Lambda}(M) = \overline{\Lambda(M)} = \bar{\lambda}\ \Lambda(M)$$

Let $u,v \in \bar{M}$ and $a \in A$. If $d:M \to \bar{M}$ is defined by $d(m) = ua\langle v,m\rangle_M$ then $\bar{d}(m) = v\bar{a}\langle u,m\rangle_M$. For we have $\langle \bar{d}(m),n\rangle_M = \langle m,d(n)\rangle_{\bar{M}} = \langle m,ua\langle v,n\rangle_M\rangle_{\bar{M}} = \langle m,u\rangle_M\ a\ \langle v,n\rangle_M$ $= \overline{\langle u,m\rangle_M}\ a\ \langle v,n\rangle_M = \langle v\bar{a}\langle u,m\rangle_M,n\rangle_M$. Further we have $\langle d(m),m\rangle_M = \overline{\langle u,m\rangle_M}\ a\ \langle v,m\rangle_M$, which lies in $\bar{\Lambda}$ if $u = v$ and $a \in \bar{\Lambda}$. It follows that

$$m \longmapsto u\langle v,m\rangle - v\lambda\langle u,m\rangle - ua\langle u,m\rangle$$

is an element of $\Lambda(M)$ for all $u,v \in \bar{M}$ and $a \in \bar{\Lambda} = \bar{\lambda}\Lambda$.

If M is finitely generated and projective then these elements generate $\Lambda(M)$ (additively). For this is a property inherited by direct sums and summands, as one sees easily, and it clearly holds for $M = A$, whence the assertion.

Lef $f:M' \to M$ be an A-linear map. It induces a C-linear map $f^*: \mathrm{Sesq}(M) \to \mathrm{Sesq}(M')$ defined by

$$(f^*h)(m',n') = h(fm',fn').$$

It is easily verified that

$$f^*\bar{h} = \overline{f^*h}\ ,\quad f^*S_\lambda(h) = S_\lambda(f^*h),$$

and $(f^*h)(m',m') = h(fm',fm')$ for $m' \in M'$. It follows that f^* maps $\mathrm{Sesq}_\Lambda^{-\lambda}(M)$ into $\mathrm{Sesq}_\Lambda^{-\lambda}(M')$ and hence induces a homomorphism

$$f^*\colon \mathrm{Sesq}(M)/\mathrm{Sesq}_\Lambda^{-\lambda}(M) \longrightarrow \mathrm{Sesq}(M')/\mathrm{Sesq}_\Lambda^{-\lambda}(M').$$

Denoting the class modulo $\mathrm{Sesq}_\Lambda^{-\lambda}(M)$ of h by $[h]$, and similarly for M', we thus have $f^*[h] = [f^*h]$.

We also have a homomorphism $\mathrm{Hom}_A(M,\bar{M}) \to \mathrm{Hom}_A(M',\overline{M'})$, $d \mapsto \bar{f}df$, and this is compatible with the above, in the following sense:

$$d_{f^*h} = \bar{f}d_h f \text{ and } {}_{f^*h}d = \bar{f}\,{}_h df.$$

Suppose $M = A^m$ with standard basis $e_1,\ldots,e_m$. Then $\mathrm{Sesq}(A^m)$ can be identified with the set $M_m(A)$ of $m \times m$ matrices over A, via the map $h \mapsto (h_{ij})$ where $h_{ij} = h(e_i,e_j)$. This identification is C-linear and converts $h \mapsto \bar{h}$ into conjugate transpose, $(h_{ij}) \mapsto \overline{(h_{ij})} = (\overline{h_{ji}})$ (see (2.7)). It follows that $\Lambda(M)$ is identified with the module

$$\Lambda_m = \Lambda(A^m)$$

of $m \times m$ matrices (h_{ij}) satisfying $h_{ij} = -\lambda\,\overline{h_{ji}}$ and $h_{ii} \in \Lambda$ for all i,j.

Relative to the basis above and the dual basis for $\bar{M}$ the matrix of $d_h: M \to \bar{M}$ coincides with that of h, and the matrix of $_hd$ is its conjugate transpose (see (2.7)).

If $\alpha: A^m \to A^m$ is A-linear then $(\alpha^*h)_{ij} = h(\alpha e_i, \alpha e_j)$ $= h(\sum_u e_u \alpha_{ui}, \sum_v e_v \alpha_{vj}) = \sum_{u,v} \overline{\alpha_{ui}}\, h_{uv}\, \alpha_{vj}$, so

$$((\alpha^*h)_{ij}) = \overline{(\alpha_{ij})}\ (h_{ij})\ (\alpha_{ij})$$

(4.4) A (λ,Λ)-<u>quadratic</u> <u>module</u> is a pair $(M,[q])$ where M is an A-module, $q \in \mathrm{Sesq}(M)$, and $[q]$ is the class of q modulo $\mathrm{Sesq}_\Lambda^{-\lambda}(M)$ (see (4.3)).

With $(M,[q])$ there is an <u>associated</u> λ-<u>hermitian</u> <u>form</u> $S_\lambda[q] = S_\lambda(q) = q + \lambda\bar{q}$, which is well defined because $\mathrm{Sesq}_\Lambda^{-\lambda}(M) \subset \mathrm{Sesq}^{-\lambda}(M) = \mathrm{Ker}(S_\lambda)$, and a <u>quadratic</u> <u>function</u> $[q]: m \mapsto [q(m,m)]$, $M \to A/\Lambda$, where $[a]$ denotes the class modulo Λ of $a \in A$. If $a \in A$ the map $x \mapsto \bar{a}\, x\, a$ stabilizes Λ, so we can define $\bar{a}[x]a = [\bar{a}\, x\, a]$. This done we have

(4.4.1) $[q](ma) = \bar{a}([q](m))a$ for $m \in M$ and $a \in A$.

We can relate the associated λ-hermitian form and quadratic function as follows: If $m,n \in M$ then

$$\begin{aligned} & q(m+n,m+n) - q(m,m) - q(n,n) \\ &= q(m,n) + q(n,m) \\ &= q(m,n) + \lambda\,\overline{q(n,m)} - \lambda\,\overline{q(n,m)} + q(n,m) \\ &= S_\lambda[q](m,n) + S_{-\lambda}(q(n,m)). \end{aligned}$$

Since $S_{-\lambda}(A) \subset \Lambda$ we obtain

(4.4.2) $[q](m+n) - [q](m) - [q](n) = [S_\lambda[q](m,n)]$

A useful consequence of (4.4.1) and (4.4.2) is the equivalence of the condition: $q \in \mathrm{Sesq}_\Lambda^{-\lambda}(M)$ (i.e. (i) $S_\lambda(q) = 0$ and (ii) $[q](m) = 0$ for all $m \in M$), with the conditions:
(i) $S_\lambda(q) = 0$ and (ii') $[q](m) = 0$ for all m in some generating set of the A-module M.

(4.5) A <u>morphism</u> $f: (M',[q']) \to (M,[q])$, of quadratic modules (we often suppress the prefix (λ,Λ) -) is an A-linear map $f: M' \to M$ such that $f^*[q] = [q']$, i.e. such that $q' - f^*q \in \mathrm{Sesq}_\Lambda^{-\lambda}(M')$.

Such an f is also a morphism $(M',S_\lambda[q']) \to (M,S_\lambda[q])$ of the associated λ-hermitian modules, and it preserves the quadratic functions: $[q](fm') = [q'](m')$ for all $m' \in M'$. It follows conversely from the end of (4.4) that f is a morphism of quadratic modules provided that $f^*S_\lambda[q] = S_\lambda[q']$ and that $[q](fm') = [q'](m')$ for all m' in some generating set of M'.

The group of automorphisms of (M,[q]) will be denoted

$$U(M,[q])$$

and called the <u>unitary group</u> of (M,[q]).

In the special cases distinguished in (4.2)(4) we use more standard notation and terminology for U(M,[q]). Thus suppose

A is commutative with trivial involution. In the orthogonal case (A,1,0) we write O(M,[q]) and call it an orthogonal group. In the symplectic case (A,-1,A) we write Sp(M,[q]) and call it a symplectic group.

(4.6) The orthogonal direct sum of quadratic modules (M,[q]) and (M',[q']) is defined to be

$$(M,[q]) \perp (M',[q']) = (M \oplus M',[q \oplus q']).$$

It is clear that $[q \oplus q']$ depends only on [q] and [q'].

Orthogonality in a quadratic module (M,[q]) is defined with reference to $(M,S_\lambda[q])$. Suppose, $M = M_0 \oplus M_1$ where M_0 and M_1 are orthogonal with respect to $S_\lambda(q)$. Put $q_i = q|M_i \times M_i$ $(i = 0,1)$. Then we claim that $[q] = [q_0 \oplus q_1]$, so that $(M,[q]) = (M_0,[q_0]) \perp (M_1,[q_1])$. That is to say orthogonal decompositions of quadratic modules are equivalent to those of the associated λ-hermitian modules.

To verify the claim put $q' = q_0 \oplus q_1$ and $t = q - q'$. By hypothesis $S_\lambda(t) = 0$. By construction $[t](m) = 0$ if $m \in M_0$ or if $m \in M_1$. Since M_0 and M_1 generate M it follows (cf. end of (4.4)) that $t \in \mathrm{Sesq}_\Lambda^{-\lambda}(M)$ as claimed.

(4.7) Hyperbolic forms. For any A-module M we put

$$H(M) = (M \oplus \bar{M}, [q_M])$$

where $q_M((m,f),(m',f')) = \langle f,m'\rangle_M$, and call this the <u>hyperbolic (quadratic) module</u> on M. Note that $h_M = S_\lambda(q_M)$ is the λ-hermitian form $h_{(M,0)}$ of (3.5). The corresponding homomorphism $D_M = d_{(h_M)}: M \oplus \bar{M} \to \overline{(M \oplus \bar{M})} = \bar{M} \oplus \bar{\bar{M}}$ is represented by the matrix $\begin{pmatrix} 0 & 1_{\bar{M}} \\ \lambda d_M & 0 \end{pmatrix}$, which is invertible if and only if M is reflexive. When this is the case we often identify $\bar{\bar{M}}$ with M via d_M, so that the matrix for D_M assumes the form $\begin{pmatrix} 0 & 1_{\bar{M}} \\ \bar{\lambda} 1_M & 0 \end{pmatrix}$.

An isomorphism $\alpha: M \to N$ of A-modules induces an isomorphism $H(\alpha) = \alpha \oplus \bar{\alpha}^{-1}: H(M) \to H(N)$ of quadratic modules. Indeed

$$\begin{aligned} & q_N(H(\alpha)(m,f), H(\alpha)(m',f')) \\ &= q_N((\alpha m, \bar{\alpha}^{-1} f), (\alpha m', \bar{\alpha}^{-1} f')) \\ &= \langle \bar{\alpha}^{-1} f, \alpha m'\rangle_N = \langle f,m'\rangle_M \\ &= q_M((m,f),(m',f')), \end{aligned}$$

so that $H(\alpha)^* q_N = q_M$. We call H the <u>hyperbolic functor</u> from A-modules to (λ,Λ)-quadratic modules.

Note that

$$\begin{array}{ccc} H(M) \perp H(M') & \longrightarrow & H(M \oplus M') \\ ((m,f),(m',f')) & \longmapsto & ((m,m'),(f,f')) \end{array}$$

is an isomorphism of quadratic modules (cf. Prop. (3.6), part e)).

(4.8) PROPOSITION. Let $(M,[q])$ be a (λ,Λ)-quadratic module. Define

$$\sigma: (M,[q]) \perp (M,[-q]) \longrightarrow H(M)$$

by $\sigma(m,n) = (m-n, {}_q dm + \lambda_{\bar{q}} dn)$. Then σ is a morphism. It is an isomorphism if and only if $h = S_\lambda[q]$ is non-singular.

As a homomorphism $M \oplus M \to M \oplus \bar{M}$, σ is represented by the matrix

$$\begin{pmatrix} 1_M & -1_M \\ {}_q d & \lambda_{\bar{q}} d \end{pmatrix} = \begin{pmatrix} 1_M & 0 \\ {}_q d & 1_{\bar{M}} \end{pmatrix} \begin{pmatrix} 1_M & -1_M \\ 0 & {}_h d \end{pmatrix}$$

This factorization shows that σ is invertible if and only if ${}_h d$ is, whence the last assertion.

To show σ is a morphism we calculate:

$$q_M(\sigma(m,n),\sigma(m',n'))$$

$$= q_M((m - n, {}_q dm + \lambda_{\bar{q}} dn), (m' - n', {}_q dm' + \lambda_{\bar{q}} dn'))$$

$$= \langle {}_q dm + \lambda_{\bar{q}} dn, m' - n' \rangle_M$$

$$= q(m,m') - \lambda\bar{q}(n,n') - q(m,n') + \lambda\bar{q}(n,m') - q(n,n') + q(n,n')$$

Subtracting $(q \oplus (-q))((m,n),(m',n')) = q(m,m') - q(n,n')$ from $q_M(\sigma(m,n),\sigma(m',n'))$ we therefore obtain $S_{-\lambda}(k)((m,n),(m',n'))$, where $k((m,n),(m',n')) = q(n,n') - q(m,n')$. Thus $\sigma^* q_M = (q \oplus (-q)) + S_{-\lambda}(k)$ so $\sigma^*[q_M] = [q \oplus (-q)]$, as claimed.

(4.9) COROLLARY. *A* (λ,Λ)-*quadratic module* $(M,[q])$ *is non-singular if and only if it is an orthogonal direct summand of a reflexive hyperbolic module* $H(N)$; *in this case we can even take* $N = M$. *If* M *is a finitely generated projective module we may also take* $N = A^n$ *for some* $n \geq 0$.

(4.10) *Totally isotropic direct summands*. Let $(M,[q])$ be a non-singular (λ,Λ)-quadratic module and put $h = S_\lambda[q]$. Let M_0 be a direct summand of M which is totally isotropic for h, i.e. $h|M_0 = 0$. Then as in Prop. (3.10) we have $M = M_0^{\perp} \oplus M_1$ and $N = M_1 \oplus M_0$ is a non-singular submodule of (M,h); we have $M = N \oplus N^{\perp}$. It follows therefore from (4.6) that $(M,[q]) = (N,[q|N]) \perp (N^{\perp},[q|N^{\perp}])$ as well.

We further have from Prop. (3.10) an isomorphism $1_{M_1} \oplus d\colon (N,h|N) \to H(M_1,h|M_1)$ of λ-hermitian modules, where $\langle dm_0,m_1\rangle_{M_1} = h(m_0,m_1)$.

(4.10.1) *Claim*: *Suppose* M_0 *is totally isotropic in* $(M,[q])$, *i.e. that* $[q|M_0] = 0$. *Then* M_1 *above can be chosen totally isotropic also, so that there is a morphism* $H(M_0) \to (M,[q])$ *inducing the identity* $M_0 \to M_0$.

Put $L = M_1$, identify M_0 with $\bar{L}$ via d as above, so that $N = \bar{L} \oplus L$, and put $q' = q|N$. Modifying q' module $\mathrm{Sesq}_\Lambda^{-\lambda}(N)$ the condition $[q'|\bar{L}] = 0$ can be strengthened to $q'|\bar{L} = 0$. Further replace q' by $q' - S_{-\lambda}(t)$, where $t((f,x),(f',x')) = q'(x,f')$, and we make

q' vanish on $L \times \bar{L}$. Since $S_\lambda(q') = h_{(L,0)} = S_\lambda(q_L)$, where $q_L((f,x),(f',x')) = \langle f,x'\rangle_L$, it follows that

$$q'((f,x),(f',x')) = \langle f,x'\rangle_L + k(x,x')$$

for some $k \in \mathrm{Sesq}(L)$. Now map L into N by $x \mapsto (-_k dx, x)$. The image, L', still satisfies $N = \bar{L} \oplus L'$. We establish the claim by showing that L' is totally isotropic. In fact $q'((-_k dx,x), (-_k dy,y)) = -\langle_k dx,y\rangle + k(x,y) = 0$.

(4.10.2) COROLLARY. Suppose $e \in M$ is unimodular, i.e. a basis for a free direct summand of M, and that $[q](e) = 0$. Then there is an element $f \in M$ such that we have

$$[q](e) = 0 \ , \ [q](f) = 0$$

(4.10.3)

$$S_\lambda(q)(f,e) = 1.$$

These conditions imply e,f is a basis for a hyperbolic plane ($\cong H(A)$) in M.

Put $M_0 = eA$. Since $q(ea,ea) = \bar{a}q(e,e)a \in \Lambda$ we have $[q](x) = 0$ for all $x \in M_0$. Hence we can apply (4.10.1). Moreover $M_1 \cong \bar{M}_0$ is a free module with basis f such that $S_\lambda(q)(f,e) = 1$. Thus we have conditions (4.10.3).

Let e,f be two elements of M satisfying (4.10.3). Define $g\colon H(A) \to (M,[q])$ by $g(a,b) = ea + fb$. Then (4.10.3) implies that g is a morphism. Since $H(A)$ is nonsingular (3.2) implies g is a monomorphism onto a direct summand of M, whence the corollary.

(4.10.4) If $(M,[q])$ is any quadratic module (not necessarily non-singular) a pair $e,f \in M$ satisfying (4.10.3) will be called a <u>hyperbolic pair</u>. We will call an element $e \in M$ a <u>hyperbolic element</u> if it can be completed to a hyperbolic pair e,f. Cor. (4.10.2) above shows that the obvious necessary condition for e to be a hyperbolic element (i.e. that e be unimodular and $[q](e) = 0$) is also sufficient provided the module $(M,[q])$ is non-singular.

(4.11) By <u>unitary module</u> or <u>(λ,Λ)-unitary module</u> we mean a non-singular (λ,Λ)-quadratic module $(P,[q])$ where P is a finitely generated projective A-module, i.e. an object of the category Ⓟ(A) of such modules and their isomorphisms. The unitary modules and their isomorphisms form a <u>category</u>

$$Ⓠ^{\lambda}(A,\Lambda)$$

The direct sums $\oplus$ in Ⓟ(A) and $\perp$ in $Ⓠ^{\lambda}(A,\Lambda)$ make these "categories with product" in the sense of [3], Ch. VII, §1 . Moreover the <u>hyperbolic functor</u> (4.7)

$$H: Ⓟ(A) \longrightarrow Ⓠ^{\lambda}(A,\Lambda)$$

is product preserving. It is also "cofinal" in the sense of [3], Ch. VII, §2; i.e. every unitary module $(P,[q])$ is an orthogonal summand of some $H(Q)$ with $Q \in Ⓟ(A)$. In fact we can take $Q = P$ or Q a sufficiently large free module if we like

(Prop. (4.8)).

(4.12) The Grothendieck, Whitehead, and Witt groups. The Grothendieck group K_0 and the Whitehead group K_1 of a category with product are defined in [3], Ch. VII, §1. For $i = 0$ or 1 we put

$$K_i(A) = K_i\textcircled{P}(A),$$

$$KU_i^{\lambda}(A,\Lambda) = K_i\textcircled{Q}^{\lambda}(A,\Lambda),$$

and we define the Witt group $W_i^{\lambda}(A,\Lambda)$ by the exact sequence

(4.12.1) $$K_i(A) \xrightarrow{H} KU_i^{\lambda}(A,\Lambda) \longrightarrow W_i^{\lambda}(A,\Lambda) \longrightarrow 0$$

A principal aim of this work is the calculation of these groups when $i = 1$ for a number of specific unitary rings (A,λ,Λ).

The forgetful functor $F: \textcircled{Q}^{\lambda}(A,\Lambda) \to \textcircled{P}(A)$, $F(P,[q]) = P$, induces homomorphisms $F: KU_i^{\lambda}(A,\Lambda) \to K_i(A)$. We define involutions $x \mapsto \bar{x}$ on $K_i(A)$ by $[P] \mapsto [\bar{P}]$ for $i = 0$ and by $[\alpha] \mapsto [\bar{\alpha}]$ for $i = 1$. Then it is clear that $F(H(x)) = x + (-1)^i\bar{x}$ for $x \in K_i(A)$. It follows that

(4.12.2) $$\operatorname{Ker}(K_i(A) \xrightarrow{H} KU_i^{\lambda}(A,\Lambda)) \subset \{x \mid x = (-1)^{i+1}\bar{x}\}.$$

Since $H(P) \cong H(\bar{P})$ and since $H(\alpha)$ is conjugate to $H(\bar{\alpha}^{-1})$ for $\alpha \in GL(P)$ it follows further that

(4.12.3) $\quad \mathrm{Ker}(K_i(A) \longrightarrow KU_i^{\lambda}(A,\Lambda)) \supset \{x + (-1)^{i+1}\bar{x}\}$.

In the special cases distinguished in (4.2)(4) we shall use a corresponding special notation for $KU_i^{\lambda}(A,\Lambda)$ and $W_i^{\lambda}(A,\Lambda)$. Thus suppose A is commutative with trivial involution. In the <u>orthogonal case</u> (A,1,0) we write $KO_i(A)$ and $WO_i(A)$. In the <u>restricted symplectic case</u> $(A,-1,\Lambda)$ we write $KSp_i(A,\Lambda)$ and $WSp_i(A,\Lambda)$. In the <u>symplectic case</u> (A,-1,A) we write $KSp_i(A)$ and $WSp_i(A)$.

Similarly in these three cases we shall refer to unitary modules instead as <u>orthogonal modules</u>, <u>restricted symplectic modules</u>, and <u>symplectic modules</u>, respectively.

(4.13) <u>Scaling</u> refers to multiplication by a central unit $\mu \in A$. Suppose (A,λ,Λ) is a unitary ring. Put $\lambda' = \lambda\mu/\bar{\mu}$ and $\Lambda' = \mu\Lambda$. Then (A,λ',Λ') is again a unitary ring.

Let (M,h) be a sesquilinear module. Then $\mu S_{\lambda}(h) = S_{\lambda'}(\mu h)$. It follows that μh is λ'-hermitian if and only if h is λ-hermitian. Moreover $\mu h \in \mathrm{Sesq}_{\Lambda'}^{-\lambda'}(M) \Leftrightarrow h \in \mathrm{Sesq}_{\Lambda}^{-\lambda}(M)$.

Let (N,g) be another sesquilinear module, and let $f: N \to M$ be an A-linear map. Then $f^*(\mu h) = \mu f^*h$. It follows that f is a morphism $(N,\mu g) \to (M,\mu h) \Leftrightarrow$ f is a morphism $(N,g) \to (M,h)$.

From these remarks one deduces that the functor $(M,h) \mapsto (M,\mu h)$, which is the identity on morphisms, induces an isomorphism from the category of λ-hermitian modules to the category of λ'-hermitian modules. Similarly $(M,[q]) \mapsto (M,[\mu q])$ defines an isomorphism of categories $Ⓠ^{\lambda}(A,\Lambda) \to Ⓠ^{\lambda'}(A,\Lambda')$. In particular we obtain isomorphisms

$$KU_i^{\lambda}(A,\Lambda) \cong KU_i^{\lambda'}(A,\Lambda'),$$

and

$$W_i^{\lambda}(A,\Lambda) \cong W_i^{\lambda'}(A,\Lambda')$$

for $i = 0,1$.

(4.3.1) __Example__. If $\mu = \bar{\lambda}$ then $\lambda' = \bar{\lambda}$ and $\Lambda' = \bar{\lambda}\Lambda = \bar{\Lambda}$. Moreover $\mathrm{Sesq}_{\Lambda}^{-\lambda}(M) = \bar{\lambda}\ \mathrm{Sesq}_{\Lambda}^{-\lambda}(M) = \mathrm{Sesq}_{\Lambda}^{-\lambda}(M)$ for any A-module M.

(4.3.2) __Example__. Suppose there is a central unit μ such that $\bar{\mu} = -\mu$. Then the categories $Ⓠ^{\lambda}(A,\Lambda)$ and $Ⓠ^{-\lambda}(A,\mu\Lambda)$ are isomorphic via scaling by μ.

Cases where such a μ exists are the following:

1) Let C = Center (A). Then any invertible $\mu = c - \bar{c}$ $(c \in C)$ works. Such a c exists if A is a simple ring with involution of the second kind.

2) Suppose $A = B\tau$, the group ring over a ring with involution B of a group $\tau = \{1,t\}$ of order 2, the involution on A being $\overline{a + bt} = \bar{a} - \bar{b}t$. Then we can take $\mu = t$. Note that if B is itself a group ring, say $B = R\pi$, then $A = R[\pi \times \tau]$.

§5. Unitary Transvections

In this section we introduce certain kinds of automorphisms of quadratic modules over a unitary ring (A,λ,Λ). The results here are later used only for the stablity theorems in Chapter IV.

(5.1) **Transvections** $\sigma_{u,a,v}$. Let $(M,[h])$ be a (λ,Λ)-quadratic module. We shall abbreviate

$$\langle x,y\rangle = S_\lambda(h)(x,y) = h(x,y) + \lambda\overline{h(y,x)}$$

for $x,y \in M$.

Suppose $u,v \in M$ and $a \in A$ satisfy

$$(5.1.1)\qquad \begin{cases} h(u,u) \in \Lambda & \text{and hence } \langle u,u\rangle = 0) \\ \langle u,v\rangle = 0 & \\ h(v,v) \equiv a \bmod \Lambda & \end{cases}$$

Then we define

$$\sigma = \sigma_{u,a,v}: M \longrightarrow M$$

by

$$\sigma(x) = x + u\langle v,x\rangle - v\bar{\lambda}\langle u,x\rangle - u\bar{\lambda}a\langle u,x\rangle .$$

(5.2) PROPOSITION. *Keep the notations and hypotheses (5.1.1) above.*

(a) $\sigma_{u,a,v} \in U(M,[h])$ *and, for any* $\alpha \in U(M,[h])$ *we have*

(5.2.1) $$\alpha\sigma_{u,a,v}\alpha^{-1} = \sigma_{\alpha u,a,\alpha v}.$$

(b) *If* u,v',a' *also satisfy conditions like those of* (5.1.1) *then*

(5.2.2) $$\sigma_{u,a',v'} \circ \sigma_{u,a,v} = \sigma_{u,a'+\langle v',v\rangle+a,v+v'}$$

Consequently

(5.2.3) $$(\sigma_{u,a,v})^{-1} = \sigma_{u,\langle v,v\rangle-a,-v}$$

(c) *Suppose further that* $h(v,v) \in \Lambda$, *whence* $a \in \Lambda$ *and* $\langle v,v\rangle = 0$. *Then*

(5.2.4) $$\sigma_{u,a,v} = \sigma_{u,a,o} \circ \sigma_{u,o,v}$$

and

(5.2.5) $$\sigma_{u,o,v} = \sigma_{v,o,-u\lambda}.$$

We first show the vanishing of $\langle\sigma x,\sigma x'\rangle - \langle x,x'\rangle = \alpha + \beta + \gamma$, where

$$\alpha = \langle x,\sigma x' - x'\rangle$$
$$= \langle x,u\rangle\langle v,x'\rangle - \langle x,v\rangle\bar{\lambda}\langle u,x'\rangle - \langle x,u\rangle\bar{\lambda}a\langle u,x'\rangle,$$

$$\beta = \langle\sigma x - x,x'\rangle$$
$$= \overline{\langle v,x\rangle}\langle u,x'\rangle - \lambda\overline{\langle u,x\rangle}\langle v,x'\rangle - \overline{\langle u,x\rangle}\lambda\bar{a}\langle u,x'\rangle,$$

and

$$\gamma = \langle\sigma x - x,\ \sigma x' - x'\rangle$$
$$= \langle -v\bar{\lambda}\langle u,x\rangle\ ,\ -v\bar{\lambda}\langle u,x'\rangle\rangle$$
$$= \overline{\langle u,x\rangle}\langle v,v\rangle\langle u,x'\rangle.$$

The simplification of γ uses the conditions $\langle u,u\rangle = 0$ and $\langle u,v\rangle = 0$ of (5.1.1). The first term of α (resp., β) cancels the second term of β (resp., α). The third terms of α and β add up to $-\overline{\langle u,x\rangle}(a + \lambda\bar{a})\langle u,x'\rangle$ which, since $a + \lambda\bar{a} = \langle v,v\rangle$ (a consequence of (5.1.1)), is $-\gamma$. Hence $\alpha + \beta + \gamma = 0$ so $\langle \sigma x,\sigma x'\rangle = \langle x,x'\rangle$ for all $x,x' \in M$.

Next we show that σ preserves $[h]: M \to A/\Lambda$. Recall that $[h](x) = [h(x,x)]$ where $[s]$ = the class modulo Λ of $s \in A$. Let $x \in M$; we have $\sigma x = x + vr + us$ where $r = -\bar{\lambda}\langle u,x\rangle$, so $\bar{r} = -\langle x,u\rangle$, and $s = \langle v,x\rangle - \bar{\lambda}a\langle u,x\rangle = \langle v,x\rangle + ar$. From (4.4.2) we have

$$[h](\sigma x) - [h](x) - [h](\sigma x - x) = [\langle x,\sigma x - x\rangle].$$

Therefore we must show that

$$h(vr + us, vr + us) + \langle x,vr + us\rangle \tag{5.2.6}$$

lies in Λ. From (5.1.1) and (4.4.2) again we have

$$\begin{aligned} h(vr + us,vr + us) &\equiv h(vr,vr) + h(us,us) \\ &\equiv \bar{r}h(v,v)r \equiv \bar{r}\,a\,r \end{aligned}$$

mod Λ. Hence (5.2.6) is congruent mod Λ to $\bar{r}ar + \langle x,v\rangle r + \langle x,u\rangle(\langle v,x\rangle + ar) = \bar{r}ar + \langle x,v\rangle r - \bar{r}(\langle v,x\rangle + ar) = \langle x,v\rangle r - \lambda\overline{\bar{r}\langle x,v\rangle} \in \Lambda$. Thus σ preserves $[h]$.

If $\alpha \in U(M,[h])$ then

$\alpha\ \sigma\ \alpha^{-1}(x)$

$= \alpha(\alpha^{-1}x + u\langle v,\alpha^{-1}x\rangle - v\bar{\lambda}\langle u,\alpha^{-1}x\rangle - u\bar{\lambda}a\langle u,\alpha^{-1}x\rangle)$

$= x + (\alpha u)\langle \alpha v,x\rangle - (\alpha v)\bar{\lambda}\langle \alpha u,x\rangle - (\alpha u)\bar{\lambda}a\langle \alpha u,x\rangle$

$= \sigma_{\alpha u,a,\alpha v}(x).$

We now go to the proof of (b). This will show that $\sigma = \sigma_{u,a,v}$ is an isomorphism and hence also complete the proof of (a). Put $\sigma' = \sigma_{u,a',v'}$, so that

$\sigma' \circ \sigma(x)$

$= \sigma x + u\langle v',\sigma x\rangle - v'\bar{\lambda}\langle u,\sigma x\rangle - u\bar{\lambda}a'\langle u,\sigma x\rangle.$

We have $\langle v',\sigma x\rangle = \langle v',x\rangle - \langle v',v\rangle\bar{\lambda}\langle u,x\rangle$ and $\langle u,\sigma x\rangle = \langle u,x\rangle$ in view of (5.1.1). Hence

$\sigma' \circ \sigma(x)$

$= x + u\langle v,x\rangle - v\bar{\lambda}\langle u,x\rangle - u\bar{\lambda}a\langle u,x\rangle$

$\quad + u\langle v',x\rangle \qquad - u\langle v',v\rangle\bar{\lambda}\langle u,x\rangle$

$\qquad - v'\bar{\lambda}\langle u,x\rangle - u\bar{\lambda}a'\langle u,x\rangle$

$= x + u\langle v + v',x\rangle - (v + v')\bar{\lambda}\langle u,x\rangle$

$\quad - u\bar{\lambda}b\langle u,x\rangle$

where $b = a + \langle v'v\rangle + a'$, whence (5.2.2).

In case $v' = -v$ we can take $a' = \langle v,v\rangle - a$, for then

$a' \equiv h(v,v) + \lambda\overline{h(v,v)} - h(v,v) \equiv h(v,v) = h(-v,-v) \bmod \Lambda$. By the formula just proved $\sigma_{u,a'-v} \circ \sigma_{u,a,v} = \sigma_{u,o,o} = 1_M$. This completes the proof of (b).

Formula (5.2.4) of (c) results from (5.2.2) of (b). Further

$$\begin{aligned}\sigma_{v,o,-u\lambda}(x) &= x + v\langle -u\lambda, x\rangle - (-u\lambda)\bar{\lambda}\langle v,x\rangle \\ &= x - v\bar{\lambda}\langle u,x\rangle + u\langle v,x\rangle \\ &= \sigma_{u,o,v}(x).\end{aligned}$$

This completes the proof of (c) and hence of Prop. (5.2).

(5.3) Transvections in $M = V \perp H(P)$. In this case $M = V \oplus (P \oplus \bar{P})$, and, if $x = (v;p,q) \in M$, we have $h(x,x) = h(v,v) + \langle q,p\rangle_P$.

Suppose P has a unimodular element p_0, i.e. there is a $q_0 \in \bar{P}$ such that $\langle q_0,p_0\rangle_P = 1$. Then for any $v \in V$ the element $x = (v;-p_0 h(v,v),q_0)$ is unimodular and isotropic, in fact $h(x,x) = 0$. These elements together with P and $\bar{P}$ clearly generate M. Thus

(5.3.1) *If P contains a unimodular element then for any quadratic module V, $V \perp H(P)$ is generated as a module by its unimodular isotropic elements.*

For any elements $p_0 \in P$, $w_0 \in V$ and $a_0 \in A$ such that $a_0 \equiv h(w_0,w_0) \bmod \Lambda$, the conditions of (5.1.1) hold, so that we may define σ_{p_0,a_0,w_0}. If $x = (v;p,q)$ then $\sigma_{p_0,a_0,w_0}(x) =$ $x + p_0\langle w_0,x\rangle - w_0\bar{\lambda}\langle p_0,x\rangle - p_0\bar{\lambda}a_0\langle p_0,x\rangle$. Now $\langle w_o,x\rangle = \langle w_o,v\rangle$ and $\langle p_0,x\rangle = \langle p_0,q\rangle$, whence

(5.3.2) $$\sigma_{p_0,a_0,w_0}(v;p,q)$$

$$= (v - w_0\bar{\lambda}\langle p_0,q\rangle \; ; \; p + p_0\langle w_0,v\rangle - p_0\bar{\lambda}a_0\langle p_0,q\rangle, \; q)$$

(5.4) PROPOSITION. *Keep the notation of* (5.3). *Let* $p_i \in P$, $w_i \in V$ *and* $a_i \in A$ *satisfy* $a_i \equiv h(w_i,w_i) \bmod A$ $(1 \leq i \leq m)$. *Let*

$$f = \sum_{i=1}^{m} p_i\langle w_i, \;\; \rangle : V \longrightarrow P$$

and

$$g = \sum_{i=1}^{m} -\bar{\lambda}w_i\langle p_i, \;\; \rangle : \bar{P} \longrightarrow V$$

Put $\sigma = \sigma_{p_1,a_1,w_1} \circ \cdots \circ \sigma_{p_m,a_m,w_m}$. *Then for a certain* A-*homomorphism* $t: \bar{P} \to P$ *we have*

$$\sigma(v;p,q) = (v + g(q) \, ; \, p + f(v) + t(q), q).$$

In particular $\sigma|\bar{P}$ *is the identity.*

(5.4.1) *Remark.* Suppose the quadratic module M is non singular. If V is a finitely generated projective A-module then every A-homomorphism $V \to P$ is of the form of f above, for suitable p_i and w_i. Similarly, if P is finitely generated and projective then every A-homomorphism $\bar{P} \to V$ has the form of g above for suitable p_i and w_i.

We prove Prop. (5.4) by induction on m. The case m = 0 is immediate. Assume now the formula holds for m and that we are given an $(m + 1)^{st}$ triple p_0, a_0, w_0. With σ as in the proposition and $\sigma_0 = \sigma_{p_0, a_o, w_o}$ we have, using (5.3.2),

$$\sigma_0\sigma(v;p,q) = \sigma_o(v + g(q); p + f(v) + t(q), q)$$

$$= (v + g'(q);\ p + f'(v) + t'(q), q)$$

where

$$g'(q) = g(q) - w_0\bar{\lambda}\langle p_0, q\rangle,$$

$$f'(v) = f(v) + p_0\langle w_0, v\rangle$$

and

$$t'(q) = p_0\langle w_0, g(q)\rangle - p_0\bar{\lambda}a_0\langle p_0, q\rangle,$$

whence the proposition.

(5.5) COROLLARY. Suppose $p_0 \in P$ and $q_0 \in \bar{P}$ satisfy $\langle q_o, p_o\rangle_P = 1$, so that p_0, q_0 is a hyperbolic pair in $M = V \perp H(P)$. Let $x = (v;p,q) \in M$.

a) If $q = q_0$ then for any $a \equiv h(v,v) \bmod \Lambda$ we have $\sigma_{p_0,a,v}(x) = (0;p', q_0)$, where $p' = p + p_0(\langle v,v\rangle - a)$.

b) Suppose $P = p_0A$, and hence $\bar{P} = q_0A$, and that p_0, x is a hyperbolic pair. Then $q = q_0$ and $p' = p_0 r$ with $r \in \Lambda$. We have $\sigma x = q_0$ and $\sigma p_0 = p_0$, where $\sigma = \sigma_{p_0,r,o} \circ \sigma_{p_0,a,v}$ $= \sigma_{p_0, a + r, v}$.

According to (5.3.2) $\sigma_{p_0,a,v}(x) = (v - v\lambda\langle p_0,q\rangle;$
$p + p_0\langle v,v\rangle - p_0\bar{\lambda}a\langle p_0,q\rangle,q)$. If $q = q_0$ then $\langle p_0,q\rangle = \lambda$ whence a).

If p_0,x is a hyperbolic pair then $1 = \langle x,p_0\rangle = \langle q,p_0\rangle_P$. Writing $q = q_0b$ with $b \in A$ we see that $b = 1$. Applying a) we obtain $\sigma_{p_0,a,v}(x) = (0;\ p_0r,q_0)$, and clearly $\sigma_{p_0,a,v}(p_0) = p_0$. We have $0 \equiv h(x,x) \equiv h(\sigma_{p_0,a,v}(x),\sigma_{p_0,a,v}(x)) = \langle q_0,p_0r\rangle_P$ $\equiv r \bmod \Lambda$, since x is isotropic and $\sigma_{p_0,a,v} \in U(M,[h])$. Now $\sigma_{p_0,r,0}(p_0) = p_0$ and $\sigma_{p_0,r,0}(0;\ p_0r,q_0) = (0;p_0r,q_0) - p_0\bar{\lambda}r\langle p_0,q_0\rangle$ $= (0;0,q_0) = q_0$. In view of (5.2.2) this completes the proof of b).

(5.6) COROLLARY. <u>Assume</u> P <u>has</u> <u>a</u> <u>unimodular</u> <u>element</u> p_0. <u>Let</u> G <u>be</u> <u>a</u> <u>subgroup</u> <u>of</u> U(M,[h]). <u>Suppose</u> <u>that</u> G <u>acts</u> <u>transitively</u> <u>on</u> <u>the</u> <u>set</u> <u>of</u> <u>hyperbolic</u> <u>elements</u> (see (4.10.4)) <u>in</u> M, <u>and</u> <u>that</u> $\sigma_{p_0,a,v} \in G$ <u>for</u> <u>all</u> $v \in (p_0)^{\perp}$ <u>and</u> $a \in A$ <u>with</u> $a \equiv h(v,v) \bmod \Lambda$. <u>Then</u> G <u>acts</u> <u>transitively</u> <u>on</u> <u>the</u> <u>set</u> <u>of</u> <u>hyperbolic</u> <u>pairs</u> <u>in</u> M <u>and</u> <u>hence</u> <u>also</u> <u>on</u> <u>the</u> <u>set</u> <u>of</u> <u>hyperbolic</u> <u>planes</u> <u>in</u> M.

Write $P = P' \oplus p_0A$ and $\bar{P} = \bar{P}' \oplus q_0A$ with $\langle q_0,p_0\rangle_P = 1$. Thus $M = V \perp H(P) = (V \perp H(P')) \perp H(p_0A)$, and p_0,q_0 is a hyperbolic pair spanning $H(p_0A)$.

Let y,x be another hyperbolic pair in M. Transforming it by

a suitable element of G we can arrange that $y = p_0$. Then Cor. (5.5) gives us an element σ of G such that $\sigma p_0 = p_0$ and $\sigma x = q_0$, whence the corollary.

§6. Change of rings

(6.1) Base change of sesquilinear forms. Let $u: A \to A'$ be a homomorphism of rings with involution. If $f: N \to M$ is a homomorphism of A-modules we shall write $f': N' \to M'$ in place of $f \otimes_A 1_{A'}: N \otimes_A A' \to M \otimes_A A'$.

Let $h: N \times M \to A$ be a sesquilinear form. Then (see Bourbaki [4], §1, n° 4, Prop. 2, p. 15) there is a sesquilinear form $h': N' \times M' \to A$ defined by

$$h'(n \otimes a', m \otimes b') = \overline{a'} \cdot uh(n,m) \cdot b \qquad (6.1.1)$$

A direct calculation shows that $h \mapsto h'$ is an additive map satisfying,

$$\bar{h}' = \overline{h'} \qquad (6.1.2)$$

Applied to $\langle\ ,\ \rangle_M: \bar{M} \times M \to A$ this defines a pairing $\langle\ ,\ \rangle'_M: \bar{M}' \times M' \to A'$, whence a homomorphism

$$j_M: \bar{M}' \longrightarrow \overline{M'}$$

defined by $\langle j_M(f \otimes a'),(m \otimes b')\rangle_{M'} = \bar{a}' \cdot u\langle f,m\rangle_M \cdot b'$. It is clear that j_A is an isomorphism. By additivity it follows that

(6.1.3) j_M is an isomorphism if M is finitely generated and projective.

In case A' is a flat left A-module the functor $M \mapsto M'$ is exact, so the functors $M \mapsto \bar{M}'$ and $M \mapsto \overline{M'}$ are contravariant left exact. It follows therefore from (6.1.3) and a 5-lemma argument that:

(6.1.4) *If A' is a flat left A-module then j_M is an isomorphism for all finitely presented A-modules M.*

From formula (6.1.1) we obtain

$$\langle_{h'} d(n \otimes a'), m \otimes b'\rangle_{M'} = \overline{a'}u\langle_h dn, m\rangle_M b'$$
$$= \langle j_M(_h dn \otimes a'), m \otimes b'\rangle_{M'} ,$$

whence

(6.1.5) $$_{(h')}d = j_M \circ (_h d)'$$

Similarly one shows that

(6.1.6) $$d_{(h')} = j_N \circ (d_h)'$$

(6.2) PROPOSITION. *Let $h: N \times M \to A$ be a non-singular sesquilinear form. Then $h': N' \times M' \to A'$ is non-singular if and only if j_N and j_M are isomorphisms. This is the case under the conditions of (6.1.3) or of (6.1.4) on N and M.*

Since h is non-singular d_h and hence also $(d_h)'$ are isomorphisms. It follows then from (6.1.6) that $d_{(h')}$ is an isomorphism if and only if j_N is one. Similarly (6.1.5) implies

$(h')^d$ is an isomorphism if and only if j_M is one.

Example. Let k be a field and let $A = k[x,y] = k[X,Y]/(X^2,Y^2)$, a commutative ring with trivial involution. Put $M = A/(Ax + Ay)$ and $A' = A/Axy$. Let $h: \bar{M} \times M \to A$ be the form $\langle\ ,\ \rangle_M$. Then it is easily checked that M is reflexive, so h is non-singular. However the A'-module $M'(= M)$ is not reflexive, and $h' = 0$, so h' is far from being non-singular.

(6.3) Base change of quadratic forms. Let $u:(A,\lambda,\Lambda) \to (A',\lambda',\Lambda')$ be a morphism of unitary rings. For any A-module M we have the additive map $\mathrm{Sesq}(M) \to \mathrm{Sesq}(M')$, $h \mapsto h'$ satisfying $\bar{h}' = \overline{h'}$. It follows, since $u\lambda = \lambda'$, that $S_{\pm\lambda}(h)' = S_{\pm\lambda'}(h')$. In particular we have induced maps $\mathrm{Sesq}_{-\lambda}(M) \to \mathrm{Sesq}_{-\lambda'}(M')$ and $\mathrm{Sesq}^{-\lambda}(M) \to \mathrm{Sesq}^{-\lambda'}(M')$. We claim further that $\mathrm{Sesq}_{\Lambda}^{-\lambda}(M)$ maps into $\mathrm{Sesq}_{\Lambda'}^{-\lambda'}(M')$. For suppose $h \in \mathrm{Sesq}_{\Lambda}^{-\lambda}(M)$. Then $h \in \mathrm{Sesq}^{-\lambda}(M)$ so $h' \in \mathrm{Sesq}^{-\lambda'}(M')$ as we saw above. To show that $h'(m',m') \in \Lambda'$ for all $m' \in M'$ it therefore suffices, in view of (4.4), to check this for a set of generators of M'. Thus we can restrict to elements $m' = m \otimes 1$, $m \in M$, in which case $h'(m',m') = uh(m,m) \in u\Lambda \subset \Lambda'$.

The conclusions above furnish a homomorphism

$$\mathrm{Sesq}(M)/\mathrm{Sesq}_{\Lambda}^{-\lambda}(M) \longrightarrow \mathrm{Sesq}(M')/\mathrm{Sesq}_{\Lambda'}^{-\lambda'}(M')$$

and hence a base change,

(6.3.1) $$(M,[q]) \longmapsto (M',[q'])$$

for quadratic modules.

Suppose $f: N \to M$ is an A-linear map. Then a simple calculation shows that

(6.3.2) $$(f^*h)' = (f'^*)h'$$

for $h \in \mathrm{Sesq}(M)$. It follows that if $f: (N,[q]) \to (M,[h])$ is a morphism of quadratic modules then so also is $f': (N',[q']) \to (M',[h'])$. Thus (6.3.1) is a functor, and it preserves orthogonal sums in an obvious sense. By Prop. (6.2) it induces a functor

$$\textcircled{Q}^{\lambda}(A,\Lambda) \longrightarrow \textcircled{Q}^{\lambda'}(A',\Lambda')$$

and, in turn, homomorphisms

$$KU_i^{\lambda}(A,\Lambda) \longrightarrow KU_i^{\lambda'}(A',\Lambda')$$

and

$$W_i^{\lambda}(A,\Lambda) \longrightarrow W_i^{\lambda'}(A',\Lambda')$$

for $i = 0,1$.

(6.4) PROPOSITION. <u>Suppose</u> $u:(A,\lambda,\Lambda) \to (A',\lambda',\Lambda)$ <u>is a unitary surjection</u> (see (4.1)), <u>i.e. that</u> $uA = A'$ <u>and</u> $u\Lambda = \Lambda'$. <u>Let</u> M <u>be a projective</u> A-<u>module</u>. <u>Then the maps</u> $\mathrm{Sesq}(M) \to \mathrm{Sesq}(M')$ <u>and</u> $\mathrm{Sesq}_{\Lambda}^{-\lambda}(M) \to \mathrm{Sesq}_{\Lambda'}^{-\lambda'}(M')$ <u>described above are surjective</u>.

Suppose M is free with ordered bases $(e_i)_{i\in I}$. Given $h' \in \mathrm{Sesq}(M')$ choose $a_{ij} \in A$ so that $ua_{ij} = h'(e_i', e_j')$ and define $h \in \mathrm{Sesq}(M)$ by $h(e_i, e_j) = a_{ij}$. Now suppose further that $h' \in \mathrm{Sesq}_{\Lambda'}^{-\lambda'}(M')$. Then $h(e_j, e_i) = -\lambda\, h(e_i, e_j)$ and $h(e_i, e_i) \in \Lambda' = u\Lambda$. Choose the $a_{ii} \in \Lambda$ and define $g(e_i, e_j) = a_{ij}$ if $i \leq j$ and $g(e_i, e_j) = -\lambda\, \overline{a_{ji}}$ if $i > j$. Then $g \in \mathrm{Sesq}_{\Lambda}^{-\lambda}(M)$ and $g' = h'$.

In the general case choose N so that $M \oplus N$ is free. Given h' on M' we can lift $h' \oplus 0$ to some g on $M \oplus N$ and then $g|M$ is the required lifting of h'. If $h' \in \mathrm{Sesq}_{\Lambda'}^{-\lambda'}(M')$ then we can choose $g \in \mathrm{Sesq}_{\Lambda}^{-\lambda}(M \oplus N)$ and hence $g|M \in \mathrm{Sesq}_{\Lambda}^{-\lambda}(M)$.

(6.5) <u>Conservation of hyperbolic and metabolic forms.</u> Let $u:(A,\lambda,\Lambda) \to (A',\lambda',\Lambda')$ be a morphism of unitary rings. Let M be an A-module and consider

$$\alpha = 1_{M'} \oplus j_M : M' \oplus \bar{M}' \longrightarrow M' \oplus \overline{M'}$$

(see (6.1)). Then

$$\begin{aligned} &q_{M'}(\alpha(m \otimes 1, f \otimes 1), \alpha(m_1 \otimes 1, f_1 \otimes 1)) \\ &\quad = \langle j_M(f \otimes 1), m_1 \otimes 1\rangle_{M'} \\ &\quad = u\langle f, m_1\rangle_M = u\, q_M((m,f),(m_1,f_1)) \\ &\quad = q_M'((m \otimes 1, f \otimes 1), (m_1 \otimes 1, f_1 \otimes 1)). \end{aligned}$$

Hence

> α: $\mathbb{H}(M)' \longrightarrow \mathbb{H}(M')$ _is a morphism of quadratic modules. It is an isomorphism if and only if_ j_M _is one._

One checks similarly that, for any $g \in Sesq^{\lambda}(M)$, $\alpha: \mathbb{H}(M,g)' \to \mathbb{H}(M',g')$ is a morphism of λ-hermitian modules.

(6.6) _Hyperbolic rings._ Let (A,λ,Λ) be a unitary ring and suppose that the ring with involution A is hyperbolic (see (1.4)), say $A = \mathbb{H}(B) = B \times B^{op}$ for a ring B. The opposite ring B^{op} has the same additive group as B and multiplication $a \cdot b = ba$. The involution on A is $\overline{(a,b)} = (b,a)$.

It follows easily that λ must be of the form (u,u^{-1}) with u a unit in the center of B, and

$$S_{-\lambda}(A) = \Lambda = S^{-\lambda}(A) = \{(b, -u^{-1}b) \mid b \in B\}.$$

Put $e_0 = (1,0)$ and $e_1 = (0,1)$; these are central indempotents in A and each A-module M canonically decomposes as $M = M_0 \times M_1$ with $M_i = Me_i$. These decompositions are preserved by A-linear maps. Thus an A-module M is built of a B-module M_0 and a B^{op}-module M_1, or, equivalently, from a right B-module M_0 and a left B-module M_1. From this point of view the pair of dual modules M_0^* and M_1^* form a new A-module $M_1^* \times M_0^*$, and this is the module we have denoted $\overline{M}$.

Let N be any B-module. Then $N \times 0$ is an A-module and $\overline{N \times 0} = 0 \times N^*$. Consider the functor

$$N \longrightarrow H(N \times 0)$$

from B-modules (and their isomorphisms) to (λ,Λ)-quadratic A-modules. Then it is not difficult to verify that it induces an <u>equivalence</u> from the category of reflexive B-modules to the category of non-singular (λ,Λ)-quadratic modules. Thus every non singular quadratic module is of the form $H(N \times 0)$, and we have a group isomorphism $GL(N) \cong U(H(N \times 0))$. Moreover we have isomorphisms

$$K_i(B) \cong KU_i^{\lambda}(A,\Lambda) \qquad (i = 0,1).$$

(6.7) <u>Contravariant base change</u>. Let $u:(A,\lambda,\Lambda) \to (A',\lambda',\Lambda')$ be a morphism of unitary rings. We shall systematically use u to view A'-modules as A-modules. In particular A' will be considered a two-sided A-module. Put $C = \{c \in \mathrm{Center}(A) \mid uc \in \mathrm{Center}(A')\}$.

Let $t: A' \to A$ be a homomorphism of A-bimodules, i.e.,

(6.7.1) $\quad$ t *is additive and* $t(a\,x\,b) = at(x)b$ *for all* $a,b \in A$ *and all* $x \in A'$.

Let M' be an A'-module and let $h' \in \mathrm{Sesq}_{A'}(M')$. Then $(th')(m,n) = t(h(m,n))$ defines an element $th' \in \mathrm{Sesq}_A(M')$. For if a,b $\in$ A we have $th'(ma,nb) = t(\bar{a}h'(m,n)b) = \bar{a}(th')(m,n)b$, by (6.7.1).

We thus have an additive map

$$t\colon \mathrm{Sesq}_{A'}(M') \longrightarrow \mathrm{Sesq}_{A}(M');$$

It is clearly C-linear, i.e. $t(ch') = c(th')$ for $c \in C$.

Suppose $f\colon N' \to M'$ is a homomorphism of A'-modules. If $h' \in \mathrm{Sesq}_{A'}(M')$ then $f^*h' \in \mathrm{Sesq}_{A'}(N')$ and we have $t(f^*h')(n_1,n_2) = t(h'(fn_1,fn_2)) = f^*(th')(n_1,n_2)$, i.e. $t(f^*h') = f^*(th')$, where f is viewed on the right as a homomorphism of A-modules. It follows that $(M',h') \mapsto (M',th')$ <u>defines a functor from sesquilinear modules over</u> A' <u>to those over</u> A.

Consider next the condition:

(6.7.2) $$t(\bar{a}) = \overline{t(a)} \qquad \text{for } a \in A'.$$

Then we have $\overline{(th')}(m,n) = \overline{(th')(n,m)} = \overline{t(h'(n,m))} = t(\overline{h'(n,m)})$ (by (6.7.2)) $= t(\overline{h'}(m,n)) = (t\overline{h'})(m,n)$. Thus (6.7.2) implies $\overline{(th')} = t\overline{h'}$. It follows that $tS_\mu(h') = S_\mu(th')$ for $\mu \in C$. Hence th' is (even) μ-hermitian whenever h' is so.

Suppose further that:

(6.7.3) $$t\Lambda' \subset \Lambda.$$

Then if $h' \in \mathrm{Sesq}_{\Lambda'}^{-\lambda'}(M')$ we have $th' \in \mathrm{Sesq}_{\Lambda}^{-\lambda}(M')$. For

$S_\lambda(th') = tS_{\lambda'}(h') = 0$, as we saw above. Further $(th')(m,m)$ $= t(h'(m,m)) \in t\Lambda' \subset \Lambda$ by (6.7.3). It follows that t induces an additive map

$$\mathrm{Sesq}_{\Lambda'}(M')/\mathrm{Sesq}_{\Lambda'}^{-\lambda'}(M') \longrightarrow \mathrm{Sesq}_A(M')/\mathrm{Sesq}_\Lambda^{-\lambda}(M') .$$

Thus we have <u>a functor</u> $(M',[h']) \mapsto (M',[th'])$ <u>from</u> (λ',Λ')<u>-quadratic</u> A'<u>-modules to</u> (λ,Λ)<u>-quadratic</u> A<u>-modules</u>.

We now investigate conservation of non-singularity. Define

$$t_{M'}: \mathrm{Hom}_{A'}(M',A') \longrightarrow \mathrm{Hom}_A(M',A)$$

by $t_{M'}(d) = t \circ d$. Then for $h' \in \mathrm{Sesq}_{A'}(M')$ we have:

(6.7.4) $$_{(th')}d = t_{M'} \circ {_{(h')}d} \text{ and } d_{(th')} = t_{M'} \circ d_{(h')} .$$

For $\langle {_{(th')}dm}, n\rangle_{M'/A} = (th')(m,n) = t(\langle {_{(h')}dm}, n\rangle_{M'/A'})$ $= \langle t_{M'}({_{(h')}dm}), n\rangle_{M'/A}$. The second equality follows similarly. Consequently:

(6.7.5) th' <u>is non-singular provided that</u> h' <u>is non-singular and</u> $t_{M'}$ <u>is an isomorphism</u>.

Consider a hyperbolic module $H_{A'}(P) = (P \oplus \bar{P},[q_P])$ where P is an A'-module, $\bar{P} = \mathrm{Hom}_{A'}(P,A')$, and $q_P((x,f),(y,g)) =$ $\langle f,y\rangle_{P/A'}$. Put $tH_{A'}(P) = (P \oplus \bar{P},[tq_P])$. We have $(tq_P)((x,f),(y,g)) = t\langle f,y\rangle_{P/A'} = \langle t_P f,y\rangle_{P/A} = q_{P/A}((x,t_P f),(y,t_P g))$. It follows that:

(6.7.6) $1_P \oplus t_P : tM_{A'}(P) \longrightarrow M_A(P)$ is a morphism of quadratic modules. It is an isomorphism provided t_P is one.

Taking $M' = A'$ and $h'(x,y) = \bar{x}y$ we obtain the sesquilinear form $\tau = th'$ defined by

$$\tau : A' \times A' \longrightarrow A, \qquad \tau(x,y) = t(\bar{x}y).$$

Identifying $Hom_{A'}(A',A')$ with A' we find that ${}_{\tau}d = t_{A'} = d_{\tau}$ (because ${}_{(h')}d = 1_{A'} = d_{(h')}$). Thus:

(6.7.7) $t_{A'}$ is an isomorphism if and only if $\tau : A' \times A' \rightarrow A$ is non-singular.

Since $t_{M'\oplus N'} = t_{M'} \oplus t_{N'}$ we see that, once $t_{A'}$ is an isomorphism, so also is $t_{M'}$ for all finitely generated projective A'-modules M'.

Consider the condition:

(6.7.8) A' is a finitely generated projective right A-module.

The same is then true of all finitely generated projective A'-modules. Hence we obtain the following conclusions from the above discussion.

(6.8) PROPOSITION. *Suppose* $t: A' \to A$ *satisfies* (6.7.i) $(i = 1,2,3)$ *and that* $\tau: A' \times A' \to A$, $\tau(x,y) = t(\bar{x}y)$, *is non-singular*. *Further assume* (6.7.8). *Then* $(M,[h']) \mapsto (M',[th'])$ *defines a functor* $t: \mathcal{Q}^{\lambda'}(A',\Lambda') \to \mathcal{Q}^{\lambda}(A,\Lambda)$. *If* P *is a finitely generated projective* A'*-module then* $1_P \oplus t_P: tH_{A'}(P) \to H_A(P)$ *is an isomorphism*. *Hence* t *induces homomorphisms* $KU_i^{\lambda'}(A',\Lambda') \to KU_i^{\lambda}(A,\Lambda)$ *and* $W_i^{\lambda'}(A',\Lambda') \to W_i^{\lambda}(A,\Lambda)$ $(i = 0,1)$.

§7. Unitary ideals and unitary relativization.

We introduce in (7.2) the notion, due to Bak[1], of a unitary ideal in a unitary ring. Then we extend the procedure of Stein [11], for relativizing group valued functors on rings to ideals, to the present setting.

(7.1) The ring $A(\mathfrak{a})$. As in Stein [11] we attach to a ring A and a two sided ideal $\mathfrak{a}$ of A the fibre product diagram

(7.1.1)
$$\begin{array}{ccc} A(\mathfrak{a}) & \xrightarrow{p_1} & A \\ {\scriptstyle p_2}\downarrow & & \downarrow \\ A & \longrightarrow & A/\mathfrak{a} \end{array}$$

where $A(\mathfrak{a}) = \{(a_1,a_2) \in A \times A \mid a_1 \equiv a_2 \bmod \mathfrak{a}\}$ and $p_i(a_1,a_2) = a_i$ $(i = 1,2)$. We also have the diagonal map

$$\Delta: A \longrightarrow A(\mathfrak{a}), \qquad \Delta(a) = (a,a),$$

which is a section of both p_1 and p_2.

Suppose now that A is a ring with involution, $a \mapsto \bar{a}$, and that the ideal $\mathfrak{a}$ is involutory, i.e. $\bar{\mathfrak{a}} = \mathfrak{a}$. Then we can (and shall) view (7.1.1) as a diagram of rings with involution, and Δ as a homomorphism of such rings.

(7.2) Unitary ideals. Let $\mathfrak{a}$ be an involutory (i.e. $\mathfrak{a} = \bar{\mathfrak{a}}$) two sided ideal of A. Denote by $\Lambda_{\mathfrak{a}}$ the additive group generated by $S_{-\lambda}(\mathfrak{q}) = \{q - \lambda\bar{q} \mid q \in \mathfrak{q}\}$ together with all $qr\bar{q}$ $(q \in \mathfrak{q},\ r \in \Lambda)$.

A <u>unitary ideal</u> $(\mathfrak{q},\Gamma)$ consists of an ideal $\mathfrak{q}$ as above and an additive subgroup Γ of A such that

(7.2.1) $$\Lambda_{\mathfrak{q}} \subset \Gamma \subset \Lambda \cap \mathfrak{q}$$

and

(7.2.2) $$a\,r\,\bar{a} \in \Gamma \quad \text{for all } a \in A,\ r \in \Gamma.$$

<u>Remarks.</u>

(7.2.3) If $u:(A,\lambda,\Lambda) \to (A',\lambda',\Lambda')$ is a morphism of unitary rings then $\mathfrak{q} = \mathrm{Ker}(u)$ and $\Gamma = \Lambda \cap \mathfrak{q}$ form a unitary ideal, called the <u>unitary kernel</u> of u.

(7.2.4) If $\mathfrak{q} = A$ then $r = 1\,r\,\bar{1} \in \Gamma$ for all $r \in \Lambda$ so Γ necessarily equals Λ.

(7.2.5) If 2 is invertible in A, more generally if 2 acts invertibly on $\mathfrak{q}$, then $S_{-\lambda}(\mathfrak{q}) = \Lambda \cap \mathfrak{q} = S^{-\lambda}(\mathfrak{q})$, so again determines Γ.

In general $(\mathfrak{q},\Lambda_{\mathfrak{q}})$ and $(\mathfrak{q},\Lambda \cap \mathfrak{q})$ are unitary ideals, but they may well be distinct if 2 is not invertible.

(7.2.6) Suppose A is commutative with trivial involution. For any ideal $\mathfrak{q}$ let $Sq(\mathfrak{q})$ denote the additive group generated by all q^2 $(q \in \mathfrak{q})$. Note that $2\mathfrak{q}^2 \subset Sq(\mathfrak{q}) \subset \mathfrak{q}^2$. Moreover $Sq(A)$ is a subring of A and Λ is an $Sq(A)$-module. We further have $\Lambda_{\mathfrak{q}} = (1-\lambda)\mathfrak{q} + Sq(\mathfrak{q})\cdot\Lambda$.

Suppose $\lambda = -1$, so we are in the restricted symplectic case. Then if $\mathfrak{q} \subset 2A$ we have $\Lambda \cap \mathfrak{q} = \mathfrak{q}$, $\Lambda_{\mathfrak{q}} = 2\mathfrak{q}$ (because $1 - \lambda = 2$), and $Sq(\mathfrak{q}) \subset 2\mathfrak{q}$. The unitary ideals $(\mathfrak{q},\Gamma)$ thus correspond to $Sq(A)$-modules Γ between $2\mathfrak{q}$ and $\mathfrak{q}$.

(7.3) <u>The unitary ring</u> $A(\mathfrak{q},\Gamma) = (A(\mathfrak{q}),\lambda,\Lambda(\Gamma))$. Let $(\mathfrak{q},\Gamma)$ be a unitary ideal in (A,λ,Λ). We define $\Lambda(\Gamma)$ by the cartesian square of groups

(7.3.1)
$$\begin{array}{ccc} \Lambda(\Gamma) & \xrightarrow{p_1} & \Lambda \\ {\scriptstyle p_2}\downarrow & & \downarrow \\ \Lambda & \longrightarrow & \Lambda/\Gamma \end{array}$$

Thus $\Lambda(\Gamma) = \{(r_1,r_2) \in \Lambda \times \Lambda \mid r_1 \equiv r_2 \bmod \Gamma\}$ and $p_i(r_1,r_2) = r_i$ $(i = 1,2)$. Note that $\Lambda(\Gamma) \subset \Lambda \times \Lambda \subset A \times A$.

<u>Claim</u>: $\Lambda(\Gamma) \subset A(\mathfrak{q})$ and, if we write λ also for $(\lambda,\lambda) \in A(\mathfrak{q})$, then $(A(\mathfrak{q}),\lambda,\Lambda(\Gamma))$ is a unitary ring.

We shall denote it $A(\mathfrak{q},\Gamma)$. Note that it fits into the commutative square

(7.3.2)
$$\begin{array}{ccc} A(\mathfrak{q},\Gamma) & \xrightarrow{p_1} & (A,\lambda,\Lambda) \\ {\scriptstyle p_2}\downarrow & & \downarrow \\ (A,\lambda,\Lambda) & \longrightarrow & (A/\mathfrak{q},\lambda',\Lambda') \end{array}$$

where $\lambda' = \lambda \bmod \mathfrak{q}$ and $\Lambda' = \Lambda/(\mathfrak{q} \cap \Lambda)$, and that $\Delta : (A,\lambda,\Lambda) \to A(\mathfrak{q},\Gamma)$

is a morphism of unitary rings which is a section of p_1 and p_2. However the induced square $\begin{array}{ccc} \Lambda(\Gamma) & \longrightarrow & \Lambda \\ \downarrow & & \downarrow \\ \Lambda & \longrightarrow & \Lambda' \end{array}$ is <u>not</u> cartesian unless $\Gamma = \mathfrak{q} \cap \Lambda$.

We now prove the above claim. If $(r_1,r_2) \in \Lambda(\Gamma)$ then $r_1 - r_2 \in \Gamma \subset \mathfrak{q}$ so $(r_1,r_2) \in A(\mathfrak{q})$. It now remains to show that

(i) $$S_{-\lambda}(A(\mathfrak{q})) \subset \Lambda(\Gamma) \subset S^{-\lambda}(A(\mathfrak{q}))$$

and

(ii) $a\ r\ \bar{a} \in \Lambda(\Gamma)$ for all $a = (a_1,a_2) \in A(\mathfrak{q})$ and $r = (r_1,r_2) \in \Lambda(\Gamma)$.

<u>Proof of (i)</u>. If $a = (a_1,a_2) \in A(\mathfrak{q})$ then $S_{-\lambda}(a) = (S_{-\lambda}(a_1),S_{-\lambda}(a_2))$ and $S_{-\lambda}(a_1) - S_{-\lambda}(a_2) = S_{-\lambda}(a_1 - a_2) \in S_{-\lambda}(\mathfrak{q}) \subset \Gamma$, so $S_{-\lambda}(a) \in \Lambda(\Gamma)$.

If $r \in \Lambda(\Gamma) \subset \Lambda \times \Lambda$ then $S_\lambda(r) \in S_\lambda(\Lambda) \times S_\lambda(\Lambda) = 0 \times 0$, so $r \in S^{-\lambda}(A(\mathfrak{q}))$.

<u>Proof of (ii)</u>. Write $a_2 = a_1 + q$ with $q \in \mathfrak{q}$ and $r_2 = r_1 + s$ with $s \in \Gamma$. Then $a\ r\ \bar{a} = (a_1r_1\bar{a}_1,\ a_2r_2\bar{a}_2) \in \Lambda \times \Lambda$ and we must show that $d = a_2r_2\bar{a}_2 - a_1r_1\bar{a}_1$ lies in Γ. Putting $a = a_1$ and $r = r_1$ we have

$$\begin{aligned} d &= (a + q)(r + s)(\bar{a} + \bar{q}) - a\ r\ \bar{a} \\ &= a(r + s)\bar{q} + q(r + s)\bar{a} \\ &\quad + q(r + s)\bar{q} + a\ s\ \bar{a} \\ &= S_{-\lambda}(a(r + s)\bar{q}) + q\ r_2\ \bar{q} + a\ s\ \bar{a} \\ &\in S_{-\lambda}(\mathfrak{q}) + q\ \Lambda\ \bar{q} + a\ \Gamma\ \bar{a} \subset \Gamma. \end{aligned}$$

(See (7.2.1) and (7.2.2).)

(7.4) <u>Relativizing a unitary functor.</u> Given a functor

$$G: (A,\lambda,\Lambda) \longmapsto G^{\lambda}(A,\Lambda)$$

from unitary rings to (non-abelian) groups we propose to define $G^{\lambda}(\mathfrak{a},\Gamma)$ for a unitary ideal $(\mathfrak{a},\Gamma)$ in (A,λ,Λ). Despite the notation, $G^{\lambda}(\mathfrak{a},\Gamma)$ will depend also on (A,λ,Λ), so we shall write $G^{\lambda}(\mathfrak{a},\Gamma)_{(A,\Lambda)}$ when, to avoid confusion, this dependence must be made explicit.

The definiton is furnished by the exact commutative diagram

(7.4.1)
$$\begin{array}{ccccccc} 1 \longrightarrow & G^{\lambda}(\mathfrak{a},\Gamma) & \longrightarrow & G^{\lambda}(A(\mathfrak{a},\Gamma)) & \xrightarrow{p_1} & G^{\lambda}(A,\Lambda) \\ & & \searrow^{\partial} & \downarrow{\scriptstyle p_2} & & \downarrow \\ & & & G^{\lambda}(A,\Lambda) & \longrightarrow & G^{\lambda'}(A/\mathfrak{a},\Lambda') \end{array}$$

obtained by applying G to (7.3.2). The diagram is natural in an obvious sense with respect to the pair $((A,\lambda,\Lambda),(\mathfrak{a},\Gamma))$. Further the composite

$$G^{\lambda}(\mathfrak{a},\Gamma) \xrightarrow{\partial} G^{\lambda}(A,\Lambda) \longrightarrow G^{\lambda}(A/\mathfrak{a},\Lambda')$$

is trivial.

Since $p_1 \circ \Delta = 1_A$ we obtain a semi-direct product decomposition

$$G^{\lambda}(A(\mathfrak{a},\Gamma)) = \Delta G^{\lambda}(A,\Lambda) \ltimes G^{\lambda}(\mathfrak{a},\Gamma) \tag{7.4.2}$$

Since p_2, like p_1, is a split epimorphism, the normal subgroup $G^{\lambda}(\mathfrak{a},\Gamma)$ of $G^{\lambda}(A(\mathfrak{a},\Gamma))$ projects via ∂ (the restriction of p_2) onto a normal subgroup of $G^{\lambda}(A,\Lambda)$. Further ∂ is $G^{\lambda}(A,\Lambda)$-equivariant with respect to the action in (7.4.2) and conjugation on itself.

In case $(\mathfrak{a},\Gamma) = (A,\Lambda)$ we have $A(\mathfrak{a},\Gamma) = (A,\lambda,\Lambda) \times (A,\lambda,\Lambda)$. If G preserves products therefore the map $\partial: G^{\lambda}(\mathfrak{a},\Gamma) \to G^{\lambda}(A,\Lambda)$ is an isomorphism in this case, so the notation is unambiguous. We shall apply the above construction only to functors which preserve products.

Taking G to be one of the functors KU_i or W_i $(i = 0,1)$ we obtain definitions of relative Grothendieck and Whitehead groups $KU_i^{\lambda}(\mathfrak{a},\Gamma)$ and Witt groups $W_i^{\lambda}(\mathfrak{a},\Gamma)$.

§8 Remarks on sesquilinear duality

The results of this section will be used here only in Ch. IV, §2.

(8.1) The data. They are: a ring Λ; an antiautomorphism j of A--we shall also write $\bar{a}$ for $j(a)$; an A-module M; and a map $h: M \times M \to A$ which is j-sesquilinear, i.e. h is biadditive and, abbreviating $[m,n]$ for $h(m,n)$ we have

$$[ma, nb] = \bar{a}[m,n]b$$

for $a,b \in A$.

(8.2) The form g_h. Put $B = \mathrm{End}_A(M)$ and define $g_h = \{\ ,\ \}: M \times M \to B$ by

(8.2.1) $$\{m,n\}x = m[n,x]$$

for $m,nx \in M$. If $a \in A$ then $\{ma,n\}x = ma[n,x] = m[n\bar{a},x] = \{m,n\bar{a}\}x$, so

(8.2.2) $$\{ma,n\} = \{m,n\bar{a}\}.$$

If $b \in B$ then $\{bm,n\}x = (bm)[n,x] = b(m[n,x])$ (b is A-linear) $= b(\{m,n\}x)$, so

(8.2.3) $$\{bm,n\} = b\{m,n\}.$$

To go further we assume henceforth that h is non-singular.

(8.3) <u>The antiautomorphism</u> j_h. The map $b \mapsto \bar{b}$ is an anti-isomorphism from $B = \mathrm{End}_A(M)$ to $\mathrm{End}_A(\bar{M})$. Thus

(8.3.1) $b \longmapsto \tilde{b} = ({}_h d)^{-1}\bar{b}({}_h d)$ <u>is an anti-automorphism</u> j_h <u>of</u> B.

We have $[\tilde{b}m,n] = \langle {}_h d\tilde{b}m,n\rangle_M = \langle \bar{b}\,{}_h d\,m,n\rangle_M = \langle {}_h dm,bn\rangle_M = [m,bn]$. Thus

(8.3.2) $$[\tilde{b}m,n] = [m,bn]$$

for $b \in B$, $m,n \in M$, and this also characterises j_h.

Define $\bar{h}$ by $\bar{h}(m,n) = j^{-1}h(n,m)$, i.e. $\overline{\bar{h}(m,n)} = h(n,m)$. Then $\bar{h}$ is (j^{-1})-sesquilinear and non-singular. Thus $j_{\bar{h}}$ is defined and we have $\bar{h}(j_{\bar{h}}(b)n,m) = \bar{h}(n,bm)$, i.e. $j^{-1}h(m,j_{\bar{h}}(b)n) = j^{-1}h(bm,n)$. Hence $h(bm,n) = h(m,j_{\bar{h}}(b)n)$ $= h(j_h(j_{\bar{h}}(b))m,n)$, by (8.3.2), and so

(8.3.3) $$j_{\bar{h}} = j_h^{-1}\ .$$

(8.4) <u>Relation to</u> C = Center (A). The homothetic map $C \to B$, $c \mapsto c_M$ is defined by

$$c_M m = mc$$

for $c \in C$ and $m \in M$. We claim this is compatible with the involutions on A and B. For if $c \in C$ then $[\widetilde{(c_M)}n,m] = [n,c_M m]$ $= [n,mc] = [n,m]c = [n\bar{c},m] = [(\bar{c})_M n,m]$, so

(8.4.1) $$(\bar{c})_M = \widetilde{(c_M)} \quad \text{for all } c \in C.$$

If c is a unit of C then ch is also non-singular and we have $(ch)(j_{ch}(b)n,m) = (ch)(n,bm) = ch(n,bm) = ch(j_h(b)n,m)$, whence

(8.4.2) $$j_{ch} = j_h \text{ for } c \text{ a unit of } C.$$

It follows from this and (8.3.3) that:

(8.4.3) <u>If</u> $h = \lambda\bar{h}$ <u>for some unit</u> λ <u>in</u> C <u>then</u> $j_h^2 = 1_B$.

The condition $h = \lambda\bar{h}$ almost implies that the automorphism $\alpha = j^2$ of A is also the identity. For a simple calculation shows then that $\alpha[m,n] = (\lambda\bar{\lambda})[m,n]$ for all $m,n \in M$. If these elements $[m,n]$ generate the ring A therefore then indeed we must have $\lambda\bar{\lambda} = 1$ and $\alpha = 1_A$.

(8.5) <u>Sesquilinearity of</u> g_h. By (8.2.3) $g_h(m,n) = \{m,n\}$ is B-linear in the first variable. Moreover $\{m,\bar{b}n\}x = m[\bar{b}n,x] = m[n,bx] = \{m,n\}bx$, so

(8.5.1) $$\{m,\bar{b}n\} = \{m,n\}b$$

for $b \in B$ and $m,n \in M$. It follows that

(8.5.2) g_h <u>is sesquilinear for</u> $j_h^{-1} = j_{\bar{h}}$.

Hence we can define $\tilde{g}_h$ by $\tilde{g}_h(m,n) = j_h g(n,m) = \widetilde{g(n,m)}$.

(8.6) <u>The relation</u> $g_{\bar{h}} = \widetilde{g_h}$. This follows from the calculation,

$$[g_{\bar{h}}(m,n)x,y] = [m\bar{h}(n,x),y]$$

$$= [m\overline{[x,n]},y] = [x,n][m,y]$$

$$= [x,n[m,y]] = [x,g_h(n,m)y]$$

$$= [\widetilde{g_h(n,m)}x,y] = [\tilde{g}_h(m,n)x,y].$$

If $\lambda \in C$ it is clear that $g_{\lambda h} = \lambda_M g_h$.

(8.6.1) <u>If</u> $h = \lambda\bar{h}$ <u>for some</u> $\lambda \in C$ <u>then</u> $g_h = \lambda_M \widetilde{g_h}$.

(8.7) <u>Non-singularity of</u> g_h. Put $g = g_h$. Then

$$_g d(m) = 0 \iff \{m,n\} = 0 \quad \text{for all } n$$

$$\iff m[n,x] = 0 \quad \text{for all } x,n$$

$$\iff m\,\mathfrak{a}_h = 0,$$

where $\mathfrak{a}_h$ is the two sided ideal of all sums of elements of the form $[n,x]$ in A.

(8.7.1) $_g d$ <u>is injective if</u> $\mathfrak{a}_h = A$.

$$d_g(n) = 0 \iff \{m,n\} = 0 \quad \text{for all } m$$

$$\iff m[n,x] = 0 \quad \text{for all } m,x$$

$$\iff [n,x] \in ann_A(M) \text{ for all } x.$$

(8.7.2) d_g <u>is injective if</u> M <u>is a faithful</u> A-<u>module</u>.

To prove non-singularity of g we must assume that M is <u>balanced</u>, i.e. that the natural homomorphism

$$A \longrightarrow \mathrm{End}_B(M)^{op}$$

is an isomorphism. Note that the faithfulness of M is just the injectivity of this map.

Given $f \in \mathrm{Hom}_B(M,B)$ and $x \in M$ the map $n \mapsto f(n)x$ is B-linear so, since M is balanced $f(n)x = n(f,x)$ for a unique element $(f,x) \in A$. If $a \in A$ then $n(f,xa) = f(n)xa = n(f,x)a$, so the map $M \to A$, $x \mapsto (f,x)$ is A-linear. Since $h = [\,,\,]$ is assumed non-singular we have $(f,x) = [m,x]$ for some (in fact unique) $m \in M$. Thus $f(n)x = n(f,x) = n[m,x] = (n,m)x = d_g(m)(n)x$, so $f = d_g(m)$. This proves the surjectivity of $d_g : M \to \mathrm{Hom}_B(M,B)$. In view of (8.7.2) we have therefore proved:

(8.7.3) <u>If the module</u> M <u>is balanced then</u> g_h <u>is non-singular</u>.

(8.8) <u>Going backwards</u>. Once $g = g_h$ is non-singular we can take B, j_h, M, and g as starting point and reverse the whole process. If M is balanced (the criterion we obtained for non-singularity of g) then A reappears as $\mathrm{End}_B(M)^{op}$; we use the opposite ring because M is a <u>left</u> B-module. Formula (8.2.1) shows that h is the form derived from g, and formula (8.2.2) shows that j is then the involution induced on A by g.

(8.10) $\mathrm{Sesq}_A(M)$ <u>as B-module</u>. This structure can be seen via the isomorphisms $\mathrm{Sesq}_A(M) \cong \mathrm{Hom}_A(M,\bar{M})$, or directly via the definition

$$(hb)(m,n) = h(bm,n)$$

for $h \in \mathrm{Sesq}_A(M)$ and $b \in B$.

Suppose h is non-singular and b is invertible. Then hb is non-singular because ${}_{hb}d = {}_{h}d \circ b$. We now compare j_{hb} with j_h: $(hb)(j_{hb}(b')m,n) = h(bj_{hb}(b')m,n)$, and it also equals $(hb)(m,b'n) = h(bm,b'n) = h(j_h(b')bm,n)$. It follows that $bj_{hb}(b') = j_h(b')b$, i.e. that

(8.10.1) $$j_{hb} = \mathrm{Int}(b^{-1}) \circ j_h,$$

where $\mathrm{Int}(b^{-1})(b') = b^{-1}b'b$ for $b' \in B$.

Now we compute $\overline{hb}$: $\overline{hb}(m,n) = j^{-1}(hb(n,m)) = j^{-1}(h(bn,m))$ $= \bar{h}(m,bn) = \bar{h}(j_{\bar{h}}(b)m,n) = \bar{h}j_{\bar{h}}(b)(m,n)$, so

(8.10.2) $$\overline{hb} = \bar{h}j_{\bar{h}}(b),$$

or $h\tilde{b} = \bar{h}b$. Recall that $j_{\bar{h}} = j_h^{-1}$ (8.3.3).

(8.11) <u>Simple rings.</u> Keep the data of (8.1) and assume that A is a simple ring, that $j^2 = 1_A$, and that M is a simple

A-module. Then B is a division ring and $Sesq(M) \cong Hom_A(M,\bar{M})$ is a one dimensional B-module. If $h \neq 0$ then Schur's lemma implies that h is non-singular, and we have $Sesq(M) = hB$.

Let $\lambda \in C$ = Center (A) be such that $\lambda\bar{\lambda} = 1$. One of the following cases must occur:

(i) h is λ-hermitian, i.e. $h = \lambda\bar{h}$. In this case the λ-hermitian forms on M are those of the form hb with $b \in B$ and $\bar{b} = b$.

(ii) $g = S_{-\lambda}(h)$ is a non-singular $(-\lambda)$-hermitian form on M.

(One uses (8.10.2) for the last part of (i).)

Note that we can always take $\lambda = 1$, in which case we conclude that M supports a non-singular hermitian or anti-hermitian form. In either case the involution on A permits us to construct a corresponding one on B: B _admits an involution compatible on_ C _with that induced from_ A.

If case (ii) above never occurs then any form h is λ-hermitian and it follows from (i) that $B = S^1(B)$. This implies that B is a field with trivial involution, and that B = C = Center (A). Moreover λ must be ± 1, and the involution on $A = End_C(M)$ is j_g for some non-singular λ-hermitian form g on the C-module M, unique up to scalar multiples. We rephrase part of this for the record:

The simple A-*module* M *admits a non-singular even* $(-\lambda)$-*hermitian form except in the following case: The involution is trivial on* $C = \text{Center}\,(A)$, $\lambda = \pm 1$,

(8.11.1) $A = \text{End}_C(M)$, *and the involution on* A *arises from a non-singular* λ-*hermitian form on the* C-*module* M. *In this case any* $h \neq 0$ *in* $\text{Sesq}_A(M)$ *is* λ-*hermitian, and* $\text{Sesq}_A(M) = hC$.

Chapter II. The groups $U(H(P))$

Here we initiate a detailed study of the unitary groups $U(H(P))$ of hyperbolic modules over a unitary ring (A,λ,Λ). When $P = A^n$ we obtain groups $U_{2n}^{\lambda}(A,\Lambda)$ whose union is the infinite unitary group $U^{\lambda}(A,\Lambda)$. We classify the normal subgroups of $U^{\lambda}(A,\Lambda)$, following ideas of Bak [1]. For each unitary ideal $(\mathfrak{q},\Gamma)$ there is the normal "congruence group" $U^{\lambda}(\mathfrak{q},\Gamma)$ and the group $EU^{\lambda}(\mathfrak{q},\Gamma) = (U^{\lambda}(A,\Lambda),U^{\lambda}(\mathfrak{q},\Gamma))$ for which we give a set of generators. Every normal subgroup of $U^{\lambda}(A,\Lambda)$ lies between $EU^{\lambda}(\mathfrak{q},\Gamma)$ and $U^{\lambda}(\mathfrak{q},\Gamma)$ for a unique $(\mathfrak{q},\Gamma)$. Thus the normal subgroups of $U^{\lambda}(A,\Lambda)$ are determined by the abelian groups

$$KU_1^{\lambda}(\mathfrak{q},\Gamma) = U^{\lambda}(\mathfrak{q},\Gamma)/EU^{\lambda}(\mathfrak{q},\Gamma).$$

A principal tool of this chapter is the unitary Whitehead lemma and other formulas in §3 which are taken from Wasserstein [15]. They give the commutator formulas in §5 and §6, also due to Wasserstein. (Cf. also Wall [14].)

The classification of normal subgroups of $U_{2n}^{\lambda}(A,\Lambda)$ under finiteness assumptions on A, and with n "sufficiently large" is due to Bak [1]. His proofs, and even the construction of the groups $U_{2n}^{\lambda}(\mathfrak{q},\Gamma)$ involve extensive calculations. Our setting is much simpler in that we work in the infinite unitary group. However it should be noted that the use of the "Stein relativization method to construct the groups of level $(\mathfrak{q},\Gamma)$ is a substantial simplification of Bak's more direct approach.

§1. Defining equations for $U(\mathbb{H}(P))$

(1.1) Notation. Let (A,λ,Λ) be a unitary ring (Ch. I, (4.1)). Let P be a reflexive A-module and use d_P to identify P with $\bar{\bar{P}}$ (Ch. I, (2.2)). Recall (Ch. I, (4.7)) that $\mathbb{H}(P)$ denotes the (λ,Λ)-quadratic module $(P \oplus \bar{P}, [q_P])$ where $q_P((x;f),(y,g)) = \langle f,y\rangle_P$. The associated λ-hermitian form is $h_P = S_\lambda(q_P) = q_P + \lambda\,\overline{q_P}$. The homomorphism

$$(h_P)^d\colon P \oplus \bar{P} \longrightarrow \overline{(P \oplus \bar{P})} = \bar{P} \oplus P$$

is represented by the matrix

$$(1.1.1) \qquad \mathbb{H}_P = \begin{pmatrix} 0 & 1_{\bar{P}} \\ \bar{\lambda} 1_P & 0 \end{pmatrix}$$

Here we treat elements of $P \oplus \bar{P}$ as column vectors with matrices multiplying on the left. Thus elements of $\mathrm{End}(P \oplus \bar{P})$ are represented by matrices of the type

$$\begin{pmatrix} \mathrm{Hom}(P,P) & \mathrm{Hom}(\bar{P},P) \\ \mathrm{Hom}(P,\bar{P}) & \mathrm{Hom}(\bar{P},\bar{P}) \end{pmatrix}$$

If $\sigma = \begin{pmatrix} \alpha & \beta \\ \gamma & \delta \end{pmatrix} \in \mathrm{End}(P \oplus \bar{P})$ then $\bar{\sigma} = \begin{pmatrix} \bar{\alpha} & \bar{\gamma} \\ \bar{\beta} & \bar{\delta} \end{pmatrix} \in \mathrm{End}(\bar{P} \oplus P)$. The letter I will be used indiscriminately to denote various identity endomorphisms. For example we shall often write $\mathbb{H}_P = \begin{pmatrix} 0 & I \\ \bar{\lambda} I & 0 \end{pmatrix}$.

(1.2) <u>Automorphisms of</u> $(P \oplus \bar{P}, h_P)$ are elements $\sigma \in GL(P \oplus \bar{P})$ such that $\bar{\sigma}\, H_P\, \sigma = H_P$, i.e. such that $\sigma^{-1} = H_P^{-1} \bar{\sigma} H_P$. Now for any endomorphism $\sigma = \begin{pmatrix} \alpha & \beta \\ \gamma & \delta \end{pmatrix}$ we have

$$H_P^{-1}\, \bar{\sigma}\, H_P = \begin{pmatrix} 0 & \lambda I \\ I & 0 \end{pmatrix} \begin{pmatrix} \bar{\alpha} & \bar{\gamma} \\ \bar{\beta} & \bar{\delta} \end{pmatrix} \begin{pmatrix} 0 & I \\ \bar{\lambda} I & 0 \end{pmatrix}$$

$$= \begin{pmatrix} \bar{\delta} & \lambda \bar{\beta} \\ \bar{\lambda}\bar{\gamma} & \bar{\alpha} \end{pmatrix}$$

so

(1.2.1) $$\sigma\, H_P^{-1}\, \bar{\sigma}\, H_P = \begin{pmatrix} \alpha\bar{\delta} + \bar{\lambda}\beta\bar{\gamma} & \lambda\alpha\bar{\beta} + \beta\bar{\alpha} \\ \gamma\bar{\delta} + \bar{\lambda}\delta\bar{\gamma} & \delta\bar{\alpha} + \lambda\gamma\bar{\beta} \end{pmatrix}$$

and

(1.2.2) $$H_P^{-1}\, \bar{\sigma}\, H_P\, \sigma = \begin{pmatrix} \bar{\delta}\alpha + \lambda\bar{\beta}\gamma & \bar{\delta}\beta + \lambda\bar{\beta}\delta \\ \bar{\alpha}\gamma + \bar{\lambda}\bar{\gamma}\alpha & \bar{\alpha}\delta + \bar{\lambda}\bar{\gamma}\beta \end{pmatrix}$$

We thus obtain:

(1.3) PROPOSITION. $\sigma = \begin{pmatrix} \alpha & \beta \\ \gamma & \delta \end{pmatrix} \in \mathrm{End}(P \oplus \bar{P})$ <u>is an automorphism of</u> $(P \oplus \bar{P}, h_P)$ <u>if and only if any two of the following three conditions hold</u>:

(1.3.1) $$\sigma \in GL(P \oplus \bar{P})$$

(1.3.2) $$\begin{cases} \alpha\bar{\delta} + \bar{\lambda}\beta\bar{\gamma} = I \\ S_\lambda(\beta\bar{\alpha})\ (= \beta\bar{\alpha} + \lambda\alpha\bar{\beta}) = 0 \\ S_{\bar{\lambda}}(\gamma\bar{\delta})\ (= \gamma\bar{\delta} + \bar{\lambda}\delta\bar{\gamma}) = 0 \end{cases}$$

$$(1.3.3)\qquad \begin{cases} \bar{\alpha}\delta + \bar{\lambda}\bar{\gamma}\beta = I \\ S_{\lambda}(\bar{\delta}\beta) = 0 \\ S_{\bar{\lambda}}(\bar{\alpha}\gamma) = 0 \end{cases}$$

In this case we have

$$(1.3.4)\qquad \sigma^{-1} = \begin{pmatrix} \bar{\delta} & \lambda\bar{\beta} \\ \bar{\lambda}\bar{\gamma} & \bar{\alpha} \end{pmatrix}$$

Remark. Suppose left invertible elements of End$(P \oplus \bar{P})$ are invertible. This is the case, for example, if $P \oplus \bar{P}$ is finitely generated and the ring A is either commutative or right noetherian. Then either of conditions (1.3.2) or (1.3.3) implies the other conditions.

(1.4) The group $U(H(P))$ consists of those σ which are automorphisms of $(P \oplus \bar{P}, h_P)$ and which, in addition, preserve the quadratic function $[q_P](x,f) = [\langle f,x\rangle_P]$, where $[a]$ denotes the class modulo Λ of $a \in A$. Once σ preserves h_P to check that it preserves $[q_P](x,f)$ it suffices, by Ch. I., (4.4), to treat only elements of the forms $(x,0)$ and $(0,f)$, on which $[q_P]$ vanishes. We have $\sigma(x,0) = (\alpha x, \gamma x)$ so $[q_P](\sigma(x,0)) = [\langle \gamma x, \alpha x\rangle_P]$ $= [\langle \bar{\alpha}\gamma x, x\rangle_P]$. By (1.3.3) $S_{\bar{\lambda}}(\bar{\alpha}\gamma) = 0$, so the vanishing of $[\langle \bar{\alpha}\gamma x, x\rangle_P]$ for all x amounts to requiring that $\bar{\alpha}\gamma \in \bar{\Lambda}(P)$. (See Ch. I, (4.3).) Next we have $\sigma(0,f) = (\beta f, \delta f)$ so that $[q_P](\sigma(0,f)) = [\langle \delta f, \beta f\rangle_P] = [\overline{\langle \beta f, \delta f\rangle_P}] = [\overline{\langle \bar{\delta}\beta f, f\rangle_{\bar{P}}}]$. Now

$[\bar{a}] = 0 \Leftrightarrow \bar{a} \in \Lambda \Leftrightarrow a \in \bar{\Lambda}$, so we have $[q_p](\sigma(0,f)) = 0$ $\Leftrightarrow \langle \bar{\delta}\beta f, f\rangle_{\bar{P}} \in \bar{\Lambda}$. By (1.3.3) $S_\lambda(\bar{\delta}\beta) = 0$. Thus adding the previous condition amounts to the requirement (Ch. I., (4.3)) that $\bar{\delta}\beta \in \Lambda(\bar{P})$

Applying these considerations to σ^{-1} we arrive at the conditions $\beta\bar{\alpha} \in \Lambda(\bar{P})$ and $\gamma\bar{\delta} \in \bar{\Lambda}(P)$. Thus from Prop. (1.3) we obtain:

(1.5) PROPOSITION. $\sigma = \begin{pmatrix}\alpha & \beta\\ \gamma & \delta\end{pmatrix} \in \mathrm{End}(P \oplus \bar{P})$ _lies in_ $U(H(P))$ _if and only if any two of the following three conditions holds:_

(1.5.1) $\sigma \in GL(P \oplus \bar{P})$

(1.5.2) $\alpha\bar{\delta} + \bar{\lambda}\beta\bar{\gamma} = I$
$\beta\bar{\alpha} \in \Lambda(\bar{P})$, _and_ $\gamma\bar{\delta} \in \bar{\Lambda}(P)$

(1.5.3) $\bar{\alpha}\delta + \bar{\lambda}\bar{\gamma}\beta = I$,
$\bar{\delta}\beta \in \Lambda(\bar{P})$, _and_ $\bar{\alpha}\gamma \in \bar{\Lambda}(P)$.

§2. The homomorphisms $X_{\pm}$ and H, the group EU(H(P)), and the elements $\omega_P(\varphi)$.

(2.1) The homomorphisms $X_{\pm}$ are the following:

$$X_-: \bar{\Lambda}(P) \longrightarrow U(H(P))$$

(2.1.1)
$$X_-(\gamma) = \begin{pmatrix} I & 0 \\ \gamma & I \end{pmatrix}$$

$$X_+: \Lambda(\bar{P}) \longrightarrow U(H(P))$$

(2.1.2)
$$X_+(\beta) = \begin{pmatrix} I & \beta \\ 0 & I \end{pmatrix}$$

These are clearly group homomorphisms from additive to multiplicative groups. It follows from (1.5.2) that X_+ maps $\bar{\Lambda}(P)$ onto the set of elements of U(H(P)) of the form $\begin{pmatrix} I & 0 \\ * & I \end{pmatrix}$. Similarly X_+ maps $\Lambda(\bar{P})$ onto the set of elements of U(H(P)) of the form $\begin{pmatrix} I & * \\ 0 & I \end{pmatrix}$.

(2.2) The group EU(H(P)) is defined to be the group generated by the images of X_+ and X_-.

(2.3) PROPOSITION. Let u: $(A,\lambda,\Lambda) \to A',\lambda',\Lambda')$ be a unitary surjection. Suppose the A-module P is projective. Then the base change homomorphism $U(H(P)) \to U(H(P'))$, where $P' = P \otimes_A A'$, induces an epimorphism $EU(H(P)) \to EU(H(P'))$.

This results from the fact (Ch. I, (6.4)) that $\bar{\Lambda}(P) \to \overline{\Lambda'}(P')$ is surjective, and similarly for $\Lambda(\bar{P}) \to \Lambda'(\overline{P'})$

(2.4) The <u>hyperbolic homomorphism</u>

$$H: GL(P) \longrightarrow U(H(P))$$

$$H(\alpha) = \begin{pmatrix} \alpha & 0 \\ 0 & \bar{\alpha}^{-1} \end{pmatrix}$$

arises from the hyperbolic functor (Ch. I, (4.7)).

(2.5) PROPOSITION.

a) *For* $\alpha \in GL(P)$, $\beta \in \Lambda(\bar{P})$, *and* $\gamma \in \bar{\Lambda}(P)$ *we have*

(2.5.1) $$H(\alpha)X_+(\beta)H(\alpha)^{-1} = X_+(\alpha\beta\bar{\alpha})$$

and

(2.5.2) $$H(\alpha)X_-(\gamma)H(\alpha)^{-1} = X_-(\bar{\alpha}^{-1}\gamma\alpha^{-1}).$$

Hence $H(GL(P))$ *normalizes the groups* $\mathrm{Im}(X_+)$, $\mathrm{Im}(X_-)$, *and* $EU(H(P))$.

b) *The map*

$$\bar{\Lambda}(P) \times GL(P) \times \Lambda(\bar{P}) \longrightarrow U(H(P))$$

$$(\gamma,\alpha,\beta) \longmapsto X_-(\gamma)H(\alpha)X_+(\beta) = \begin{pmatrix} \alpha & \alpha\beta \\ \gamma\alpha & \bar{\alpha}^{-1}+\gamma\alpha\beta \end{pmatrix}$$

is injective, and its image consists of all $\sigma = \begin{pmatrix} \alpha & \beta' \\ \gamma' & \delta \end{pmatrix} \in U(H(P))$ *for which* $\alpha \in GL(P)$. (*We have* $\gamma = \gamma'\alpha^{-1}$ *and* $\beta = \alpha^{-1}\beta'$.)

$$\begin{pmatrix}\alpha & 0\\ 0 & \bar{\alpha}^{-1}\end{pmatrix}\begin{pmatrix}I & \beta\\ 0 & I\end{pmatrix}\begin{pmatrix}\alpha^{-1} & 0\\ 0 & \bar{\alpha}\end{pmatrix} = \begin{pmatrix}I & \alpha\beta\bar{\alpha}\\ 0 & I\end{pmatrix}$$

whence (2.5.1), and a similar calculation gives (2.5.2). These formulas clearly imply the last assertion of a).

If $X_-(\gamma)H(\alpha)X_+(\beta) = \begin{pmatrix}\alpha' & \beta'\\ \gamma' & \delta'\end{pmatrix}$ then $\alpha = \alpha'$, $\beta = \alpha^{-1}\beta'$, and $\gamma = \gamma'\alpha^{-1}$, so the map in b) is injective.

Suppose $\sigma = \begin{pmatrix}\alpha & \beta'\\ \gamma' & \delta\end{pmatrix} \in U(H(P))$ and $\alpha \in GL(P)$. Put $\beta = \alpha^{-1}\beta$; and $\gamma = \gamma'\alpha^{-1}$. Since $H(\alpha)^{-1}\sigma = \begin{pmatrix}I & \beta\\ * & *\end{pmatrix}$ and $\sigma\ H(\alpha)^{-1} = \begin{pmatrix}I & *\\ \gamma & *\end{pmatrix}$ we see that $\beta \in \Lambda(\bar{P})$ and $\gamma \in \bar{\Lambda}(P)$. Now $X_-(\gamma)H(\alpha)X_+(\beta) = \begin{pmatrix}\alpha & \beta'\\ \gamma' & \delta'\end{pmatrix}$ for some δ', and we have $\bar{\alpha}\delta' + \lambda\overline{\gamma'\beta'} = I = \bar{\alpha}\delta + \lambda\overline{\gamma'\beta'}$. Since α is invertible we have $\delta' = \delta$. This concludes the proof of b).

(2.6) COROLLARY. $EU(H(P)) \cap H(GL(P))$ *is a normal subgroup of* $H(GL(P))$.

For, by part a) of Prop. (2.5), it is the intersection of groups normalized by $H(GL(P))$.

(2.7) PROPOSITION. *Suppose* $\beta \in \Lambda(\bar{P})$ *and* $\gamma \in \bar{\Lambda}(P)$ *satisfy* $\beta\gamma\beta = 0$. *Then*

(2.7.1) $$(X_+(\beta), X_-(\gamma)) = X_-(\gamma\beta\gamma)H(I+\beta\gamma),$$

and so $H(I + \beta\gamma) \in EU(H(P))$. *If* $\gamma\beta\gamma = 0$ *also then* $H(I + \beta\gamma)$ *is a commutator in* $EU(H(P))$.

We have $X_+(\beta)X_-(\gamma) = \begin{pmatrix}I+\beta\gamma & \gamma\\ \beta & I\end{pmatrix}$ so $(X_+(\beta), X_-(\gamma))$
$= \begin{pmatrix}I+\beta\gamma & \gamma\\ \beta & I\end{pmatrix}\begin{pmatrix}I+\beta\gamma & -\gamma\\ -\beta & I\end{pmatrix} = \begin{pmatrix}I+\beta\gamma & 0\\ \gamma\beta\gamma & I-\beta\gamma\end{pmatrix}$ (since $\beta\gamma\beta = 0$) $= X_-(\gamma\beta\gamma)H(I+\beta\gamma)$.

The other assertions are immediate from this.

(2.8) <u>The element</u> $\omega_P(\varphi)$ attached to an isomorphism $\varphi: \bar{P} \to P$ is

$$\omega_P(\varphi) = \begin{pmatrix} 0 & \varphi \\ \bar{\lambda}\bar{\varphi}^{-1} & 0 \end{pmatrix} \in U(H(P)).$$

Note that

(2.8.1) $$\omega_P(\varphi)^2 = H(\bar{\lambda}\varphi\bar{\varphi}^{-1})$$

and

(2.7.2) $$\omega_P(\varphi)^{-1} = \begin{pmatrix} 0 & \lambda\bar{\varphi} \\ \varphi^{-1} & 0 \end{pmatrix} = \omega_P(\lambda\bar{\varphi})$$

For $\alpha \in GL(P)$, $\beta \in \Lambda(\bar{P})$, and $\gamma \in \bar{\Lambda}(P)$ we have

(2.8.3) $$\omega_P(\varphi)H(\alpha)\omega_P(\varphi)^{-1} = H(\varphi\bar{\alpha}^{-1}\varphi^{-1}),$$

(2.8.4) $$\omega_P(\varphi)X_+(\beta)\omega_P(\varphi)^{-1} = X_-(\bar{\lambda}\bar{\varphi}^{-1}\beta\varphi^{-1}),$$

and

(2.8.5) $$\omega_P(\varphi)X_-(\gamma)\omega_P(\varphi)^{-1} = X_+(\lambda\varphi\gamma\bar{\varphi})$$

Hence

(2.8.6) $\omega_P(\varphi)$ <u>normalizes</u> $H(GL(P))$ <u>and</u> $EU(H(P))$.

<u>Suppose there is an isomorphism</u> $\gamma: P \to \bar{P}$ <u>such that</u> $\gamma \in \bar{\Lambda}(P)$. Then

(2.8.7) $$\omega_P(\gamma^{-1}) = X_-(-\gamma)X_+(\gamma^{-1})X_-(-\gamma),$$

as a direct calculation shows. Moreover

$$(2.8.8) \qquad \omega_P(\varphi) = H(\varphi\gamma)\,\omega_P(\gamma^{-1}) \in H(GL(P)) \cdot EU(H(P))$$

for all φ as above.

§3. Passage from $U(H(P))$ to $U(H(P \oplus P'))$.

(3.1) The homomorphism $U(H(P)) \times U(H(P')) \to U(H(P \oplus P'))$, $(\sigma,\sigma') \mapsto \sigma \perp \sigma'$, arises from the identification of $H(P \oplus P')$ with $H(P) \perp H(P')$ through switching the middle terms of $(P \oplus \bar{P}) \oplus (P' \oplus \overline{P'})$ to obtain $(P \oplus P') \oplus (\bar{P} \oplus \overline{P'})$.

In our matrix notation, if $\sigma = \begin{pmatrix} \alpha & \beta \\ \gamma & \delta \end{pmatrix}$ and $\sigma' = \begin{pmatrix} \alpha' & \beta' \\ \gamma' & \delta' \end{pmatrix}$ then

$$\sigma \perp \sigma' = \begin{pmatrix} \alpha & 0 & \beta & 0 \\ 0 & \alpha' & 0 & \beta' \\ \gamma & 0 & \delta & 0 \\ 0 & \gamma' & 0 & \delta' \end{pmatrix}$$

We shall record below some special cases of this formula.

(3.2) The module $\bar{\Lambda}(P \oplus P')$ can be described as the set of matrices $\begin{pmatrix} \gamma & \epsilon \\ -\bar{\lambda}\bar{\epsilon} & \gamma' \end{pmatrix}$ with $\gamma \in \bar{\Lambda}(P)$, $\gamma' \in \bar{\Lambda}(P')$, and $\epsilon \in \mathrm{Hom}(P',\bar{P})$. We then have

$$X_-(\gamma) \perp X_-(\gamma') = X_-\begin{pmatrix} \gamma & 0 \\ 0 & \gamma' \end{pmatrix}$$

Analogous considerations and notation for $\Lambda(\bar{P} \oplus \overline{P'})$ yield

$$X_+(\beta) \perp X_+(\beta') = X_+\begin{pmatrix} \beta & 0 \\ 0 & \beta' \end{pmatrix}$$

for $\beta \in \Lambda(\bar{P})$ and $\beta' \in \Lambda(\overline{P'})$. It follows that

$$EU(H(P)) \perp EU(H(P')) \subset EU(H(P \oplus P'))$$

(3.3) <u>The homomorphism</u> $\oplus$: $GL(P) \times GL(P') \to GL(P \oplus P')$, $(\alpha,\alpha') \mapsto \alpha \oplus \alpha' = \begin{pmatrix} \alpha & 0 \\ 0 & \alpha' \end{pmatrix}$, and that of (3.1) are compatible via the hyperbolic homomorphism:

$$H(\alpha) \perp H(\alpha') = H \begin{pmatrix} \alpha & 0 \\ 0 & \alpha' \end{pmatrix}$$

(3.4) If $\varphi: \bar{P} \to P$ and $\varphi': \overline{P'} \to P'$ are isomorphisms then we have

$$\omega_P(\varphi) \perp \omega_{P'}(\varphi') = \omega_{P\oplus P'}(\varphi \oplus \varphi').$$

(3.5) <u>The subgroup</u> $E(P,P')$ <u>of</u> $GL(P \oplus P')$ is defined to be the group generated by the subgroups $\begin{pmatrix} 1_P & \mathrm{Hom}(P',P) \\ 0 & 1_{P'} \end{pmatrix}$ and $\begin{pmatrix} 1_P & 0 \\ \mathrm{Hom}(P,P') & 1_{P'} \end{pmatrix}$ (cf. [3], Ch. IV, §3).

(3.6) PROPOSITION. <u>Let</u> $\sigma = \begin{pmatrix} \alpha & \beta \\ \gamma & \delta \end{pmatrix} \in U(H(P))$. <u>Then in</u> $U(H(P \oplus P'))$ <u>we have the formulas</u>

(3.6.1)
$$(\sigma \perp I) H\begin{pmatrix} I & 0 \\ x & I \end{pmatrix} X_+\begin{pmatrix} 0 & y \\ -\lambda\bar{y} & z \end{pmatrix} (\sigma \perp I)^{-1}$$
$$= H\begin{pmatrix} I & 0 \\ x\bar{\delta}-\bar{y}\gamma & I \end{pmatrix} \cdot X_+ \begin{pmatrix} 0 & \alpha y - \beta\bar{x} \\ \lambda(x\bar{\beta}-\bar{y}\bar{\alpha}) & z+xy+(\bar{y}\bar{\gamma}-x\bar{\delta})(\alpha y-\beta\bar{x}) \end{pmatrix}$$

<u>and</u>

(3.6.2)
$$(\sigma \perp I) H\begin{pmatrix} I & x \\ 0 & I \end{pmatrix} X_-\begin{pmatrix} 0 & y \\ -\bar{\lambda}\bar{y} & z \end{pmatrix} (\sigma \perp I)^{-1}$$
$$= X_-\begin{pmatrix} 0 & \gamma x + \delta y \\ -\bar{\lambda}\,\overline{(\gamma x+\delta y)} & z-\bar{x}y+\bar{\lambda}\,\overline{(\gamma x+\delta y)}(\alpha x+\beta y) \end{pmatrix} H\begin{pmatrix} I & \alpha x + \beta y \\ 0 & I \end{pmatrix}$$

$$\begin{pmatrix} \alpha & 0 & \beta & 0 \\ 0 & I & 0 & 0 \\ \gamma & 0 & \delta & 0 \\ 0 & 0 & 0 & I \end{pmatrix} \begin{pmatrix} I & 0 & 0 & 0 \\ x & I & 0 & 0 \\ 0 & 0 & I & -\bar{x} \\ 0 & 0 & 0 & I \end{pmatrix} \begin{pmatrix} I & 0 & 0 & y \\ 0 & I & -\lambda\bar{y} & z \\ 0 & 0 & I & 0 \\ 0 & 0 & 0 & I \end{pmatrix} \begin{pmatrix} \delta & 0 & \lambda\bar{\beta} & 0 \\ 0 & I & 0 & 0 \\ \bar{\lambda}\bar{\gamma} & 0 & \bar{\alpha} & 0 \\ 0 & 0 & 0 & I \end{pmatrix}$$

$$= \begin{pmatrix} \alpha & 0 & \beta & -\beta\bar{x} \\ x & I & 0 & 0 \\ \gamma & 0 & \delta & -\delta\bar{x} \\ 0 & 0 & 0 & I \end{pmatrix} \begin{pmatrix} \delta & 0 & \lambda\bar{\beta} & y \\ -\bar{y}\bar{\gamma} & I & -\lambda\bar{y}\bar{\alpha} & z \\ \bar{\lambda}\bar{\gamma} & 0 & \bar{\alpha} & 0 \\ 0 & 0 & 0 & I \end{pmatrix}$$

$$= \begin{pmatrix} I & 0 & 0 & \alpha y-\beta\bar{x} \\ x\bar{\delta}-\bar{y}\bar{\gamma} & I & \lambda(x\bar{\beta}-\bar{y}\bar{\alpha}) & z+xy \\ 0 & 0 & I & \gamma y-\delta\bar{x} \\ 0 & 0 & 0 & I \end{pmatrix}$$

$$= H\begin{pmatrix} I & 0 \\ x\bar{\delta}-\bar{y}\bar{\gamma} & I \end{pmatrix} X_+(\varepsilon),$$

where

$$\varepsilon = \begin{pmatrix} I & 0 \\ \bar{y}\bar{\gamma}-x\bar{\delta} & I \end{pmatrix} \begin{pmatrix} 0 & \alpha y-\beta\bar{x} \\ \lambda(x\bar{\beta}-\bar{y}\bar{\alpha}) & z+xy \end{pmatrix}$$

$$= \begin{pmatrix} 0 & \alpha y - \beta\bar{x} \\ \lambda(x\bar{\beta}-\bar{y}\bar{\alpha}) & z+xy+(\bar{y}\bar{\gamma}-x\bar{\delta})(\alpha y-\beta\bar{x}) \end{pmatrix}$$

whence (3.6.1). Formula (3.6.2) follows from a similar direct calculation.

For reference we shall record here some special cases of the above formulas.

$$(3.6.3)\quad (\sigma \perp I)H\begin{pmatrix} I & 0 \\ x & I \end{pmatrix}(\sigma \perp I)^{-1} = H\begin{pmatrix} I & 0 \\ x\bar{\delta} & I \end{pmatrix} X_{+}\begin{pmatrix} 0 & -\beta\bar{x} \\ \lambda x\bar{\beta} & x\bar{\delta}\beta\bar{x} \end{pmatrix}$$

$$(3.6.4)\quad (\sigma \perp I)H\begin{pmatrix} I & x \\ 0 & I \end{pmatrix}(\sigma \perp I)^{-1} = X_{-}\begin{pmatrix} 0 & \gamma x \\ -\bar{\lambda}\bar{x}\bar{\gamma} & \bar{\lambda}\bar{x}\bar{\gamma}\alpha x \end{pmatrix} H\begin{pmatrix} I & \alpha x \\ 0 & I \end{pmatrix}$$

$$(3.6.5)\quad (\sigma \perp I)X_{+}\begin{pmatrix} 0 & y \\ -\lambda\bar{y} & 0 \end{pmatrix}(\sigma \perp I)^{-1} = H\begin{pmatrix} I & 0 \\ -\bar{y}\gamma & I \end{pmatrix} X_{+}\begin{pmatrix} 0 & \alpha y \\ -\lambda\bar{y}\bar{\alpha} & \bar{y}\bar{\gamma}\alpha y \end{pmatrix}$$

$$(3.6.6)\quad (\sigma \perp I)X_{-}\begin{pmatrix} 0 & y \\ -\bar{\lambda}\bar{y} & 0 \end{pmatrix}(\sigma \perp I)^{-1} = X_{-}\begin{pmatrix} 0 & \delta y \\ -\bar{\lambda}\bar{y}\bar{\delta} & \bar{\lambda}\bar{y}\bar{\delta}\beta y \end{pmatrix}$$

We now prove the "unitary Whitehead lemma," which is due to Wasserstein [15] (Lemma 1.1).

(3.7) PROPOSITION. Let $\sigma, \tau \in U(H(P))$. Then $\sigma \perp \sigma^{-1}$ and $(\sigma, \tau) \perp I$ belong to $H(E(P,P)) \cdot EU(H(P \oplus P))$. In fact

(3.7.1) $\quad \sigma \perp \sigma^{-1}$

$$= X_-\begin{pmatrix} 0 & \gamma \\ -\bar{\lambda}\bar{\gamma} & \bar{\lambda}\bar{\gamma}\alpha \end{pmatrix} H\begin{pmatrix} I & \alpha \\ 0 & I \end{pmatrix} H\begin{pmatrix} I & 0 \\ -\bar{\delta} & I \end{pmatrix} X_+\begin{pmatrix} 0 & \beta \\ -\lambda\bar{\beta} & \bar{\delta}\beta \end{pmatrix} X_-\begin{pmatrix} 0 & \gamma \\ -\bar{\lambda}\bar{\gamma} & \bar{\lambda}\bar{\gamma}\alpha \end{pmatrix} H\begin{pmatrix} I & \alpha \\ 0 & I \end{pmatrix} H\begin{pmatrix} 0 & -I \\ I & 0 \end{pmatrix}$$

<u>and</u>

(3.7.2) $\quad (\sigma,\tau) \perp I = ((\sigma\tau) \perp (\sigma\tau)^{-1})(\sigma^{-1} \perp \sigma)(\tau^{-1} \perp \tau).$

Formula (3.7.2) is immediate. To prove (3.7.1) we introduce the notation $e^x_{12} = \begin{pmatrix} I & x \\ 0 & I \end{pmatrix}$ and $e^x_{21} = \begin{pmatrix} I & 0 \\ x & I \end{pmatrix}$ for $x \in \mathrm{End}(P)$. These elements generate $E(P,P)$ (see (3.5)). Put

$$w = e^I_{12}\, e^{-I}_{21}\, e^I_{12} = \begin{pmatrix} 0 & I \\ -I & 0 \end{pmatrix} \in E(P,P).$$

Then a simple calculation shows that $H(w)(\sigma \perp I)H(w)^{-1} = I \perp \sigma$ (cf. Prop. (4.4) below). Hence $\sigma \perp \sigma^{-1} = (\sigma \perp I)(I \perp \sigma^{-1})$ $= (\sigma \perp I)H(w)(\sigma^{-1} \perp I)H(w)^{-1}$. Using (3.6.3) and (3.6.4) we have

$$(\sigma \perp I)H(w)(\sigma \perp I)^{-1} = (\sigma \perp I)H(e^I_{12})H(e^{-I}_{21})H(e^I_{12})(\sigma \perp I)^{-1}$$

$$= X_-\begin{pmatrix} 0 & \gamma \\ -\bar{\lambda}\bar{\gamma} & \bar{\lambda}\bar{\gamma}\alpha \end{pmatrix} H(e^{\alpha}_{12})H(e^{-\bar{\delta}}_{21}) X_+\begin{pmatrix} 0 & \beta \\ \lambda\bar{\beta} & \bar{\delta}\beta \end{pmatrix} X_-\begin{pmatrix} 0 & \gamma \\ -\bar{\lambda}\bar{\gamma} & \bar{\lambda}\bar{\gamma}\alpha \end{pmatrix} H(e^{\alpha}_{12})$$

This proves (3.7.2) and shows that $\sigma \perp \sigma^{-1} \in H(E(P,P)) \cdot EU(H(P \oplus P))$. The fact that $(\sigma,\bar{\tau}) \perp I$ also lies in this group follows now from (3.7.2), thus proving the proposition.

(3.8) <u>The element σ_u</u>. We next record an identity due to Wall [14] (Lemma (6.2)).

If $\sigma \in U(H(P))$, and if u is a central unit of A such that $u = \bar{u}$, we put σ_u = the conjugate of σ by $\begin{pmatrix} uI & 0 \\ 0 & I \end{pmatrix}$; thus if $\sigma = \begin{pmatrix} \alpha & \beta \\ \gamma & \delta \end{pmatrix}$ then $\sigma_u = \begin{pmatrix} \alpha & u\beta \\ u^{-1}\gamma & \delta \end{pmatrix}$. It is easily seen from the defining equations (1.5) that $\sigma_u \in U(H(P))$.

(3.9) PROPOSITION. <u>Let</u> $\sigma = \begin{pmatrix} \alpha & \beta \\ \gamma & \delta \end{pmatrix} \in U(H(P))$. <u>In</u> $U(H(P \oplus P))$ <u>put</u> $M = X_{-}\begin{pmatrix} 0 & -I \\ \bar{\lambda}I & 0 \end{pmatrix} \cdot (I \perp \omega(\lambda I))$. <u>Then</u>

$$M(\sigma \perp \sigma_{-1})M^{-1} = H(\sigma)X_{+}\begin{pmatrix} \bar{\delta}\beta & -\lambda^2\bar{\beta}\gamma \\ \bar{\lambda}\bar{\gamma}\beta & \bar{\gamma}\alpha \end{pmatrix}.$$

We have

$$M = \begin{pmatrix} I & 0 & 0 & 0 \\ 0 & I & 0 & 0 \\ 0 & -I & I & 0 \\ \bar{\lambda}I & 0 & 0 & I \end{pmatrix}\begin{pmatrix} I & 0 & 0 & 0 \\ 0 & 0 & 0 & \lambda I \\ 0 & 0 & I & 0 \\ 0 & I & 0 & 0 \end{pmatrix} = \begin{pmatrix} I & 0 & 0 & 0 \\ 0 & 0 & 0 & \lambda I \\ 0 & 0 & I & -\lambda I \\ \bar{\lambda}I & I & 0 & 0 \end{pmatrix}$$

and

$$M^{-1} = \begin{pmatrix} I & 0 & 0 & 0 \\ 0 & 0 & 0 & I \\ 0 & 0 & I & 0 \\ 0 & \bar{\lambda}I & 0 & 0 \end{pmatrix}\begin{pmatrix} I & 0 & 0 & 0 \\ 0 & I & 0 & 0 \\ 0 & I & I & 0 \\ -\bar{\lambda}I & 0 & 0 & I \end{pmatrix} = \begin{pmatrix} I & 0 & 0 & 0 \\ -\bar{\lambda}I & 0 & 0 & I \\ 0 & I & I & 0 \\ 0 & \bar{\lambda}I & 0 & 0 \end{pmatrix}$$

Thus $M(\sigma \perp \sigma_{-1})M^{-1} =$

$$\begin{pmatrix} I & 0 & 0 & 0 \\ 0 & 0 & 0 & \lambda I \\ 0 & 0 & I & -\lambda I \\ \bar{\lambda} I & I & 0 & 0 \end{pmatrix} \begin{pmatrix} \alpha & 0 & \beta & 0 \\ 0 & \alpha & 0 & -\beta \\ \gamma & 0 & \delta & 0 \\ 0 & -\gamma & 0 & \delta \end{pmatrix} \begin{pmatrix} I & 0 & 0 & 0 \\ -\bar{\lambda} I & 0 & 0 & I \\ 0 & I & I & 0 \\ 0 & \bar{\lambda} I & 0 & 0 \end{pmatrix}$$

$$= \begin{pmatrix} \alpha & 0 & \beta & 0 \\ 0 & -\lambda\gamma & 0 & \lambda\delta \\ \gamma & \lambda\gamma & \delta & -\lambda\delta \\ \bar{\lambda}\alpha & \alpha & \bar{\lambda}\beta & -\beta \end{pmatrix} \begin{pmatrix} I & 0 & 0 & 0 \\ -\bar{\lambda} I & 0 & 0 & I \\ 0 & I & I & 0 \\ 0 & \bar{\lambda} I & 0 & 0 \end{pmatrix}$$

$$= \begin{pmatrix} \alpha & \beta & \beta & 0 \\ \gamma & \delta & 0 & -\lambda\gamma \\ 0 & 0 & \delta & \lambda\gamma \\ 0 & 0 & \bar{\lambda}\beta & \alpha \end{pmatrix} = H(\sigma) X_+(\beta'),$$

where

$$\beta' = \sigma^{-1} \begin{pmatrix} \beta & 0 \\ 0 & -\lambda\gamma \end{pmatrix} = \begin{pmatrix} \bar{\delta} & \lambda\bar{\beta} \\ \bar{\lambda}\bar{\gamma} & \bar{\alpha} \end{pmatrix} \begin{pmatrix} \beta & 0 \\ 0 & -\lambda\gamma \end{pmatrix}$$

$$= \begin{pmatrix} \bar{\delta}\beta & -\lambda^2\bar{\beta}\gamma \\ \bar{\lambda}\bar{\gamma}\beta & -\lambda\bar{\alpha}\gamma \end{pmatrix}.$$

Since $\bar{\alpha}\gamma + \bar{\lambda}\bar{\gamma}\alpha = 0$ this proves Prop. (3.9).

(3.10) <u>The transvections</u> $\sigma_{u,a,v}$. Let $(M,[h])$ be a quadratic module and write $\langle x,y\rangle = (S_\lambda h)(x,y)$ for $x,y \in M$. If $u,v \in M$

and $a \in A$ satisfy $h(u,u) \in \Lambda$, $\langle u,v\rangle = 0$, and $a \equiv h(v,v) \bmod \Lambda$ then(Ch. I, (5.1)) one has the element $\sigma_{u,a,v} \in U(M,[h])$ defined by

$$\sigma_{u,a,v}(x) = x + u\langle v,x\rangle - v\bar{\lambda}\langle u,x\rangle - u\bar{\lambda}a\langle u,x\rangle.$$

We proposed to relate these $\sigma_{u,a,v}$ to the elements introduced above when $(M,[h]) = H(Q)$ with $Q = P \oplus P'$.

(3.10.1) <u>The case</u> $u,v \in \bar{Q}$: Define $d,e\colon Q \to \bar{Q}$ by $d(x) = u\langle v,x\rangle_Q$ and $e(x) = u\bar{\lambda}a\langle u,x\rangle_Q$. We have $a \in \Lambda$ so $e \in \bar{\Lambda}(Q)$. Moreover $\bar{d}(x) = v\langle u,x\rangle_Q$ (cf. Ch. I, (4.3)). One sees therefore that $\sigma_{u,a,v} = X_-(\gamma)$ where $\gamma = d - \bar{\lambda}\bar{d} - e \in \bar{\Lambda}(Q)$. If $u \in \bar{P}$ and $v \in \overline{P'}$ then γ is represented by the matrix $\begin{pmatrix} -e & d \\ -\bar{\lambda}\bar{d} & 0 \end{pmatrix}$. If $u \in \overline{P'}$ and $v \in \bar{P}$ the representing matrix is $\begin{pmatrix} 0 & -\bar{\lambda}\bar{d} \\ d & -e \end{pmatrix}$. If Q is finitely generated and projective such elements γ as above additively generate $\bar{\Lambda}(Q)$.

(3.10.2) <u>The case</u> $u,v, \in Q$: Define $d,e\colon \bar{Q} \to Q$ by $d(f) = v\langle u,f\rangle_{\bar{Q}} = v\overline{\langle f,u\rangle}_Q$ and $e(f) = ua\langle u,f\rangle_{\bar{Q}}$. Then as above we see that $\sigma_{u,a,v} = X_+(\beta)$ where $\beta = -d + \lambda\bar{d} - e \in \Lambda(\bar{Q})$. If $u \in P$ and $v \in P'$ then β is represented by the matrix $\begin{pmatrix} -e & \lambda\bar{d} \\ -d & 0 \end{pmatrix}$; if $u \in P'$ and $v \in P$ the representing matrix is $\begin{pmatrix} 0 & -d \\ \lambda\bar{d} & -e \end{pmatrix}$. If Q is finitely generated and projective such elements β as above additively generate $\Lambda(\bar{Q})$.

(3.10.3) <u>The case</u> $u \in P$, $v \in \overline{P'}$: We have $a \in \Lambda$ and $\sigma_{u,a,v} = \sigma_{u,a,o} \cdot \sigma_{u,o,v}$ (Ch. I, (5.2.4)); the first factor is covered by case (3.10.2) above. Define $d: P' \to P$ by $d(x) = u\langle v,x\rangle_{P'}$. Then $\begin{pmatrix} I & d \\ 0 & I \end{pmatrix} \in E(P,P')$ (see (3.5)) and $\sigma_{u,o,v} = H\begin{pmatrix} I & d \\ 0 & I \end{pmatrix}$. If Q is finitely generated and projective then elements d as above additively generate $\mathrm{Hom}_A(P',P)$.

(3.10.4) <u>The case</u> $u \in \bar{P}$, $v \in P'$: Then we have $\sigma_{u,o,v} = H\begin{pmatrix} I & 0 \\ -\lambda d & I \end{pmatrix}$ where $d: P \to P'$ is defined by $d(x) = v\langle u,x\rangle_P$.

(3.11) PROPOSITION. <u>Suppose</u> $Q = P \oplus P'$ <u>with</u> $P' \cong A$. <u>Let</u> G <u>be a subgroup of</u> $U(H(Q))$ <u>containing</u> $E = H(E(P,P')) \cdot EU(H(Q))$.

a) <u>If</u> G <u>acts transitively on the set of unimodular isotropic elements in</u> $H(Q)$ <u>then</u> G <u>acts transitively on the set of hyperbolic pairs and on the set of hyperbolic planes in</u> $H(Q)$.

b) <u>If, in addition,</u> Q <u>is finitely generated and projective and if</u> G <u>normalizes</u> E <u>then</u> E <u>is a normal subgroup of</u> $U(H(Q))$.

Write $P' = p_0 A$ with p_0 unimodular. Part a) follows from (Ch. I, Cor. (5.6)) once we show that $\sigma_{p_0,a,v} \in E$ for all

$v \in P'^{\perp}$ and $a \in A$ such that $a \equiv q_Q(v,v) \bmod \Lambda$. Now $H(Q) = H(P) \perp H(P')$ and $P'^{\perp} = H(P) \oplus P'$. Using formulas (5.2.2) and (5.2.4) of Ch. I we see that it suffices to show that $\sigma_{p_0,a,o} \in E$ for all $a \in \Lambda$ and $\sigma_{p_0,o,v} \in E$ for v in the additive generating set $P \cup \bar{P} \cup P'$ of $P'^{\perp}$. We have $\sigma_{p_0,a,v} \in X_+(\Lambda(\bar{Q})) \subset E$ if $v \in P \oplus P' = Q$, by (3.10.2). If $v \in \bar{P}$ then $\sigma_{p_0,o,v} \in H(E(P,P'))$ $\subset E$ by (3.10.3). This proves a).

Note that Q is generated by unimodular elements. If Q is finitely generated and projective then $\mathrm{Hom}_A(Q,\bar{Q})$ is generated by homomorphisms of the form $x \mapsto u\langle v,x\rangle_Q$, and we can even restrict u or v to be unimodular. The same applies to $\mathrm{Hom}_A(\bar{Q},Q)$, $\mathrm{Hom}_A(P,P')$, etc. It follows easily (cf. (8.6)) that E is generated by elements of the form $\sigma_{u,a,v}$ with u unimodular and isotropic. Such an element is, by assumption and formula (5.2.1) of Ch. I, G-conjugate to an element of the form $\sigma_{p_0,a,v}$ (with $v \in (P')^{\perp}$). To prove b) it therefore suffices to show that the elements $\alpha\sigma_{p_0,a,v}\alpha^{-1} = \sigma_{\alpha p_0,a,\alpha v}$ lie in E for each $\alpha \in U(H(Q))$. By assumption $\alpha p_0 = \beta p_0$ for some $\beta \in G$. Writing $\alpha = \beta\gamma$ we have $\gamma p_0 = p_0$ so $\alpha\sigma_{p_0,a,v}\alpha^{-1} = \beta\sigma_{p_0,a,\gamma v}\beta^{-1}$. By assumption G normalizes E, and we have seen above that $\sigma_{p_0,a,w} \in E$ for all $w \in P'^{\perp}$, whence b).

§4. The groups $U^{\lambda}_{2n}(A,\Lambda) = U(H(A^n))$.

(4.1) Notation. We use the standard basis $e_1,\ldots,e_n$ of A^n and dual basis $\bar{e}_1,\ldots,\bar{e}_n$ of $\overline{A^n}$ to identify the four Hom's, Hom (A^n or $\overline{A^n}$, A^n or $\overline{A^n}$) with the set $M_n(A)$ of $n \times n$ matrices over A. The maps $\varepsilon \mapsto \bar{\varepsilon}$ of Hom's are converted to conjugate-transpose of matrices (Ch. I, (2.7)). Moreover we put

$$\Lambda_n = \Lambda(\overline{A^n}) = \{\beta \in M_n(A) \,|\, \beta = \lambda\bar{\beta} \text{ and } \beta_{ii} \in \Lambda (1 \le i \le n)\}$$

and

$$\bar{\Lambda}_n = \bar{\Lambda}(A^n) = \{\gamma \in M_n(A) \,|\, \gamma = -\bar{\lambda}\bar{\gamma} \text{ and } \gamma_{ii} \in \bar{\Lambda}(1 \le i \le n)\}$$

(see Ch. I, (4.3)). We thus have

(4.1.1) $$\bar{\Lambda}_n = \overline{(\Lambda_n)} = \bar{\lambda}\Lambda_n$$

It follows from Prop. (1.5) that

(4.1.2) $$U^{\lambda}_{2n}(A,\Lambda)$$

$$= \left\{ \sigma = \begin{pmatrix} \alpha & \beta \\ \gamma & \delta \end{pmatrix} \in GL_{2n}(A) \;\middle|\; \begin{matrix} \alpha\bar{\delta} + \bar{\lambda}\beta\bar{\gamma} = I, \\ \beta\bar{\alpha} \in \Lambda_n, \gamma\bar{\delta} \in \bar{\Lambda}_n \end{matrix} \right\}$$

$$= \left\{ \sigma = \begin{pmatrix} \alpha & \beta \\ \gamma & \delta \end{pmatrix} \in GL_{2n}(A) \;\middle|\; \begin{matrix} \bar{\alpha}\delta + \bar{\lambda}\bar{\gamma}\beta = I \\ \bar{\delta}\beta \in \Lambda_n, \bar{\alpha}\gamma \in \bar{\Lambda}_n \end{matrix} \right\}$$

If $\sigma = \begin{pmatrix}\alpha & \beta \\ \gamma & \delta\end{pmatrix} \in U_{2n}^{\lambda}(A,\Lambda)$ then

$$\sigma^{-1} = \begin{pmatrix} \bar{\delta} & \lambda\bar{\beta} \\ \bar{\lambda}\bar{\gamma} & \bar{\alpha} \end{pmatrix}.$$

The group

$$(4.1.3) \qquad EU_{2n}^{\lambda}(A,\Lambda) = EU(H(A^n))$$

is generated by the images of the homomorphisms

$$(4.1.4) \qquad X_+ : \Lambda_n \longrightarrow U_{2n}^{\lambda}(A,\Lambda), \quad X_+(\beta) = \begin{pmatrix} I & \beta \\ 0 & I \end{pmatrix}$$

and

$$(4.1.5) \qquad X_- : \bar{\Lambda}_n \longrightarrow U_{2n}^{\lambda}(A,\Lambda), \quad X_-(\gamma) = \begin{pmatrix} I & 0 \\ \gamma & I \end{pmatrix}.$$

By Prop. (2.5) $EU_{2n}^{\lambda}(A,\Lambda)$ is normalized by the image of

$$(4.1.6) \qquad H: GL_n(A) \longrightarrow U_{2n}^{\lambda}(A,\Lambda), \quad H(\alpha) = \begin{pmatrix} \alpha & 0 \\ 0 & \bar{\alpha}^{-1} \end{pmatrix}.$$

If $\varphi \in GL_n(A)$ we have

$$(4.1.7) \qquad \omega_n(\varphi) = \begin{pmatrix} 0 & \varphi \\ \bar{\lambda}\bar{\varphi}^{-1} & 0 \end{pmatrix} \in U_{2n}^{\lambda}(A,\Lambda).$$

(4.2) <u>The involution</u> $\sigma \mapsto \bar{\sigma}$ stabilized $U_{2n}^{\lambda}(A,\Lambda)$. For if $\sigma = \begin{pmatrix}\alpha & \beta \\ \gamma & \delta\end{pmatrix}$ satisfies the first set of defining equations (4.1.2) for $U_{2n}^{\lambda}(A,\Lambda)$ then $\bar{\sigma} = \begin{pmatrix}\bar{\alpha} & \bar{\gamma} \\ \bar{\beta} & \bar{\delta}\end{pmatrix}$ satisfies the second set.

If $\beta \in \Lambda_n$ then

(4.2.1) $$\overline{X_+(\beta)} = X_-(\bar{\beta}) ,$$

so $\overline{X_+(\Lambda_n)} = X_-(\bar{\Lambda}_n)$.

If $\alpha \in GL_n(A)$ then

(4.2.2) $$\overline{H(\alpha)} = H(\bar{\alpha})$$

If $\varphi \in GL_n(A)$ then

(4.2.3) $$\overline{\omega_n(\varphi)} = \omega_n(\bar{\lambda}\ \varphi^{-1})$$

(4.3) PROPOSITION. *The groups* $EU_{2n}^{\lambda}(A,\Lambda)$, $H(GL_n(A))$, *and*

$$G_n = \{\alpha \in GL_n(A) \mid H(\alpha) \in EU_{2n}^{\lambda}(A,\Lambda)\}$$

are stable under the involution (conjugate transpose). *Moreover* G_n *is a normal subgroup of* $GL_n(A)$.

Stability of $EU_{2n}^{\lambda}(A,\Lambda)$ results from (4.2.1), and that of $H(GL_n(A))$ and G_n from (4.2.2). Normality of G_n in $GL_n(A)$ follows from Cor. (2.6).

The following technical result is often useful. As in [3], we write

$$E_n(A)$$

for the subgroup of $GL_n(A)$ generated by all elementary matrices $I + ae_{ij}$ $(a \in A,\ i \neq j)$.

(4.4) PROPOSITION. *Let* $\sigma \in U^{\lambda}_{2n}(A,\Lambda)$ *and* $\sigma' \in U^{\lambda}_{2m}(A,\Lambda)$. *Let* $\pi = \begin{pmatrix} 0 & tI_m \\ I_n & 0 \end{pmatrix}$ *where* $t = (-1)^{nm}$. *Then* $\pi \in E_{n+m}(A)$ *and*

$$H(\pi)(\sigma \perp \sigma')H(\pi)^{-1} = \sigma' \perp \sigma . \tag{4.4.1}$$

In case $n = m$ *we have*

$$(H(\pi),(\sigma^{-1} \perp I_{2n})) = \sigma \perp \sigma^{-1} \tag{4.4.2}$$

and

$$(\sigma,\sigma') \perp I_{2n} = ((\sigma\sigma') \perp (\sigma\sigma')^{-1})(\sigma^{-1} \perp \sigma)(\sigma'^{-1} \perp \sigma') . \tag{4.4.3}$$

Since π is an integer matrix of determinant 1 and $SL_{n+m}(\mathbb{Z}) = E_{n+m}(\mathbb{Z})$ ($\mathbb{Z}$ is euclidean) we have $\pi \in E_{n+m}(A)$.

$$H(\pi)(\sigma \perp \sigma')H(\pi)^{-1}$$

$$= \begin{pmatrix} 0 & tI_m & 0 & 0 \\ I_n & 0 & 0 & 0 \\ 0 & 0 & 0 & tI_m \\ 0 & 0 & I_n & 0 \end{pmatrix} \begin{pmatrix} \alpha & 0 & \beta & 0 \\ 0 & \alpha' & 0 & \beta' \\ \gamma & 0 & \delta & 0 \\ 0 & \gamma' & 0 & \delta' \end{pmatrix} \begin{pmatrix} 0 & I_n & 0 & 0 \\ tI_m & 0 & 0 & 0 \\ 0 & 0 & 0 & I_n \\ 0 & 0 & tI_m & 0 \end{pmatrix}$$

$$= \begin{pmatrix} \alpha' & 0 & \beta' & 0 \\ 0 & \alpha & 0 & \beta \\ \gamma' & 0 & \delta' & 0 \\ 0 & \gamma & 0 & \delta \end{pmatrix} ,$$

whence (4.4.1). If $n = m$ then $(H(\pi), (\sigma^{-1} \perp I_{2n}))$
$= H(\pi)(\sigma^{-1} \perp I_{2n})H(\pi)^{-1}(\sigma^{-1} \perp I_{2n})^{-1} = (I_{2n} \perp \sigma^{-1})(\sigma \perp I_{2n})$
(by (4.4.1)) $= \sigma \perp \sigma^{-1}$. Formula (4.4.3) is immediate.

(4.5) The element ω_n is defined by

(4.5.1) $$\omega_n = \omega_n(I) = \begin{pmatrix} 0 & I \\ \bar{\lambda}I & 0 \end{pmatrix} \in U_{2n}^{\lambda}(A,\Lambda).$$

We have by (3.4)

(4.5.2) $$\omega_n = \omega_1 \perp \cdots \perp \omega_1 \qquad \text{(n terms)}.$$

If $\varphi \in GL_n(A)$ then evidently

(4.5.3) $$\omega_n(\varphi) = H(\varphi)\omega_n .$$

Moreover

(4.5.4) $$\omega_n \begin{pmatrix} \alpha & \beta \\ \gamma & \delta \end{pmatrix} \omega_n^{-1} = \begin{pmatrix} \delta & \lambda\gamma \\ \bar{\lambda}\beta & \alpha \end{pmatrix}$$

whence

(4.5.5) $$\begin{aligned} \omega_n H(\alpha)\omega_n^{-1} &= H(\bar{\alpha})^{-1} \\ \omega_n X_+(\beta)\omega_n^{-1} &= X_-(\bar{\lambda}\beta) = X_-(-\bar{\beta}) \\ \omega_n X_-(\gamma)\omega_n^{-1} &= X_+(\lambda\gamma) = X_+(-\bar{\gamma}) \end{aligned}$$

for $\alpha \in GL_n(A)$, $\beta \in \Lambda_n$, and $\gamma \in \bar{\Lambda}_n$. Hence

(4.5.6) $$\omega_n X_+(\Lambda_n)\omega_n^{-1} = X_-(\bar{\Lambda}_n)$$

It follows that:

(4.5.7) *A subgroup of* $U^{\lambda}_{2n}(A,\Lambda)$ *containing* $X_+(\Lambda_n)$ *or* $X_-(\bar{\Lambda}_n)$ *and normalized by* ω_n *must contain* $EU^{\lambda}_{2n}(A,\Lambda)$.

From (4.4.1) of Prop. (4.4) we obtain:

(4.5.8) *The group generated by* $H(E_n(A))$ *and* $\omega_1 \perp I_{2(n-1)}$ *contains all the elements* $I_{2r} \perp \omega_s \perp I_{2t}$ *for which* $r + s + t = n$.

(4.6) We *stablize* $U^{\lambda}_{2n}(A,\Lambda)$ by identifying it with $U^{\lambda}_{2n}(A,\Lambda) \perp I_{2m}$ in $U_{2(n+m)}(A,\Lambda)$ and passing to the limit to obtain

(4.6.1) $$U^{\lambda}(A,\Lambda) = \bigcup_{n\geq 1} U^{\lambda}_{2n}(A,\Lambda)$$

Similarly

(4.6.2) $$EU^{\lambda}(A,\Lambda) = \bigcup_{n\geq 1} EU^{\lambda}_{2n}(A,\Lambda).$$

Analogous stabilizations can be defined for the groups $GL_n(A)$ and $E_n(A)$, and we obtain a stabilization of H,

(4.6.3) $$H: GL(A) \longrightarrow U^{\lambda}(A,\Lambda).$$

Just at the modules A^n are cofinal in Ⓟ(A), the hyperbolic modules $H(A^n)$ are cofinal in $Ⓠ^{\lambda}(A,\Lambda)$ (Ch. I, Prop (4.8); see also (4.11)). It therefore follows from [3], Ch. VII, Cor. (2.3) that

(4.6.4) $KU_1^\lambda(A,\Lambda) = U^\lambda(A,\Lambda)/(U^\lambda(A,\Lambda),U^\lambda(A,\Lambda))$

and that

(4.6.5) $K_1(A) \xrightarrow{H} KU_1^\lambda(A,\Lambda)$ _is the abelianization of_ (4.6.3).

We shall see in §5 that

$$(U^\lambda(A,\Lambda),U^\lambda(A,\Lambda)) = EU^\lambda(A,\Lambda).$$

This result is used in Chapter III to verify the "E-surjectivity" conditions of [3], Ch. VII needed to construct exact sequences for the functors KU_i.

(4.7) When A _is commutative_ we have the homomorphisms $\det: U_{2n}^\lambda(A,\Lambda) \to A^\cdot$ (units of A) which induce a homomorphism $\det: KU_1^\lambda(A,\Lambda) \to A^\cdot$.

If $\sigma \in U_{2n}^\lambda(A,\Lambda)$ then $\bar{\sigma}\,\omega_n\,\sigma = \omega_n$, where $\omega_n = \begin{pmatrix} 0 & I \\ \bar{\lambda}I & 0 \end{pmatrix}$. It follows that:

(4.7.1) _If_ $\sigma \in U_{2n}^\lambda(A,\Lambda)$ _then_ $\det(\sigma) \cdot \overline{\det(\sigma)} = 1.$

This gives an upper bound for Im(det). Lower bounds are furnished by the following readily verified formulas:

(4.7.2) $\det \omega_n = (-\bar{\lambda})^n$

(4.7.3) _If_ $\alpha \in GL_n(A)$ _then_ $\det H(\alpha) = \det(\alpha)/\overline{\det(\alpha)}.$

(4.8) PROPOSITION. *Suppose* A *is commutative.* *Let* A_1 *denote the fixed ring of the involution in* A. *Let* $\sigma = \begin{pmatrix} a & b \\ c & d \end{pmatrix} \in U_2^{\lambda}(A,\Lambda)$.

a) $\sigma^{-1} = \begin{pmatrix} \bar{d} & \lambda\bar{b} \\ \overline{\lambda c} & \bar{a} \end{pmatrix}$ *and* $\bar{a}b$, $a\bar{c}$, $b\bar{d}$, $\bar{c}d \in \Lambda$. *Moreover* $a\bar{d}$, $\lambda\bar{b}c \in A_1$.

b) *Put* $u = \det(\sigma) = ad - bc$. *Then* $u\bar{u} = 1$ *and we have:*

$$u\bar{a} = a \quad , \quad u\bar{b} = -\bar{\lambda}b,$$

$$u\bar{c} = -\lambda c \ , \quad u\bar{d} = d \quad .$$

Hence if a,b,c *or* d *is a unit then* u *lies in the group generated by* $-\lambda$ *together with all* $v/\bar{v}$ $(v \in A^{\cdot})$.

c) *Suppose* $\lambda = -1$, *so that* $\Lambda \subset A_1$. *If* $u = 1$ *then* $a,b,c,d \in A_1$ *so that* $\sigma \in Sp_2(A_1,\Lambda)$. *Conversely, if* $a,b,c,d \in A_1$ *then* $u = 1$.

d) *Suppose we are in the orthogonal case,* $\lambda = 1$, $A = A_1$, *and* $\Lambda = 0$. *Then* ad *and* bc *are orthogonal idempotents with sum* 1.

The first part of a) follows from (4.1.2). The last assertion of a) results from a comparison of the equations $a\bar{d} = \bar{\lambda}\, b\bar{c} = 1$ and $\bar{a}d + \bar{\lambda}\, b\bar{c} = 1$.

That $u\bar{u} = 1$ follows from (4.7.1) above. We have $u\bar{a} = ad\bar{a} - bc\bar{a} = ad\bar{a} + \lambda\,\bar{b}ca = (\bar{a}d + \lambda\bar{b}c)a = a$. Similarly

$u\bar{d} = d$. Next $u\bar{b} = ad\bar{b} - bc\bar{b} = -\bar{\lambda}a\bar{d}b - bc\bar{b} = -\bar{\lambda}b(a\bar{d} + \lambda\bar{b}c) = -\bar{\lambda}b$. Finally $u\bar{c} = ad\bar{c} - bc\bar{c} = -\lambda\bar{a}dc - bc\bar{c} = -\lambda c(\bar{a}d + \bar{\lambda}b\bar{c}) = -\lambda c$. The last assertion of b), and part c), result immediately from these equations.

In the orthogonal case we have ad + bc = 1 and ab = ac = bd = cd = 0, whence d).

§5. The derived group of $U^{\lambda}(A,\Lambda)$.

(5.1) PROPOSITION.

a) *We have* $H(E_n(A)) \subset EU^{\lambda}_{2n}(A,\Lambda)$ *provided either* $n \neq 2$ *or* $A = A\Lambda + \Lambda A$.

b) *Suppose either* $n \geq 3$ *or that* $n = 2$ *and* $A = AS_{-\lambda}(A) + S_{-\lambda}(A)A$. *Then* $(H(E_n(A)), X_+(\Lambda_n)) = X_+(\Lambda_n)$ *and* $(H(E_n(A)), X_-(\bar{\Lambda}_n)) = X_-(\bar{\Lambda}_n)$. *Hence* $EU^{\lambda}_{2n}(A,\Lambda)$ *is its own derived group.*

Proof of a). We may assume $n \geq 2$ since $E_1(A) = \{1\}$. Put

$$G_n = \{\alpha \in GL_n(A) \mid H(\alpha) \in EU^{\lambda}_{2n}(A,\Lambda)\}$$

and $J_n = \{a \in A \mid I + ae_{12} \in G_n\}$. Clearly J_n is an additive subgroup of A. According to Prop. (4.3) G_n is a normal subgroup of $GL_n(A)$ stable under $\alpha \mapsto \bar{\alpha}$. Conjugation by permutation matrices thus shows that $I + J_n e_{ij}$ and $I + \overline{J_n} e_{ji}$ lie in G_n for all $i \neq j$. Hence $J_n = \overline{J_n}$ and we have $E_n(A) \subset G_n \Leftrightarrow J_n = A$. Thus a) will follow once we show that (i) $\Lambda A \subset J_2$, and (ii) $J_n = A$ if $n \geq 3$.

It follows from Prop. (2.7) that

$$(5.1.1) \qquad (X_+(re_{ii}), X_-(ce_{ik} - \bar{\lambda}\bar{c}e_{ki})) = X_-(-\bar{\lambda}\bar{c}rce_{kk})H(I + rce_{ik})$$

and

$$(5.1.2) \qquad (X_+(be_{ij} - \lambda\bar{b}e_{ji}), X_-(ce_{jk} - \bar{\lambda}\bar{c}e_{kj})) = H(I + bce_{ik})$$

for $r \in \Lambda, b,c \in A$, and i,j,k distinct. It follows from (5.1.1) that $I + rce_{ik} \in G_n$ for $n \geq 2$, whence (i), and from (5.1.2) that $I + bce_{ik} \in G_n$ for $n \geq 3$, whence (ii).

Proof of b). It suffices to thow that $(H(E_n(A)),X_+(\Lambda^n)) = X_+(\Lambda^n)$. For then the conjugate transpose of this gives $(H(E_n(A)), X_-(\bar{\Lambda}^n)) = X_-(\bar{\Lambda}^n)$, whence $(H(E_n(A)), EU^{\lambda}_{2n}(A,\Lambda)) = EU^{\lambda}_{2n}(A,\Lambda)$. But under the assumption of b) it follows from a) proved above that $H(E_n(A)) \subset EU^{\lambda}_{2n}(A,\Lambda)$.

The action of $H(E_n(A))$ on $X_+(\Lambda^n)$ is given by the following formulas: Let $a,b \in A$, $r \in \Lambda$, $i \neq j$, and $u \neq v$. Then

(5.1.3) $$(H(I + ae_{ij}),X_+(re_{uu})) = \begin{cases} X_+(are_{iu}+r\bar{a}e_{ui}+ar\bar{a}e_{ii}) & \text{if } j = u \\ I & \text{if } j \neq u \end{cases}$$

(5.1.4) $$(H(I + ae_{ij}),X_+(be_{uv} - \lambda\bar{b}e_{vu}))$$

$$= \begin{cases} X_+(abe_{iv} - \lambda\bar{b}\bar{a}e_{vi}) & \text{if } j = u \\ X_+(-\lambda a\bar{b}e_{iu} + b\bar{a}e_{ui}) & \text{if } j = v \\ I & \text{if } j \neq u \text{ and } j \neq v. \end{cases}$$

It suffices to verify these formulas in $U^{\lambda}_4(A,\Lambda)$ and $U^{\lambda}_6(A,\Lambda)$, respectively; we omit the details.

Let $\Gamma_n = \{\beta \in \Lambda_n | X_+(\beta) \in (H(E_n(A)), X_+(\Lambda_n))\}$ and put $S = S_{-\lambda}(A)$. From (5.1.4) we have $Se_{ii} \subset \Gamma_n$ for all i. Taking $r \in S$ in (5.1.3) we further obtain $be_{ij} - \lambda\bar{b}e_{ji} \in \Gamma_n$ for all $b \in AS + SA$. If $AS + SA = A$ therefore, (5.1.3) further shows that $\Lambda e_{ii} \subset \Gamma_n$ for all i, whence $\Gamma_n = \Lambda_n$.

If $n \geq 3$ then (5.1.4) shows that $be_{ij} - \lambda\bar{b}e_{ji} \in \Gamma_n$ for all $b \in A$ and all $i \neq j$. Then, as above, (5.1.3) shows further that $\Lambda e_{ii} \subset \Gamma_n$ for all i, so $\Gamma_n = \Lambda_n$. This concludes the proof of b), and hence of the proposition.

(5.2) THEOREM (Wasserstein [15]).

$$(U^\lambda(A,\Lambda), U^\lambda(A,\Lambda)) = EU^\lambda(A,\Lambda) = (EU^\lambda(A,\Lambda), EU^\lambda(A,\Lambda)).$$

<u>Hence</u>

$$KU_1^\lambda(A,\Lambda) = U^\lambda(A,\Lambda)/EU^\lambda(A,\Lambda)$$

<u>and</u>

$$W_1^\lambda(A,\Lambda) = U^\lambda(A,\Lambda)/H(GL(A))\cdot EU^\lambda(A,\Lambda).$$

In fact this theorem results from the following assertions, by letting $n \to \infty$:

<u>For all</u> $n \geq 1$,

$$(U_{2n}^\lambda(A,\Lambda), U_{2n}^\lambda(A,\Lambda)) \subset H(E_{2n}(A))\cdot EU_{4n}^\lambda(A,\Lambda).$$

<u>For all</u> $n \geq 3$

$$H(E_n(A)) \subset EU_{2n}^\lambda(A,\Lambda)$$

<u>and</u>

$$(EU_{2n}^\lambda(A,\Lambda), EU_{2n}^\lambda(A,\Lambda)) = EU_{2n}^\lambda(A,\Lambda).$$

The first assertion follows from Prop. (3.7), and the second one from Prop. (5.1).

(5.3) COROLLARY. *If* $u:(A,\lambda,\Lambda) \to (A',\lambda',\Lambda')$ *is a unitary surjection then the homomorphism* $U^{\lambda}(A,\Lambda) \to U^{\lambda'}(A',\Lambda')$ *induces an epimorphism of derived groups.*

This results from Thm. (5.2) and Prop. (2.3).

We conclude this section with some results on the normalizer of $EU^{\lambda}_{2n}(A,\Lambda)$.

(5.4) *Conjugation by* $\omega_1 \perp I_2$. Let $\alpha = \begin{pmatrix} a & b \\ c & d \end{pmatrix} \in GL_2(A)$ and put $\alpha^{-1} = \begin{pmatrix} d_1 & b_1 \\ c_1 & a_1 \end{pmatrix}$. (For example if A is commutative then $\begin{pmatrix} d_1 & b_1 \\ c_1 & a_1 \end{pmatrix} = \frac{1}{ad-bc}\begin{pmatrix} d & -b \\ -c & a \end{pmatrix}$.) Then

$$(\omega_1 \perp I_2)H(\alpha)\ (\omega_1 \perp I_2)^{-1}$$

$$= \begin{pmatrix} 0 & 0 & 1 & 0 \\ 0 & 1 & 0 & 0 \\ \bar{\lambda} & 0 & 0 & 0 \\ 0 & 0 & 0 & 1 \end{pmatrix} \begin{pmatrix} a & b & 0 & 0 \\ c & d & 0 & 0 \\ 0 & 0 & \overline{d_1} & \overline{c_1} \\ 0 & 0 & \overline{b_1} & \overline{a_1} \end{pmatrix} \begin{pmatrix} 0 & 0 & \lambda & 0 \\ 0 & 1 & 0 & 0 \\ 1 & 0 & 0 & 0 \\ 0 & 0 & 0 & 1 \end{pmatrix}$$

$$= \begin{pmatrix} \bar{d}_1 & 0 & 0 & \overline{c_1} \\ 0 & d & \lambda c & 0 \\ 0 & \bar{\lambda} b & a & 0 \\ \overline{b_1} & 0 & 0 & \bar{a}_1 \end{pmatrix}$$

As special cases we obtain

(5.4.1) $$(\omega_1 \perp I_2)H\begin{pmatrix}1 & b\\ 0 & 1\end{pmatrix}(\omega_1 \perp I_2)^{-1} = X_-\begin{pmatrix}0 & \bar{\lambda}b\\ -\bar{b} & 0\end{pmatrix}$$

and

(5.4.2) $$(\omega_1 \perp I_2)H\begin{pmatrix}1 & 0\\ c & 1\end{pmatrix}(\omega_1 \perp I_2)^{-1} = X_+\begin{pmatrix}0 & -\bar{c}\\ \lambda c & 0\end{pmatrix}$$

(These formulas are also special cases of (3.6.3) and (3.6.4).) Moreover it follows from Prop. (4.4) that

(5.4.3) $$H\begin{pmatrix}0 & 1\\ -0 & 0\end{pmatrix}(\omega_1 \perp I_2)H\begin{pmatrix}0 & -1\\ 1 & 0\end{pmatrix} = I_2 \perp \omega_1.$$

(5.5) PROPOSITION. <u>The group generated by</u> $EU^{\lambda}_{2n}(A,\Lambda)$ <u>and</u> $\omega_1 \perp I_{2n-2}$ <u>contains</u> $H(E_n(A)) \cdot EU^{\lambda}_{2n}(A,\Lambda)$ <u>as a normal subgroup.</u>

If $n = 1$ then $E_n(A) = \{I\}$ and the result follows from (2.8.6).

If $n = 2$ the result follows from formulas (5.4.1) and (5.4.2) above.

Suppose $n \geq 3$. Then (Prop. (5.1) $H(E_n(A)) \subset EU^{\lambda}_{2n}(A,\Lambda)$, so we need only show that $\omega_1 \perp I_{2n-2}$ normalizes $H(E_n(A)) \cdot EU^{\lambda}_{2n}(A,\Lambda)$. As generating set for the latter group we take the following elements: $H(\alpha)$, $X_+(\beta)$, and $X_-(\bar{\beta})$, where $\alpha = I + ae_{ij}$ ($a \in A$, $i \neq j$), and $\beta = re_{ii}$ ($r \in \Lambda$) or $\beta = ae_{ij} - \lambda\bar{a}e_{ji}$ ($a \in A$, $i \neq j$). If we write $H(A^n) = H_1 \perp \cdots \perp H_n$ where H_i is the hyperbolic plane spanned by e_i, $\overline{e_i}$ then each of the above elements lies in some $U(H_i)$ or $U(H_i \perp H_j)$, these being identified with the subgroups of $U_{2n}^{\lambda}(A,\Lambda)$ fixing $H_i^{\perp}$, respectively, $(H_i \perp H_j)^{\perp}$. With this notation $\omega_1 \perp I_{2n-2}$ lies in $U(H_1)$ so it centralizes $U(H_i)$ and $U(H_i \perp H_j)$ if $i,j \geq 2$. Thus we need only consider conjugates by $\omega_1 \perp I_{2n-2}$ of those generators above which lie in $U(H_1)$ or in $U(H_1 \perp H_i)$ for some $i \geq 2$. The fact that these conjugates remain in $H(E_n(A)) \cdot EU_{2n}^{\lambda}(A,\Lambda)$ follows from the cases $n = 1$ and $n = 2$ already treated above.

(5.6) COROLLARY. Assume that $H(E_n(A)) \subset EU_{2n}^{\lambda}(A,\Lambda)$. (This is automatic if $n \geq 3$ or if $A = A\Lambda + \Lambda A$ by (5.1).) Then the normalizer N of $EU_{2n}^{\lambda}(A,\Lambda)$ contains $H(GL_n(A))$ and $\omega_1 \perp I_{2n-2}$. If N acts transitively on the set of unimodular isotropic elements in $H(A^n)$ then $N = U_{2n}^{\lambda}(A,\Lambda)$.

Normalization by $H(GL_n(A))$ follows from Prop. (2.5); normalization by $\omega_1 \perp I_{2n-2}$ follows from Prop. (5.5) in view of the assumption $H(E_n(A)) \subset EU_{2n}^{\lambda}(A,\Lambda)$. The last assertion follows from Prop. (3.11), part b).

§6. Relativization.

(6.1) Relative unitary groups. Let $(\mathfrak{a},\Gamma)$ be a unitary ideal in (A,λ,Λ) (Ch. I, (6.2)). Recall (Ch. I, (7.3)) that we associate to it a commutative diagram

$$
(6.1.1)\qquad
\begin{array}{ccccc}
A(\mathfrak{a},\Gamma) & & \xrightarrow{\ p_1\ } & & (A,\lambda,\Lambda) \\
 & \nwarrow^{\Delta} & & // & \\
{\scriptstyle p_2}\big\downarrow & & (A,\lambda,\Lambda) & & \big\downarrow \\
 & // & & & \\
(A,\lambda,\Lambda) & & \longrightarrow & & (A/\mathfrak{a},\lambda',\Lambda/\mathfrak{a}\cap\Lambda)
\end{array}
$$

of unitary rings. If

$$G:\ (A,\lambda,\Lambda)\longmapsto G^{\lambda}(A,\Lambda)$$

is a group valued functor on unitary rings then $G^{\lambda}(\mathfrak{a},\Gamma)$ is defined by the exact commutative diagram (Ch. I, (7.4.1)):

$$
(6.1.2)\qquad
\begin{array}{ccccccc}
1\longrightarrow & G^{\lambda}(\mathfrak{a},\Gamma) & \longrightarrow & G^{\lambda}(A(\mathfrak{a},\Gamma)) & \xrightarrow{\ p_1\ } & G^{\lambda}(A,\Lambda) \\
 & & \searrow^{\partial} & {\scriptstyle p_2}\big\downarrow & & \big\downarrow \\
 & & & G^{\lambda}(A,\Lambda) & \longrightarrow & G^{\lambda'}(A/\mathfrak{a},\Lambda/\mathfrak{a}\cap\Lambda)
\end{array}
$$

Taking $G = U_{2n}, EU_{2n},\ldots$ we thus define groups $U^{\lambda}_{2n}(\mathfrak{a},\Gamma)$, $EU^{\lambda}_{2n}(\mathfrak{a},\Gamma),\ldots$. We also introduce

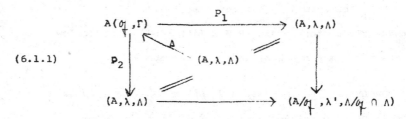

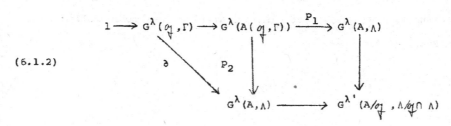

$$(\mathfrak{q},\Gamma)_n = \{\gamma \in \Lambda_n \mid \gamma \equiv 0 \bmod \mathfrak{q},\ \gamma_{ii} \in \Gamma (1 \le i \le n)\}$$

and its conjugate transpose $\overline{(\mathfrak{q},\Gamma)_n} = (\mathfrak{q},\bar{\Gamma})_n$.

(6.2) THEOREM.

a) When $G = U_{2n}$, ∂ defines an isomorphism from $U_{2n}^{\lambda}(\mathfrak{q},\Gamma)$ onto the group of $\sigma = \begin{pmatrix}\alpha & \beta\\ \gamma & \delta\end{pmatrix} \in U_{2n}^{\lambda}(A,\Lambda)$ such that $\sigma \equiv I_{2n} \bmod \mathfrak{q}$ and $\beta\bar{\alpha},\ \delta\bar{\gamma} \in (\mathfrak{q},\Gamma)_n$. The latter is a normal subgroup of $U_{2n}^{\lambda}(A,\Lambda)$. The kernel of $U_{2n}^{\lambda}(A,\Lambda) \to U_{2n}^{\lambda}(A/\mathfrak{q},\ \Lambda/\mathfrak{q} \cap \Lambda)$ is $U_{2n}^{\lambda}(\mathfrak{q},\ \mathfrak{q} \cap \Lambda)$.

b) When $G = EU_{2n}$, ∂ defines an isomorphism from $EU_{2n}^{\lambda}(\mathfrak{q},\Gamma)$ onto the least normal subgroup of $EU_{2n}^{\lambda}(A,\Lambda)$ containing $X_+((\mathfrak{q},\Gamma)_n)$ and $X_-((\mathfrak{q},\bar{\Gamma})_n)$.

c) For all $n \ge 1$,

$$(U_{2n}^{\lambda}(A,\Lambda),\ U_{2n}^{\lambda}(\mathfrak{q},\Gamma)) \subset H(E_{2n}(\mathfrak{q})) \cdot EU_{4n}^{\lambda}(\mathfrak{q},\Gamma)$$

For all $n \ge 3$,

$$H(E_n(\mathfrak{q})) \subset EU_{2n}^{\lambda}(\mathfrak{q},\Gamma)$$

and

$$(EU_{2n}^{\lambda}(A,\Lambda),\ EU_{2n}^{\lambda}(\mathfrak{q},\Gamma)) = EU_{2n}^{\lambda}(\mathfrak{q},\Gamma)$$

Moreover

$$(U^{\lambda}(A,\Lambda),U^{\lambda}(\mathfrak{q},\Gamma)) = EU^{\lambda}(\mathfrak{q},\Gamma) = (EU^{\lambda}(A,\Lambda),EU^{\lambda}(\mathfrak{q},\Gamma)).$$

We shall use the isomorphisms in a) and b) to identify $U_{2n}^{\lambda}(\mathfrak{q},\Gamma)$ and $EU_{2n}^{\lambda}(\mathfrak{q},\Gamma)$ with the indicated subgroups of

$U^{\lambda}_{2n}(A,\Lambda)$. It is with these identifications that we interpret the commutator formulas in c).

To prove the theorem recall first that $A(\mathfrak{q},\Gamma) = (A(\mathfrak{q}),\lambda,\Lambda(\Gamma))$ where $A(\mathfrak{q}) \subset A \times A$ is the fibre product $A \times_{A/\mathfrak{q}} A$, $\Lambda(\Gamma) = \Lambda \times_{\Lambda/\Gamma}\Lambda$, and λ stands for $(\lambda,\lambda) \in A(\mathfrak{q})$. Let $\Delta : A \to A(\mathfrak{q})$, $\Delta a = (a,a)$. Put $\mathfrak{q}_2 = 0 \times \mathfrak{q} \subset A(\mathfrak{q})$ and $q_2 = (0,q) \in \mathfrak{q}_2$ for $q \in \mathfrak{q}$. Then $A(\mathfrak{q}) = \Delta A \oplus \mathfrak{q}_2$ and $p_1 : A(\mathfrak{q}) \to A$ is the map $\Delta a + q_2 \mapsto a$, with kernel $\mathfrak{q}_2$. Further, $p_2 : \Delta a + q_2 \mapsto a + q$. We similarly have $\Lambda(\Gamma) = \Delta\Lambda \oplus \Gamma_2$ where $\Gamma_2 = 0 \times \Gamma = \mathfrak{q}_2 \cap \Lambda(\Gamma)$.

For any functor G we have

$$G^{\lambda}(\mathfrak{q},\Gamma) = \operatorname{Ker}(G^{\lambda}(A(\mathfrak{q},\Gamma)) \xrightarrow{p_1} G^{\lambda}(A,\Lambda))$$

and $\partial : G^{\lambda}(\mathfrak{q},\Gamma) \to G^{\lambda}(A,\Lambda)$ is induced by $p_2 : G^{\lambda}(A(\mathfrak{q},\Gamma)) \to G^{\lambda}(A,\Lambda)$; the image of ∂ is a normal subgroup of $G^{\lambda}(A(\mathfrak{q},\Gamma)$ (Ch. I, (7.4)). Since Δ is a section of p_1 we have the semi-direct product decomposition

$$G^{\lambda}(A(\mathfrak{q},\Gamma)) = G^{\lambda}(\Delta A,\Delta\Lambda) \times G^{\lambda}(\mathfrak{q},\Gamma)$$

so that on the first factor p_1 induces the inverse to the isomorphism $\Delta : G^{\lambda}(A,\Lambda) \to G^{\lambda}(\Delta A,\Delta\Lambda)$.

Suppose $G = U_{2n}$. Then $U^{\lambda}_{2n}(\mathfrak{q},\Lambda)$ consists of the elements of $U^{\lambda}_{2n}(A(\mathfrak{q},\Gamma))$ which are $\equiv I_{2n} \bmod \mathfrak{q}_2$. Since p_2 maps $\mathfrak{q}_2$

isomorphically to $\mathfrak{q}$, ∂ maps $U^{\lambda}_{2n}(\mathfrak{q},\Gamma)$ isomorphically onto the group of matrices $\sigma = \begin{pmatrix}\alpha & \beta\\ \gamma & \delta\end{pmatrix} \in U^{\lambda}_{2n}(A,\Lambda)$ which are $\equiv I_{2n} \bmod \mathfrak{q}$ and such that $\beta\bar{\alpha}$ and $\delta\bar{\gamma}$ lie in the projection (under p_2) of $M_n(\mathfrak{q}_2) \cap \Lambda(\Gamma)_n = (\mathfrak{q}_2,\Gamma_2)_n$. This latter projects isomorphically to $(\mathfrak{q},\Gamma)_n$, whence the first assertions of a). When $\Gamma = \mathfrak{q} \cap \Lambda$ it is clear that $(\mathfrak{q},\Gamma)_n = M_n(\mathfrak{q}) \cap \Lambda_n$, whence the last assertion of a).

To prove b) we write

$$EU^{\lambda}_{2n}(A(\mathfrak{q},\Gamma)) = EU^{\lambda}_{2n}(\Delta A,\Delta\Lambda) \ltimes EU^{\lambda}_{2n}(\mathfrak{q},\Gamma)$$

as above, where $EU^{\lambda}_{2n}(\mathfrak{q},\Gamma)$ consists of the elements $\equiv I_{2n} \bmod \mathfrak{q}_2$. As above we see that p_2 maps $EU^{\lambda}_{2n}(\mathfrak{q},\Gamma)$ isomorphically to a normal subgroup of $EU^{\lambda}_{2n}(A,\Lambda)$. This isomorphism is equivariant with respect to the isomorphism $p_2 : EU^{\lambda}_{2n}(\Delta A,\Delta\Lambda) \to EU^{\lambda}_{2n}(A,\Lambda)$, the first group acting in the semi-direct product and the second by conjugation on itself.

The group $EU^{\lambda}_{2n}(A(\mathfrak{q},\Gamma))$ is generated by $X_+(\Lambda(\Gamma)_n)$ and $X_-(\overline{\Lambda(\Gamma)}_n)$. The decomposition $\Lambda(\Gamma) = \Delta\Lambda \oplus \Gamma_2$ induces a decomposition $\Lambda(\Gamma)_n = \Delta\Lambda_n \oplus (\mathfrak{q}_2,\Gamma_2)_n$; similarly $\overline{\Lambda(\Gamma)}_n = \Delta\bar{\Lambda}_n \oplus (\mathfrak{q}_2,\bar{\Gamma}_2)_n$. Thus $S_1 = X_+(\Delta\Lambda_n) \cup X_-((\mathfrak{q}_2,\bar{\Gamma}_2)_n) \subset EU^{\lambda}_{2n}(\mathfrak{q},\Gamma)$ generates the semi-direct product $EU^{\lambda}_{2n}(A(\mathfrak{q},\Gamma))$. The assertions of b) and c) now follow from the next lemma, in view of Theorem (5.2) and the assertions displayed in its proof.

(6.3) LEMMA. *Let* $G = G_1 \ltimes G_2$ *be a semi-direct product of groups. Let* S_i *be a subset of* G_i $(i = 1,2)$.

a) *The group* $\langle S_1,S_2\rangle$ *generated by* $S_1 \cup S_2$ *is the semi-direct product of* $\langle S_1\rangle$ *with* $\langle S_2\rangle_{\langle S_1\rangle}$, *the least* $\langle S_1\rangle$*-normalized subgroup of* G_2 *containing* S_1.

b) $(G,G) = (G_1,G_1) \ltimes G,G_2)$.

Since (by definition) $\langle S_1\rangle$ normalizes $\langle S_2\rangle_{\langle S_1\rangle}$ the product $\langle S_1\rangle \cdot \langle S_2\rangle_{\langle S_1\rangle}$ is a group. It clearly containes $S_1 \cup S_2$ and is contained in $\langle S_1,S_2\rangle$, whence a). Part b) is well known (cf. for example [3], Ch. VII, Lemma (2.6)).

(6.4) COROLLARY. *We have*

$$KU_1^\lambda(\mathfrak{q},\Gamma) = U^\lambda(\mathfrak{q},\Gamma)/EU^\lambda(\mathfrak{q},\Gamma),$$

and the sequence

$$KU_1^\lambda(\mathfrak{q}, \mathfrak{q} \cap \Lambda) \longrightarrow KU_1^\lambda(A,\Lambda) \longrightarrow KU_1^{\lambda'}(A/\mathfrak{q}, \Lambda/\mathfrak{q} \cap \Lambda)$$

is exact.

$KU_1^\lambda(\mathfrak{q},\Gamma)$ is the kernel of the abelianization of $U^\lambda(A(\mathfrak{q}),\Lambda(\Gamma)) \to U^\lambda(A,\Lambda)$. Using the semi-direct product decomposition of the first group, part c) of Theorem (6.2), and part b) of Lemma (6.3), we obtain the above description of $KU_1^\lambda(\mathfrak{q},\Gamma)$.

The exact sequence results, in view of Cor. (5.3) and part a) of Theorem (6.2), from the following easily verified fact:

If $1 \to G_0 \to G_1 \xrightarrow{p} G_2$ is an exact sequence of groups such that $p(G_1,G_1) = (G_2, G_2)$ then the sequence

$$G_0/(G_1,G_0) \longrightarrow G_1/(G_1,G_1) \longrightarrow G_2/(G_2,G_2)$$

is exact (cf. Ch. III, Lemma (A.10)).

(6.5) COROLLARY. Let $u:(A,\lambda,\Lambda) \to (A',\lambda',\Lambda')$ be a unitary surjection. If $(\mathfrak{q},\Gamma)$ is a unitary ideal in (A,λ,Λ) then $(u\mathfrak{q},u\Gamma)$ is a unitary ideal in (A',λ',Λ') and for all $n \geq 1$, the homomorphism $EU_{2n}^{\lambda}(\mathfrak{q},\Gamma) \to EU_{2n}^{\lambda'}(u\mathfrak{q},u\Gamma)$ is surjective.

The fact that $(u\mathfrak{q},u\Gamma)$ is a unitary ideal is immediate. We have seen in Prop. (2.3) that $EU_{2n}^{\lambda}(A,\Lambda) \to EU_{2n}^{\lambda'}(A',\Lambda')$ is surjective. It follows that the image of $EU_{2n}^{\lambda}(\mathfrak{q},\Gamma)$ is normalized by $EU_{2n}^{\lambda'}(A',\Lambda')$. This image visibly contains $X_+((u\mathfrak{q},u\Gamma)_n)$ and $X_-((u\mathfrak{q},u\bar{\Gamma})_n)$. By Thm. (6.2) part b) therefore it contains $EU_{2n}^{\lambda}(u\mathfrak{q},u\Gamma)$, whence the corollary.

(6.6) PROPOSITION. Let $(\mathfrak{q},\Gamma)$ be a unitary ideal of (A,λ,Λ). The multiplication map

(6.6.1) $$(\mathfrak{q},\bar{\Gamma})_n \times GL_n(\mathfrak{q}) \times (\mathfrak{q},\Gamma)_n \longrightarrow U_{2n}^{\lambda}(\mathfrak{q},\Gamma)$$
$$(\gamma,\alpha,\beta) \longmapsto X_-(\gamma)H(\alpha)X_+(\beta) = \begin{pmatrix}\alpha & \alpha\beta \\ \gamma\alpha & \bar{\alpha}^{-1}+\gamma\alpha\beta\end{pmatrix}$$

is injective, and its image consists of all $\sigma = \begin{pmatrix}\alpha & \beta' \\ \gamma' & \delta\end{pmatrix} \in U_{2n}^{\lambda}(\mathfrak{q},\Gamma)$ for which α is invertible. If $\mathfrak{q} \subset \operatorname{rad} A$ then (6.6.1) is bijective.

The first assertion follows from Prop. (2.5), part b). The last assertion results from it, since $\alpha \equiv I \bmod \operatorname{rad} A \Rightarrow \alpha$ is invertible.

§7. The normal subgroups of $U^{\lambda}(A,\Lambda)$

Let (A,λ,Λ) be a unitary ring. We shall here describe all of the normal subgroups of $U^{\lambda}(A,\Lambda)$. They fall into disjoint classes (called "levels") parametrized by the set of unitary ideals $(\mathfrak{q},\Gamma)$ of (A,λ,Λ) (Thm. (7.3)).

Under finiteness assumptions on A an analogous, and much more difficult, theorem was first proved by Bak [1], Ch. IV.

In Prop (7.8) we compute the level of a commutator of two normal subgroups. In Thms. (7.9) and (7.11) we give some precise information about $KU_1^{\lambda}(\mathfrak{q},\Gamma)$ when $\mathfrak{q} \subset \text{rad } A$, Thm. (7.11) dealing with the restricted symplectic case.

(7.1) Subgroups of level $(\mathfrak{q},\Gamma)$. Let $(\mathfrak{q},\Gamma)$ be a unitary ideal of (A,λ,Λ). A subgroup G of $U^{\lambda}(A,\Lambda)$ is said to be of level $(\mathfrak{q},\Gamma)$ if

$$EU^{\lambda}(\mathfrak{q},\Gamma) \subset G \subset U^{\lambda}(\mathfrak{q},\Gamma).$$

It then follows from Thm. (6.2) part c) that

(7.1.1) $$(U^{\lambda}(A,\Lambda),G) = EU^{\lambda}(\mathfrak{q},\Gamma) = (EU^{\lambda}(A,\Lambda),G).$$

Since this group is contained in G we conclude that G is a normal subgroup of $U^{\lambda}(A,\Lambda)$.

Indeed the lattice of subgroups of level $(\mathfrak{q},\Gamma)$ is isomorphic to the lattice of subgroups of $KU_1^\lambda(\mathfrak{q},\Gamma) = U^\lambda(\mathfrak{q},\Gamma)/EU^\lambda(\mathfrak{q},\Gamma)$, the latter being a central subgroup of $U^\lambda(A,\Lambda)/EU^\lambda(\mathfrak{q},\Gamma)$.

Suppose $u\colon (A,\lambda,\Lambda) \to (A',\lambda',\Lambda')$ is a unitary surjection. Then $(u\mathfrak{q},u\Gamma)$ is a unitary ideal of (A',λ',Λ') and the homomorphism $EU^\lambda(\mathfrak{q},\Gamma) \to EU^\lambda(u\mathfrak{q},u\Gamma)$ is surjective (Cor. (6.5)). It follows that _if_ G _is a subgroup of_ $U^\lambda(A,\Lambda)$ _of level_ $(\mathfrak{q},\Gamma)$ _then_ uG _is a subgroup of_ $U^{\lambda'}(A',\Lambda')$ _of level_ $(u\mathfrak{q},u\Gamma)$.

(7.2) _Uniqueness of level._ Suppose $(\mathfrak{q}_1,\Gamma_1)$ and $(\mathfrak{q}_2,\Gamma_2)$ are unitary ideals such that $EU^\lambda(\mathfrak{q}_1,\Gamma_1) \subset U^\lambda(\mathfrak{q}_2,\Gamma_2)$. Then $\mathfrak{q}_1 \subset \mathfrak{q}_2$ and $\Gamma_1 \subset \Gamma_2$.

For we have $X_+((\mathfrak{q}_1,\Gamma_1)_n) \subset U_{2n}^\lambda(\mathfrak{q}_2,\Gamma_2)$ for all $n \geq 1$, whence, by Thm. (6.2) part a), $(\mathfrak{q}_1,\Gamma_1)_n \subset (\mathfrak{q}_2,\Gamma_2)_n$ for all $n \geq 1$. For $n = 1$ this says $\Gamma_1 \subset \Gamma_2$ and for $n \geq 2$ it implies that $\mathfrak{q}_1 \subset \mathfrak{q}_2$.

The above conclusion implies that if a subgroup G of $U^\lambda(A,\Lambda)$ has level $(\mathfrak{q}_1,\Gamma_1)$ and level $(\mathfrak{q}_2,\Gamma_2)$ then $(\mathfrak{q}_1,\Gamma_1) = (\mathfrak{q}_2,\Gamma_2)$.

(7.3) THEOREM. _Let_ G _be a subgroup of_ $U^\lambda(A,\Lambda)$ _normalized by_ $EU^\lambda(A,\Lambda)$. _Then there is a (unique) unitary ideal_ $(\mathfrak{q},\Gamma)$ _such that_ G _has level_ $(\mathfrak{q},\Gamma)$.

The proof of this theorem will be based on the following two propositions, together with the analogue of Theorem (7.3) for the group GL(A) ([3], Ch. V, Thm. (2.1)).

(7.4) PROPOSITION. Let $GL_n(A)$ act on Λ_n by $\alpha[\beta] = \alpha\beta\bar{\alpha}$. For each unitary ideal $(\mathfrak{q},\Gamma)$ the subgroup $(\mathfrak{q},\Gamma)_n$ of Λ_n is $GL_n(A)$-invariant. If $n \geq 2$ then every $E_n(A)$-invariant subgroup of Λ_n is of the form $(\mathfrak{q},\Gamma)_n$ for some unitary ideal $(\mathfrak{q},\Gamma)$.

Remark. In $U_{2n}^{\lambda}(A,\Lambda)$ we have $H(\alpha)X_+(\beta)H(\alpha)^{-1} = X_+(\alpha[\beta])$. Thus $H(GL_n(A)) \cdot X_+(\Lambda_n)$ is isomorphic to the semi-direct product defined by the above action.

Since $H(GL_n(A))$ normalizes $X_+(\Lambda_n)$ and $U_{2n}^{\lambda}(\mathfrak{q},\Gamma)$ it normalizes their intersection, $X_+((\mathfrak{q},\Gamma)_n)$, whence the first assertion of the proposition.

To prove the last assertion we note first that if $(\mathfrak{q}_i,\Gamma_i)$ $(i \in I)$ is a family of unitary ideals then $(\mathfrak{q},\Gamma) = (\Sigma_i\mathfrak{q}_i,\Sigma_i\Gamma_i)$ is a unitary ideal, and $(\mathfrak{q},\Gamma)_n = \Sigma_i(\mathfrak{q}_i,\Gamma_i)_n$. Therefore to show that an $E_n(A)$-invariant subgroup L of Λ_n is of the form $(\mathfrak{q},\Gamma)_n$ it suffices to do so in case L is the sub $E_n(A)$-module of Λ_n generated by a single element $\beta = (b_{ij}) \in \Lambda_n$. We claim $L = (\mathfrak{q},\Gamma)_n$ where $\mathfrak{q}$ is the two sided ideal generated by all b_{ij} (note that $b_{ij} = -\lambda\,\overline{b_{ji}}$ so $\mathfrak{q} = \overline{\mathfrak{q}}$) and where Γ is the group generated by Λ together with all elements $ab_{ii}\bar{a}$

$(a \in A,\ 1 \leq i \leq n)$. Recall (Ch. I. (7.2)) that:

(7.4.1) $\Lambda_{\mathfrak{q}}$ is the additive group generated by $S_{-\lambda}(\mathfrak{q})$ together with all $qr\bar{q}$ $(q \in \mathfrak{q},\ r \in \Lambda)$.

It is readily seen that $(\mathfrak{q},\Gamma)$ is indeed a unitary ideal, and that $L \subset (\mathfrak{q},\Gamma)_n$.

We have $\beta = \Sigma_{i,j} b_{ij} e_{ij}$. Let $a \in A$ and $u \neq v$ and put

$$\beta' = H(I + ae_{uv})[\beta] - \beta = ae_{uv}\beta + \beta\bar{a}e_{vu} + ae_{uv}\beta e_{vu}\bar{a}$$

$$= \Sigma_j (ab_{vj}e_{uj} + b_{jv}\bar{a}e_{ju}) + ab_{vv}\bar{a}e_{uu}.$$

Let $c \in A$ and $s \neq r$ or u, and put

$$\beta'' = H(I + ce_{rs})[\beta'] - \beta' = ce_{rs}\beta' + \beta'\bar{c}e_{sr}$$

$$= cb_{sv}\bar{a}\, e_{ru} + ab_{vs}\bar{c}\, e_{ur}$$

$$= S_{-\lambda}(ab_{vs}\bar{c}\, e_{ur})$$

From this it follows that L contains $S_{-\lambda}(qe_{ij})$ for all i,j and $q \in \mathfrak{q}$. Since $\beta' \in L$ it further follows that $ab_{ii}\bar{a}\, e_{jj} \in L$ for all i,j and all $a \in A$. The same calculation shows that the set of $r \in A$ such that $re_{jj} \in L$ for all j is stable under

$r \mapsto ar\bar{a}$ ($a \in A$). These conclusions establish the proposition.

<u>Remark</u>. The proof shows that L is the $E_n(A)$-module generated by all $\epsilon\beta\bar{\epsilon} - \beta$ with $\epsilon \in E_n(A)$.

(7.5) <u>Products of unitary ideals</u>. For the next proposition on commutators we introduce some further notation.

Let $(\mathfrak{q},\Gamma)$ and $(\mathfrak{q}',\Gamma')$ be unitary ideals. Then $\mathfrak{q}'' = \mathfrak{q}\mathfrak{q}' + \mathfrak{q}'\mathfrak{q}$ is a two sided involutory ideal. Denote by

$$\Gamma'_{\mathfrak{q}}$$

the additive group generated by $\Lambda_{\mathfrak{q}''}$ (see (7.4.1) above) together with all elements $qr'\bar{q}$ ($q \in \mathfrak{q}$, $r' \in \Gamma'$). It is clear that $(\mathfrak{q}'', \Gamma'_{\mathfrak{q}})$ is a unitary ideal. Similarly we have the unitary ideal $(\mathfrak{q}'', \Gamma_{\mathfrak{q}'})$ and hence also $(\mathfrak{q}'', \Gamma'_{\mathfrak{q}} + \Gamma_{\mathfrak{q}'})$. (It might be natural to call the latter the "unitary product" of $(\mathfrak{q},\Gamma)$ and $(\mathfrak{q}',\Gamma')$.)

(7.6) PROPOSITION. <u>Let</u> $(\mathfrak{q},\Gamma)$ <u>and</u> $(\mathfrak{q}',\Gamma')$ <u>be unitary ideals, and put</u> $\mathfrak{q}'' = \mathfrak{q}\mathfrak{q}' + \mathfrak{q}'\mathfrak{q}$. <u>Let</u> $n \geq 2$.

a) $(H(E_n(\mathfrak{q})), X_+((\mathfrak{q}',\Gamma')_n)) = X_+((\mathfrak{q}'',\Gamma'_{\mathfrak{q}})_n)$ <u>and</u> $(H(E_n(\mathfrak{q})), X_-((\mathfrak{q}',\Gamma')_n)) = X_-((\mathfrak{q}'',\overline{\Gamma'_{\mathfrak{q}}})_n)$. <u>Hence</u> $(H(E_n(\mathfrak{q})), EU^{\lambda}_{2n}(\mathfrak{q}',\Gamma')) \supset EU^{\lambda}_{2n}(\mathfrak{q}'',\Gamma'_{\mathfrak{q}})$.

b) <u>If</u> $n \geq 3$ <u>then</u> $H(E_n(\mathfrak{q})) \subset EU^{\lambda}_{2n}(\mathfrak{q},\Gamma)$ <u>and</u> $(EU^{\lambda}_{2n}(\mathfrak{q},\Gamma), EU^{\lambda}_{2n}(\mathfrak{q}',\Gamma')) \supset EU^{\lambda}_{2n}(\mathfrak{q}'',\Gamma'_{\mathfrak{q}} + \Gamma_{\mathfrak{q}'})$.

In part a) the last assertion follows from the first two, and the second is the conjugate transpose of the first.

It is clear that $(H(E_n(\mathfrak{q})), X_+((\mathfrak{q}', \Gamma')_n) = X_+(L)$ for some $E_n(A)$-invariant subgroup L of Λ_n. By Prop. (7.4) therefore $L = (\mathfrak{q}_1, \Gamma_1)_n$ for some unitary ideal $(\mathfrak{q}_1, \Gamma_1)$. That $\mathfrak{q}_1 \subset \mathfrak{q}''$ follows from the general fact: $(GL_m(\mathfrak{q}), GL_m(\mathfrak{q}')) \subset GL_m(\mathfrak{q}\mathfrak{q}' + \mathfrak{q}'\mathfrak{q})$ for any two sided ideals $\mathfrak{q}, \mathfrak{q}'$ in any ring A. To verify this we may pass to $A/(\mathfrak{q}\mathfrak{q}' + \mathfrak{q}'\mathfrak{q})$ and so assume $\mathfrak{q}\mathfrak{q}' = 0 = \mathfrak{q}'\mathfrak{q}$. Then if $\sigma \in GL_m(\mathfrak{q})$ and $\sigma' \in GL_m(\mathfrak{q}')$ we have $(\sigma - I)(\sigma' - I) = 0 = (\sigma' - I)(\sigma - I)$ so σ and σ' commute.

For later use we record the following consequence of the above.

(7.6.1) $$(U_{2n}^{\lambda}(\mathfrak{q}, \Gamma), U_{2n}^{\lambda}(\mathfrak{q}', \Gamma')) \subset U_{2n}^{\lambda}(\mathfrak{q}'', \Lambda \cap \mathfrak{q}'').$$

Back to the proof of a), we now have $L = (\mathfrak{q}_1, \Gamma_1)_n$ with $\mathfrak{q}_1 \subset \mathfrak{q}''$. The reverse inclusion follows from formula (5.1.4), whence $\mathfrak{q}_1 = \mathfrak{q}''$. Further formula (5.1.3) shows that $qr'\bar{q} \in \Gamma_1$ for all $q \in \mathfrak{q}$, $r' \in \Gamma'$, whence $\Gamma'_{\mathfrak{q}} \subset \Gamma_1$. Formulas (5.1.3) and (5.1.4) show in addition that $(H(\varepsilon), X_+((\mathfrak{q}', \Gamma')_n)) \subset X_+((\mathfrak{q}'', \Gamma')_n)$ whenever ε is of the form $I + qe_{ij}$ $(q \in \mathfrak{q},\ i \neq j)$. Now $E_n(\mathfrak{q})$ is generated by elements of the form $\alpha\varepsilon\alpha^{-1}$ with ε as above and $\alpha \in E_n(A)$. We have $(H(\alpha\varepsilon\alpha^{-1}), X_+((\mathfrak{q}', \Gamma')_n))$

$= H(\alpha)(H(\epsilon), H(\alpha)^{-1}X_+((\mathfrak{q}', \Gamma')_n)H(\alpha))H(\alpha)^{-1}$

$= H(\alpha)(H(\epsilon), X_+((\mathfrak{q}', \Gamma')_n))H(\alpha)^{-1} \subset H(\alpha)X_+((\mathfrak{q}'', \Gamma')_n)H(\alpha)^{-1}$

$= X_+((\mathfrak{q}'', \Gamma')_n)$. This shows that $\Gamma_1 = \Gamma'_{\mathfrak{q}}$ and thus completes the proof of a).

The first assertion of b) follows from formula (5.1.2). The last assertion follows from the first together with part a).

(7.7) Proof of Theorem (7.3). We shall write $\Lambda_\infty = \varinjlim \Lambda_n$ where $\beta \in \Lambda_n$ maps to $\begin{pmatrix}\beta & 0\\ 0 & 0\end{pmatrix} \in \Lambda_{n+m}$. Similarly, for any unitary ideal $(\mathfrak{q}, \Gamma)$ we put $(\mathfrak{q}, \Gamma)_\infty = \varinjlim (\mathfrak{q}, \Gamma)_n$.

Let G be a subgroup of $U^\lambda(A, \Lambda)$ normalized by $EU^\lambda(A, \Lambda)$. Then $G \cap X_+(\Lambda_\infty) = X_+(L)$ where L is an E(A)-invariant subgroup of Λ_∞. It follows therefore from Prop. (7.4) that $L = (\mathfrak{q}, \Gamma)_\infty$ for some unitary ideal $(\mathfrak{q}, \Gamma)$. By (4.5.5) we have $\omega_n X_+((\mathfrak{q}, \Gamma)_n)\omega_n^{-1} = X_-((\mathfrak{q}, \bar{\Gamma})_n)$. Passing to $U^\lambda_{4n}(A, \Lambda)$ and replacing ω_n by $\omega_n \perp \omega_n^{-1} \in EU^\lambda_{4n}(A, \Lambda)$ (Prop. (3.7)) we see that a group containing $X_+((\mathfrak{q}, \Gamma)_n)$ and normalized by $EU^\lambda_{4n}(A, \Lambda)$ also contains $X_-((\mathfrak{q}, \Gamma)_n)$, and hence $EU^\lambda_{2n}(\mathfrak{q}, \Gamma)$. Letting $n \to \infty$ we conclude above that G contains $EU^\lambda(\mathfrak{q}, \Gamma)$. It follows further from Prop. (7.6) that $H(E(\mathfrak{q})) \subset G$.

Theorem (7.3) will be proved by showing that $G \subset U^\lambda(\mathfrak{q}, \Gamma)$.

Put $G_+ = G \cap (H(GL(A)) \cdot X_+(\Lambda_\infty))$. We have $H(E(\mathfrak{q})) \cdot X_+((\mathfrak{q}, \Gamma)_\infty) \subset G_+$. Let H(K) denote the projection of G_+ into H(GL(A)) in the semi-direct product above; K is a subgroup of GL(A) normalized by E(A). Hence by [3], Ch. V, Thm. (2.1) there is a two sided ideal $\mathfrak{q}'$ in A such that $E(\mathfrak{q}') \subset K \subset GL(\mathfrak{q}')$. We claim $\mathfrak{q}' = \mathfrak{q}$. Clearly $\mathfrak{q} \subset \mathfrak{q}'$. Moreover $\mathfrak{q}' = \overline{\mathfrak{q}'}$ because K is stable under conjugate transpose. (In U_{2n} we have $\omega_n H(\alpha)\omega_n^{-1} = H(\bar{\alpha})^{-1}$ so in U_{4n} we have $(\omega_n \perp \omega_n^{-1})(H(\alpha \oplus I))(\omega_n \perp \omega_n^{-1}) = H(\bar{\alpha} \oplus I)^{-1}$, and $\omega_n \perp \omega_n^{-1} \in EU_{4n}^{\lambda}(A,\Lambda)$.) Now G contains the commutator group $(G_+, X_+(\Lambda_\infty))$. Since $X_+(\Lambda_\infty)$ is an abelian normal subgroup of the semi-direct product $H(GL(A)) \cdot X_+(\Lambda_\infty)$ it follows that $(G_+, X_+(\Lambda_\infty))$ $= (H(K), X_+(\Lambda_\infty)) \supset (H(E(\mathfrak{q}')), X_+(\Lambda_\infty)) = X_+((\mathfrak{q}', \Lambda_{\mathfrak{q}'})_\infty)$ (Prop. (7.6)). By definition of $(\mathfrak{q}, \Gamma)$ it follows that $(\mathfrak{q}', \Lambda_{\mathfrak{q}'})$ $\subset (\mathfrak{q}, \Gamma)$ whence $\mathfrak{q}' \subset \mathfrak{q}$. This completes the proof that $\mathfrak{q}' = \mathfrak{q}$.

<u>Claim</u>. $G_+ \subset H(GL(\mathfrak{q})) \cdot X_+((\mathfrak{q},\Gamma)_\infty)$. Suppose $H(\alpha)X_+(\beta) \in G_+$. We have just shown that $\alpha \in GL(\mathfrak{q})$. Moreover in case $\alpha \in E(\mathfrak{q})$ then $H(\alpha) \in G_+$ whence $X_+(\beta) \in G \cap X_+(\Lambda_\infty) = X_+((\mathfrak{q},\Gamma)_\infty)$. In the general case we take the commutator with a general element $H(\epsilon) \in H(E(A))$: $(H(\epsilon), H(\alpha)X_+(\beta)) = (H(\epsilon), H(\alpha)) \cdot {}^{H(\alpha)}(H(\epsilon), X_+(\beta))$ (where we write ${}^{x}y = xyx^{-1}$) $= H((\epsilon,\alpha))X_+(\alpha(\epsilon\beta\bar{\epsilon} - \beta)\bar{\alpha})$. Since $(\epsilon,\alpha) \in (E(A), GL(\mathfrak{q})) = E(\mathfrak{q})$ ([3], Ch. V, Thm. (2.1)) we see from the case above that $\alpha(\epsilon\beta\bar{\epsilon} - \beta)\bar{\alpha} \in (\mathfrak{q},\Gamma)_\infty$ and hence $\epsilon\beta\bar{\epsilon} - \beta \in (\mathfrak{q},\Gamma)_\infty$, for all $\epsilon \in E(A)$. The remark after the

proof of Prop. (7.4) shows that these elements generate the same sub $E(A)$-module of Λ_∞ as β, whence $\beta \in (\mathfrak{q},\Gamma)_\infty$. This proves the claim.

Finally we shall prove that $G \subset U^\lambda(\mathfrak{q},\Gamma)$. Let $\sigma = \begin{pmatrix}\alpha & \beta\\ \gamma & \delta\end{pmatrix} \in G$. Say $\sigma \in U^\lambda_{2n}(A,\Lambda)$. From formula (3.6.3) we obtain

$$\left(H\begin{pmatrix} I & 0\\ I & I\end{pmatrix};\ (\sigma \perp I)\right) = H\begin{pmatrix} I & 0\\ I-\bar{\delta} & I\end{pmatrix} X_+\begin{pmatrix} 0 & \beta\\ -\lambda\bar{\beta} & \bar{\delta}\beta\end{pmatrix} \in G_+$$

It follows therefore from the claim above that $\delta \equiv I \bmod \mathfrak{q}$, $\beta \equiv 0 \bmod \mathfrak{q}$, and $\bar{\delta}\beta \in (\mathfrak{q},\Gamma)_n$. From (3.6.5) we have

$$\left((\sigma \perp I),\ X_+\begin{pmatrix} 0 & I\\ -\lambda I & 0\end{pmatrix}\right) = H\begin{pmatrix} 0 & \alpha-I\\ -\lambda(\bar{\alpha}-I) & \bar{\gamma}\alpha\end{pmatrix} \in G_+.$$ Again it

follows from the claim above that $\alpha \equiv I \bmod \mathfrak{q}$, $\gamma \equiv 0 \bmod \mathfrak{q}$, and $\bar{\gamma}\alpha \in (\mathfrak{q},\Gamma)_n$. This completes the proof of Theorem (7.3).

(7.8) PROPOSITION. *Let $(\mathfrak{q},\Gamma)$ and $(\mathfrak{q}',\Gamma')$ be unitary ideals. Let G,G' be subgroups of $U^\lambda(A,\Lambda)$ of levels $(\mathfrak{q},\Gamma)$ and $(\mathfrak{q}',\Gamma')$, respectively. Then (G,G') is a subgroup of level* $(\mathfrak{q}\mathfrak{q}' + \mathfrak{q}'\mathfrak{q},\ \Gamma'_{\mathfrak{q}} + \Gamma_{\mathfrak{q}'})$.

(See (7.5) for the above notation.)

Put $\mathfrak{q}'' = \mathfrak{q}\mathfrak{q}' + \mathfrak{q}'\mathfrak{q}$ and $\Gamma'' = \Gamma'_{\mathfrak{q}} + \Gamma_{\mathfrak{q}'}$. We have seen in Prop. (7.6) that $(EU^\lambda_{2n}(\mathfrak{q},\Gamma),\ EU^\lambda_{2n}(\mathfrak{q}',\Gamma'))$ contains $EU^\lambda_{2n}(\mathfrak{q}'',\Gamma'')$ for all $n \geq 3$. It therefore suffices to show that

$K = (U^{\lambda}(\mathfrak{q},\Gamma), U^{\lambda}(\mathfrak{q}',\Gamma'))$ is contained in $U^{\lambda}(\mathfrak{q}'',\Gamma'')$. By Thm. (7.3) K has some level $(\mathfrak{q}_1,\Gamma_1)$, clearly containing $(\mathfrak{q}'',\Gamma'')$. By (7.6.1) we even have $\mathfrak{q}_1 = \mathfrak{q}''$. It remains to show that the inclusion $\Gamma'' \subset \Gamma_1$ is not strict.

From (7.1.1) we have $EU^{\lambda}(\mathfrak{q}'',\Gamma_1) = (EU^{\lambda}(A,\Lambda),K)$. If X,Y,Z are normal subgroups of a group then (X,(Y,Z)) is contained in (Y,(Z,X)) · (Z,(X,Y)) (cf. [6], Ch. 10, Thm. 10.3.5). Taking $X = EU^{\lambda}(A,\Lambda)$, $Y = U^{\lambda}(\mathfrak{q},\Gamma)$, and $Z = U^{\lambda}(\mathfrak{q}',\Gamma')$ we find that $EU^{\lambda}(\mathfrak{q}'',\Gamma_1)$ is contained in $(EU^{\lambda}(\mathfrak{q},\Gamma), U^{\lambda}(\mathfrak{q}',\Gamma')) \cdot (U^{\lambda}(\mathfrak{q},\Gamma), EU^{\lambda}(\mathfrak{q}',\Gamma'))$. Therefore it suffices to show, for example, that $(EU^{\lambda}(\mathfrak{q},\Gamma), U^{\lambda}(\mathfrak{q}',\Gamma')) \subset U^{\lambda}(\mathfrak{q}'',\Gamma'')$. For the same will then follow for $(U^{\lambda}(\mathfrak{q},\Gamma), EU^{\lambda}(\mathfrak{q}',\Gamma'))$, whence $EU^{\lambda}(\mathfrak{q}'',\Gamma_1) \subset U^{\lambda}(\mathfrak{q}'',\Gamma'')$, and so $\Gamma_1 \subset \Gamma''$ as required.

Now $EU^{\lambda}(\mathfrak{q},\Gamma)$ is the normal subgroup of $U^{\lambda}(\mathfrak{q},\Gamma)$ generated by $X_+((\mathfrak{q},\Gamma)_\infty)$ (Cf. (4.5.5)). It thus suffices to show that for $\epsilon \in (\mathfrak{q},\Gamma)_\infty$ and $\sigma = \begin{pmatrix}\alpha & \beta\\ \gamma & \delta\end{pmatrix} \in U^{\lambda}(\mathfrak{q}',\Gamma')$ we have $(\sigma, X_+(\epsilon)) \in U^{\lambda}(\mathfrak{q}'',\Gamma'')$.

$$\begin{pmatrix}\alpha & \beta\\ \gamma & \delta\end{pmatrix}\begin{pmatrix}I & \epsilon\\ 0 & I\end{pmatrix}\begin{pmatrix}\bar\delta & \lambda\bar\beta\\ \bar\lambda\bar\gamma & \bar\alpha\end{pmatrix}\begin{pmatrix}I & -\epsilon\\ 0 & I\end{pmatrix}$$

$$= \begin{pmatrix}\alpha & \alpha\epsilon + \beta\\ \gamma & \gamma\epsilon + \delta\end{pmatrix}\begin{pmatrix}\bar\delta & -\bar\delta\epsilon + \lambda\bar\beta\\ \bar\lambda\bar\gamma & -\bar\lambda\bar\gamma\epsilon + \bar\alpha\end{pmatrix}$$

$$= \begin{pmatrix}I + \bar\lambda\alpha\epsilon\bar\gamma & \alpha(\lambda\bar\beta - \bar\delta\epsilon) + (\alpha\epsilon + \beta)(\bar\alpha - \bar\lambda\bar\gamma\epsilon)\\ \bar\lambda\gamma\epsilon\bar\gamma & \gamma(\lambda\bar\beta - \bar\delta\epsilon) + (\gamma\epsilon + \delta)(\bar\alpha - \bar\lambda\bar\gamma\epsilon)\end{pmatrix}.$$

Denoting this matrix $\begin{pmatrix} a & b \\ c & d \end{pmatrix}$ we are required by Thm. (6.2) to show that $b\bar{a}$ and $d\bar{c}$ belong to $(\mathfrak{q}'', \Gamma'')_\infty$. Since $\epsilon = -\lambda\bar{\epsilon}$ we have $\bar{a} = I - \gamma\epsilon\bar{\alpha}$. Using $\beta\bar{\alpha} + \lambda\alpha\bar{\beta} = 0$ we further find that $b = \alpha\epsilon\bar{\alpha} - \alpha\bar{\delta}\epsilon - \bar{\lambda}\beta\bar{\gamma}\epsilon - \bar{\lambda}\alpha\epsilon\bar{\gamma}\epsilon = \alpha\epsilon\bar{\alpha} - \epsilon - \bar{\lambda}\alpha\epsilon\bar{\gamma}\epsilon$. Thus

$$\begin{aligned} b\bar{a} = {} & \alpha\epsilon\bar{\alpha} - \epsilon - \bar{\lambda}\,\alpha\epsilon\,\bar{\gamma}\,\epsilon \\ & - \alpha\epsilon\bar{\alpha}\gamma\epsilon\bar{\alpha} + \epsilon\gamma\epsilon\bar{\alpha} + \bar{\lambda}\alpha\epsilon\bar{\gamma}\epsilon\gamma\epsilon\bar{\alpha} \end{aligned}$$

Writing $\alpha = I + \alpha_0$ where α_0 has coordinates in $\mathfrak{q}'$ we have $\alpha\epsilon\bar{\alpha} - \epsilon = \alpha_0\epsilon + \epsilon\bar{\alpha}_0 + \alpha_0\epsilon\bar{\alpha}_0 = S_{-\lambda}(\alpha_0\epsilon) + \alpha_0\epsilon\bar{\alpha}_0$. Since $\alpha_0\epsilon$ has coordinates in $\mathfrak{q}'\mathfrak{q}$ we have $S_{-\lambda}(\alpha_0\epsilon) \in (\mathfrak{q}'', \Lambda_{\mathfrak{q}''})$. Further it is clear that $\alpha_0\epsilon\bar{\alpha}_0 \in (\mathfrak{q}'', \Gamma_{\mathfrak{q}'})_\infty$.

We have $\epsilon\gamma\epsilon\bar{\alpha} - \lambda\alpha\epsilon\bar{\gamma}\epsilon = S_{-\lambda}(\epsilon\gamma\epsilon\bar{\alpha}) \in (\mathfrak{q}'', \Lambda_{\mathfrak{q}''})_\infty$ because $\epsilon \in (\mathfrak{q}, \Gamma)_\infty$ and γ has coordinates in $\mathfrak{q}'$. We have $-\alpha\epsilon\bar{\alpha}\gamma\epsilon\bar{\alpha} = (\alpha\epsilon)\lambda\bar{\alpha}\gamma\overline{(\alpha\epsilon)} \in (\mathfrak{q}'', \Gamma'_{\mathfrak{q}})_\infty$ because $\alpha\epsilon$ has coordinates in $\mathfrak{q}$ and $\lambda\bar{\alpha}\gamma \in (\mathfrak{q}', \Gamma'_{\mathfrak{q}})_\infty$. Finally $\lambda\alpha\epsilon\bar{\gamma}\epsilon\gamma\epsilon\bar{\alpha} = -(\alpha\epsilon\bar{\gamma})\epsilon\overline{(\alpha\epsilon\bar{\gamma})} \in (\mathfrak{q}'', \Gamma_{\mathfrak{q}''})_\infty$ since $\alpha\epsilon\bar{\gamma}$ has coordinates in $\mathfrak{q}\mathfrak{q}'$ and $\epsilon \in (\mathfrak{q}, \Gamma)_\infty$. This shows that $b\bar{a} \in (\mathfrak{q}'', \Gamma'')_\infty$. The proof that $d\bar{c} \in (\mathfrak{q}'', \Gamma'')_\infty$ from similar calculations, which we omit.

(7.8.1) <u>The restricted symplectic case</u>. Here the notation simplifies slightly, as follows. For any ideal $\mathfrak{q}$ write $\mathrm{sq}(\mathfrak{q})$ for the additive group generated by all q^2 $(q \in \mathfrak{q})$. Then, in the setting of Prop. (7.8), the commutator group (G, G') has level $(\mathfrak{q}'', \Gamma'')$ where $\mathfrak{q}'' = \mathfrak{q}\mathfrak{q}'$ and where

$\Gamma'' = 2\mathfrak{q}\mathfrak{q}' + sq(\mathfrak{q}\mathfrak{q}')\Lambda + sq(\mathfrak{q})\Gamma' + sq(\mathfrak{q}')\Gamma$

(7.9) THEOREM. *Let* $(\mathfrak{q},\Gamma)$ *be a unitary ideal in* (A,λ,Λ) *such that* $\mathfrak{q} \subset \mathrm{rad}\,(A)$.

a) *The multiplication map*

$$X_-((\mathfrak{q},\bar{\Gamma})_n) \times H(GL_n(\mathfrak{q})) \times X_+((\mathfrak{q},\Gamma)_n) \longrightarrow U^\lambda_{2n}(\mathfrak{q},\Gamma)$$

is bijective for all n, $1 \leq n \leq \infty$.

b) *The homomorphism* $H{:}K_1(\mathfrak{q}) \to KU^\lambda_1(\mathfrak{q},\Gamma)$ *is surjective. Its kernel is the subgroup* L *of* $K_1(\mathfrak{q})$ *generated by all elements* $[I + \beta\bar{\gamma}]$ *and* $[I + \gamma\bar{\beta}]$ *with* $\beta \in \Lambda_\infty$ *and* $\gamma \in (\mathfrak{q},\Gamma)_\infty$. *The inverse isomorphism* $KU^\lambda_1(\mathfrak{q},\Gamma) \to K_1(\mathfrak{q})/L$ *is induced by* $\sigma = \begin{pmatrix}\alpha & \beta\\ \gamma & \delta\end{pmatrix} \mapsto [\alpha] \bmod L$.

If $\sigma = \begin{pmatrix}\alpha & \beta\\ \gamma & \delta\end{pmatrix} \in U^\lambda_{2n}(\mathfrak{q},\Gamma)$ then $\alpha \equiv I \bmod \mathfrak{q}$ so, since $\mathfrak{q} \subset \mathrm{rad}\,(A)$, α is invertible. Therefore a) follows from Prop. (2.5) part b); in fact $\sigma = X_-(\gamma\alpha^{-1})H(\alpha)X_+(\alpha^{-1}\beta)$. Moreover, the surjectivity of $H{:}K_1(\mathfrak{q}) \to KU^\lambda_1(\mathfrak{q},\Gamma)$ is an immediate consequence of this.

Suppose $\beta \in \Lambda_\infty$ and $\gamma \in (\mathfrak{q},\Gamma)_\infty$. Then $\begin{pmatrix}I & \beta\\ 0 & I\end{pmatrix}\begin{pmatrix}I & 0\\ \bar{\gamma} & I\end{pmatrix}\begin{pmatrix}I & -\beta\\ 0 & I\end{pmatrix} = \begin{pmatrix}I+\beta\bar{\gamma} & -\beta\bar{\gamma}\beta\\ \bar{\gamma} & I-\bar{\gamma}\beta\end{pmatrix}$. Using the factorization in part a) we conclude that $H(I + \beta\bar{\gamma}) \in EU^\lambda_{2n}(\mathfrak{q},\Gamma)$ so $[I + \beta\bar{\gamma}] \in \mathrm{Ker}(H)$. A similar argument with $\begin{pmatrix}I & 0\\ -\bar{\beta} & I\end{pmatrix}\begin{pmatrix}I & \gamma\\ 0 & I\end{pmatrix}\begin{pmatrix}I & 0\\ \bar{\beta} & I\end{pmatrix}$ shows that $[I + \gamma\bar{\beta}] \in \mathrm{Ker}\,(H)$. Therefore $L \subset \mathrm{Ker}(H)$.

If $\alpha \in GL(\mathfrak{q})$ let $\langle\alpha\rangle$ denote its class in $K_1(\mathfrak{q})/L$. Define a map $f{:}U^\lambda(\mathfrak{q},\Gamma) \to K_1(\mathfrak{q})/L$ by sending $\sigma = \begin{pmatrix}\alpha & \beta\\ \gamma & \delta\end{pmatrix}$ to $\langle\alpha\rangle$. We claim that (i) f is a homomorphism, and (ii) $f(\varepsilon\sigma\varepsilon^{-1}) = f(\sigma)$ for any $\varepsilon \in EU^\lambda(A,\Lambda)$. Once (i) and (ii) are established it will follow that f induces a homorphism $KU^\lambda_1(\mathfrak{q},\Gamma) \to K_1(\mathfrak{q})/L$, and this is clearly inverse to the homomorphism defined by H, whence the theorem.

Proof of (i). If $\sigma' = \begin{pmatrix}\alpha' & \beta' \\ \gamma' & \delta'\end{pmatrix}$ then $\sigma\sigma' = \begin{pmatrix}\alpha\alpha'+\beta\gamma' & * \\ * & *\end{pmatrix}$. We have $\alpha\alpha' + \beta\gamma' = \alpha(I + \alpha^{-1}\beta\gamma'\alpha'^{-1})\alpha'$. Further $\alpha^{-1}\beta = \alpha^{-1}(\beta\bar{\alpha})\overline{\alpha^{-1}} \in (\mathfrak{q},\Gamma)_\infty$ and $\gamma'\alpha'^{-1} = \overline{\alpha'^{-1}}(\overline{\alpha'}\gamma')\alpha'^{-1} \in (\mathfrak{q},\bar{\Gamma})_\infty$ (see Thm. (6.2)) so that $[I + \alpha^{-1}\beta\gamma'\alpha'^{-1}] \in L$. It follows that $\langle\alpha\alpha' + \beta\gamma'\rangle = \langle\alpha\rangle + \langle\alpha'\rangle$, so $f(\sigma\sigma') = f(\sigma) + f(\sigma')$ as claimed.

Proof of (ii). It suffices to verify (ii) for $\epsilon = \begin{pmatrix}I & \varphi \\ 0 & I\end{pmatrix}$ or $\epsilon = \begin{pmatrix}I & 0 \\ \bar{\varphi} & I\end{pmatrix}$ with $\varphi \in \Lambda_\infty$, since such elements ϵ generate $EU^\lambda(A,\Lambda)$. We have $\epsilon\sigma\epsilon^{-1} = \begin{pmatrix}\alpha + \varphi\gamma & * \\ * & *\end{pmatrix}$, resp., $\begin{pmatrix}\alpha + \beta\bar{\varphi} & * \\ * & *\end{pmatrix}$, in the two cases. Further $\alpha + \varphi\gamma = (I + \varphi\gamma\alpha^{-1})\alpha$ and $\gamma\alpha^{-1} = \overline{\alpha^{-1}}(\bar{\alpha}\gamma)\alpha^{-1} \in (\mathfrak{q},\bar{\Gamma})_\infty$, so that $[I + \varphi\gamma\alpha^{-1}] \in L$. It follows that $\langle\alpha + \varphi\gamma\rangle = \langle\alpha\rangle$. A similar argument shows that $\langle\alpha + \beta\bar{\varphi}\rangle = \langle\alpha\rangle$, thus proving (ii).

(7.9.1) Remark. If $\beta \in \Lambda_\infty$ and $\gamma \in (\mathfrak{q},\Gamma)_\infty$ then $\overline{I + \beta\bar{\gamma}} = I + \gamma\bar{\beta} = I+(-\lambda\bar{\gamma})(-\bar{\lambda}\beta) = I + \bar{\gamma}\beta$. Moreover we have $[I + \bar{\gamma}\beta] = [I + \beta\bar{\gamma}]$. This can easily be verified directly, but it follows from Ch. I, (4.12.2) since $[I + \beta\bar{\gamma}] \in \mathrm{Ker}(H)$.

(7.10) When A is commutative, and $\mathfrak{q} \subset$ rad A we have an isomorphism det: $K_1(\mathfrak{q}) \to 1 + \mathfrak{q}$, [3], Ch. V, Cor. (9.2). It follows therefore from Thm. (7.9) that $KU_1^{\lambda}(\mathfrak{q},\Gamma) \cong (1+\mathfrak{q})/\det(L)$.

Let A_1 denote the fixed ring of the involution on A and put $\mathfrak{q}_1 = \mathfrak{q} \cap A_1$. By Ch. I, (4.12.2) and (4.12.3) we have $\{x+\bar{x} \mid x \in K_1(\mathfrak{q})\} \subset L \subset \{x \mid x = \bar{x}\}$. It follows that $\det L \subset 1 + \mathfrak{q}_1$ and that det L contains $(1+q)\overline{(1+q)} = 1 + q + \bar{q} + q\bar{q}$ for $q \in \mathfrak{q}$.

If $r \in \Lambda$ and $s \in \Gamma$ then $1 + r\bar{s} = 1 + \bar{r}s \in \det L$ by Thm. (7.9) (part b), with $\beta = r$ and $\gamma = s$).

We do not know what additional elements are required to generate det L in general. However we can give a complete answer in certain restricted symplectic cases.

(7.11) THEOREM. *Suppose that A is commutative with trivial involution and that $\lambda = -1$ (the restricted symplectic case). Let $(\mathfrak{q},\Gamma)$ be a unitary ideal in $(A,-1,\Lambda)$ such that $\Gamma = \mathfrak{q} \subset \Lambda$. Then the map*

(7.11.1) $$\sigma = \begin{pmatrix} \alpha & \beta \\ \gamma & \delta \end{pmatrix} \longmapsto \mathrm{Tr}(\alpha - I) \bmod \Lambda\mathfrak{q}$$

induces an epimorphism

(7.11.2) $$KU_1^{-1}(\mathfrak{q},\Gamma) \longrightarrow \mathfrak{q}/\Lambda\mathfrak{q} .$$

If $\mathfrak{q} \subset \mathrm{rad}\, A$ this is an isomorphism.

We first treat the case $\Lambda\mathfrak{q} = 0$. Then $\mathfrak{q}^2 \subset \Lambda\mathfrak{q} = 0$. It follows that for $\alpha = I + \alpha_0 \in GL(\mathfrak{q})$ we have $\det \alpha = 1 + \mathrm{Tr}(\alpha_0)$, and $\alpha \mapsto Tr(\alpha_0)$ induces an isomorphism $t: K_1(\mathfrak{q}) \to \mathfrak{q}$. We must show that $t(L) = 0$, where L is generated by the elements $[I + \beta\bar{\gamma}]$ and $[I + \gamma\bar{\beta}]$ with $\beta \in \Lambda_\infty$ and $\gamma \in (\mathfrak{q},\Gamma)_\infty$. Say $\beta = (b_{ij})$ and $\bar{\gamma} = (c_{ij})$. Then $t[I + \beta\bar{\gamma}] = Tr(\beta\bar{\gamma}) = \sum_i \sum_j b_{ij}c_{ji}$

$= \sum_{1\leq i<j\leq n} (b_{ij}c_{ji} + b_{ji}c_{ij}) + \sum_{1\leq i\leq n} b_{ii}c_{ii}$. We have $b_{ii}c_{ii} \in \Lambda\mathfrak{q}$

$= 0$ and $b_{ij}c_{ji} + b_{ji}c_{ij} = 2b_{ij}c_{ij} \in \Lambda\mathfrak{q} = 0$ because $2A \subset \Lambda$. The proof that $t[I + \gamma\bar{\beta}] = 0$ is similar.

In the general case the homomorphism (7.11.2) is constructed by passage from A to $A/\Lambda\mathfrak{q}$, as the composite $KU_1^{-1}(\mathfrak{q},\Gamma) \to KU_1^{-1}(\mathfrak{q}/\Lambda\mathfrak{q},\Gamma/\Lambda\mathfrak{q}) \to \mathfrak{q}/\Lambda\mathfrak{q}$, the second arrow being that treated in the case above.

As above put $L = Ker(K_1(\mathfrak{q}) \to KU_1^{-1}(\mathfrak{q},\Gamma))$. The homomorphism (7.11.2) shows that $\det(L) \subset 1 + \Lambda\mathfrak{q}$.

The fact that, when $\mathfrak{q} \subset$ rad A, (7.11.2) is injective is equivalent to the equality $\det(L) = 1 + \Lambda\mathfrak{q}$. This equality follows from Lemma (7.13) below by virtue of the fact (see (7.10)) that det(L) contains $1 + r\mathfrak{q}$ for all $r \in \Lambda$.

It remains to show that (7.11.2) is always surjective. This is clear when $\mathfrak{q} \subset$ rad A, for the element $[H(1 + q)] \in KU_1^{-1}(\mathfrak{q},\Gamma)$ maps onto q mod $\Lambda\mathfrak{q}$ for $q \in \mathfrak{q}$.

Note that $(A,-1,\Lambda)$ is contained in the "symplectic ring" $(A,-1,A)$ and that $(\mathfrak{q},\Gamma) = (\mathfrak{q},\mathfrak{q})$ is also a unitary ideal of

$(A,-1,A)$. Write $Sp(\mathfrak{q},\Gamma)$ and $Ep(\mathfrak{q},\Gamma)$ for the groups $U^{-1}(\mathfrak{q},\Gamma)$ and $EU^{-1}(\mathfrak{q},\Gamma)$ defined relative to the symplectic ring $(A,-1,A)$, reserving the latter notation for these groups defined relative to $(A,-1,\Lambda)$. Then it is clear that $Sp(\mathfrak{q},\Gamma) = U^{-1}(\mathfrak{q},\Gamma)$; it is just the principal congruence group of level $\mathfrak{q}$ in $Sp(A)$ $(= U^{-1}(A,A))$. Moreover $EU^{-1}(\mathfrak{q},\Gamma) \subset Ep(\mathfrak{q},\Gamma)$.

Suppose $\mathfrak{q} \subset \text{rad } A$. The isomorphism (7.11.2) then shows that $KSp_1(\mathfrak{q},\Gamma) \cong \mathfrak{q}/A\mathfrak{q} = 0$, i.e. that $Sp(\mathfrak{q},\Gamma) = Ep(\mathfrak{q},\Gamma)$.

In the general case we have $\mathfrak{q}/\Lambda\mathfrak{q} \subset \text{rad}(A/\Lambda\mathfrak{q})$, since $(\mathfrak{q}/\Lambda\mathfrak{q})^2 = 0$, so $Sp(\mathfrak{q}/\Lambda\mathfrak{q},\Gamma/\Lambda\mathfrak{q}) = Ep(\mathfrak{q}/\Lambda\mathfrak{q},\Gamma/\Lambda\mathfrak{q})$. Since $Ep(\mathfrak{q},\Gamma) \to Ep(\mathfrak{q}/\Lambda\mathfrak{q},\Gamma/\Lambda\mathfrak{q})$ is surjective (Cor. (6.5)) it follows that $Sp(\mathfrak{q},\Gamma) \to Sp(\mathfrak{q}/\Lambda\mathfrak{q},\Gamma/\Lambda\mathfrak{q})$ is surjective. As we observed above this means that $U^{-1}(\mathfrak{q},\Gamma) \to U^{-1}(\mathfrak{q}/\Lambda\mathfrak{q},\Gamma/\Lambda\mathfrak{q})$ is surjective, whence the surjectivity of $KU_1^{-1}(\mathfrak{q},\Gamma) \to KU_1^{-1}(\mathfrak{q}/\Lambda\mathfrak{q},\Gamma/\Lambda\mathfrak{q})$. This establishes the surjectivity of (7.11.2), and hence completes the proof of Thm. (7.11).

(7.12) COROLLARY. In the symplectic case, $\Lambda = A$, of Thm. (7.11) we have $KSp_1(\mathfrak{q},\Gamma) = 0$ whenever $\Gamma = \mathfrak{q} \subset \text{rad } A$.

The next lemma was used in the proof of Thm. (7.11) above.

(7.13) LEMMA. Let B be a ring and let (J_i) be a family of two sided ideals in rad B. Put $J = \Sigma J_i$. Then the multiplicative *group* $1 + J$ *is generated by the family of subgroups* $1 + J_i$.

It clearly suffices to treat the case when there is only a finite number, $J_1, \ldots, J_n$, of J_i's. We argue by induction on n, the case n = 1 being immediate. Let G denote the group generated by $1 + J_1, \ldots, 1 + J_n$. Put $J' = J_2 + \ldots + J_n$. By induction G contains $1 + J'$. By the case n = 1 G projects onto $1 + (J/J')$, whence $G = 1 + J$.

Chapter III. Unitary exact sequences

In Chapter VII of [3] it is shown how to construct exact sequences of K_o's and K_1's in an axiomatic setting. In an appendix to this chapter we add some complements to these constructions which, in particular, allow the extension of the exact sequences to some K_2 terms.

In the body of this chapter we simply verify the appropriate axioms for the unitary categories under discussion here, and then quote the appendix to obtain certain exact sequences. These occur in Theorem (1.1) and Corollary (2.3).

§1. The exact sequence of a unitary surjection.

It is the exact sequence (1.1.1) below.

(1.1) THEOREM. *Let* $f:(A,\lambda,\Lambda) \to (A',\lambda',\Lambda')$ *be a unitary surjection* (see Ch. I, (4.1)) *with unitary kernel* $(\mathfrak{q},\Gamma) = (\mathrm{Ker}\, f, \Lambda \cap \mathrm{Ker}\, f)$. *Then the base change functor* $F: \textcircled{Q}^{\lambda}(A,\Lambda) \to \textcircled{Q}^{\lambda}(A',\Lambda')$ *is cofinal and* E-*surjective so it induces an exact sequence*

$$(1.1.1) \qquad KU_2^{\lambda}(A,\Lambda) \longrightarrow KU_2^{\lambda'}(A',\Lambda') \longrightarrow KU_1^{\lambda}(\mathfrak{q},\Gamma) \longrightarrow$$
$$\longrightarrow KU_1^{\lambda}(A,\Lambda) \longrightarrow KU_1^{\lambda'}(A',\Lambda') \longrightarrow KU_0^{\lambda}(\mathfrak{q},\Gamma) \longrightarrow$$
$$\longrightarrow KU_0^{\lambda}(A,\Lambda) \longrightarrow KU_0^{\lambda'}(A',\Lambda') .$$

This sequence is natural in f.

Proof of theorem (1.1). It follows from Ch. I, Cor. (4.9) that the hyperbolic plane H(A) is a basic object in $\textcircled{Q}^{\lambda}(A,\Lambda)$ (in the sense of (A.4) below). Moreover $F: H(A) \mapsto H(A')$ in $\textcircled{Q}^{\lambda'}(A',\Lambda')$. In the notation of (A.4) we have $G(H(A)^{\infty}) = U^{\lambda}(A,\Lambda)$ and similarly $G(H(A')^{\infty}) = U^{\lambda'}(A',\Lambda')$. According to Ch. II, Cor. (5.3) the homomorphism $U^{\lambda}(A,\Lambda) \overset{F}{\to} U^{\lambda'}(A',\Lambda')$ induces an epimorphism of derived groups. Therefore it follows from Prop. (A.7) below that the functor F is cofinal and E-surjective. In view of (A.8) and Ch. II, Thm. (5.2) the exact sequence

(1.1.1) now follows from Thm. (A.9) below. Its naturality in f follows from that of the functor F.

(1.2) <u>The hyperbolic homomorphism</u>. The involution on A induces a (contravariant) automorphism of order two on Ⓟ(A): $P \mapsto \bar{P}$, $\alpha \mapsto \bar{\alpha}$. This induces involutions on the groups $K_i(A) = K_i(Ⓟ(A))$ which we shall denote $x \mapsto \bar{x}$.

We have functors

(1.2.1) $$Ⓟ(A) \xrightarrow{H} Ⓠ^{\lambda}(A,\Lambda) \xrightarrow{\phi} Ⓟ(A),$$

$\phi(P,[h]) = P$ ("forget the form"), and their composite is $P \mapsto P \oplus \bar{P}$, $\alpha \mapsto \alpha \oplus \bar{\alpha}^{-1}$. They induce homomorphisms

$$K_i(A) \xrightarrow{H} KU_i^{\lambda}(A,\Lambda) \xrightarrow{\phi} K_i(A)$$

whose composite is $x \mapsto x + (-1)^i \bar{x}$ $(i = 0,1)$. Thus

(1.2.2) $$\mathrm{Ker}\,(H) \subset \{x \mid x + (-1)^i \bar{x} = 0\}.$$

On the other hand $H(P) \cong H(\bar{P})$ and $H(\alpha)$ is conjugate to $H(\bar{\alpha}^{-1})$ in $U(H(P))$, for $\alpha \in GL(P)$. Thus $H(x) = H((-1)^i \bar{x})$ for $x \in K_i(A)$, so

(1.2.3) $$\mathrm{Ker}(H) \supset \{x - (-1)^i \bar{x}\}$$

for $i = 0,1$.

In the setting of Thm. (1.1) the functor Ⓟ(A) → Ⓟ(A') induces an exact sequence

$$(1.2.4) \qquad K_2A \longrightarrow K_2A' \longrightarrow K_1 \longrightarrow$$
$$\longrightarrow K_1A \longrightarrow K_1A' \longrightarrow K_0 \longrightarrow$$
$$\longrightarrow K_0A \longrightarrow K_0A'$$

analogous to (1.1.1) (cf. Thm. (A.9)). The functors H and Φ induce homomorphisms between the exact sequences (1.1.1) and (1.2.4).

§2. The unitary Mayer-Vietoris sequence.

(2.1) Unitary fibre products. Let

$$\text{(2.1.1)}\qquad \begin{array}{ccc} (A,\lambda,\Lambda) & \xrightarrow{h_1} & (A_1,\lambda_1,\Lambda_1) \\ {\scriptstyle h_2}\downarrow & & \downarrow{\scriptstyle f_1} \\ (A_2,\lambda_2,\Lambda_2) & \xrightarrow[f_2]{} & (A',\lambda',\Lambda') \end{array}$$

be a commutative square of morphisms of unitary rings. We shall call it cartesian if, up to isomorphism, (A,λ,Λ) is the fibre product of $(A_1,\lambda_1,\Lambda_1)$ with $(A_2,\lambda_2,\Lambda_2)$ over (A',λ',Λ'), i.e. if:

$$A = \{(a_1,a_2) \in A_1 \times A_2 \mid f_1a_1 = f_2a_2\}$$

$$\lambda = (\lambda_1,\lambda_2)$$

$$\Lambda = \{(r_1,r_2) \in \Lambda_1 \times \Lambda_2 \mid f_1r_1 = f_2r_2\}$$

The square (2.1.1) induces a square of base change functors of categories with product,

$$\text{(2.1.2)}\qquad \begin{array}{ccc} Ⓠ^{\lambda}(A,\Lambda) & \xrightarrow{H_1} & Ⓠ^{\lambda_1}(A_1,\Lambda_1) \\ {\scriptstyle H_2}\downarrow & & \downarrow{\scriptstyle F_1} \\ Ⓠ^{\lambda_2}(A_2,\Lambda_2) & \xrightarrow[F_1]{} & Ⓠ^{\lambda'}(A',\Lambda') \end{array}$$

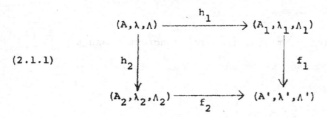

equipped with a natural isomorphism $\beta\colon F_1H_1 \to F_2H_2$.

Let

$$
(2.1.3)\qquad
\begin{array}{ccc}
\textcircled{Q} & \xrightarrow{\;G_1\;} & \textcircled{Q}^{\lambda_1}(A_1,\Lambda_1) \\
{\scriptstyle G_2}\big\downarrow & & \big\downarrow{\scriptstyle F_1} \\
\textcircled{Q}^{\lambda_2}(A_2,\Lambda_2) & \xrightarrow[\;F_2\;]{} & \textcircled{Q}^{\lambda'}(A',\Lambda')
\end{array}
\qquad \alpha\colon F_1G_1 \to F_2G_2
$$

be the cartesian square defined by (F_1,F_2) as in [3], Ch. VII, §3. Then there is a unique functor

$$(2.1.4)\qquad T\colon \textcircled{Q}^{\lambda}(A,\Lambda) \longrightarrow \textcircled{Q}$$

such that $G_iT = H_i$ $(i = 1,2)$ and so that $\beta = \alpha T$.

We now extend Milnor's theorem [3], Ch. IV, Thm. (5.3) to the present unitary setting.

(2.2) THEOREM. *Suppose the commutative square* (2.1.1) *of morphisms of unitary rings is cartesian, and that* f_1 *is a unitary surjection. Then the functor* $T\colon \textcircled{Q}^{\lambda}(A,\Lambda) \to \textcircled{Q}$ *of* (2.1.4) *is an equivalence of categories. Thus the square* (2.1.2) *is cartesian and* E-*surjective* (in the sense of [3], Ch. VII, §3).

The last assertions follow from the first together with Thm. (1.1), which implies the E-surjectivity of F_1, whence that of (2.1.2) ([3], Ch. VII, Prop. (3.4)(c)).

To prove that T is an equivalence we use Milnor's theorem cited above, which asserts that the corresponding functor Ⓟ$(A) \to$ Ⓟ$(A_1) \times_{Ⓟ(A')}$ Ⓟ(A_2) is an equivalence. To formulate this explicitly we introduce some notation: If M is an A_i-module put $M' = M \otimes_A A' = F_i M$; if $x \in M$ put $x' = x \otimes 1 \in M'$. If $f: M \to N$ is an A_i-homomorphism put $f' = F_i f: M' \to N'$.

Now let (P_1, σ, P_2) be a triple with $P_i \in$ Ⓟ(A_i) and $\sigma: P_1' \to P_2'$ an A'-isomorphism. Define P to be the fibre product

$$P = \{(x_1, x_2) \in P_1 \times P_2 \mid \sigma x_1' = x_2'\};$$

it is an A-module. If $(v_1, v_2): (Q_1, \tau, Q_2) \to (P_1, \sigma, P_2)$ is a morphism of such triples, i.e. $v_i: Q_i \to P_i$ is an A_i-homomorphism $(i = 1,2)$ and $\sigma v_1' = v_2' \tau$, then $v = [v_1, v_2]: Q \to P$, defined by $v(x_1, x_2) = (v_1(x_1), v_2(x_2))$, is an A-homomorphism. Now Milnor's theorem asserts that $(P_1, \sigma, P_2) \mapsto P$, $(v_1, v_2) \mapsto v$ defines a functor Ⓟ$(A_1) \times_{Ⓟ(A')}$ Ⓟ$(A_2) \to$ Ⓟ(A), that it is an equivalence, and that it is inverse to the canonical functor in the reverse sense. The latter includes the assertion that the A_i-homomorphism $P \otimes_A A_i \to P_i$ induced by the coordinate projection $P \to P_i$ is an isomorphism. We shall use this to identify $P \otimes_A A_i$ with P_i.

Let P arise from (P_1, σ, P_2) as above. Let $h_i \in \mathrm{Sesq}_{A_i}(P_i)$ and define $h = [h_1, h_2]$ on $P \times P$ by $h((x_1, x_2), (y_1, y_2))$

$= (h_1(x_1,y_1), h_2(x_2,y_2)) \in A_1 \times A_2$. Then h takes values in $A \Leftrightarrow h_1(x_1,y_1)' = h_2(x_2,y_2)'$ for all $x,y \in P$, where we put $a_i' = f_i a_i \in A'$ for $a_i \in A_i$ $(i = 1,2)$. We have $h_i(x_i,y_i)'$ $= h_i'(x_i',y_i')$ $(i = 1,2)$ and $h_2'(x_2',y_2') = h_2'(\sigma x_1',\sigma y_1')$. Thus h takes values in $A \Leftrightarrow \sigma\colon (P_1',h_1') \to (P_2',h_2')$ is a morphism of sesquilinear modules, i.e. $\Leftrightarrow \sigma^* h_2' = h_1'$. In this case $h\colon P \times P \to A$ is clearly sesquilinear.

Conversely, suppose $h \in \mathrm{Sesq}_A(P)$. By base change h defines forms $h_i \in \mathrm{Sesq}_{A_i}(P_i)$, such that $\sigma\colon (P_1',h_1') \to (P_2',h_2')$ is a morphism. In summary then we have established an identification

(2.2.1) $$\mathrm{Sesq}_A(P) = \mathrm{Sesq}_{A_1}(P_1) \underset{\sigma^*}{\times} \mathrm{Sesq}\,(P_2)$$
$$= \{(h_1,h_2) \mid \sigma^* h_2' = h_1'\};$$

The dual module $\bar{P}$ arises from the triple $(\bar{P}_1, \bar{\sigma}^{-1}, \bar{P}_2)$ so that $v = [v_1,v_2] \in \bar{P}$ satisfies

$$\langle v,x\rangle_P = (\langle v_1,x_1\rangle_{P_1}\ ,\ \langle v_2,x_2\rangle_{P_2}).$$

If $h \in \mathrm{Sesq}_A(P)$ then $\langle {}_h dx,y\rangle_P = h(x,y) = (h_1(x_1,y_1), h_2(x_2,y_2))$ $= (\langle {}_{h_1}dx_1,y_1\rangle_{P_1}, \langle {}_{h_2}dx_2,y_2\rangle_{P_2}))$, whence

(2.2.2) $h^d = ({}_{h_1}d, {}_{h_2}d)$, <u>so</u> h <u>is non-singular</u>
$\Leftrightarrow h_1$ <u>and</u> h_2 <u>are non-singular</u>.

If $v = [v_1, v_2]: Q \to P$ is an A-homomorphism as above then one sees easily that $v^*h = [v_1^*h_1, v_2^*h_2]$. Thus $v: (Q,g) \to (P,h)$ is a morphism of sesquilinear modules if and only if $v_i: (Q_i, g_i) \to (P_i, h_i)$ is such a morphism $(i = 1,2)$.

To investigate quadratic modules we first note that, for $h \in \mathrm{Sesq}_A(P)$, we have $\bar{h} = [\bar{h}_1, \bar{h}_2]$ and $ch = [c_1h_1, c_2h_2]$ for $c = (c_1, c_2) \in$ Center (A). Hence $S_\lambda(h) = [S_{\lambda_1}(h_1), S_{\lambda_2}(h_2)]$; taking $\mathrm{Ker}(S_\lambda)$ we find that

(2.2.3) $$\mathrm{Sesq}^{-\lambda}(P) = \mathrm{Sesq}^{-\lambda_1}(P_1) \underset{\sigma^*}{\times} \mathrm{Sesq}^{-\lambda_2}(P_2).$$

Further $h(x,x) \in \Lambda$ for all $x \in P \Leftrightarrow h_1(x_1,x_1) \in \Lambda_1$ and $h_2(x_2,x_2) \in \Lambda_2$ for all $(x_1,x_2) \in P$, whence

(2.2.4) $$\mathrm{Sesq}_\Lambda^{-\lambda}(P) = \mathrm{Sesq}_{\Lambda_1}^{-\lambda_1}(P_1) \underset{\sigma^*}{\times} \mathrm{Sesq}_{\Lambda_2}^{-\lambda_2}(P_2)$$

Recall that $[h]$ denotes the class of h modulo $\mathrm{Sesq}_\Lambda^{-\lambda}(P)$, and $[h_i]$ that of h_i modulo $\mathrm{Sesq}_{\Lambda_i}^{-\lambda_i}(P_i)$ $(i = 1,2)$. If $h,g \in \mathrm{Sesq}_A(P)$ then it follows from (2.2.4) that $[h] = [g] \Leftrightarrow [h_1] = [g_1]$ and $[h_2] = [g_2]$.

Suppose $(Q,[g])$ and $(P,[h])$ are unitary A-modules and that $v:Q \to P$ is an A-linear map. Then v is a morphism $\Leftrightarrow g - v^*h \in \mathrm{Sesq}_\Lambda^{-\lambda}(P) \Leftrightarrow g_i - v_i^* h_i \in \mathrm{Sesq}_{\Lambda_i}^{-\lambda}(P_i)$ $(i = 1,2) \Leftrightarrow v_i: (Q_i,[g_i]) \to (P_i,[h_i])$ is a morphism $(i = 1,2)$.

The conclusions so far imply that the functor $T: \textcircled{Q}^{\lambda}(A,\Lambda) \to \textcircled{Q}$ is fully faithful, i.e. it is bijective on Hom's. It remains to show that each object of $\textcircled{Q}$ is isomorphic to one coming from $\textcircled{Q}^{\lambda}(A,\Lambda)$. Now an object of $\textcircled{Q}$ is a triple $((P_1,[h_1]),\sigma,(P_2,[h_2]))$ where $(P_i,[h_i]) \in \textcircled{Q}^{\lambda_i}(A_i,\Lambda_i)$ $(i = 1,2)$ and where $\sigma: (P_1',[h_1']) \to (P_2',[h_2'])$ is an isomorphism of unitary modules. Let P be the module associated to (P_1,σ,P_2). We have $\sigma^*h_2' - h_1' \in \mathrm{Sesq}_{\Lambda'}^{-\lambda'}(P_1')$. Since $f_1:(A_1,\lambda_1,\Lambda_1) \to (A',\lambda',\Lambda')$ is, by hypothesis, a unitary surjection it follows from Ch. I, Prop. (5.4) that $\mathrm{Sesq}_{\Lambda_1}^{-\lambda_1}(P_1) \to \mathrm{Sesq}_{\Lambda'}^{-\lambda'}(P_1')$ is surjective. Hence there is a $k \in \mathrm{Sesq}_{\Lambda_1}^{-\lambda_1}(P_1)$ such that $k' = \sigma^*h_2' - h_1'$. Replacing h_1 by $h_1 + k$, which does not affect $[h_1]$, we may thus assume that $h_1' = \sigma^*h_2'$. This done we have $h = (h_1,h_2) \in \mathrm{Sesq}_A(P)$, and clearly $(P,[h]) \in \textcircled{Q}^{\lambda}(A,\Lambda)$ and $T(P,[h]) = ((P_1,[h_1]),\sigma,(P_2,[h_2]))$. This completes the proof of Theorem (2.2).

Remark. The proof of Thm. (2.2) establishes an equivalence analogous to T for the categories of (non-singular) λ-hermitian forms on finitely generated projective modules.

In view of Thm. (2.2) we can apply [3], Ch. VII, Thm. (4.3) and Thm. (A.13) to obtain:

(2.3) COROLLARY. ("Mayer-Vietoris Sequence") *In the setting of Thm (2.2) there is an exact sequence*

(2.3.1) $KU_1^{\lambda}(A,\Lambda) \longrightarrow KU_1^{\lambda 1}(A_1,\Lambda_1) \oplus KU_1^{\lambda 2}(A_2,\Lambda_2) \longrightarrow KU_1^{\lambda'}(A',\Lambda') \longrightarrow$

$\longrightarrow KU_0^{\lambda}(A,\Lambda) \longrightarrow KU_0^{\lambda 1}(A_1,\Lambda_1) \oplus KU_0^{\lambda 2}(A_2,\Lambda_2) \longrightarrow KU_0^{\lambda'}(A',\Lambda')$.

If both f_1 _and_ f_2 _are unitary surjections this sequence extends to_

(2.3.2) $KU_2^{\lambda 1}(A_1,\Lambda_1) \oplus KU_2^{\lambda 2}(A_2,\Lambda_2) \longrightarrow KU_2^{\lambda'}(A',\Lambda') \longrightarrow$

$\longrightarrow KU_1^{\lambda}(A,\Lambda) \longrightarrow \ldots$

These sequences are natural with respect to morphisms of cartesian squares (2.1.1).

(2.4) _The relative groups_ $KU_0(f)$. Let $f:(B,\lambda,\Lambda) \to (B',\lambda',\Lambda')$ be a morphism of unitary rings, and let $F: Ⓠ^{\lambda}(B,\Lambda) \to Ⓠ^{\lambda'}(B',\Lambda')$ be the base change functor. Then from [3], Ch. VII, Thm. (5.3) we have an exact sequence which we shall denote

(2.4.1) $KU_1^{\lambda}(B,\Lambda) \longrightarrow KU_1^{\lambda}(B',\Lambda') \longrightarrow KU_0(f) \longrightarrow KU_0^{\lambda}(B,\Lambda) \longrightarrow KU_0^{\lambda}(B',\Lambda')$.

In case f is a unitary surjection with unitary kernel $(\mathfrak{q},\Gamma)$ then it is easily checked that $KU_0(f)$ can be naturally identified with the group $KU_0^{\lambda}(\mathfrak{q},\Gamma)$ as defined in Ch. I, (7.4). Moreover the sequence (2.4.1) is then a portion of the exact sequence (1.1.1) above.

(2.5) COROLLARY. ("Excision isomorphisms") _In the setting of Thm._ (2.2) _the homomorphisms_ $KU_0(h_2) \to KU_0(f_1)$ _and_

$KU_0(h_1) \to KU_0(f_2)$, *obtained by applying* (2.4) *to the cartesian square* (2.1.1), *are isomorphisms. If both* f_1 *and* f_2 *are surjective then* $KU_1(h_2) \to KU_1(f_1)$ *and* $KU_1(h_1) \to KU_1(f_2)$ *are isomorphisms.*

In view of Thm. (2.2) the first assertion follows from [3], Ch. VII, Thm. (6.1), and the second assertion from Thm. (A.13).

Remark. Let $(\mathfrak{q}_1,\Gamma_1)$ be the unitary kernel of the unitary surjection $f_1\colon (A_1,\lambda_1,\Lambda_1) \to (A',\lambda',\Lambda')$. It is easily seen that $h_2\colon(A,\lambda,\Lambda) \to (A_2,\lambda_2,\Lambda_2)$ is also a unitary surjection; denote its unitary kernel $(\mathfrak{q},\Gamma)$. Then the first in each pair of isomorphisms of Cor. (2.5) can be rewritten as an isomorphism

$$KU_i^{\lambda}(\mathfrak{q},\Gamma) \longrightarrow KU_i^{\lambda_1}(\mathfrak{q}_1,\Gamma_1).$$

§A. Appendix: Some complements on categories with product.

(A.1) A central functor H from groups to abelian groups is one such that the action of G on H(G) induced by inner automorphisms is always trivial. If H is not necessarily central then $G \mapsto H_0(G,H(G))$ is the largest central quotient functor of H.

Examples of central functors are the homology groups

$$H_n(G) = H_n(G,\mathbb{Z}),$$

where G acts trivially on $\mathbb{Z}$. For n = 1 we have $H_1(G) = G/G'$, where G' denotes the derived group (G,G) of a group G.

(A.2) The functors K_H. Consider a category Ⓒ with product $\perp$ in the sense of [3], Ch. VII, §1. If A and B are objects of Ⓒ then G(A) denotes the automorphism group of A, and $G(A) \to G(A \perp B)$ the homomorphism $\alpha \mapsto \alpha \perp 1_B$.

Let H be a central functor as in (A.1) above. Then H(G(A)) depends only on the isomorphism class of A, since two isomorphisms $A \to A'$ induce isomorphisms $G(A) \to G(A')$ differing only by an inner automorphism. In this way the groups H(G(A)) and the homomorphisms $H(G(A)) \to H(G(A \perp B))$ induced by the homomorphisms $\alpha \mapsto \alpha \perp 1_B$ above, form a filtered inductive system indexed by the isomorphism classes (A) of ojbects A of ©. We put

$$K_H(\copyright) = \varinjlim_{(A)} H(G(A))$$

If $F: \copyright \to \copyright'$ is a product preserving functor then the homomorphisms $H(G(A)) \to H(G(FA))$ induced by F induce in turn a homomorphism of inductive limits $F_H: K_H(\copyright) \to K_H(\copyright')$, thus making K_H a functor.

(A.3) <u>The functors</u> K_1 <u>and</u> K_2. Taking $H(G) = H_1(G) = G/G'$ we obtain the functor $K_H = K_1$ of [3], Ch. VII, §1 and Prop. (2.1).

The functor $G \mapsto H_2(G')$ is not central, so we make it so by forming $H(G) = H_0(G, H_2(G'))$. The resulting functor K_H is denoted K_2. We shall see below (Cor. A.6)) that this definition of K_2 is consistent with that of Milnor [9].

(A.4) A <u>basic object</u> A in a Category © with product is one such that, for any object B, $B \perp B' \cong A^n = A \perp \cdots \perp A$ for some B' and some $n \geq 0$. The inductive limit of the groups $G(A^n)$ $(n \geq 1)$ with respect to the homomorphisms $G(A^n) \to G(A^{n+m})$, $\alpha \mapsto \alpha \perp 1_{A^m}$ will be denoted $G(A^\infty)$.

Let H be a central functor as in (A.2). The canonical homomorphisms $H(G(A^n)) \to K_H(\copyright)$ induce a homomorphism

$\varinjlim_n H(G(A^n)) \to K_H(Ⓒ)$. If H commutes with sequential inductive limits then $\varinjlim_n H(G(A^n)) = H(G(A^\infty))$, whence a canonical homomorphism

$$(A.4.1) \qquad H(G(A^\infty)) \longrightarrow K_H(Ⓒ).$$

(A.5) PROPOSITION. *Let* Ⓒ *be a category with product. Let* A *be a basic object of* Ⓒ *and put* $G = G(A^\infty)$. *Let* U *be a subgroup of* G *normalized by* $G' = (G,G)$.

a) U *is a normal subgroup of* G. *If* $V = (G,U)$ *then* $V = (G',V)$. *In particular* $G'' = G'$.

Let H *be a functor from groups to abelian groups which commutes with filtered inductive limits.*

b) *The canonical epimorphism* $H_0(G',H(U)) \to H_0(G,H(U))$ *is an isomorphism.*

c) *If* H *is a central functor then the canonical homomorphism* (A.4.1) : $H(G) \to K_H(Ⓒ)$, *is an isomorphism.*

Proof of a). Put $G_n = G(A^n)$, and let U_n and V_n denote the inverse images of U and V, respectively, under $G_n \to G = \varinjlim_n G_n$.

Let $\tau \in G_{2n}$ be the transposition of $A^{2n} = A^n \perp A^n$. If $\alpha,\beta \in G_n$ then $\tau(\alpha \perp \beta)\tau^{-1} = \beta \perp \alpha$. Thus the commutator $(\tau, 1_{A^n} \perp \alpha)$ equals $\alpha \perp \alpha^{-1}$.

If $\beta \in U_n$ then $\alpha\beta\alpha^{-1} \perp 1_{A^n} = (\alpha \perp \alpha^{-1})(\beta \perp 1_{A^n})(\alpha^{-1} \perp \alpha) \in U_{2n}$

because $\alpha \perp \alpha^{-1} \in G'_{2n}$ and G' normalizes U. This holds for all n, whence $U = \varinjlim_n U_n$ is normal in G.

With $\alpha \in G_n$ and $\beta \in U_n$ we have $h_{12}(\alpha) = \alpha \perp \alpha^{-1} \perp 1_{A^n} \in G'_{3n}$ and $h_{13}(\beta) = \beta \perp 1_{A^n} \perp \beta^{-1} \in V_{3n}$, as we have seen above. Moreover $(h_{12}(\alpha), h_{13}(\beta)) = (\alpha,\beta) \perp 1_{A^{2n}}$, clearly.

Suppose $\gamma \in V_n$; say $\gamma = (\alpha_1,\beta_1)\ldots(\alpha_r,\beta_r)$ with $\alpha_i \in G_n$ and $\beta_i \in U_n$ $(1 \le i \le r)$. Then we have $\gamma \perp 1_{A^{2n}} = (h_{12}(\alpha_1), h_{13}(\beta_1))\ldots$ $\ldots(h_{12}(\alpha_r), h_{13}(\beta_r))$, by the last paragraph. Thus $\gamma \perp 1_{A^{2n}} \in (G'_{3n}, V_{3n})$. Letting n go to ∞ we obtain $V = (G',V)$.

In the case $U = G$ we have $V = G'$ and so $G' = (G',G') = G''$. This completes the proof of a).

<u>Proof of b)</u>. Suppose $\alpha \in G(A^n)$. Let $I(\alpha)$ denote conjugation by α in U_n, so that α acts on $H(U_n)$ via $H(I(\alpha))$. For $m \ge n$ put $\alpha_m = \alpha \perp 1_{A^{m-n}} \in G(A^m)$. If $x \in H(U_n)$ let x_m denote the image of x under $H(U_n) \to H(U_m)$ for $m \ge n$.

The homomorphism $U_n \to U_{2n}$ is equivariant with respect to $I(\alpha_n)$ on U_n and $I(\alpha_{2n})$ on U_{2n}. However the same is true if we replace $I(\alpha_{2n})$ by $I(\alpha \perp \alpha^{-1})$. It follows that if $x \in H(U_n)$ then $(I(\alpha)x)_{2n} = I(\alpha \perp \alpha^{-1})(x_{2n})$. Since $\alpha \perp \alpha^{-1} \in G(A^{2n})'$ it follows, on letting $n \to \infty$, that trivializing the action of $G(A^\infty)$ on $H(U)$ is equivalent to trivializing that of $G(A^\infty)'$. This proves b).

Proof of c). This results directly from [3], Ch. I, Prop. (8.6) (cf. also Ch. VII, Cor. (2.3)).

(A.6) COROLLARY. The natural homomorphisms

(A.6.1) $$H_1(G(A^\infty)) \longrightarrow K_1(\copyright)$$

and

(A.6.2) $$H_2(G(A^\infty)') \longrightarrow K_2(\copyright)$$

are isomorphisms.

That (A.6.1) is an isomorphism is the special case $H(G) = H_1(G)$ of part c) of Prop. (A.5). In the case $H(G) = H_0(G,H_2(G'))$ the same result implies that

$$H_0(G(A^\infty), H_2(G(A^\infty)')) \longrightarrow K_2(\copyright)$$

is an isomorphism. Since H_2 is a central functor part b) of Prop. (A.5) implies $G(A^\infty)$ acts trivially on $H_2(G(A^\infty)')$, whence the isomorphism (A.6.2).

(A.7) PROPOSITION. Let $\copyright$ be a category with product with a basic object A. Let $F: \copyright \to \copyright_1$ be a product preserving functor.

a) F is cofinal if and only if $A_1 = FA$ is a basic object of $\copyright_1$.

b) If F is cofinal then F is E-surjective (in the sense of [3], Ch. VII, Defn. (2.4)) if and only if the homomorphism

$G(A^\infty) \to G(A_1{}^\infty)$ <u>defined by</u> F <u>induces an epimorphism</u> $G(A^\infty)' \to G(A_1{}^\infty)'$ <u>of derived groups</u>.

<u>Proof of a)</u>. Suppose F is cofinal. We must show each object B_1 of C_1 is a $\perp$-summand of some A_1^n. But this follows since B_1 is a $\perp$-summand of some FB, and B is a $\perp$-summand of some A^n.

Suppose conversely that A_1 is basic. Then each B_1 in C_1 is a $\perp$-summand of some $A_1^n = FA^n$, so F is cofinal.

<u>Proof of b)</u>. We first show that $G(A^\infty)' \to G(A_1{}^\infty)'$ is surjective if F is E-surjective. Say $\epsilon_1 \in G(A_1^n)'$. By E-surjectivity there is an object B of Ⓒ and an $\epsilon \in G(A^n \perp B)$ such that $F\epsilon = \epsilon_1 \perp 1_{FB}$. Writing $B \perp C \cong A^m$ for some C and m, and then replacing B by $B \perp C$ and ϵ by $\epsilon \perp 1_C$ we may assume $B = A^m$. Then $\epsilon \in G(A^{n+m})'$ and $F\epsilon = \epsilon_1 \perp 1_{A_1^m}$, whence our assertion.

Suppose conversely that $G(A^\infty)' \to G(A_1{}^\infty)'$ is surjective (and that F is cofinal); we claim that F is E-surjective. Suppose given an object B_1 of C_1 and an $\epsilon_1 \in G(B_1)'$. Replacing B_1 by $B_1 \perp C_1$ and ϵ_1 by $\epsilon_1 \perp 1_{C_1}$ we may arrange, since A_1 is basic, that $B_1 = A_1^n$. Our hypothesis implies that, for some m, $\epsilon_1 \perp 1_{A_1^m}$ lifts to some $\epsilon \in G(A^{n+m})'$. This shows that F is E-surjective.

(A.8) <u>The relative group</u> $K_1(F)$. Let $F: © \to ©_1$ be a product preserving functor. If A is an object of © put

$$G(A,F) = \mathrm{Ker}(G(A) \xrightarrow{F} G(FA))$$

and

$$H_F(A) = G(A,F)/(G(A),G(A,F)).$$

The latter depends only on the isomorphism class (A) of A. The homomorphisms $G(A) \to G(A \perp B)$, $\alpha \mapsto \alpha \perp 1_B$, induce homomorphisms $H_F(A) \to H_F(A \perp B)$, and the group $\varinjlim_{(A)} H_F(A)$ is easily seen to be the group $K_1(©, F)$ defined in [3], Ch. VII, §1.

Suppose A is a basic object of ©, and put

$$G(A^\infty,F) = \mathrm{Ker}(G(A^\infty) \xrightarrow{F} G((FA)^\infty)).$$

Then just as in the proof of Prop. (A.5) part c), it follows from [3], Ch. I, Prop. (8.6) that the canonical homomorphism

(A.8.1) $$G(A^\infty(F))/(G(A^\infty),G(A^\infty,F)) \longrightarrow K_1(©, F)$$

is an isomorphism. In case F is cofinal and E-surjective then $K_1(©,F)$ coincides with the group $K_1(F)$ of [3], Ch. VII, §5 and Thm. (5.3).

(A.9) THEOREM. <u>Let</u> $F: © \to ©_1$ <u>be a product preserving functor which is cofinal and</u> E-<u>surjective</u>. <u>Then there is an exact sequence</u>

(A.9.1) $K_2 © \to K_2 ©_1 \to K_1F \to K_1 © \to K_1 ©_1 \to K_0F \to K_0 © \to K_0 ©_1$

<u>which is natural in</u> F.

The portion of the sequence from K_1F to $K_0 ©_1$ is just the exact sequence of [3], Ch. VII, Thm. (5.3). For the remainder we shall assume © has a basic object A. To see that this assumption is no essential restriction consider, for each object A, the full subcategory $©_A$ of © consisting of objects isomorphic to a $\perp$-summand of some A^n. In $©_A$ clearly A is a basic object. Then © is the filtered union of the subcategories $©_A$, and F induces functors $F_A: ©_A \to (©_1)_{FA}$. Now the exact sequence above can be obtained as the inductive limit of those for the functors F_A.

Let A be a basic object in ©. Consider the exact sequence

$$1 \longrightarrow G(A^{\infty},F) \longrightarrow G(A^{\infty}) \xrightarrow{F} G((FA)^{\infty}),$$

which we shall denote $1 \to U \to \Gamma \xrightarrow{F} \Gamma_1$. The following conditions hold:

(i) F induces a surjection $\Gamma' \to \Gamma_1'$ of derived groups.

(ii) $\Gamma'' = \Gamma'$ and (hence, in view of (i),) $\Gamma_1'' = \Gamma_1'$.

(iii) If $V = (\Gamma,U)$ then $(\Gamma',V) = V$.

Property (i) follows from Prop. (A.7), since the functor F is E-surjective. Properties (ii) and (iii) follow from Prop. (A.5), part a).

It will be convenient to introduce the following *definition*: A homomorphism of groups $F: \Gamma \to \Gamma_1$ will be called *stable* if properties (i), (ii), and (iii) above hold, with $U = \mathrm{Ker}(F)$.

With this terminology, Thm. (A.9) follows now from the next lemma.

(A.10) LEMMA. *Let* $F: \Gamma \to \Gamma_1$ *be a stable group homomorphism* (see defn. above) *with kernel* U. *Then there is an exact sequence*

(A.10.1) $$H_2(\Gamma') \longrightarrow H_2(\Gamma_1') \xrightarrow{\ \delta\ } H_0(\Gamma, H_1(U)) \longrightarrow H_1(\Gamma) \longrightarrow H_1(\Gamma_1)$$

which is natural with respect to F.

Consider the commutative diagram

$$\begin{array}{ccccccccc}
 & & & & & & \Gamma_1 & & \\
 & & & & & & \cup & & \\
1 & \longrightarrow & U & \longrightarrow & \Gamma & \longrightarrow & \Gamma_2 & \longrightarrow & 1 \\
 & & \cup & & \cup & & \cup & & \\
1 & \longrightarrow & U_0 & \longrightarrow & \Gamma' & \longrightarrow & \Gamma_1' & \longrightarrow & 1
\end{array}$$

where $\Gamma_2 = F(\Gamma)$ and $U_0 = U \cap \Gamma'$. The exact rows of this diagram yield the exact rows of the following one:

$$
\begin{array}{ccccccccc}
 & & & & & & & & H_1(\Gamma_1) \\
 & & & & & & & & \uparrow{\scriptstyle i} \\
 & & & & H_0(\Gamma,H_1(U)) & \xrightarrow{g} & H_1(\Gamma) & \longrightarrow & H_1(\Gamma_2) \longrightarrow 0 \\
\text{(A.10.2)} & & & {\scriptstyle\delta}\nearrow & \uparrow{\scriptstyle j} & & & & \\
H_2(\Gamma') & \rightarrow & H_2(\Gamma_1') & \xrightarrow{\partial} & H_0(\Gamma',H_1(U_0)) & \longrightarrow & 0 & &
\end{array}
$$

The "0" corresponds to $H_1(\Gamma')$; see condition (ii) in the definition of stable. This same condition and the fact that $\Gamma_1' \subset \Gamma_2 \subset \Gamma_1$ imply that $\Gamma_2' = \Gamma_1'$, whence $i: \Gamma_2/\Gamma_2' \rightarrow \Gamma_1/\Gamma_1'$ is injective. Thus the right hand three terms of (A.10.1) form an exact sequence.

We define δ in (A.10.1) to be the composite $j \circ \partial$ in (A.10.2). Then the exactness of (A.10.1) results once we show that j is injective and Im (j) = Ker (g). (Its naturality in F is clear from that of diagram (A.10.2).) We have $U_0 = \Gamma' \cap U \supset (\Gamma,U) = V$, and so $(\Gamma',U_0) \supset (\Gamma',V) = V$, by condition (iii) in the definition of stable. Thus $(\Gamma',U_0) = V = (\Gamma,U)$ so $j: U_0/(\Gamma,U_0) \rightarrow U/(\Gamma,U)$ is injective. Its image, $U_0/(\Gamma,U) = (\Gamma' \cap U)/(\Gamma,U)$, is clearly the kernel of $g: U/(\Gamma,U) \rightarrow \Gamma/\Gamma'$. This completes the proof of Lemma (A.10).

(A.11) <u>Cartesian squares</u>. Let

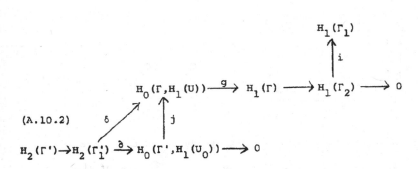

$$\text{(A.11.1)}\qquad \begin{array}{ccc} \mathbb{C} & \xrightarrow{G_1} & \mathbb{C}_1 \\ {\scriptstyle G_2}\downarrow & & \downarrow{\scriptstyle F_1} \\ \mathbb{C}_2 & \xrightarrow[F_2]{} & \mathbb{C}' \end{array} \quad , \qquad \alpha : F_1G_1 \to F_2G_2$$

be a cartesian square of categories with product, in the sense of [3], Ch. VII, §3. Recall that the objects of $\mathbb{C}$ are of the form $A = (A_1, \alpha_A, A_2)$ with $\alpha_A : F_1A_1 \to F_2A_2$ an isomorphism.

(A.12) PROPOSITION. <u>Suppose that</u> F_1 <u>is</u> E-<u>surjective. Suppose</u> $A = (A_1, \alpha_A, A_2) \in \mathbb{C}$ <u>is such that</u> A_i <u>is a basic object of</u> $\mathbb{C}_i$ $(i = 1,2)$. <u>Then</u> A <u>is a basic object of</u> $\mathbb{C}$. <u>Moreover the functor</u> G_2 <u>is</u> E-<u>surjective</u>.

Let $B = (B_1, \beta_B, B_2) \in \mathbb{C}$. We can solve $B_i \perp C_i \cong A_i^{n_i}$ with $C_i \in \mathbb{C}_i$ $(i = 1,2)$. We can further arrange that $n_1 = n_2 = n$. Then $F_1(C_1 \perp A_1^n) \cong F_1C_1 \perp F_2A_2^n$ (via α_A^n) $\cong F_1C_1 \perp F_2B_2 \perp F_2C_2$ $\cong F_1C_1 \perp F_1B_1 \perp F_2C_2$ (via α_B) $\cong F_1A_1^n \perp F_2C_2 \cong F_2(A_2^n \perp C_2)$. It follows that there is an isomorphism $\gamma : F_1(C_1 \perp A_1^n) \to F_2(C_2 \perp A_2^n)$. Hence $B \perp (C_1 \perp A_1^n, \gamma, C_2 \perp A_2^n) \cong (A_1^{2n}, \delta, A_2^{2n})$ for some δ. Put $\delta = (\alpha_A^{2n}) \cdot \varepsilon$, with $\varepsilon \in G(F_1A_1^{2n})$, so that $(A_1^{2n}, \delta, A_2^{2n}) = A^{2n} \cdot \varepsilon$ (in the notation of [3], p. 360). Thus B is a $\perp$-summand of $A^{2n} \cdot \varepsilon \perp A^{2n} \cdot \varepsilon^{-1} = A^{4n}(\varepsilon \perp \varepsilon^{-1})$. Moreover $\varepsilon \perp \varepsilon^{-1} \in G(F_1A_1^{4n})'$. Since F_1 is E-surjective we can, after enlarging n to n + m and replacing ε by $\varepsilon \perp 1_{F_1A_1^{2m}}$, write $\varepsilon \perp \varepsilon^{-1} = F_1\varphi$ for some $\varphi \in G(A_1^{4m})$. But then $(\varphi, 1_{A_2^{4n}})$ is an isomorphism from

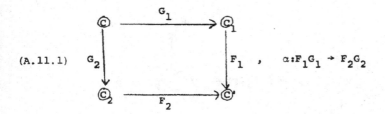

$A^{4m}(\epsilon \perp \epsilon^{-1})$ to A^{4m}. This proves the first assertion of the proposition.

Next consider the commutative diagram

$$
\text{(A.12.1)} \qquad
\begin{array}{ccccc}
G(A^{\infty}) & & \xrightarrow{\quad G_2 \quad} & & G(A_2^{\infty}) \\
{\scriptstyle G_1}\downarrow & & & & \downarrow{\scriptstyle F_2} \\
G(A_1^{\infty}) & \xrightarrow[F_1]{} & G(F_1A_1^{\infty}) & \xrightarrow[\alpha_A^{\infty}]{\cong} & G(F_2A_2^{\infty})
\end{array} .
$$

The corners form a cartesian square

$$
\begin{array}{ccc}
\Gamma & \longrightarrow & \Gamma_2 \\
\downarrow & & \downarrow \\
\Gamma_1 & \longrightarrow & \Gamma_0
\end{array}
$$

of group homomorphisms in which $\Gamma_1' \to \Gamma_0'$ is surjective (E-surjectivity of F_1). It follows that $\mathrm{Im}(\Gamma \to \Gamma_2)$ contains Γ_2', and hence $\mathrm{Im}(\Gamma' \to \Gamma_2')$ contains $\Gamma_2'' = \Gamma_2'$ (Prop. (A.5)), whence the E-surjectivity of G_2. This proves Prop. (A.12).

(A.13) THEOREM. *In the cartesian square* (A.11.1) *suppose that* F_1 *and* F_2 *are* E-*surjective*. *Further assume there is an object* $A = (A_1, \alpha_A, A_2) \in$ Ⓒ *such that* A_i *is basic in* $\mathbb{C}_i$ $(i = 1,2)$. *Then there is an exact Mayer-Vietoris sequence*

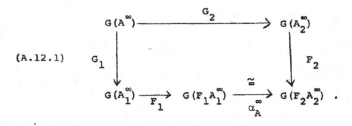

(A.13.1)

$$K_2\textcircled{C}_1 \oplus K_2\textcircled{C}_2 \longrightarrow K_2\textcircled{C}' \longrightarrow$$

$$\longrightarrow K_1\textcircled{C} \longrightarrow K_1\textcircled{C}_1 \oplus K_1\textcircled{C}_2 \longrightarrow K_1\textcircled{C}' \longrightarrow$$

$$\longrightarrow K_0\textcircled{C} \longrightarrow K_0\textcircled{C}_1 \oplus K_0\textcircled{C}_2 \longrightarrow K_0\textcircled{C}'$$

<u>Moreover the homomorphism</u> $K_1(G_2) \to K_1(F_1)$ <u>induced by</u> (A.11.1) <u>is an isomorphism</u> (<u>excision</u>).

The sequence of K_o's and K_1's follows from [3], Ch. VII, Thm. (4.3). To obtain the sequence of K_1's and K_2's and the excision isomorphism we consider the following commutative diagram:

(A.13.2)

$$\begin{array}{ccccc} G(A^\infty) & & \longrightarrow & & G(A_2^\infty) \\ \downarrow & & & & \downarrow F_2 \\ G(A_1^\infty) & \xrightarrow{F_1} & G(F_1A_1^\infty) & \xrightarrow[G(\alpha_A^\infty)]{\cong} & G(F_2A_2^\infty) \end{array}$$

Its four corners form a cartesian square which we shall denote

(A.13.3)

$$\begin{array}{ccc} \Gamma & \xrightarrow{G_2} & \Gamma_2 \\ {\scriptstyle G_2}\downarrow & & \downarrow{\scriptstyle F_2} \\ \Gamma_1 & \xrightarrow[F_1]{} & \Gamma_0 \end{array}$$

The functors F_1 and F_2 are E-surjective. It follows therefore from Prop. (A.12) that the functors G_1 and G_2 are also

E-surjective. As noted in the proof of Thm. (A.9) this implies that the group homomorphisms in (A.13.3) are all stable (in the sense of (A.9)). Thm. (A.13) is thus a consequence of the next lemma, parts c) and a).

(A.14) LEMMA. *Let*

(A.14.1)

$$\begin{array}{ccc} \Gamma & \xrightarrow{g_2} & \Gamma_2 \\ {\scriptstyle g_1}\downarrow & & \downarrow{\scriptstyle f_2} \\ \Gamma_1 & \xrightarrow[f_1]{} & \Gamma_0 \end{array}$$

be a cartesian square of group homomorphisms in which f_1 *and* f_2 *are stable* (in the sense of (A.9)).

a) *Put* $U = \mathrm{Ker}(g_2)$, $U_0 = (\Gamma, U)$, $W = \mathrm{Ker}(f_1)$, *and* $W_0 = (\Gamma_1, W)$. *Then* g_1 *induces isomorphisms* $U \to W$ *and* $U_0 \to W_0$, *and hence also an isomorphism* $H_0(\Gamma, H_1(U)) \to H_0(\Gamma_1, H_1(W))$. *Moreover* $(\Gamma', U_0) = U_0$.

b) *We have* $g_i\Gamma' = \Gamma_i'$ $(i = 1,2)$ *and* $\Gamma''' = \Gamma''$.

c) *If* $\Gamma'' = \Gamma'$ *then the homomorphisms* g_1 *and* g_2 *are stable, and there is an exact sequence*

(A.14.2) $$H_2(\Gamma_1') \oplus H_2(\Gamma_2') \longrightarrow H_2(\Gamma_0') \longrightarrow H_1(\Gamma) \longrightarrow H_1(\Gamma_1) \oplus H_1(\Gamma_2) \longrightarrow H_1(\Gamma_0)$$

which is natural with respect to the diagram (A.14.1).

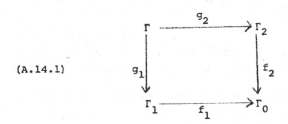

Since the square is cartesian the inclusion $f_2\Gamma_2 \supset \Gamma_0'$ implies $g_1\Gamma \supset \Gamma_1'$, whence $g_1\Gamma' \supset \Gamma_1'' = \Gamma_1'$. Thus $g_1\Gamma' = \Gamma_1'$, and similarly $g_2\Gamma' = \Gamma_2'$; this is the first part of b).

Since the square is cartesian $g_1 : U \to W$ is an isomorphism, and clearly $g_1U_0 \subset W_0$. But $g_1U_0 = (g_1\Gamma, g_1U) = (g_1\Gamma, W) \supset (\Gamma_1', W)$ (by the last paragraph) $\supset (\Gamma_1', W_0) = W_0$ (because f_1 is stable). Thus $g_1 : U_0 \to W_0$ is an isomorphism. Further $g_1(\Gamma', U_0) = (g_1\Gamma', g_1U_0) = (\Gamma_1', W_0) = W_0 = g_1U_0$, whence $(\Gamma', U_0) = U_0$. These conclusions establish part a) of the lemma. Moreover the last argument plus the fact that $\Gamma_1'' = \Gamma_1'$, shows that $(\Gamma^{(n)}, W_0) = W_0$ for any term $\Gamma^{(n)}$ of the derived series of Γ.

Put $W_1 = W \cap \Gamma'$, so that we have an exact sequence $1 \to W_1 \to \Gamma' \to \Gamma_2' \to 1$. Since $\Gamma_2'' = \Gamma_2'$ it follows that $\Gamma' = \Gamma'' \cdot W_1$ and hence $\Gamma'' = \Gamma''' \cdot (\Gamma'', W_1) \cdot W_1'$. Since $W_0 \subset W_1 \subset W$ both W_1' and (Γ'', W_1) belong to $W_0 = (\Gamma, W)$. Since $W_0 = (\Gamma''', W_0) \subset \Gamma'''$, as we have shown above, we have $\Gamma'' = \Gamma'''$, thus proving the last assertion of part b).

We have now established all the properties for stability of g_1 except the condition, $\Gamma'' = \Gamma'$. By symmetry the same applies to g_2, whence the first assertion of c).

If f_1 and g_2 are stable then we associate to them the exact sequences of Lemma (A.10), and the map between these exact sequences induced by the diagram (A.14.1):

$$\begin{array}{ccccccccc}
H_2(\Gamma') & \longrightarrow & H_2(\Gamma_2') & \longrightarrow & H_0(\Gamma, H_2(U)) & \longrightarrow & H_1(\Gamma) & \longrightarrow & H_1(\Gamma_2) \\
\downarrow & & \downarrow & & \downarrow & & \downarrow & & \downarrow \\
H_2(\Gamma_1') & \longrightarrow & H_2(\Gamma_0') & \longrightarrow & H_0(\Gamma_1, H_1(W)) & \longrightarrow & H_1(\Gamma_1) & \longrightarrow & H_1(\Gamma_0)
\end{array}$$

The middle vertical arrow is, by part a), an isomorphism. Further, since (A.14.1) is cartesian, and since $f_1\Gamma_1' = \Gamma_0'$, it follows easily that $\mathrm{Coker}(H_1(\Gamma) \to H_1(\Gamma_2))$ injects into $\mathrm{Coker}(H_1(\Gamma_1) \to H_1(\Gamma_0))$. Thus the exact sequence (A.14.2) results from the following well known construction, which completes the proof of Lemma (A.14).

(A.15) LEMMA. *Let*

$$\begin{array}{ccccccccc}
A_1 & \xrightarrow{a_1} & A_2 & \xrightarrow{a_2} & A_3 & \xrightarrow{a_3} & A_4 & \xrightarrow{a_4} & A_5 \\
\downarrow{\scriptstyle c_1} & & \downarrow{\scriptstyle c_2} & & \downarrow{\scriptstyle c_3} & & \downarrow{\scriptstyle c_4} & & \downarrow{\scriptstyle c_5} \\
B_1 & \xrightarrow[b_1]{} & B_2 & \xrightarrow[b_2]{} & B_3 & \xrightarrow[b_3]{} & B_4 & \xrightarrow[b_4]{} & B_5
\end{array}$$

be a commutative exact diagram of abelian groups in which c_3 *is an isomorphism. Then the sequence*

$$B_1 \oplus A_2 \xrightarrow{(b_1, c_2)} B_2 \xrightarrow{a_3 c_3^{-1} b_2} A_4 \xrightarrow{(c_4, -a_4)} B_4 \oplus A_5$$

is exact. If $\mathrm{Coker}(a_4) \to \mathrm{Coker}(b_4)$ *is injective then*

$$A_4 \xrightarrow{(c_4, -a_4)} B_4 \oplus A_5 \xrightarrow{(b_4, c_5)} B_5$$

is exact. If $\mathrm{Ker}(a_1) \to \mathrm{Ker}(b_1)$ *is surjective then*

$$A_1 \xrightarrow{(c_1, -a_1)} B_1 \oplus A_2 \xrightarrow{(b_1, c_2)} B_2$$

is exact.

The proof is standard diagram chasing; we omit the details.

Chapter IV. Some stability theorems

We wish to prove that certain groups of isometries of certain unitary modules act transitively on their sets of hyperbolic planes. Such results imply Witt cancellation theorems ($V \perp W \cong V' \perp W \Rightarrow V \cong V'$, for V "sufficiently large") and that the $s_n : U^{\lambda}_{2n}(A,\Lambda) \to KU^{\lambda}_1(A,\Lambda)$ are surjective for n "sufficiently large". Such results have been proved by Bak [1] and by Wasserstein [15]. Moreover Wasserstein has even proved the much more difficult fact that the homomorphisms s_n are eventually injective as well; we do not treat this matter here.

In attempting to compute the groups $KU^{\lambda}_1(A,\Lambda)$ it is important to know how soon s_n becomes surjective. One assumes, say, that A is a finite algebra over a commutative ring R and that max(R) is a noetherian space of dimension d. Then Bak and Wasserstein show that s_n is surjective for $n \geq d + 1$. However, it appears in the sympletic case (cf. [2], Prop. 13.2) that $2n \geq d + 1$ already suffices. This has also been proved by Wagner [12]. In trying to find a general explanation for this difference it appeared that the "size" of Λ should also influence these considerations, a larger Λ giving surjectivity of s_n sooner. (In the symplectic case $\Lambda = A$!)

These considerations lead to the notion of the "ampleness of Λ" introduced and investigated in §2. The stability theorems proper are then treated in §§3-4.

§1. Unitary (R,λ)-algebras (A,Λ)

(1.1) A unitary base ring (R,λ) consists of a commutative ring R with involution, $t \mapsto \bar{t}$, and an element $\lambda \in R$ such that $\lambda\bar{\lambda} = 1$.

The subring R_0 of R is defined to be the subring generated by all $t\bar{t}$ $(t \in R)$. It is contained in the fixed ring R_1 of the involution.

If $t \in R$ put $P(X) = (X - t)(X - \bar{t}) = X^2 - (t + \bar{t})X + t\bar{t}$. Setting $X = 1$ we see that $t + \bar{t} \in R_0$, so $P(X) \in R_0[X]$. It follows, that $2R_1 \subset R_0 \subset R_1$, and that R is integral over R_0.

Hence spec $(R) \to$ spec (R_0), $\mathfrak{q} \mapsto \mathfrak{q}_0 = \mathfrak{q} \cap R_0$, is surjective, and injective on chains of prime ideals. If $\mathfrak{q} \in$ spec (R) then $\mathfrak{q}$ and $\bar{\mathfrak{q}}$ are the only primes lying over $\mathfrak{q}_0$. For if $\mathfrak{q}'$ were another we could find $t \in \mathfrak{q}'$, $t \notin \mathfrak{q} \cup \bar{\mathfrak{q}}$, whence $t\bar{t} \in \mathfrak{q}_0' = \mathfrak{q}_0$ and $t\bar{t} \notin \mathfrak{q}$; contradiction. It follows that $\mathfrak{q} \cap \bar{\mathfrak{q}}$ is the radical of the ideal $\mathfrak{q}_0 R$.

The ideal $I(R)$ of R generated by all $t - \bar{t} (t \in R)$ will be called the inertial ideal of R. It is the smallest ideal modulo which the involution is trivial. We also put $I_0(R) = I(R) \cap R_0$.

In case the involution on R is trivial R_0 is the subring of R generated by all squares. Taking for R a suitable imperfect field of characteristic 2 one thus finds examples where

R is not finitely generated over R_0.

(1.2) <u>Examples</u>.

(1.2.1) <u>The hyperbolic case</u>: Suppose R is a hyperbolic ring, say $R = H(S) = S \times S$, with $\overline{(s,t)} = (t,s)$. If $\Delta : S \to R$ is the diagonal then ΔS is the fixed ring of the involution. If $s \in S$ then $\Delta s = (s,s) = (s,1)(1,s) = (s,1)\overline{(s,1)}$ so $R_0 = \Delta S$. The inertial ideal I(R) contains $(1,0) - \overline{(1,0)} = (1,-1)$, which is a unit. so $I(R) = R$.

(1.2.2) <u>The local case</u>: Suppose R_0 is a local ring with maximal ideal $\mathfrak{m}_0$. If $\mathfrak{m}$ is a maximal ideal of R above $\mathfrak{m}_0$ then R is semi-local with maximal ideals $\mathfrak{m}, \bar{\mathfrak{m}}$. If $\mathfrak{m} \neq \bar{\mathfrak{m}}$ and if R is complete in the $(\mathfrak{m}_0 R)$-adic topology then R splits as a product of two local rings so R is a hyperbolic ring on a local ring, and we are again in case (1.2.1).

Suppose $\mathfrak{m} = \bar{\mathfrak{m}}$. If $I(R) \subset \mathfrak{m}$ then $\lambda \equiv \pm 1 \bmod \mathfrak{m}$ and $R_0/\mathfrak{m}_0$ is the subfield of $R/\mathfrak{m}$ generated by all squares, i.e. $R/\mathfrak{m}$ itself if $\mathrm{char}(R/\mathfrak{m}) \neq 2$, and the image of Frobenius if $\mathrm{char}(R/\mathfrak{m}) = 2$.

If $I(R) \not\subset \mathfrak{m}$ then $I(R) = R$, the involution on $R/\mathfrak{m}$ is non trivial, and $R_0/\mathfrak{m}_0$ is its fixed field.

(1.2.3) <u>Abelian groups rings</u>. Suppose $R = k\pi$ where k is a commutative ring with involution, π is an abelian group, and R is given the involution $\overline{\sum_{x\in\pi} a_x x} = \sum_{x\in\pi} \bar{a}_x x^{-1}$. Then one checks easily that $R/I(R) = (k/I(k))[\pi/\pi^2]$.

(1.3) <u>A unitary</u> (R,λ)-<u>algebra</u> (A,Λ) consists of an R-algebra-with-involution A such that (A,λ,Λ) is a unitary ring. By R-algebra-with-involution we mean that A is a ring with involution, $a \mapsto \bar{a}$, that A is an R-algebra, and that the structure map $R \to A$ (with image in Center (A)) is compatible with the respective involutions in R and in A. Further, in writing (A,λ,Λ) we allow ourselves to confuse λ with its image in Center (A).

If $t \in R$ and $r \in \Lambda$ then we have $tr\bar{t} = t\bar{t}r \in \Lambda$. It follows that:

(1.3.1) Λ <u>is an</u> R_0-<u>module</u>

For any homomorphism $(R,\lambda) \to (R',\lambda')$ of unitary base rings (defined in the obvious way) we obtain a unitary (R',λ')-algebra (A',Λ') where $A' = A \otimes_R R'$, with involution $\overline{a \otimes r'} = \bar{a} \otimes \overline{r'}$, and where Λ' is the additive group generated by $S_{-\lambda'}(A')$ together with all $r \otimes s'$, $r \in \Lambda$ and $s' \in R_0'$ = the subring of R' generated by all norms, $s\bar{s}$ $(s \in R')$.

As special cases we can reduce modulo involution invariant ideals of R, or we can localize with respect to involution invariant multiplicative sets in R (e.g. any multiplicative set in R_0).

(1.3.2) Example. Suppose $R = H(S)$ is hyperbolic. If $e = (1,0)$ and $\bar{e} = (0,1)$ in R then in A these give rise to central idempotents $e, \bar{e}$ with sum 1, so it follows that A is also hyperbolic; in fact $A = H(B)$ where B is an S-algebra.

The R_0 algebra A is said to be quasi-finite if, for each $\mathfrak{m} \in \max(R_0)$, we have $\mathfrak{m}A_{\mathfrak{m}} \subset \operatorname{rad} A_{\mathfrak{m}}$ and the ring $A_{\mathfrak{m}}$ is semi-local. The latter means that the ring $A_{\mathfrak{m}}/\operatorname{rad} A_{\mathfrak{m}}$, which we shall denote $A[\mathfrak{m}]$, is semi-simple (artinian).

We call (A,Λ) a quasi-finite (R,λ)-algebra if A is a quasi-finite R_0-algebra. If $\mathfrak{m} \in \max(R_0)$ we denote by $\Lambda[\mathfrak{m}]$ the image of Λ (or of $\Lambda_{\mathfrak{m}}$) in $A[\mathfrak{m}]$. Thus $(A[\mathfrak{m}], \Lambda[\mathfrak{m}])$ is a semi-simple unitary $(R/\mathfrak{m}R, \lambda)$-algebra.

§2. Ampleness of Λ in A

This is a condition which intervenes in the hypotheses of the stability theorems in §3. We describe below when this condition holds.

(2.1) The semi-simple case. Let (A,λ,Λ) be a unitary ring such that the ring A is semi-simple (artinian). Then we say Λ is ample in A if the following condition holds:

(2.1.1) Given $a,b \in A$ there exists an $r \in \Lambda$ such that
$$A(a + rb) = Aa + Ab.$$

Remark: To illustrate one useful consequence of this condition we observe that it implies that $U_2^\lambda(A,\Lambda)$ is generated by $EU_2^\lambda(A,\Lambda)$ and $H(A^\cdot)$. Indeed suppose $\sigma = \begin{pmatrix} a & b \\ c & d \end{pmatrix} \in U_2^\lambda(A,\Lambda)$. Since $Aa + Ac = A$ we can find $r \in \Lambda$ so that $u = a + rc$ is a unit. Then we have

$$X_+(r)\sigma = \begin{pmatrix} u & b' \\ c & d \end{pmatrix} = X_-(cu^{-1})H(u)X_+(u^{-1}b')$$

(see Ch. II, Prop. (2.5)), whence the assertion.

In order to see when this condition holds we decompose A into a product of simple rings-with-involution A_i, so that (A,λ,Λ) similarly decomposes into a product of simple unitary rings $(A_i,\lambda_i,\Lambda_i)$ (see Ch. I, (1.5) and (4.2.5)). It is easily seen that Λ is ample in A if and only if Λ_i is ample in A_i for all i.

Thus we are led to consider the simple case. We begin by giving some counterexamples. Then we shall prove (Theorem (2.3)) that they are essentially the only ones.

(2.2) Counterexamples. Let C be a field with trivial

involution. We shall show that $\Lambda = S_{-\lambda}(A)$ is <u>not</u> ample in A in the following two cases:

(2.2.1) $$A = M_n(C),\ \lambda = 1, \text{ and } \quad \bar{\alpha} = \delta\alpha^*\delta^{-1}, \text{ where } \quad \delta = \operatorname{diag}(d_1,\dots,d_n).$$

(The * signifies transpose.)

(2.2.2) $$A = M_{2n}(C),\ \lambda = -1, \text{ and } \quad \bar{\sigma} = \epsilon\sigma^*\epsilon^{-1}, \text{ where } \quad \epsilon = \begin{pmatrix} 0 & I \\ -I & 0 \end{pmatrix}.$$

<u>Case (2.2.1)</u>. $\rho \in \Lambda = S_{-1}(A) \Leftrightarrow \rho\delta$ is an alternating matrix, as a simple calculation shows. Consider the elements

$$\alpha = \left(\begin{array}{c|cc} I_{n-2} & 0 & \\ \hline 0 & 1 & 0 \\ & 1 & 0 \end{array}\right), \quad \beta = \left(\begin{array}{c|cc} 0 & 0 & \\ \hline 0 & 0 & -d_{n-1} \\ & 0 & d_n \end{array}\right)$$

Clearly $A\alpha + A\beta = A$. We will show however that for any $\rho \in \Lambda$ the element $\gamma = \alpha + \rho\beta$ is not invertible. We have

$$\rho = \left(\begin{array}{c|cc} * & * & \\ \hline * & 0 & td_n^{-1} \\ & -td_{n-1}^{-1} & 0 \end{array}\right)$$ for some $t \in C$, whence

$$\rho\beta = \left(\begin{array}{c|cc} 0 & & * \\ \hline 0 & 1 & t \\ & 1 & t \end{array}\right). \quad \text{It follows that}$$

$$\gamma = \left(\begin{array}{c|cc} I & & * \\ \hline 0 & 1 & t \\ & 1 & t \end{array}\right), \text{ which has zero determinant.}$$

Case (2.2.2). If $\sigma = \begin{pmatrix} \alpha & \beta \\ \gamma & \delta \end{pmatrix}$ we have $\bar{\sigma} = \begin{pmatrix} \delta^* & -\beta^* \\ -\gamma^* & \alpha^* \end{pmatrix}$. It follows that $\Lambda = S_1(A)$ consists of those σ for which $\delta = \alpha^*$ and β and γ are alternating. Consider the elements

$$a = \left(\begin{array}{c|c} I_n & 0 \\ \hline \begin{array}{c|c} 0 & 0 \\ \hline 0 & 1 \end{array} & \begin{array}{c|c} I_{n-1} & 0 \\ \hline 0 & 0 \end{array} \end{array}\right), \quad b = \left(\begin{array}{c|c} 0 & \begin{array}{c|c} 0 & 0 \\ \hline 0 & 1 \end{array} \\ \hline 0 & \begin{array}{c|c} 0 & 0 \\ \hline 0 & 1 \end{array} \end{array}\right)$$

Clearly $Aa + Ab = A$. We claim $a + rb$ ($r \in \Lambda$) is never invertible. If $r = \begin{pmatrix} \alpha & \beta \\ \gamma & \alpha^* \end{pmatrix} \in \Lambda$ then rb is zero except in the last column, which is $\begin{pmatrix} \alpha_n + \beta_n \\ \gamma_n + \alpha^*_n \end{pmatrix}$, where α_n denotes the last column of α and similarly for $\beta_n, \gamma_n, \alpha^*_n$. Since β and γ are alternating we have $\beta_{nn} = 0 = \gamma_{nn}$, whence the last column of rb has the form

$\begin{pmatrix} * \\ \alpha_{nn} \\ * \\ \alpha_{nn} \end{pmatrix}$. It follows that

$$a + rb = \left(\begin{array}{c|c} I & \begin{array}{c|c} 0 & * \\ \hline 0 & \alpha_{nn} \end{array} \\ \hline \begin{array}{c|c} 0 & 0 \\ \hline 0 & 1 \end{array} & \begin{array}{c|c} I & * \\ \hline 0 & \alpha_{nn} \end{array} \end{array}\right)$$

The last column of a + rb is easily seen to be a linear combination of the others; hence a + rb is not invertible, as claimed.

(2.3) THEOREM. *Let* (A,λ,Λ) *be a unitary ring such that* A *is a simple ring and* $\Lambda = S_{-\lambda}(A)$. *Then* Λ *is ample in* A *if and only if* (A,λ,Λ) *is not isomorphic to one of the examples* (2.2.1) *or* (2.2.2).

Let M be a simple A-module. Suppose M does not admit a non singular even $(-\lambda)$-hermitian form. Then according to Ch. I, (7.11.1), $C = \text{Center }(A) = \text{End}_A(M)$, C has trivial involution, $\lambda = \pm 1$, and the involution on $A = \text{End}_C(M)$ arises from a λ-hermitian form g on the C-module M. If g is diagonalisable then $\lambda = 1$ and, relative to a suitable basis for M, $A \cong M_n(C)$ and $\bar{\alpha} = \delta\alpha^*\delta^{-1}$ where $\delta = \text{diag }(d_1,\ldots,d_n)$ is a diagonal matrix representing g (see Ch. I, (7.3)). This is just example (2.2.1). If g is not diagonalizable then (see Bourbaki, [4], §6, Théorème 1) g is alternating. Relative to a symplectic basis of M we have $A \cong M_{2n}(C)$ and $\bar{\sigma} = \epsilon\sigma^*\epsilon^{-1}$ where $\epsilon = \begin{pmatrix} 0 & I \\ -I & 0 \end{pmatrix}$ is the matrix representing g (loc. cit.). This is just example (2.2.2). In each of these cases we have seen above that Λ is not ample in A.

It remains therefore to show that Λ is ample in A whenever there exists a non-singular even $(-\lambda)$-hermitian form

$$h = [\ ,\ ] : M \times M \longrightarrow A$$

Put $B = \mathrm{End}_A(M)$, a division ring. By sesquilinear duality theory (Ch. I, §7) we have an involution $b \mapsto \bar{b}$ on B defined by $[\bar{b}n,m] = [n,bm]$ for $n,m \in M$, and a non-singular, $(-\lambda)$-hermitian form $g = \{\ ,\ \} : M \times M \to B$ defined by $\{m,n\}x = m[n,x]$ for $m,n,x \in M$.

Given $a,b \in A$ we seek $r \in \Lambda$ such that, if $c(r) = a + rb$, we have $Ac(r) = Aa + Ab$. For $u,v \in A = \mathrm{End}_B(M)$ we have $Au \subset Av \Leftrightarrow Mu \subset Mv$. Hence we can restate the above condition as $Mc(r) = Ma + Mb$.

We argue by induction on $t = \dim_B(Mb) - \dim_B(Mb \cap Ma)$. If $t = 0$ we have $Mb \subset Ma$ and we can take $r = 0$. Suppose now that $t > 0$, so $Mb \not\subset Ma$. Then a is not invertible so $K_a = \{x \in M \mid xa = 0\} \neq 0$. Put $K_a^{\perp} = \{x \in M \mid \{K_a, x\} = 0\}$. Then

$K_a^{\perp}$ and also $H = \{x \in M | xb \in Ma\}$ are proper B-submodules of M. It follows that M is not their union. This is clear if one of them contains the other. Otherwise choose $u \in K_a^{\perp} - H$ and $v \in H - K_a^{\perp}$ and then $u + v \notin K_a^{\perp} \cup H$.

It follows that we can find an $x_0 \in M$ such that $x_0 b \notin Ma$ and $\{K_a, x_0\} \neq 0$. The latter condition implies that $x_0^{\perp} = \{x \in M | \{x, x_0\} = 0\}$ is a B-hyperplane in M not containing K_a so, $M = (x_0^{\perp}) + K_a$ and hence $(x_0^{\perp})a = Ma$.

Put $r_0 = [x_0, x_0]$. Since $[\ ,\ ]$ is even $(-\lambda)$-hermitian we have $r_0 \in S_{-\lambda}(A) = \Lambda$. Now if $x \in M$ we have $xc(r_0) = xa + x[x_0,x_0]b = xa + \{x,x_0\}x_0 b$. Hence $(x_0^{\perp})c(r_0) = (x_0^{\perp})a = Ma$, so $Ma \subset Mc(r_0)$. Since $\{K_a, x_0\} \neq 0$ we also have $K_a c(r_0) = \{K_a, x_0\}x_0 b = Bx_0 b$. It follows that $Mb \cap Mc(r_0) \supset (Mb \cap Ma) + Bx_0 b$. We can therefore apply our induction hypothesis to the pair $c(r_0)$, b in place of a,b. We obtain an $r_1 \in \Lambda$ such that $Mc = Mc(r_0) + Mb$ where $c = c(r_0) + r_1 b = a + (r_0 + r_1)b = c(r_0 + r_1)$. Since $Mc(r_0) + Mb = Ma + Mb$ (clearly) this completes the proof.

(2.4) Involutions of second kind. Let (A, λ, Λ) be a unitary ring such that A is a simple ring-with-involution. Let C_0 be the fixed ring of the involution on C = Center (A).

If A is a simple ring then C and C_0 are fields and $[C:C_0] \leq 2$.

If $A = H(B)$ is hyperbolic then B is simple with center C_0 diagonally embedded in $C = C_0 \times C_0$; in particular C_0 is a field and $[C:C_0] = 2$.

In all cases then C_0 is a field and $[C:C_0] \leq 2$. One says the involution is of <u>second kind</u> if $[C:C_0] = 2$; this case includes hyperbolic rings but excludes the examples in (2.2). We can then find a unit $\mu \in C$ such that $\lambda\mu/\bar{\mu} = -1$. If C is a field this follows from Hilbert's Theorem 90. In the hyperbolic case we have $\lambda = (u,u^{-1})$ for some $u \in C_0$ and we take $\mu = (1,-u)$. Scaling by μ (Ch. I, (4.13)) transforms (A,λ,Λ) to $(A,-1,\mu\Lambda)$, and it is clear that the ampleness in A of Λ and of $\mu\Lambda$ are equivalent. Hence: <u>For involutions of second kind there is no loss of generality in assuming</u> $\lambda = -1$ for checking the ampleness of Λ.

(2.5) <u>The hyperbolic case</u>. We have $A = H(B) = B \times B^{op}$ with B simple. We may assume $\lambda = -1$ (see (2.4)). Then $\Lambda = S_1(A) = S^1(A) = \Delta B$ where $\Delta r = (r,r)$ for $r \in B$. Let $a \cdot b = ba$ denote the multiplication in B^{op}. If $a = (a_1,a_2) \in A$ then $Aa = Ba_1 \times B^{op} \cdot a_2 = Ba_1 \times a_2B$. Thus if $b = (b_1,b_2) \in A$ then $Aa + Ab = (Ba_1 + Bb_1) \times (a_2B + b_2B)$. If $\Delta r = (r,r) \in \Lambda$ then $a + rb = (a_1 + rb_1,\ a_2 + b_2r)$. We conclude therefore that $\Lambda = \Delta B$ <u>is ample in</u> $A = H(B)$ <u>if and only if the following condition holds</u>:

<u>Given elements</u> $a_i, b_i \in B$ $(i = 1,2)$

<u>there is an element</u> $r \in B$ <u>such that</u>

(2.5.1)
$$Ba_1 + Bb_1 = B(a_1 + rb_1)$$

<u>and</u>

$$a_2B + b_2B = (a_2 + b_2r)B$$

We first note that

<u>one can find an</u> $r \in B$ <u>such that</u>

(2.5.2)
$$B(a_1 + rb_1) = Ba_1 + Bb_1$$

and similarly as $s \in B$ such that $(a_2 + b_2s)B = a_2B + b_2B$. Thus the problem lies in choosing r and s to be equal.

To prove (2.5.2) we represent B as the endomorphism ring of a vector space V over a division ring. If $a \in B$ write $K_a = \mathrm{Ker}(V \overset{a}{\rightarrow} V)$. Then Ba consists precisely of the $c \in B$ with $K_a \subset K_c$. Given $a,b \in B$ we want to find $c = a + rb$ so that $K_c = K_a \cap K_b$, for then we will have $Bc = Ba + Bb$. Put $N = K_a \cap K_b$ and write $K_a = N \oplus K_a'$ and $K_b = N \oplus K_b'$. Then $V = K_a \oplus K_b' \oplus X$ for some X. If $T = K_b' \oplus X$ then $aV = aT \cong T$ so $V = aT \oplus U$ with $\dim U = \dim K_a \geq \dim K_a'$. Define $d \in B$ to be zero on T and N and to carry K_a' isomorphically to a subspace of U. Since $K_b = N \oplus K_b' \subset N \oplus T$ we have $d \in Bb$, i.e. $d = rb$ for some $r \in B$. Moreover, if $c = a + d$ then $cT = aT$ and $cK_a' = dK_a' \subset U$.

Hence $cV = aT \oplus dK'_a$ so it follows by dimension count that $K_c = N$, as required.

In case B is a division ring the conditions on r and s above each exclude at most one possible value. Hence there is a simultaneous solution if Card $B \geq 3$. On the other hand if $B = \mathbb{F}_2$ then the case $a_1 = 0$ and $b_1 = a_2 = b_2 = 1$ admits no such solution. Thus:

(2.5.3) <u>If</u> B <u>is a division ring then</u> (2.5.1) <u>holds if and only if</u> $B \neq \mathbb{F}_2$.

Suppose now that B <u>is finite dimensional over</u> C_0 = Center (B).

Consider first the case $B = M_n(C_0)$. Put $c_1(r) = a_1 + rb_1$ and $c_2(r) = a_2 + b_2r$. The inclusion $Bc_1(r) \subset Ba_1 + Bb_1$ is an equality if and only if the matrix $c_1(r)$ has the same rank, say m_1, as a generator of $Ba_1 + Bb_1$. Similarly $c_2(r)B = a_2B + b_2B$ if and only if $c_2(r)$ has the same rank m_2 as a generator of $a_2B + b_2B$. Since each of these conditions can be achieved separately there exists an $m_1 \times m_1$ subdeterminant $P_1(r)$ of $c_1(r)$ which is not zero for some r, and an $m_2 \times m_2$ subdeterminant $P_2(r)$ of $c_2(r)$ which is non zero for some (possibly different) r. Thus P_1 and P_2 are non zero polynomials of degrees m_1 and m_2, respectively, on the C_0-vector space B. We want an $r \in B$ such that $P_1(r) \neq 0$ and $P_2(r) \neq 0$. Since P_1P_2 has degree

$m_1 + m_2 \leq 2n$ the existence of such an r follows from the next lemma, provided that Card $C_0 > 2n$.

(2.5.4) LEMMA. Let P be a non zero polynomial of degree $\leq d$ in m variables over a field extension of C_0. If Card $C_0 > d$ then there is an $x \in C_0^m$ such that $P(x) \neq 0$.

If $m = 1$ then x need only avoid the at most d roots of P. In general view P as a polynomial in the last variable x_m. There are $\leq d$ roots so some element a_m of C_0 is not one of them. Set $x_m = a_m$ to obtain a non zero polynomial Q in $(m - 1)$ variables, of degree $\leq d$. Apply induction to Q.

Now return to the general case of our problem where B is finite dimensional over C_0, but we no longer assume $B = M_n(C_0)$. If $\bar{C}_0$ is an algebraic closure of C_0 then $B \otimes_{C_0} \bar{C}_0 \cong M_n(\bar{C}_0)$, and we can apply the above discussion to $B' = B \otimes_{C_0} \bar{C}_0$. Given $a_i, b_i \in B$ $(i = 1,2)$, we obtain a non zero polynomial $P = P_1P_2$ of degree $\leq 2n$ on the $\bar{C}_0$-vector space B' such that, for $r \in B'$, $P(r) \neq 0 \Leftrightarrow B'c_1(r) = B'a_1 + B'b_1$ and $c_2(r)B' = a_2B' + b_2B'$. According to Lemma (2.5.4) we can find $r \in B$ such that $P(r) \neq 0$ provided that Card $C_0 > 2n$. In this case the condition above implies that $Bc_1(r) = Ba_1 + Bb_1$ and $c_2(r)B = a_2B + b_2B$, since these equations hold after tensoring with $\bar{C}_0$.

We have thus proved:

(2.6) THEOREM. *Let* (A,λ,Λ) *be a unitary ring such that* $A = H(B)$ *is the hyperbolic ring* $B \times B^{op}$ *on a central simple algebra* B *of dimension* n^2 *over a field* C_0. *Then* Λ *is ample in* A *provided that*

(2.6.1) $$\text{Card } C_0 > 2n = \sqrt{2[A:C_0]}$$

(2.7) COROLLARY. *Let* (A,λ,Λ) *be a unitary ring such that* A *is a simple ring-with-involution of the second kind. Let* C_0 *be the fixed field of the involution on* C = Center (A) *and assume* $[A:C_0] < \infty$. *Then* Λ *is ample in* A *provided that* Card $C_0 > \sqrt{2[A:C_0]}$.

If A is hyperbolic this follows from Thm. (2.6). If A is a simple ring then the exceptional cases of (2.2) are excluded because A is of second kind, so the result follows from Thm. (2.3).

(2.8) *Delocalisation*. Let (R,λ) be a unitary base ring (1.1). Recall that R_0 denotes the subring of R generated by all $t\bar{t}$ $(t \in R)$, I(R) denotes the (inertial) ideal generated by all $t - \bar{t}$ $(t \in R)$, and $I_0(R) = I(R) \cap R_0$. For any constant $q > 0$ we shall put

$$J_q(R_0) = \bigcap_{\text{Card}(R_0/\mathfrak{m}) \leq q} \mathfrak{m} ,$$

where $\mathfrak{m}$ varies over maximal ideals.

Let (A,Λ) be a quasi-finite (R,λ)-algebra (1.3). For $\mathfrak{m} \in \max(R_0)$ we have the semi-simple unitary $(R/\mathfrak{m}R,\lambda)$-algebra $(A[\mathfrak{m}],\Lambda[\mathfrak{m}])$ where $A[\mathfrak{m}] = A_{\mathfrak{m}}/\text{rad } A_{\mathfrak{m}} = (A/\mathfrak{m}A)/\text{rad}(A/\mathfrak{m}A)$, and where $\Lambda[\mathfrak{m}]$ denotes the image of Λ in $A[\mathfrak{m}]$. We shall say that Λ *is ample in* A *at* $\mathfrak{m}$ if $\Lambda[\mathfrak{m}]$ is ample in $A[\mathfrak{m}]$, in the sense of (2.1).

(2.9) THEOREM. *Let* (A,Λ) *be a quasi-finite unitary* (R,λ)*-algebra as above. Let* n *be an integer such that* $[A[\mathfrak{m}]:R_0/\mathfrak{m}] \leq n$ *for all* $\mathfrak{m} \in \max(R_0)$. *Let* $\mathfrak{m} \in \max(R_0)$ *and suppose that* Λ *is not ample in* A *at* $\mathfrak{m}$. *Then* $\mathfrak{m}$ *contains the ideal* $I_0(R) \cap J_{\sqrt{2n}}(R_0)$. *In case* $A = R$ (so A is commutative) $\mathfrak{m}$ *contains* $(I(R) + R(1-\lambda)) \cap J_2(R_0)$.

If $I_0(R) \not\subset \mathfrak{m}$ then the involution on $R[\mathfrak{m}] = R_{\mathfrak{m}}/\text{rad } R_{\mathfrak{m}}$ is non trivial, so each simple factor A' of the ring-with-involution $A[\mathfrak{m}]$ has an involution of second kind. Moreover the fixed field C_0 of the involution on Center (A') contains $R_0/\mathfrak{m}$ so if $J_{\sqrt{2n}}(R_0) \not\subset \mathfrak{m}$ we have $\text{Card } C_0 \geq \text{Card}(R_0/\mathfrak{m}) > \sqrt{2n} = \sqrt{2[A':C_0]}$. It follows therefore from Cor. (2.7) that if $I_0(R) \not\subset \mathfrak{m}$ and $J_{\sqrt{2n}}(R_0) \not\subset \mathfrak{m}$ then $\Lambda[\mathfrak{m}]$ is ample in $A[\mathfrak{m}]$. This proves the first assertion of the theorem.

Suppose $A = R$ and that $\Lambda[\mathfrak{m}]$ is not ample in $A[\mathfrak{m}]$ $(= R[\mathfrak{m}])$.

If $A[\mathfrak{m}]$ is hyperbolic, say $H(C_0)$, then (see (2.5.1)) we must have Card $C_0 = 2$, so $J_2(R_0) \subset \mathfrak{m}$.

Suppose $A[\mathfrak{m}]$ is a field. Then by Thm. (2.3) we must have one of the examples described in (2.2) (after replacing Λ by $S_{-\lambda}(A)$, the latter certainly being non ample whenever Λ is non ample). Example (2.2.2) is necessarily non commutative, hence impossible. Example (2.2.1) occurs precisely in the following case: $\lambda = 1$ in $A[\mathfrak{m}]$ and the involution is trivial on $A[\mathfrak{m}]$. These conditions imply that $\mathfrak{m}$ contains $(I(R) + R(1 - \lambda)) \cap R_0$, whence the last assertion of the theorem.

The presense of the terms $J_q(R_0)$ in the theorem will not be too troublesome in practice thanks to the following result.

(2.10) PROPOSITION. *Let* R *be a commutative noetherian ring. Let* q *be a constant* > 0. *Then* R *has only finitely many maximal ideals with residue class fields of cardinal* $\leq q$.

It suffices to show that, for each prime power q, the set S_q of maximal ideals $\mathfrak{m}$ with $R/\mathfrak{m} \cong \mathbb{F}_q$ is finite. Let R' denote the quotient of R modulo the intersection of all $\mathfrak{m} \in S_q$. Then $a^q = a$ for all $a \in R'$ so R' is integral over its prime ring, which is a field. Hence R' has Krull dimension zero. Since R' is reduced and noetherian it is a finite product of fields, whence the proposition.

Remark. The proposition clearly fails for $R = \mathbb{F}_q[X_1,\dots,X_n,\dots]$ or for an infinite product of copies of $\mathbb{F}_q$.

(2.11) PROPOSITION. *Let* (R,λ) *and* R_0 *be as in* (2.8). *Let* $f_i:(A_i,\Lambda_i) \to (A',\Lambda')$ $(i = 1,2)$ *be unitary surjections of quasi-finite unitary* (R,λ)*-algebras. Let* (A,Λ) *be their fibre product* (Ch. III, (2.1)), *with projection* h_i *to* (A_i,Λ_i) $(i = 1,2)$.

a) (A,Λ) *is a quasi-finite* (R,λ)*-algebra*.

b) *If* $\mathfrak{m} \in \max(R_0)$ *the square*

(2.11.1)
$$\begin{array}{ccc}
(A[\mathfrak{m}],\Lambda[\mathfrak{m}]) & \longrightarrow & (A_1[\mathfrak{m}],\Lambda_1[\mathfrak{m}]) \\
\big\downarrow & & \big\downarrow \\
(A_2[\mathfrak{m}],\Lambda_2[\mathfrak{m}]) & \longrightarrow & (A'[\mathfrak{m}],\Lambda'[\mathfrak{m}])
\end{array}$$

is cartesian.

c) Λ *is ample at* $\mathfrak{m}$ *if and only if* Λ_1 *and* Λ_2 *are ample at* $\mathfrak{m}$, *in which case* Λ' *is also ample at* $\mathfrak{m}$.

a) follows from Stein [11], Prop. (2.7). Stein treats only the case $A = A(\mathfrak{q})$, but his proofs use only the fact that all arrows in his cartesian squares are surjective. This remark applies to all of our references to [11].

To prove b) one first localizes at $\mathfrak{m}$. The square

$\begin{array}{ccc} A_{\mathfrak{m}} & \longrightarrow & A_{1\mathfrak{m}} \\ \downarrow & & \downarrow \\ A_{2\mathfrak{m}} & \longrightarrow & A'_{\mathfrak{m}} \end{array}$ is still cartesian and, by [11], Lemma (2.5), it remains cartesian upon reducing each of the rings modulo its radical. This shows that the first coordinate of the square (2.11.1) is cartesian. It follows that $\Lambda[\mathfrak{m}]$ projects onto each $\Lambda_i[\mathfrak{m}]$ $(i = 1,2)$. This forces $\Lambda[\mathfrak{m}]$ to equal $\Lambda_1[\mathfrak{m}] \times_{\Lambda'[\mathfrak{m}]} \Lambda_2[\mathfrak{m}]$ since all the rings involved are semi-simple

For suppose, $\begin{array}{ccc} (B,\Gamma) & \longrightarrow & (B_1,\Gamma_1) \\ \downarrow & & \downarrow \\ (B_2,\Gamma_2) & \longrightarrow & (B',\Gamma') \end{array}$ is a square of semi-simple unitary rings and unitary surjections such that the square of rings is cartesian. Then we can write $B_1 = B_1' \times B'$ and $B_2 = B' \times B_2'$ so that the homomorphism $B_i \to B'$ are coordinate projections. We then have $\Gamma_1 = \Gamma_1' \times \Gamma'$, $\Gamma_2 = \Gamma' \times \Gamma_2'$, $B = B_1 \times B' \times B_2'$, and $\Gamma \subset \Gamma_1' \times \Gamma' \times \Gamma_2'$. Since Γ is the product of its projections on each factor of B (Ch. I, (4.2.5)) and since Γ projects onto Γ_1 and Γ_2 it follows that $\Gamma = \Gamma_1' \times \Gamma' \times \Gamma_2'$ $= \Gamma_1 \times_{\Gamma'} \Gamma_2$.

The same considerations show that Γ is ample in $B \Leftrightarrow \Gamma_i$ is ample in B_i $(i = 1,2)$, in which case Γ' is ample in B'. This concludes the proof of b) and c), and hence of the proposition.

§3. A stability theorem.

(3.1) The setting. It will be fixed throughout §§3-4, as follows:

(R,λ) is a unitary base ring (see (1.1)).

R_0 is the subring of R generated by all norms $t\bar{t}$ $(t \in R)$.

$X = \max(R_0)$, which we assume to be a noetherian space of dimension d.

(A,Λ) is a quasi-finite unitary (R,λ)-algebra (see (1.3)).

$\mathfrak{a}_\Lambda$ is an ideal of R_0 such that Λ is ample at all $\mathfrak{m} \notin V(\mathfrak{a}_\Lambda)$

d_Λ denotes the dimension of the closed set $V(\mathfrak{a}_\Lambda)$ in X.

$O(x) = O_M(x)$, where x is an element of an A-module M, denotes $\{f(x) \mid f \in \mathrm{Hom}_A(M,A)\}$. It is a left ideal of A, and we call x unimodular in M if $O(x) = A$. Clearly $O_{M\oplus N}(x,0) = O_M(x)$. If $M = A^n$ and $x = (a_1,\ldots,a_n)$ then $O(x) = Aa_1+\ldots+Aa_n$.

(3.2) The module P and the group $E(P) \subset GL(P)$. We shall simultaneously consider two cases, in each of which P will be a finitely generated projective A-module.

Case a) $\text{f-rank}_{R_0}(P) \geq d+2$, i.e. for all $\mathfrak{m} \in X$ there is an epimorphism $P_{\mathfrak{m}} \to A^{d+2}_{\mathfrak{m}}$.

Case b) $P = A^n$ with $n \geq d + 1$ and $n \geq d_\Lambda + 2$.

In case b) we write $P = P_1 \oplus p_0A$ where p_0 is the last standard basis element of A^n and $P_1 = A^{n-1}$. In case a) we choose a decomposition $P = P_1 \oplus p_0A$ where p_0 is any unimodular element of P; such exist by a theorem of Serre [3] (Ch. IV, Cor. (2.7)). We can then write $\overline{P} = \overline{P_1} \oplus q_0A$ so that p_0, q_0 is a hyperbolic pair in $H(P) = H(P_1) \perp (p_0A \oplus q_0A)$.

Put $E(P) = E(P_1, p_0A)$ (cf. Ch. II, (3.5), for this notation). In case b) it is easily seen that $E(P) = E_n(A)$. In case a) it follows from [3] (Ch. IV, Cor. (3.7)) that E(P) is independent of the decomposition $P = P_1 \oplus p_0A$ used to define it.

If $A \to A'$ is a surjective ring homomorphism and $P' = P \otimes_A A'$ then the homomorphism $GL_A(P) \to GL_{A'}(P')$ induces an epimorphism $E_A(P) \to E_{A'}(P')$ in each of the above cases ([3], Ch. IV, Prop. (3.3)).

(3.3) THEOREM. *Let* $x = (p,q) \in H(P)$ *and let* L *be a left ideal of* A *such that* $O(x) + L = A$. *Then there is a* $\sigma \in H(E(P)) \cdot EU(H(P))$ *such that* $\sigma x = (p',q')$ *and* $O(p') + L = A$.

Before proving this theorem we shall draw several consequences.

(3.4) THEOREM. *Let* Γ *be a subgroup of* GL(P) *which contains* E(P) *and which acts transitively on the set of unimodular elements in* P. *Put* $G = H(\Gamma)\cdot EU(H(P))$.

1) *If* $x \in H(P)$ *is unimodular there is a* $\sigma \in G$ *such that* $\sigma x = p_0 + q_0 b$ *where* $\bar{b} \equiv q_P(x,x) \bmod \Lambda$.

2) G *acts transitively on the set of hyperbolic elements in* $H(P)$, *on the set of hyperbolic pairs in* $H(P)$, *and on the set of hyperbolic planes in* $H(P)$.

3) $EU(H(P))$ *is a normal subgroup of* $U(H(P))$.

In case a) *of* (3.2) *the group* Γ *can be taken to be* $E(P)$. *In case* b) *of* (3.2), *if* $d \leq 1$, $d_\Lambda \leq 0$, $n \geq 2$ *and* $A = R$, *then* Γ *can be taken to be* $SL_n(A)$.

Proof of Thm. (3.4). Say $x = (p,q) \in P \oplus \bar{P}$. Applying Thm. (3.3) with $L = 0$ we can transform x by an element of G to make p unimodular. Further applying an element of $H(\Gamma)$ we can even achieve $p = p_0$. Write $q = q_0 b - q_1 \in q_0 A \oplus \bar{P}_1$. The transvection $\sigma_{q_1,0,q_0}$ (see Ch. I, (5.1)) belongs to $X\_(\bar{\Lambda}(P)) \subset G$ and transforms x to $x + q_1\langle q_0,x\rangle - q_0\bar{\lambda}\langle q_1,x\rangle$, where $\langle\ ,\ \rangle$ is the λ-hermitian form on $H(P)$. It is easily seen that $\langle q_1,x\rangle = 0$ and that $\langle q_0,x\rangle = \langle q_0,p_0\rangle_P = 1$. Thus $\sigma_{q_1,0,q_0}x = x + q_1 = p_0 + q_0 b$. The quadratic function $[q_P]$ in $H(P)$ is invariant under $U(G(P))$. Its value on $p_0 + q_0 b$ is $[q_P(p_0 + q_0 b,\ q_0 + q_0 b)] = [\langle q_0 b, p_0\rangle] = [\bar{b}]$, where $[a]$ denotes a mod Λ for $a \in A$. This establishes assertion 1) of the theorem.

Suppose x above is isotropic. Then $[\bar{b}] = 0$, i.e. $\bar{b} \in \Lambda$. In the hyperbolic plane $p_0A + q_0A$ the element $X_-(-b)$ transforms $p_0 + q_0b$ to p_0. This shows that G can transform any hyperbolic element to p_0, whence the first assertion of 2). The remaining assertions of 2) follow from Ch. II, Prop. (3.11). Part 3) follows from the same proposition in view of the fact (Ch. II, Prop. (2.5)) that H(GL(P)), and hence also G, normalizes EU(H(P)).

We now prove the last assertions of the theorem. In case a) the condition f-rank $P \geq d + 2$ implies, by [3] (Ch. IV, Thm. 3.4) that E(P) acts transitively on the set of unimodular elements in P. Suppose $P = A^n$, $A = R$, $d \leq 1$, and $d_\Lambda \leq 0$. If we are in case b) we have $n \geq 2$. If $n \geq 3$ we are also in case a) already treated above. If $n = 2$ then we note that, for any commutative ring A, $SL_2(A)$ acts transitively on the set of unimodular elements in A^2 ([3], Ch. V, Prop. (3.4)(b)). This concludes the proof of Thm. (3.4).

(3.5) COROLLARY. Keep the assumptions and notation of Theorem (3.4). Let V be any (λ,Λ)-quadratic module (Ch. I, (4.4)) and put $M = V \perp H(P)$. Let G_1 denote the subgroup of U(M) generated by $1_V \perp G$ together with all transvections $\sigma_{p,a,v}$ ($p \in P$ or $\bar{P}$, $v \in V$). Then G_1 acts transitively on the set of hyperbolic pairs and on the set of hyperbolic planes in M.

It is easily seen that G_1 contains all transvections $\sigma_{p_0,a,v}$ with $v \in (p_0)^{\perp} = V \oplus H(P_1) \oplus p_0A$; cf. proof of Prop. (3.11) of Ch. II. It follows therefore from Ch. I, Cor. (5.6) that it suffices to show that G_1 acts transitively on the set of hyperbolic elements in M.

Let $x = (v;p,q) \in V \oplus (P \oplus \bar{P})$ be a hyperbolic element (Ch. I, (4.10.4)). If $\langle\ ,\ \rangle$ denotes the λ-hermitian form on M then $\langle x,y\rangle = 1$ for some $y \in M$. It follows that $\langle V,v\rangle + O(p) + O(q) = A$. Choose $w \in V$ so that $\langle w,v\rangle + O(p) + O(q)$ contains 1; put $c = \langle v,w\rangle$. In either case a) or case b) we have $f - \mathrm{rank}_{R_0}(P) \geq d + 1$. It follows therefore from [3], Ch. IV, Thm. (3.1), that there is a $p_1 \in P$ such that $O(p + p_1c) + O(q) = A$.

Consider the transvection $\sigma = \sigma_{p_1,a,w}$ where $a \equiv H(w,w) \bmod \Lambda$ (h being the sesquilinear form on M; $M = (M,[h])$). Then (Ch. I, Prop. (5.4)) we have $\sigma x = (v + g(q);\ p + f(v) + t(q),q)$ where $g:\bar{P} \to V$ and $t:\bar{P} \to P$ are homomorphisms, and where $f = p_1\langle w,\ \rangle : P \to V$. Thus $\sigma x = (v';\ p + p_1c + t(q),q)$. Clearly $O(p + p_1c + t(q),q) = O(p + p_1c,q) = O(p + p_1c) + O(q) = A$, by the construction of c. Thus we can reduce the proof to the case when $x = (v;p,q)$ with (p,q) unimodular, which we now assume.

In this case Thm. (3.4) permits us to transform x by an element of $1_V \perp G$ to make q unimodular. Then there is a homomorphism $g: \bar{P} \to V$ such that $g(q) = -v$. Using Prop. (5.4) of Ch. I, and Remark (5.4.1) following it, we obtain a $\sigma \in G_1$ such that $\sigma x = (v + g(q); p + f(v) + t(q), q) = (0; p', q)$. Thus we have transformed x into $H(P)$, so the proof concludes by applying Thm. (3.4).

(3.6) COROLLARY. Let W, W' be (λ, Λ)-quadratic modules. Then we have

$$W \perp N \cong W' \perp N \Longrightarrow W \cong W'$$

for all unitary modules N provided the following condition holds: $W \perp H(A)$ has an orthogonal summand isomorphic to $H(P)$, and $GL(P)$ acts transitively on the set of unimodular elements in P. The latter condition is automatic if f - rank $P \geq d + 2$ or if $A = R$, $d \leq 1$, $d_\Lambda \leq 0$, and $P = A^n$ with $n \geq 2$.

By Ch. I, Cor. (4.5) we can find a quadratic module N' such that $N \perp N' \cong H(A^r) = H(A) \perp \dots \perp H(A)$. Since the assumption on W is inherited by $W \perp H$ for any quadratic module H we see that it suffices to treat the case $N = H(A^r)$, and then, by induction on r, the case $N = H(A)$. Using the given isomorphism to identify $M = W \perp H(A)$ with $W' \perp H(A)$ we realize W and W' as the orthogonal complements of two hyperbolic planes in M. Hence it suffices to show that $U(M)$ acts transitively on the set of hyperbolic planes in M. In view of the assumption on W this follows from Cor. (3.5), taking $\Gamma = GL(P)$. The last assertion of the corollary follows from the last assertions of Thm. (3.4).

(3.7) COROLLARY. Put $n_0 = \max(d, d_\Lambda + 1)$. The natural homomorphism $U^\lambda_{2n}(A,\Lambda) \to KU^\lambda_1(A,\Lambda)$ is surjective for $n \geq n_0$.

It suffices, in view of Ch. II, Prop. (5.1) and Thm. (5.2), to show that, for $n > n_0$, $U^\lambda_{2n}(A,\Lambda) = H(E_n(A)) \cdot EU^\lambda_{2n}(A,\Lambda) \cdot U^\lambda_{2(n-1)}(A,\Lambda)$ Let $\sigma \in U^\lambda_{2n}(A,\Lambda)$. Let $H(A^n) = H_1 \perp \cdots \perp H_n$ where H_i is the hyperbolic plane $e_i A \oplus \overline{e_i} A$. By Thm. (3.4) there is an $\varepsilon \in H(E_n(A)) \cdot EU^\lambda_{2n}(A,\Lambda)$ such that $\varepsilon\sigma$ fixes e_n and $\overline{e_n}$. The fixer of H_n is just $U^\lambda_{2(n-1)}(A,\Lambda)$, so $\sigma = \varepsilon^{-1}(\varepsilon\sigma)$ is the required factorization.

(3.8) COROLLARY. Let $(\mathfrak{q}, \Gamma)$ be a unitary ideal in (A, λ, Λ). The natural homomorphism $U^\lambda_{2n}(\mathfrak{q},\Gamma) \to KU^\lambda_1(\mathfrak{q},\Gamma)$ is surjective and $(U^\lambda_{4n}(A,\Lambda), U^\lambda_{4n}(\mathfrak{q},\Gamma)) \subset EU^\lambda_{4n}(\mathfrak{q},\Gamma)$ for all $n \geq n_0$.

We have a commutative exact diagram

$$\begin{array}{ccccccccc}
1 & \longrightarrow & U^\lambda_{2n}(\mathfrak{q},\Gamma) & \longrightarrow & U^\lambda_{2n}(A(\mathfrak{q},\Gamma)) & \longrightarrow & U^\lambda_{2n}(A,\Lambda) & \longrightarrow & 1 \\
 & & \downarrow & & \downarrow & & \downarrow & & \\
0 & \longrightarrow & KU^\lambda_1(\mathfrak{q},\Gamma) & \longrightarrow & KU^\lambda_1(A(\mathfrak{q},\Gamma)) & \longrightarrow & KU^\lambda_1(A,\Lambda) & \longrightarrow & 0
\end{array}$$

with compatible splittings of the horizontal surjections. Therefore the left vertical arrow is surjective provided the middle and right ones are. That the latter is the case follows from Cor. (3.7), in view of the fact that $A(\mathfrak{q})$, just as A, is a quasi-finite R_0-algebra and that $\Lambda(\Gamma)$, just as Λ, is ample at all $\mathfrak{m} \notin V(\mathfrak{a}_\Lambda)$ (Prop. (2.11)).

Let $\sigma \in U^\lambda_{4n}(A,\Lambda)$ and $\tau \in U^\lambda_{4n}(\mathfrak{q},\Gamma)$. To show that $(\sigma,\tau) \in EU^\lambda_{4n}(\mathfrak{q},\Gamma)$ we are free to modify σ modulo $EU^\lambda_{4n}(A,\Lambda)$ and τ modulo $EU^\lambda_{4n}(\mathfrak{q},\Gamma)$. As automorphisms of $H(A^{2n}) = H(A^n) \perp H(A^n)$ we can thus arrange that σ fixes the elements of the second $H(A^n)$, and τ those of the first. This follows from the fact that $U^\lambda_{4n}(\mathfrak{q},\Gamma) = U_{2n}(\mathfrak{q},\Gamma)\cdot EU^\lambda_{4n}(\mathfrak{q},\Gamma)$, and the analogous statement for (A,Λ), which results from what has been proved above since $n \geq n_0$. But with σ and τ so modified they commute evidently, whence the Corollary.

To compute n_0 in practice it is convenient to have a simple estimate for d_Λ.

(3.9) PROPOSITION. <u>Assume</u> R_0 <u>is noetherian and that</u> A <u>is a finitely generated</u> R_0-<u>module</u>. <u>One can choose</u> $\mathfrak{a}_\Lambda$ <u>so that</u> $d_\Lambda \leq$ sup (0, dim max $(R/I(R))$). <u>If</u> $A = R$ <u>one can choose</u> $_\Lambda$ <u>so that</u> $d_\Lambda \leq$ sup (0, dim max $(R/(I(R) + R(1-\lambda)))$).

There is an integer n such that $[A[\mathfrak{m}]:R_0/\mathfrak{m}] \leq n$ for all $\mathfrak{m} \in \max(R_0)$. According to Theorem (2.9) we can choose $\mathfrak{a}_\Lambda$ to contain $I_0(R) \cap J_{\sqrt{2n}}(R_0)$ where $I_0(R) = I(R) \cap R_0$ and $J_q(R_0) = \bigcap_{\mathrm{Card}\, R_0/\mathfrak{m} \leq q} \mathfrak{m}$. According to Prop. (2.10) $R_0/J_q(R_0)$ is artinian, whence dim max $(R_0/(I_0(R) \cap J_2(R_0))) \leq$ sup (0, dim max $(R_0/I_0(R))$). Since R is integral over R_0 we have dim max $(R_0/\mathfrak{a} \cap R_0)$ = dim max $(R/\mathfrak{a})$ for any ideal $\mathfrak{a}$ of R. Taking $\mathfrak{a} = I(R)$ we obtain the first assertion of the proposition. If $A = R$ Thm. (2.9) permits us to take $\mathfrak{A}_\Lambda = (I(R) + R(1 - \lambda)) \cap J_2(R_0)$. The last assertion thus follows from an argument just like the preceding one.

(3.10) COROLLARY. *Suppose* $A = R$, A *is a finitely generated* R_0*-module,* R_0 *is noetherian,* $\lambda = -1$ *and* $\dim \max(R/2R) < \dim \max(R)$ *Then we can arrange that* $d_\Lambda < d$. *In particular we can take* $n_0 = d$ *in Corollaries* (3.7) *and* (3.8).

This follows from the hypotheses and Prop. (3.9) since $1 - \lambda = 2$.

Corollary (3.10) applies notably when A is an order in a commutative semi-simple algebra over $\mathbb{Q}$, for example $A = \mathbb{Z}\pi$ with π a finite abelian group.

We prepare for the proof of theorem (3.1) now, with a lemma, due essentially to Bak [1].

(3.11) LEMMA. *Suppose the ring* A *is semi-simple and that either* (a) f-rank $P \geq 2$ *or* (b) $P = A$ *and* Λ *is ample in* A. *Let* $x = (p,q) = (p_1 + p_0 a, q_1 + q_0 b) \in H(P)$. *Then there is a* $\sigma \in H(E(P)) \cdot EU(H(P))$ *such that* $\sigma x = (p_1' + p_0 a', q_1' + q_0 b')$ *and* $O(x) = Aa'$.

In case (b) there is an $r \in \Lambda$ such that $A(a + rb) = Aa + aB$, by definition of ampleness (2.1.1). We can then take $\sigma = X_+(r)$.

Suppose now we are in case (a). Abbreviate the coordinates of x by $\begin{pmatrix} p_1, & a \\ q_1, & b \end{pmatrix}$. If $u \in \bar{P}_1$ the element $\sigma_{p_0,0,u} \in H(E(P))$ transforms x to $\begin{pmatrix} p_1, & a + \langle u,p_1\rangle \\ q_1 - ub, & b \end{pmatrix}$. Since $\bar{P}_1$ has a direct summand isomorphic to A we can choose u so that $O(q_1 - ub) = O(q_1) + Ab$

(see (2.5.2)). Assuming this already done we reduce to the case when $O(q) = O(q_1)$.

If $u \in P_1$ then $\sigma_{p_0,0,u} \in X_+(\Lambda(P))$ and it transforms x, to $x' = \begin{pmatrix} p_1 - ub & , & a + \langle u,q_1\rangle \\ q_1 & , & b \end{pmatrix} = \begin{pmatrix} p_1' & , & a'' \\ q_1 & , & b \end{pmatrix}$. For a suitable choice of u we have $Aa'' = Aa + O(q_1)$ (see (2.5.2)). Since $b \in O(q) = O(q_1) \subset Aa''$ it follows that $O(x) = O(p_1',a'')$. Now, as in the first step, if $v \in \bar{P}_1$ then $\sigma_{q_0,0,v} \in H(E(P))$ transforms x' to $\begin{pmatrix} p_1' & , & a'' + \langle v,p_1'\rangle \\ q_1' & , & b \end{pmatrix} = \begin{pmatrix} p_1' & , & a' \\ q_1' & , & b \end{pmatrix}$. Choosing v so that $O(a') = O(p_1', a'')$ (using (2.5.2) again) we have $Aa' = O(x)$, as required.

<u>Proof of theorem</u> (3.3). We must transform $x = (p,q) = (p_1 + p_0 a, q_1 + q_0 b)$ by an element of $H(E(P)) \cdot EU(H(P))$ to achieve the condition $O(p) + L = A$. We shall argue by induction on $d = \dim X$.

Put $L_1 = Aa + L$, $\mathfrak{a} = \operatorname{ann}_{R_0}(A/L_1)$ = the largest ideal $\mathfrak{a}$ in R_0 such that $\mathfrak{a} A \subset L_1$, $X' = V(\mathfrak{a}) = \max(R_0/\mathfrak{a})$, $d' = \dim X'$, $X'_\Lambda = X' \cap X_\Lambda = V(\mathfrak{a} + \mathfrak{a}_\Lambda)$, $d'_\Lambda = \dim X'_\Lambda$, and let $\pi : A \to A' = A/\mathfrak{a} A$ be the natural projection. If M is an A-module we shall write $\pi M = M \otimes_A A'$ and $\pi m = m \otimes 1 \in \pi M$ for $m \in M$.

Consider the following conditions:

(i) $d' \leq d - 1$ and $d'_\Lambda \leq d_\Lambda - 1$.

(ii) $O(p_1, q_1) + L_1 = A$.

If $d = 0$ then (i) implies $X' = \emptyset$ whence $L_1 = Aa + L = A$, and the theorem is established. If $d > 0$ then (i) and (ii) permit us to apply induction to $x_1 = (\pi p_1, \pi q_1) \in H(\pi P_1)$ and πL_1. The result is an element $\tau' \in H(E(\pi P_1))EU(H(\pi P_1))$ which transforms x_1 to an element $((\pi p_1)', (\pi q_1)')$ such that $O((\pi p_1)') + \pi L_1 = A'$. We shall prove below that τ' can be lifted to an element $\tau_1 \in U(H(P_1))$ which, when extended to fix p_0 and q_0, yields an element $\tau \in H(E(P)) \cdot EU(H(P))$. We then have $\tau x = (p'_1 + p_0 a, q'_1 + q_0 b)$ with $\pi p'_1 = (\pi p_1)'$ and $\pi q'_1 = (\pi q_1)'$, and hence $O(p'_1) + L_1 + \mathfrak{a} A = A$. But $\mathfrak{a} A \subset L_1 = Aa + L$ and hence $O(p'_1) + Aa + L = A$. Since $O(p'_1) + Aa = O(p'_1 + p_0 a)$ this establishes the conclusion of the theorem.

To obtain the lifting of τ' promised above we first note that $EU(H(P_1)) \to EU(H(\pi P_1))$ is surjective (Ch. II, Prop. (2.3)) and that $EU(H(P_1)) \perp 1_{(p_0 A \oplus q_0 A)} \subset EU(H(P))$. To complete the proof it suffices clearly to define a group $E(P_1) \subset GL(P_1)$ so that $E(P_1)$ maps surjectively to $E(\pi P_1)$ and so that $E(P_1) \oplus 1_{p_0 A} \subset E(P)$. In case (b) we have $P_1 = A^{n-1}$ and we take $E(P_1) = E_{n-1}(A)$. In case (a) we have f-rank $(P_1) \geq d + 1$

so P_1 contains a unimodular element u, say $P_1 = P_2 \oplus uA$. Put $E(P_1) = E(P_2, uA)$. This maps surjectively to $E(\pi P_1) = E(\pi P_2, \pi uA')$ (which is automatically defined as in (3.2) because f-rank $(\pi P_1) \geq d + 1 \geq d' + 2$). Moreover it is clear that $E(P_1) \oplus 1_{P_0 A} \subset E(P)$.

In conclusion then we have shown that the theorem follows once we have conditions (i) and (ii) above.

Realisation of (i). Choose a finite set S in X meeting each irreducible component of X as well as each irreducible component of X_Λ. Put $\mathfrak{q} = \prod_{\mathfrak{m} \in S} \mathfrak{m}$ and $A[\mathfrak{q}] = (A/\mathfrak{q}A)/\mathrm{rad}(A/\mathfrak{q}A)$ $= \prod_{\mathfrak{m} \in S} A[\mathfrak{m}]$. After passing to $A[\mathfrak{q}]$ we can use Lemma (3.11) to transform x so that "$O(x) = Aa$ in $A[\mathfrak{q}]$". The transformation in that lemma lifts to one in $H(E(P)) \cdot EU(H(P))$. Assuming x already transformed by this lifting we then have $Aa + \mathfrak{q}A = O(x) + \mathfrak{q}A$, and hence $L_1 = Aa + L$ maps onto $A[\mathfrak{q}]$. Since A is a quasi-finite R_0 algebra it follows that $L_{1\mathfrak{m}} = A_{\mathfrak{m}}$ for all $\mathfrak{m} \in S$ (Nakayama's lemma). Consequently $\mathfrak{m} \notin \mathrm{supp}(A/L_1)$ $= V(\mathfrak{a}) = X'$ for all $\mathfrak{m} \in S$. Therefore X' contains no irreducible component of X. Similarly X'_Λ contains no irreducible component of X_Λ. This establishes (i).

<u>Realisation of (ii)</u>. We have $O(p_1, q_1 + q_0 b) \pm L_1 = A$ and so $O(\pi p_1, \pi(q_1 + q_0 b)) + \pi L_1 = A'$. Moreover f-rank $P_1 \geq d + 1$ (in both cases (a) and (b)) $\geq d' + 2$. It follows therefore from [3], Ch. IV, Thm. (3.1) that there is an element $u \in \pi P_1$ such that $O(\pi p_1 - ub) + O(\pi q_1) + \pi L_1 = A'$. Now u lifts to an element $v \in P_1$, and since $\mathfrak{a} + \mathfrak{q} = R_0$ (see proof of (i) above) we can choose $v \in P_1 \cdot \mathfrak{q}$. Now apply $\sigma_{p_0,0,v}$ to x to obtain $x + p_0\langle v,x\rangle - v\bar{\lambda}\langle p_0,x\rangle = (p_1 - vb + p_0(a + \langle v,q_1\rangle), q)$. Since $\pi v = u$ the relation above shows that $O(p_1 - vb) + O(q_1) + L_1 + \mathfrak{a}A = A$. But $\mathfrak{a}A \subset L_1$, so this realizes condition (ii).

This completes the proof of Theorem (3.3).

§4. The (restricted) sympletic case

We present here some stability theorems for $U^{\lambda}_{2n}(A,\Lambda)$ in which n is required to be roughly only half as large as in the theorems of §3. Some of these results are only partial, but those in the symplectic case are quite satisfactory.

(4.1) Some notation. For the next theorem we shall keep the hypotheses and notation of (3.1).

We shall use the following notation for elements $x \in H(A^n)$: If $x = \Sigma_i e_i a_i + \Sigma_i \bar{e}_i b_i$, where $e_1,\ldots,e_n$ is the standard basis of A^n, and $\bar{e}_1,\ldots,\bar{e}_n$ its dual basis, then we put

$$x = \begin{pmatrix} a_1,\ldots a_n \\ b_1,\ldots,b_n \end{pmatrix}$$

(4.2) THEOREM. Assume that $2n \geq d + 2$. Let L be a left ideal in A and let $x = \binom{a_1,\ldots,a_n}{b_1,\ldots,b_n} \in H(A^n)$ be such that

(4.2.1) $$O(x) + L = A$$

and

(4.2.2) $$Aa_n + L + \mathfrak{a}_\Lambda A = A.$$

Let d be any element of A. Then there is a $\sigma \in H(E_n(A)) \cdot EU^{\lambda}_{2n}(A,\Lambda)$ such that $\sigma x = \binom{c_1,\ldots,c_n}{d_1,\ldots,d_n}$ with

(4.2.3) $$O(c_1,\ldots,c_n,d_1,\ldots,d_{n-1}) + L = A.$$

(4.2.4) $$Ac_n + L + \mathfrak{A}_\Lambda A = A,$$

and

(4.2.5) $$d_n \equiv d \bmod (Ac_n + L).$$

Remark. In the proof, condition (4.2.4) actually is derived from the stronger condition: $c_n \equiv a_n \bmod \mathfrak{A}_\Lambda A$.

(4.3) COROLLARY. *Assume that* $2n \geq d + 2$. *Let* $x = \binom{a_1,\ldots,a_n}{b_1,\ldots,b_n} \in H(A^n)$ *be a unimodular element such that* $Aa_n + \mathfrak{A}_\Lambda A = A$. *Let* $d \in A$. *Then there is a* $\sigma \in H(E_n(A))\cdot EU^\lambda_{2n}(A,\Lambda)$ *such that* $\sigma x = \binom{c_1,\ldots,c_n}{d_1,\ldots,d_n}$ *with* $Ac_n + \mathfrak{A}_\Lambda A = A$ *and with* $d_n \equiv \bmod c_n A$.

We simply apply the theorem with $L = 0$.

Conddition (4.2.5) can be harmlessly be imposed after conditions (4.2.3) and (4.2.4) have been achieved, thanks to the following lemma.

(4.4) LEMMA. *Let* $y = \binom{c_1,\ldots,c_n}{d_1,\ldots,d_n} \in H(A^n)$ *satisfy* (4.2.3). *Let* $d \in A$. *Then there is a* $\sigma \in H(E_n(A))\cdot X_-(\bar{\Lambda}_n)$ *such that* $\sigma y = \binom{c'_1,\ldots,c'_{n-1},c_n}{d'_1,\ldots,d'_{n-1},d'_n}$ *with* $c'_i \equiv c_i \bmod Ac_n$, *and* $d'_i \equiv d_i \bmod Ac_n$ $(1 \leq i \leq n-1)$ *and* $d'_n \equiv d \bmod (Ac_n + L)$.

For (4.2.3) implies we can write $d - d_n = \sum_{i=1}^{n} r_i c_i + \sum_{i=1}^{n-1} s_i d_i + e$

with $e \in L$. Put $\sigma = H(I - \sum_{i=1}^{n-1} \bar{s}_i e_{ni}) X_-(\sum_{i=1}^{n-1} (r_i e_{ni} - \lambda\bar{r}_i e_{in}))$.

Then $\sigma Y = \binom{c_1', \ldots, c_{n-1}', c_n}{d_1', \ldots, d_{n-1}', d_n'}$ where $c_i' = c_i - \bar{s}_i c_n (1 \leq i \leq n-1)$, $d_i' = d_i - \overline{\lambda\bar{r}}_i c_n$ $(1 \leq i \leq n-1)$, and

$$d_n' = d_n + \sum_{i=1}^{n-1} r_i c_i + \sum_{i=1}^{n-1} s_i d_i'$$

$$= d_n - r_n c_n + \sum_{i=1}^{n} r_i c_i + \sum_{i=1}^{n-1} s_i (d_i - \overline{\lambda\bar{r}}_i c_n)$$

$$= d_n - r_n c_n - (\sum_{i=1}^{n-1} \bar{\lambda}\, s_i \bar{r}_i) c_n - e$$

$$+ (\sum_{i=1}^{n} r_i c_i + \sum_{i=1}^{n-1} s_i d_i + e).$$

The last term in parentheses is $d - d_n$. Hence, putting $t = r_n + \sum_{i=1}^{n-1} \bar{\lambda}\, s_i \bar{r}_i$, we have $d_n' = d - tc_n - e \in d + Ac_n + L$, whence the lemma.

<u>Proof of theorem</u> (4.2). Let Y denote the complement in $X = \max(R_0)$ of $V(\mathfrak{A}_\Lambda)$; put $\delta = \dim Y \leq d = \dim X$. We have $2n \geq \delta + 2$. Under this assumption we shall argue by induction

on δ to produce a $\sigma \in H(E_n(\mathfrak{a}_\Lambda\Lambda))EU_{2n}^{\lambda}(\mathfrak{a}_\Lambda A, \mathfrak{a}_\Lambda\Lambda)$ such that σx satisfies (4.2.3). The condition (4.2.4) will be automatic from (4.2.2) and the congruence $\sigma \equiv I \bmod \mathfrak{a}_\Lambda A$. Finally condition (4.2.5) can be imposed at the end using Lemma (4.4).

Let S consist of one element from each irreducible component of Y. Since Y is disjoint from $V(\mathfrak{a}_\Lambda)$ it follows that Λ is ample at each $\mathfrak{m} \in S$. Thus we can choose an $s_i \in \Lambda$ so that $a_i' = a_i + s_i b_i$ generates in $A[\mathfrak{m}] = (A/A\mathfrak{m})/\mathrm{rad}(A/A\mathfrak{m})$ the same left ideal as a_i and b_i $(1 \leq i \leq n)$. The different $\mathfrak{m}$'s in S, together with $\mathfrak{a}_\Lambda$, are pairwise comaximal in R_0, and Λ is an R_0-module. It follows that we can choose $s_i \in \mathfrak{a}_\Lambda\Lambda$ so that the above condition holds simultaneously for all $\mathfrak{m} \in S$ (Chinese Remainder Theorem).

Put $\sigma_1 = X_+(\sum_{i=1}^{n} s_i e_{ii}) \in EU_{2n}^{\lambda}(\mathfrak{a}_\Lambda A, \mathfrak{a}_\Lambda\Lambda)$ and $x_1 = \sigma_1 x = \binom{a_1', \ldots, a_n'}{b_1, \ldots, b_n}$. Note that we have $Aa_n' + L + \mathfrak{a}_\Lambda A = A$ and $O(x_1)[\mathfrak{m}] = O(a_1', \ldots, a_n')[\mathfrak{m}]$ for each $\mathfrak{m} \in S$. (We put $D[\mathfrak{m}]$ = the image of $D \subset A$ in $A[\mathfrak{m}]$.)

Suppose $\delta \leq 0$. Then $S = Y$ so $O(a_1', \ldots, a_n')[\mathfrak{m}] = O(x_1)[\mathfrak{m}]$ for all $\mathfrak{m}$ not containing $\mathfrak{a}_\Lambda$. If $\mathfrak{m}$ contains $\mathfrak{a}_\Lambda$ then $(Aa_n' + L)[\mathfrak{m}] = A[\mathfrak{m}]$. Thus in either case $(O(a_1', \ldots, a_n') + L)[\mathfrak{m}] = A[\mathfrak{m}]$. Since A is quasi-finite over R_0 this implies that

$O(a_1',\ldots,a_n') + L = A$, so the theorem is proved in this case.

Now assume $\delta > 0$. Then $d > 0$ so $n > 1$, because $2n \geq d + 2$. Choose $u_2,\ldots,u_n \in \mathfrak{a}_\Lambda A$ such that $a_1'' = a_1' + u_2a_2' +\ldots+ u_na_n'$ generates the same left ideal in $A[\mathfrak{m}]$ as $a_1',\ldots,a_n'$ for each $\mathfrak{m} \in S$ (cf. (2.5.2)). Put

$$\sigma_2 = H(I + \sum_{i=2}^{n} u_ie_{ii}) \in H(E_n(\mathfrak{a}_\Lambda\Lambda)) \text{ and } x_2 = \sigma x_1 = \binom{a_1'',a_2',\ldots,a_n'}{b_1,b_2',\ldots,b_n'}.$$

Then $O(x_2)[\mathfrak{m}] = (Aa_1'')[\mathfrak{m}]$ for all $\mathfrak{m} \in S$ so, putting $L_1 = Aa_1'' + L$, we have $L_1[\mathfrak{m}] = A[\mathfrak{m}]$ for all $\mathfrak{m} \in S$. Put $\mathfrak{a}_1 = \text{ann}_{R_0}(A/L_1)$. Since A is quasi-finite over R_0 it follows that S is disjoint from $V_Y(\mathfrak{a}_1) = Y \cap V(\mathfrak{a}_1)$. Thus $V_Y(\mathfrak{a}_1)$ contains no irreducible component of Y, so $\dim V_Y(\mathfrak{a}_1) < \dim Y = \delta$.

Let $T \subset Y$ consist of one element from each irreducible component of $V_Y(\mathfrak{a}_1)$. As in the construction of the elements s_i above, we can find elements $t_1,\ldots,t_{n-1} \in \mathfrak{a}_\Lambda\bar{\Lambda}$ and $t_n \in \mathfrak{a}_\Lambda\Lambda$ such that $b_1' = b_1 + t_1a_1''$, $b_2'' = b_2' + t_2a_2',\ldots,b_{n-1}'' = b_{n-1}' + t_{n-1}a_{n-1}'$, and $a_n'' = a_n' + t_nb_n'$ satisfy, for each $\mathfrak{m} \in T$, $(Ab_1')[\mathfrak{m}] = (Ab_1 + Aa_1'')[\mathfrak{m}]$, $(Ab_i'')[\mathfrak{m}] = (Ab_i' + Aa_i')[\mathfrak{m}]$ $(2 \leq i \leq n-1)$, and $(Aa_n'')[\mathfrak{m}] = (Aa_n' + Ab_n')[\mathfrak{m}]$. Put

$$\sigma_3 = X_+(t_ne_{nn})X_-(\sum_{i=1}^{n-1} t_ie_{ii}) \in EU_{2n}^{\lambda}(\mathfrak{a}_\Lambda A, \mathfrak{a}_\Lambda\Lambda) \text{ and } x_3 = \sigma_3x_2 =$$

$\binom{a_1'',a_2',\ldots,a_{n-1}',a_n''}{b_1',b_2'',\ldots,b_{n-1}'',b_n'}$. Then we have $a_n'' \equiv a_n' \equiv a_n \mod \mathfrak{a}_\Lambda A$ so

$Aa_n'' + L + \mathfrak{a}_\Lambda A = A$. Further $O(x_3)[\mathfrak{m}] = O(b_1', b_2'', \ldots, b_{n-1}'', a_n'')[\mathfrak{m}]$ for all $\mathfrak{m} \in T$.

Now choose $v_2, \ldots, v_n \in \mathfrak{a}_\Lambda A$ so that $b_1'' = b_1' + v_2 b_2'' + \ldots + v_{n-1} b_{n-1}'' + v_n a_n''$ generates in $A[\mathfrak{m}]$ the left ideal $O(b_1', b_2'', \ldots, b_{n-1}'', a_n'')[\mathfrak{m}]$ for all $\mathfrak{m} \in T$. Put

$$\sigma_4 = X_-(v_n e_{1n} - \bar{\lambda}\bar{v}_n e_{n1}) H(I - \sum_{i=2}^{n-1} \bar{v}_i e_{i1})$$

$\in H(E_n(\mathfrak{a}_\Lambda A))\, EU_{2n}^\lambda(\mathfrak{a}_\Lambda A, \mathfrak{a}_\Lambda \Lambda)$ and $x_4 = \sigma_4 x_3 = \binom{a_1'', a_2'', \ldots, a_{n-1}'', a_n''}{b_1'', b_2'', \ldots, b_{n-1}'', b_n''}$

Put $L_2 = Ab_1'' + L_1 = Ab_1'' + Aa_1'' + L$ and $\mathfrak{a}_2 = \operatorname{ann}_{R_0}(A/L_2)$. By construction $V_Y(\mathfrak{a}_2) \subset V_Y(\mathfrak{a}_1)$ and the former contains no irreducible component of the latter.

If $V_Y(\mathfrak{a}_2) = \emptyset$ then $L_2[\mathfrak{m}] = A[\mathfrak{m}]$ for all $\mathfrak{m}$ not containing $\mathfrak{a}_\Lambda$. Further $(L_2 + Aa_n'')[\mathfrak{m}] = A[\mathfrak{m}]$ for all $\mathfrak{m}$ containing $\mathfrak{a}_\Lambda$. It follows that $A = L_2 + Aa_n'' = O(b_1'', a_1'', a_n'') + L$, so the theorem is proved in this case (taking $\sigma = \sigma_4\sigma_3\sigma_2\sigma_1$).

If $V_Y(\mathfrak{a}_2) \neq \emptyset$ then $\delta_2 = \dim V_Y(\mathfrak{a}_2) < \dim V_Y(\mathfrak{a}_1) < \delta = \dim Y$, so $\delta_2 \leq \delta - 2$. Denote the natural projections $R \to R' = R/\mathfrak{a}_2 R$ and $A \to A' = A/\mathfrak{a}_2 A$ by π. Put $\Lambda' = \pi\Lambda$ and $\mathfrak{a}_{\Lambda'} = \pi\mathfrak{a}_\Lambda$ in $R_0' = \pi R_0$. Put $Y' = \max(R_0') - V(\mathfrak{a}_{\Lambda'})$; we can identify Y' in X with $V(\mathfrak{a}_2) - V(\mathfrak{a}_2 + \mathfrak{a}_\Lambda) = Y \cap V(\mathfrak{a}_2) = V_Y(\mathfrak{a}_2)$. Hence $\dim Y' \leq \delta - 2$. Further it is clear that Λ' is ample in A' for all $\mathfrak{m} \in Y'$.

Put $y = \binom{a_2'', \ldots, a_n''}{b_2'', \ldots, b_n''}$. We propose to apply our induction hypothesis to the ideal πL_2 in A', the element $\pi y \in H(A'^{n-1})$, and the ideal $\mathfrak{a}_{\Lambda'}$ in R_0. Note that since $L_2 + Ab_1'' + Aa_1'' + L$ we have $O(\pi y) + \pi L_2 = \pi(O(x_4) + L) = \pi A = A'$. Moreover $A'\pi a_n'' + \pi L + \mathfrak{a}_{\Lambda'} A' = \pi(Aa_n'' + L + \mathfrak{a}_\Lambda \Lambda) = \pi A = A'$. Finally, since $2n \geq \delta + 2$ we have $2(n-1) \geq \delta \geq \delta' + 2$.

By induction therefore there is a $\tau' \in H(E_{n-1}(\mathfrak{a}_{\Lambda'}A'))EU^{\lambda}_{2(n-1)}(\mathfrak{a}_{\Lambda'}A', \mathfrak{a}_{\Lambda'}\Lambda')$ such that $\tau'\pi y = \binom{c_2', \ldots, c_n'}{d_2', \ldots, d_n'}$ with $O(c_2', \ldots, c_n', d_2', \ldots, d_{n-1}') + \pi L_2 = A'$. Now we can lift τ' to a $\tau \in H(E_{n-1}(\mathfrak{a}_\Lambda A))EU^{\lambda}_{2(n-1)}(\mathfrak{a}_\Lambda A, \mathfrak{a}_\Lambda \Lambda)$. Put $\sigma_5 = I_2 \perp \tau$ and $x_5 = \sigma_5 x_4 = \binom{c_1, \ldots, c_n}{d_1, \ldots, d_n}$. We claim $\sigma = \sigma_5\sigma_4\sigma_3\sigma_2\sigma_1$ solves our problem. First $\sigma \in H(E_n(\mathfrak{a}_\Lambda A))EU^{\lambda}_{2n}(\mathfrak{a}_\Lambda A, \mathfrak{a}_\Lambda \Lambda)$ so $c_n \equiv a_n \bmod \mathfrak{a}_\Lambda A$. Hence $Ac_n + L + \mathfrak{a}_\Lambda A = A a_n + L + \mathfrak{a}_\Lambda A = A$, whence (4.2.4). To verify (4.2.3) note first that $c_1 = a_1''$, $d_1 = b_1''$, and $\pi\binom{c_2, \ldots, c_n}{d_2, \ldots, d_n} = \binom{c_2', \ldots, c_n'}{d_2', \ldots, d_n'}$. Hence $O(\sigma x) + L = K + L_2$ where $K = O\binom{c_2, \ldots, c_n}{d_2, \ldots, d_n}$. By the construction of τ' above we have $\pi(K + L_2) = A'$, so $K + L_2 + \mathfrak{a}_2 A = A$. But $\mathfrak{a}_2 = \mathrm{ann}_{R_0}(A/L_2)$ so $\mathfrak{a}_2 A \subset L_2$, whence $K + L_2 = A$. This concludes the proof of

Theorem (4.2).

(4.5) <u>The restricted sympletic case.</u> We assume now that $A = R$, a commutative ring with trivial involution. Then R_0 is the subring of A generated by all squares; we have $2A \subset R_0$. We further assume $\lambda = -1$. Thus Λ is an R_0-module between 2A and A, the case $\Lambda = A$ being the symplectic case. Finally we assume the ideal $\mathfrak{a}_\Lambda$ in R_0 satisfies $\mathfrak{a}_\Lambda A \subset \Lambda$. This implies that $\Lambda_{\mathfrak{m}} = A_{\mathfrak{m}}$ for all $\mathfrak{m}$ in $X = \max(R_0)$ not in $V(\mathfrak{a}_\Lambda)$. In particular Λ is ample in A outside of $V(\mathfrak{a}_\Lambda)$.

As in (3.1) we put $d = \dim X$ and $d_\Lambda = \dim V(\mathfrak{a}_\Lambda)$.

Further we shall write Sp in place of U^λ and Ep in place of EU^λ in the present setting.

(4.6) THEOREM. <u>Keep the notation of</u> (4.5). <u>Assume that either</u> (i) $\Lambda = A$ <u>and</u> $2n \geq d + 2$ <u>or</u> (ii) $2n \geq d + 3$ <u>and</u> $n \geq d_\Lambda + 2$.

1) <u>If</u> $x \in H(A^n)$ <u>is unimodular there is a</u> $\sigma \in H(E_n(A))Ep_{2n}(A,\Lambda)$ <u>such that</u> $\sigma x = \binom{0,\ldots,0,1}{0,\ldots,0,b}$, <u>where</u> $b \equiv q_{An}(x,x) \bmod \Lambda$.

2) $H(E_n(A))Ep_{2n}(A,\Lambda)$ <u>acts transitively on the set of hyperbolic elements, on the set of hyperbolic pairs, and on the set of hyperbolic planes in</u> $H(A^n)$.

3) $Ep_{2n}(A,\Lambda)$ __is a normal subgroup of__ $Sp_{2n}(A,\Lambda)$.

As in the proof of Thm. (3.4) assertion 1) implies assertions 2) and 3), so we shall prove only 1).

__Proof in case__ (i). If $x \in H(A^n)$ is unimodular we apply Cor. (4.3) to transform x to $\sigma x = \binom{c_1,\ldots,c_n}{d_1,\ldots,d_n}$ so that $d_n = 1 + bc_n$ for some $b \in A = \Lambda$. Further applying $X_-(-be_{nn})$ we can make $d_n = 1$. Then it is straightforward to transform the resulting element to $\binom{0,\ldots,0,1}{0,\ldots,0,0}$ using elements of $H(E_n(A))\cdot Ep_{2n}(A,\Lambda)$.

__Proof in case__ (ii). As a preliminary we replace $\mathfrak{A}_\Lambda$ by its intersection with a finite number of maximal ideals (this will not increase d_Λ) so that $V(\mathfrak{A}_\Lambda)$ meets each irreducible component of X. Then if $\mathfrak{A}$ is an ideal of R_0 such that $\mathfrak{A} + \mathfrak{A}_\Lambda = R_0$ the set $V(\mathfrak{A})$ is disjoint from $V(\mathfrak{A}_\Lambda)$, hence contains no irreducible component of X, and hence has dimension $<$ dim X.

Let $x \in H(A^n)$ be unimodular. We first reduce everything modulo $\mathfrak{A}_\Lambda$, whereupon the condition $n \geq d_\Lambda + 2$ allows us to invoke Thm. (3.4). This furnishes an element of $H(E_n(A/\mathfrak{A}_\Lambda A))\ Ep_{2n}(A/\mathfrak{A}_\Lambda A, \Lambda/\Lambda \cap \mathfrak{A}_\Lambda A)$, which lifts to an element $\sigma_1 \in H(E_n(A))Ep_{2n}(A,\Lambda)$, so that

$$x_1 = \sigma_1 x = \begin{pmatrix} a_1,\ldots,a_n \\ b_1,\ldots,b_n \end{pmatrix}$$

$$\equiv \begin{pmatrix} 0,\ldots,0,1 \\ 0,\ldots,0,b_n \end{pmatrix} \text{ mod } \mathfrak{a}_\Lambda A.$$

Since $2n \geq d + 3 \geq d + 2$ we can invoke Cor. (4.3) to obtain a $\sigma_2 \in H(E_n(A))Ep_{2n}(A,\Lambda)$ such that $x_2 = \sigma_2 x_1 = \begin{pmatrix} c_1,\ldots,c_n \\ d_1,\ldots,d_n \end{pmatrix}$ with $c_n \equiv a_n \equiv 1 \text{ mod } \mathfrak{a}_\Lambda A$ and $d_n \in Ac_n$. Put $A' = A/Ac_n$ and $R_0' = R_0/\mathfrak{a}$ where $\mathfrak{a} = Ac_n \cap R_0$. Since $c_n \equiv 1 \text{ mod } \mathfrak{a}_\Lambda A$ and A is integral over R_0 it follows that $\mathfrak{a} + \mathfrak{a}_\Lambda = R_0$. It follows that the image Λ' of Λ in A' is all of A' and further that $d' = \dim \max(R_0') = \dim V(\mathfrak{a}) < d$ (see the first paragraph of the proof above). We propose to apply case (i) to the image y' of $y = \begin{pmatrix} c_1,\ldots,c_{n-1} \\ d_1,\ldots,d_{n-1} \end{pmatrix}$ in $H(A'^{(n-1)})$. Since $d_n \in Ac_n$ it follows that y' is unimodular. Further we have $2(n-1) = 2n - 2 \geq d + 3 - 2 \geq d' + 2$. We therefore obtain from case (i) a $\tau' \in H(E_{n-1}(A'))Ep_{2(n-1)}(A',\Lambda')$, which lifts to a $\tau \in H(E_{n-1}(A))\cdot Ep_{2(n-1)}(A,\Lambda)$ such that $\sigma_3 = \tau \perp I_2$ transforms x_2 to $x_3 = \sigma_3 x_2 = \begin{pmatrix} c_1',\ldots,c_n' \\ d_1',\ldots,d_n' \end{pmatrix}$ with $c_n' = c_n$, $d_n' = d_n$, and $c_{n-1}' \equiv 1 \text{ mod } Ac_n$. Now it is straightforward to find an $\epsilon \in En(A)$ such that $\epsilon(c_1',\ldots,c_n') = (0,\ldots,0,1)$, whence

$x_4 = H(\epsilon)x_3 = \binom{0,\ldots,0,1}{r_1,\ldots,r_n}$. Finally $X_-(\sum_{i=1}^{n-1}(r_i e_{ni} + r_i e_{in}))$ $\in Ep_{2n}(A,\Lambda)$ transforms x_4 to $x_5 = \binom{0,\ldots,0,1}{0,\ldots,0,r_n}$. This concludes the proof of 1), and hence of Thm. (4.6).

(4.7) Remarks. Theorem (4.6) has corollaries analogous to the corollaries (3.5), (3.6), (3.7), and (3.8) of Theorem (3.4). We formulate only the latter two. The first two will be proved below in a sharper form in the symplectic case.

(4.8) COROLLARY. *Put* $n_0 = \frac{d}{2}$ *if* $\Lambda = A$, *and* $n_0 = \max(\frac{d+1}{2}, d_\Lambda + 1)$ *otherwise. Then* $Sp_{2n}(A,\Lambda) \to KSp_1(A,\Lambda)$ *is surjective for* $n \geq n_0$.

(4.9) COROLLARY. *Let* $(\mathfrak{q},\Gamma)$ *be a unitary ideal in* (A,λ,Λ). *Then* $Sp_{2n}(\mathfrak{q},\Gamma) \to KSp_1(\mathfrak{q},\Gamma)$ *is surjective for* $n \geq n_0$, *provided we take* $n_0 = \frac{d}{2}$ *only when* $\Lambda = A$ *and* $\Gamma = \mathfrak{q}$.

The proofs of these corollaries are entirely analogous to those of Corollaries (3.7) and (3.8). The proviso in (4.9) when $n_o = d/2$ is to guarantee that $\Lambda(\Gamma) = A(\mathfrak{q})$ in the fibre product unitary ring $A(\mathfrak{q},\Gamma)$.

(4.10) COROLLARY. *If* $d \leq 1$ *and* $d_\Lambda \leq 0$ *then* $Sp_2(\mathfrak{q},\Gamma) \to KSp_1(\mathfrak{q},\Gamma)$ *is surjective and* $(Sp_{2n}(A,\Lambda), Sp_{2n}(\mathfrak{q},\Gamma)) \subset Ep_{2n}(\mathfrak{q},\Gamma)$. *If* $d \leq 0$, $\Lambda = A$, *and* $\Gamma = \mathfrak{q}$ *then* $KSp_1(\mathfrak{q},\Gamma) = 0$.

All but the commutator relation follows from Cor. (4.9) since $n_0 \leq 1$, resp., $n_0 = 0$

in the two cases cited. The commutator relation results from this just as in the proof of Cor. (3.8).

Remark. In the paper of J. Milnor, J.-P. Serre, and the author, "Solution of the congruence subgroup problem for $SL_n (n \geq 3)$ and $Sp_{2n} (n \geq 2)$," Publ. IHES n°33 (1967) pp. 59-137, the following result (Prop. 13.2) is announced and used:

Let A be a Dedekind ring, let $\mathfrak{q}$ be an ideal of A, and suppose $n \geq 2$.

a) $Sp_{2n}(A, \mathfrak{q}) = Sp_2 (A, \mathfrak{q}) Ep_{2n}(A, \mathfrak{q})$

b) $Ep_{2n}(A, \mathfrak{q}) \supset (Sp_{2n}(A), Sp_{2n}(A, \mathfrak{q})) \supset Ep_{2n}(A, \mathfrak{q}'\mathfrak{q})$, *where* $\mathfrak{q}' = A$ *if* $n \geq 3$, *and* $\mathfrak{q}'$ *is generated by all* $t^2 - t \ (t \in A)$ *if* $n = 2$.

c) *Every subgroup N of finite index in* $Sp_{2n}(A)$ *contains* $Ep_{2n}(A, \mathfrak{q})$ *for some* $\mathfrak{q} \neq 0$ (if A is not finite).

The six year old manuscript in which this was proved was the starting point for its generalizations in Bak's thesis [1], but the result above does not quite follow from any yet published material. Therefore it may be worth pointing out how to deduce it from the results presented here.

Corollary (4.10) (applied with $\Lambda = A$) implies a), the first inclusion in b), and, when $n \geq 3$, the second inclusion in b).

The case $n = 2$ is easily handled using the formulas in the proof of Prop. (5.1) of Ch. II.

To prove c) one may assume N is normal in $Sp_{2n}(A)$. Then it follows from Prop. (7.4) in Ch. II that $N \cap X_+(\Lambda_n) = X_+((\mathfrak{q},\Gamma)_n)$ for some $(\mathfrak{q},\Gamma)$, and clearly $\mathfrak{q} \neq 0 \neq \Gamma$ if A is not finite. It then follows easily that N contains $Ep_{2n}(\mathfrak{q},\Gamma)$, and this contains $Ep_{2n}(A\Gamma,A\Gamma)$, whence c).

(4.11) <u>The symplectic case</u>. In addition to the assumptions of (4.5) we now assume $\Lambda = A$, so we are in the symplectic case. If $P(=(P,[h]))$ is a quadratic module we here write $Sp(P)$ in place of $U(P)$. In case $P = H(Q)$ we similarly write $Sp(H(Q))$ for $U(H(Q))$ and $Ep(H(Q))$ for $EU(H(Q))$. When $Q = A^n$ we thus obtain groups denoted $Sp_{2n}(A)$ and $Ep_{2n}(A)$. It follows from Ch. II, Prop. (5.1) that:

(4.11.1) $$H(E_n(A)) \subset Ep_{2n}(A) \text{ for all } n \geq 1.$$

With $(P,[h])$ as above, $[h]$ is determined by $S_{-1}(h) = h - \bar{h}$ together with the quadratic function $x \mapsto [h(x,x)] \in A/\Lambda$. Since $\Lambda = A$ the latter function can be ignored, so $(P,[h])$ is equivalent to the pair consisting of P together with the alternating form $g = h - \bar{h}$. Note in particular that every element of P is isotropic. Recall that $(P,[h])$ (or (P,g)) is called a <u>symplectic module</u> if P is finitely generated and projective and g is non singular.

Suppose L is a direct summand of P and that P (i.e. g) is non singular. Then if L is totally isotropic it follows from Ch. I, (4.10.1) that the inclusion $L \to P$ extends to a morphism $H(L) \to P$, so that $P \cong H(L) \perp P'$ for some P'. If L is an invertible A-module then L is locally monogene and hence L is automatically totally isotropic. We conclude therefore:

(4.11.2) <u>If</u> P <u>is non-singular and if</u> L <u>is an invertible direct summand of</u> P <u>then</u> $P \cong H(L) \perp P'$ <u>for some</u> P'.

It follows from this that:

(4.11.3) <u>If</u> L <u>and</u> L' <u>are invertible</u> A-<u>modules then</u> $H(L) \cong H(L')$ $\Leftrightarrow$ L <u>is isomorphic to a direct summand of</u> $L' \oplus \overline{L'}$. <u>In particular</u> $H(L) \cong H(A) \Leftrightarrow L$ <u>can be generated by</u> ≤ 2 <u>elements</u>.

The last condition always holds if max(A) is a noetherian space of dimension ≤ 1 [3], Ch. IV, Cor. (3.8). Whence:

(4.11.4) <u>If</u> max(A) <u>is a noetherian space of dimension</u> ≤ 1 <u>then</u> $H(L) \cong H(A)$ <u>for all invertible</u> A-<u>modules</u> L.

(4.12) Serre's theorem; symplectic version. For the rest of this section we assume max(A) is a finite union of noetherian spaces each of dimension $\leq$ d. An A-module P is said to have f-rank $\geq$ n if for all $\mathfrak{m} \in$ max(A) there is an epimorphism $P_{\mathfrak{m}} \to A^n_{\mathfrak{m}}$.

THEOREM. *Let P be a finitely presented A-module of f-rank $\geq d + 1$. Let g be a non-singular alternating form on P. Then (P,g) contains a hyperbolic plane H(A).*

It suffices, by (4.11.2), to produce a direct summand $L \cong A$ of P. Such an L is furnished by a theorem of Serre [3], Ch. IV, Cor. (2.7).

Remark. Serre's theorem requires only that P be a direct summand of a direct sum of finitely presented modules; e.g. P may be any projective module of f-rank $\geq d + 1$ in the theorem.

(4.12.1) COROLLARY. *Suppose spec (A) is connected. Let (P,g) be a non-zero symplectic module. Then P has even rank, say 2n.*

(i) *If $d \leq 1$ then $(P,g) \cong H(A^n)$.*

(ii) *If $d \leq 2$ then $(P,g) \cong (Q,h) \perp H(A^{n-1})$ where rank Q = 2 and $\Lambda^2 Q \cong A$.*

Only the fact that $\Lambda^2 Q \cong A$ is not immediate from the theorem. This fact results because h is a free basis for the dual of Q, as one sees by localizing, for example.

(4.13) Symplectic Cancellation. Let M be an A-module equipped with an alternating form $\langle\ ,\ \rangle$. We wish to show that $Sp(M)$ acts transitively on the set of hyperbolic elements in M. For this we assume that M admits a decomposition, $M = V \perp P$ where P is a projective A-module of f-rank $\geq d + 2$ such that $\langle\ ,\ \rangle$ has a non singular restriction to P. It then follows from Thm. (4.12) (and the following remark) that $P = W \perp H$ where H is a hyperbolic plane, say spanned by a hyperbolic pair p_0, q_0: $\langle q_0, p_0\rangle = 1$. We thus have $M = V \perp W \perp H = U \perp H$ where $U = V \perp W$. Moreover W is a projective module of f-rank $\geq d$ on which $\langle\ ,\ \rangle$ is non-singular.

Suppose $u, v \in M$ with $\langle u,v\rangle = 0$ and $a \in A$. Then we have the transvection $\sigma_{u,a,v}$ sending $x \in M$ to $x + u\langle v,x\rangle + v\langle u,x\rangle + ua\langle u,x\rangle$ (see Ch. I, (5.1)). Moreover we have

$\sigma_{u,a,v} = \sigma_{u,a,0} \circ \sigma_{u,0,v}$.

Let $G \subset Sp(M)$ denote the subgroup generated by all transvection $\sigma_{p,a,u}$ with $p = p_0$ or q_0, $a \in A$, and $u \in p^{\perp} = U \oplus pA$.

(4.14) THEOREM. *Keep the notation and assumptions of* (4.13). *The group* G *acts transitively on the set of hyperbolic elements, on the set of hyperbolic pairs, and on the set of hyperbolic planes in* M.

An element $x \in M$ is hyperbolic, i.e. part of a hyperbolic pair, if and only if $\langle x,M \rangle = A$. We claim we can transform such an x into p_0 by an element of G.

Put $x = x_U + p_0 a + q_0 b \in U \oplus p_0 A \oplus q_0 A$. Then $\langle q_0,x \rangle = a$ and $\langle p_0,x \rangle = -b$. If $w \in U$ and $c \in A$ then we have

$$(4.14.1) \quad \sigma_{q_0,c,w}(x) = (x_U + wa) + p_0 a + q_0(b + \langle w,x_U \rangle + ca)$$

We now prove the above claim, distinguishing various cases.

Case 1: $a = 1$. Choosing $w = -x_U$ in (4.14.1) we can further transform x into $x' = p_0 + q_0 b'$. Then $\sigma_{q_0,-b',0}$ transforms x' to p_0.

Remark. The transformations used here both fix q_0.

Case 2: $x_U + p_0 a$ is a hyperbolic element. This means that $\langle U,x_U \rangle + Aa = A$, so that we can solve $1 - a - b = \langle w,x_U \rangle + ca$ for some $w \in U$ and $c \in A$. Then by (4.14.1) we can transform x to $(x_U + wa) + p_0 a + q_0(1-a)$. Now $\sigma_{p_0,-1,0}$ transforms the latter to $(x_U + wa) + p_0 + q_0(1-a)$, which belongs to case 1.

To complete the proof we shall transform x to achieve case 2, or rather its analogue: $x_U + q_0 b$ is hyperbolic. By the symmetry between p_0 and q_0 this difference is unimportant.

Put $x_U = x_V + x_W \in V \oplus W = U$. Since x is hyperbolic we have $A = \langle M,x \rangle = \langle V,x_V \rangle + \langle W,x_W \rangle + Aa + Ab$. Hence

$A = \langle V, x_V\rangle + \langle W, x_W\rangle + Aa^2 + Ab$ also. Note that $W \oplus q_0 A$ is projective of f-rank $> d$. It follows therefore from [3], Ch. IV, Thm. (3.1) that there is a homomorphism $h: p_0 A \to W \oplus q_0 A$ such that $\langle V, x_V\rangle + O(x_W + q_0 b + h(p_0 a^2)) = A$. Put $h(p_0) = w_0 + q_0 c$ and $w = w_0 a$. Then $x_W + q_0 b + h(q_0 a^2) = (x_W + wa) + q_0(b+ca^2)$, which we abbreviate $w' + q_0 b'$. Since W is a non singular submodule of M we have $O(w' + q_0 b') = O(w') + Ab' = \langle W, w'\rangle + Ab'$. It follows therefore that $x_V + w' + q_0 b'$ is a hyperbolic element of M, so the proof will be completed if we can transform x into $x_V + w' + p_0 a + q_0 b'$. To do this we put $d = ca - \langle w_0, x\rangle$, so that $da = ca^2 - \langle w, x\rangle$, and put $\sigma = \sigma_{q_0, d, w}$. Then

$$\sigma x = x + w\langle q_0, x\rangle + q_0(\langle w, x\rangle + d\langle q, x\rangle) = x + wa + q_0(\langle w, x\rangle + da)$$
$$= x_V + (x_W + wa) + p_0 a + q_0(b + ca^2) = x_V + w' + p_0 a + q_0 b'.$$

Now let x, y be a hyperbolic pair in M. By what we have just proved we can transform by an element of G to achieve $y = q_0$. Since $\langle y, x\rangle = 1$ we must be in case 1 of the proof above, so (see the remark after case 1) we can transform x into p_0 while fixing q_0. Thus G acts transitively on the set of hyperbolic pairs, and hence also on the set of hyperbolic planes in M. Theorem (4.14) is now proved.

(4.15) COROLLARY. Let N and N' be A-modules equipped with alternating forms. Suppose that $N \perp H(A)$ contains a symplectic module of f-rank $\geq d + 2$. Then for any symplectic module Q we have:

$$N \perp Q \cong N' \perp Q \Longrightarrow N \cong N'.$$

As in the proof of Cor. (3.6) one reduces to showing that, if $M = N \perp H(A)$, the group $Sp(M)$ acts transitively on the set of hyperbolic planes in M. But this follows from Thm. (4.14).

(4.16) COROLLARY. Suppose $d \leq 2$. Let N, N' and Q be any symplectic modules. Then

$$N \perp Q \cong N' \perp Q \Longrightarrow N \cong N'.$$

We can reduce easily to the case where N has constant rank, say $2n$, with $n \geq 1$. Then $N \perp H(A)$ has rank $2n + 2 \geq 4 \geq d + 2$, so the result follows from Cor. (4.15).

References

[1] A. Bak, The stable structure of quadratic modules, Thesis, Columbia University, 1969.

[2] H. Bass, J. Milnor, and J. P. Serre, Solution of the congruence subgroup problem for $SL_n (n \geq 3)$ and $Sp_{2n} (n \geq 2)$. Publ. I.H.E.S. n°33 (1967) 59-137.

[3] H. Bass, Algebraic K-theory, W. A Benjamin, New York, 1968.

[4] N. Bourbaki, Formes sesquilinéaires et formes quadratiques (Algèbre, Ch. IX), Hermann, Paris, 1959.

[5] A. Frölich and A.M. McEvett, Forms over rings with involution, Jour. Alg. 12 (1969) 79-104.

[6] M. Hall, Theory of groups, MacMillan, New York, (1959).

[7] M. Karoubi, Périodicité de la K-théorie hermitienne. CR. Acad. Sc. Paris t 273 (1971) pp. 599-602, 802-805, and 840-843.

[8] M. Knebusch, Grothendieck und Wittringe von nichtausgearteten symmetrischen Bilinearformen, Sitzb. Heidelberg Akad. Wiss., Math. naturw, Klasse (1969/70) 3, 1-69.

[9] J. Milnor, Introduction to Algebraic K-theory, Ann. Math. Studies , 72 , Princeton (1972).

[10] R. Sharpe, On the structure of the unitary Steinberg groups, Thesis, Yale University, 1970.

[11] M. Stein, Relativizing functors on rings and algebraic K-theory, Jour. Alg. 19 (1971) 140-152.

[12] R. Wagner, Some Witt cancellation theorems, Amer. Jour. Math. (to appear).

[13] C.T.C. Wall, On the axiomatic foundations of the theory of Hermitian forms, Proc. Camb. Phil. Soc., 67 (1970) 243-250.

[14] C.T.C. Wall, Surgery on compact manifolds, Academic Press, New York (1971).

[15] L.I. Wasserstein, Stability of unitary and orthogonal groups over rings with involution, Math. Sbornik, vol 81 (123), n°3, 1970.

FOUNDATIONS OF ALGEBRAIC L-THEORY

C.T.C. Wall

In my book [17] I introduced certain algebraic functors L_n which were then used to express the obstruction to doing surgery. I did not give a full account of the algebra, which at that time I did not yet have in sufficiently good shape. This paper (intended as my definitive account) is designed to fill this gap. A more immediate reason for writing it was the need for adequate foundation material for the papers [18] , and indeed it has enabled me to make my calculations much more effective.

I have presented this work as a sequel to the short paper [16].
I am grateful to Andrew Ranicki for sending a preprint to [12] and for showing me the proof of Lemma 7 below. The relation of this paper to other foundation material on the subject is discussed in § 5 below.

§1 Preliminary definitions

For any category $\mathcal{C}$ with product (in the sense of Bass [4, p. 344] we define $k\mathcal{C}$ to be the monoid of isomorphism classes of objects of $\mathcal{C}$, $K_0(\mathcal{C})$ its universal (Grothendieck) group. Similarly, $K_1(\mathcal{C})$ is the universal group for functions on the automorphisms of $\mathcal{C}$ to (additive) abelian groups which are additive for sums and composites. If A_1, $A_1 \oplus A_2$, $A_1 \oplus A_2 \oplus A_3$, ... defines a cofinal sequence in $k(\mathcal{C})$, we can define Aut $\mathcal{C}$ as the direct limit of

$$\mathrm{Aut}_{\mathcal{C}}\, A_1 \subset \mathrm{Aut}_{\mathcal{C}}\, (A_1 \oplus A_2) \subset \dots ,$$

and $K_1(\mathcal{C})$ is then its commutator quotient group.

For any ring R, we write $\mathcal{P}(R)$ for the category of finitely generated projective (right) R -modules. There is a standard meaning for $\oplus$ here. The groups $K_0\mathcal{P}(R)$, $K_1\mathcal{P}(R)$ are written simply as $K_0(R)$, $K_1(R)$. Any automorphism of such a module thus has a 'determinant' in $K_1(R)$; in particular, so does any nonsingular matrix over R.

Now let (R, α, u) be an antistructure in the sense of [18] - i.e. α is an antiautomorphism and $u \in R^{\times}$ a unit of R such that

$$x^{\alpha^2} = u x u^{-1} \qquad \text{for all } x \in R ,$$
$$u^{\alpha} = u^{-1} .$$

For an R-module M, the space $Q_{(\alpha,u)}(M)$ of (α, u)-quadratic forms on M was defined in [16], as was the concept of nonsingular form. We will write θ for a quadratic form and (b_{θ}, q_{θ}) for the corresponding [16 Theorem 1] element of $\mathrm{Quad}_{(\alpha,u)}(M)$. We define $\mathcal{Q}(R, \alpha, u)$ to be the category whose objects are pairs (P, θ), P a finitely generated projective R-module, $\theta \in Q_{(\alpha,u)}(P)$ nonsingular; and whose morphisms $(P, \theta) \to (P', \theta')$ are the isomorphisms $P \to P'$ which carry θ to θ'. [A possible variant is to regard representatives $\phi \in S_{\alpha}(M)$ of θ as defining different objects, but still have morphisms as above.] An object of this category is called a quadratic module. There is an obvious notion of (orthogonal) direct sum. Forgetting the quadratic structure defines a functor

$$F : \mathcal{Q}(R, \alpha, u) \to \mathcal{P}(R) .$$

We also have a hyperbolic functor

$$H = H_{\alpha} : \mathcal{P}(R) \to \mathcal{Q}(R, \alpha, u) .$$

This was defined on objects on [16, p.249] : $H(M) = M \oplus M^{\alpha}$ as module; θ is the equivalence class of the pairing

$$(m, f) . (m', f') = f(m').$$

The definition on morphisms is obvious: $H(f) = f \oplus (f^{\alpha})^{-1}$ it is clear that this does define a functor. It will be important for us to recognise hyperbolic modules; as a preliminary, if (N, θ) is a quadratic module, we define a submodule $M \subset N$ to be a subkernel [17] (alias Lagrangian subspace [11] [12]) if the identity on M extends to an isomorphism of (N, θ) on $H(M)$. Clearly a necessary condition for this is that M be isotropic, i.e. that $q_\theta(M) = 0$ and $b_\theta(M \times M) = 0$. Note that M^α is also a subkernel of M. Indeed, the map

$$M \oplus M^\alpha = H(M) \to H(M^\alpha) = M^\alpha \oplus M^{\alpha\alpha}$$

given by $(x, f) \mapsto (f, \lambda_u\ \omega_{M,\alpha}(x))$ is an isometry. For the sesquilinear form in $H(M^\alpha)$ is

$$\lambda_u \omega_{M,\alpha}(x)\ (f') = u\ \omega_{M,\alpha}(x)(f')$$
$$= u(f'(x))^{\alpha^{-1}} = f'(x)^\alpha u$$

which comes from the defining form $f(x')$ for $H(M)$ by applying T_u. In general, two subkernels E, F of N are complementary (alias Hamiltonian complements) if there is an isomorphism of (N, θ) on $H(E)$ which is the identity on E and takes F to E^α. We can weaken this condition as follows.

Lemma 1 *Let* (N, θ) *be a nonsingular quadratic module, and* E, F *isotropic subspaces with* $E + F = N$. *Then* E *and* F *are complementary subkernels.*

Proof Since $E \cap F$ is orthogonal to E and to F, it is orthogonal to $E + F = N$, hence is zero by nonsingularity (it lies in $\operatorname{Ker}(Ab_\theta) = \{0\}$). Hence, additivity, $N = E \oplus F$. The isomorphism

$$E \oplus F = N \xrightarrow{Ab_\theta} N^\alpha = E^\alpha \oplus F^\alpha$$

has zero components $E \to E^\alpha$, $F \to F^\alpha$, hence yields isomorphisms $E \to F^\alpha$, $F \to E^\alpha$. Identifying F with E^α by the isomorphism yields $N = E \oplus E^\alpha = H(E)$, an additive isomorphism which (we readily verify) is also an isometry.

<u>Lemma 2</u> <u>Let</u> (N, θ) <u>be a nonsingular quadratic module,</u> $E \subset N$ <u>an isotropic projective submodule. Then</u> E <u>is a subkernel if and only if the map</u> $N/E \overset{b'}{\to} E^{\alpha}$ <u>induced by</u> $Ab_{\theta} : N \to N^{\alpha}$ <u>is an isomorphism.</u>

<u>Proof</u> The condition is clearly necessary: suppose it satisfied. Then N/E is projective, so the extension N of E by it splits, and we can find an additive complement, M say, to E, and identify E with the dual M^{α}.

Then N is additively isomorphic to $M \oplus M^{\alpha}$, and θ is given by a sesquilinear form

$$(m, f) . (m', f') = \xi(m, m') + f(m') .$$

(This can be seen from our description [16 , p.246] of $Q_{(\alpha,u)}$ of a direct sum.) We now see that if $\zeta \in S_{\alpha}(M)$, and we embed M in $M \oplus M^{\alpha}$ by the graph of ζ, the induced quadratic form comes from the sesquilinear form $\xi + \zeta$. Thus if we choose $\zeta = -\xi$, we obtain an isotropic subspace, complementary to M^{α}.

This last argument also yields the

<u>Corollary 1</u> <u>The subkernels of</u> $H(M)$ <u>complementary to</u> M^{α} <u>are the graphs of the</u> $Ab_{\theta} : M \to M^{\alpha}$ <u>corresponding to the</u> $\theta \in Q_{(\alpha,-u)}(M)$.

For here, $\xi = 0$, and ζ determines $0 \in Q_{(\alpha,u)}(M)$ if and only if

$$\zeta = \mathrm{Im}(1 - T_u) = \mathrm{Im}(1 + T_{-u})$$

is the bilinearisation of an $(\alpha, -u)$ quadratic form.

<u>Corollary 2</u> <u>Any automorphism of</u> $H(M)$ <u>leaving</u> M^{α} <u>pointwise fixed is given by</u> $x \in M \mapsto (x, Ab_{\theta}(x))$ <u>for some</u> $\theta \in Q_{(\alpha,-u)}(M)$.

For if $x \mapsto (p, q)$ we find $p = x$ since (p, q) and x have the same inner products with each element of M^{α}. The conclusion now follows from the preceding.

We recall [16 Theorem 3]; that if θ is nonsingular, $(N, \theta) \oplus (N, -\theta) \cong H(N)$. The simplest way to see this is now to use Lemma 2 to show that the diagonal $\Delta(N) \subset N \oplus N$ is a subkernel. The special case when $(N, \theta) = H(M)$ will be important below.

We will need to study based modules. A based module is a pair (M, v) where M is a free R-module and v an equivalence class of free (ordered) bases of M, two bases being equivalent if the automorphism of M taking one to the other has determinant $0 \in K_1(R)$. We can regard v as a sort of volume element on M. Now define $\mathcal{B}(R)$ as the category whose objects are based modules (M, v) and morphisms are based isomorphisms (i.e. preserving preferred classes of bases). There is an obvious definition of sum in $\mathcal{B}(R)$, but it is not commutative (permutation matrices can have determinant -1). Hence we restrict to the subcategory $\mathcal{B}_0(R)$ of modules of even rank.

More interesting is the category $\mathcal{B}\mathcal{Q}(R, \alpha, u)$ of based quadratic modules, i.e. triples (N, v, θ) where (N, v) is a based module and (N, θ) a quadratic module. A morphism here is an isomorphism class of modules respecting both structures. Again we have a direct sum, which behaves well on the subcategory $\mathcal{B}_0\mathcal{Q}(R, \alpha, u)$ of modules of even rank. There is an obvious forgetful functor $F : \mathcal{B}\mathcal{Q}(R, \alpha, u) \to \mathcal{B}(R)$, but before we can define a hyperbolic functor, we must discuss duality in $\mathcal{B}(R)$: this needs some care.

We recall from [16] that for α an antiautomorphism of R and M a right R-module, the dual module M^α is $\mathrm{Hom}_R(M, R)$ with module structure defined by

$$fr(m) = r^\alpha f(m);$$

that the natural map of M to its double dual is $\omega_{M,\alpha} : M \to (M^{\alpha})^{\alpha^{-1}}$, where

$$\omega_{M,\alpha}(m)(f) = f(m)^{\alpha^{-1}} ;$$

and that if (R, α, u) is an antistructure, there is an isomorphism $\lambda_u : M^{\alpha^{-1}} \to M^{\alpha}$ given by

$$\lambda_u(f)(m) = u^{-1} f(m) .$$

If $e_1, \dots, e_n$ is a free basis of M, the 'dual basis' $e_1^*, \dots, e_n^*$ of $\mathrm{Hom}_R(M, R)$ is defined by

$$e_i^*(e_j) = \delta_{ij} \qquad \text{(Kronecker delta).}$$

If we identify this with $M^{\alpha^{-1}}$ and with M^{α}, however, the isomorphism λ_u does not preserve the class of this basis. I thus declare that for $n = 2k$, a preferred base of M^{α} shall be $e_1^*, e_2^* u^{-1}, \dots, e_{2k-1}^*, e_{2k}^* u^{-1}$, and one of $M^{\alpha^{-1}}$ is $e_1^* u, e_2^*, \dots, e_{2k-1}^* u, e_{2k}^*$. Then λ_u preserves preferred bases and so (up to equivalence) does $\omega_{M,\alpha}$. For the case n odd, we do not define the concept of dual preferred base: ad hoc definitions can be found in special cases, but are not invariant under Morita equivalences (c.f. discussion in [18, II]).

We now define the hyperbolic functor $H : \mathcal{B}_0(R) \to \mathcal{B}\mathcal{Q}(R, \alpha, u)$: it suffices to describe the case of rank 2. If e_1, e_2 is a base of M and e_1^*, e_2^* as above, we find

$$b_\theta(e_1, e_1^*) = b_\theta(e_2, e_2^*) = u , \; b_\theta(e_1^*, e_1) = b_\theta(e_2^*, e_2) = 1 ,$$

b_θ vanishes on other pairs of basis elements and q_θ on all basis elements. We will usually use the base $f_1 = e_1^* u^{-1}$, $f_2 = e_2^* u^{-1}$, but our preferred base (for M^{α}) is $e_1^*, e_2^* u^{-1}$.

We now call (M, v) a <u>based subkernel</u> of $H(M, v)$ and M^{α} (with the above base) a <u>complementary based subkernel</u> (we see, as in the unbased case, that it __is__ a based subkernel). As before, there is a recognition principle: if E, F are complementary subkernels of (N, θ), bases for E, F are complementary iff (i) they are dual in the above sense and (ii) they combine to give a preferred base of N.

If (M, θ) is an object of $B_0\mathcal{Q}(R, \alpha, u)$, the homomorphism $Ab_\theta : M \to M^\alpha$ associated to b_0 is now a map of based modules, hence has a well defined determinant in $K_1(R)$: we call this the discriminant of (M, θ), $\delta(M, \theta)$. We write $\mathcal{L}(R, \alpha, u)$ for the full subcategory of forms with zero discriminant.

This completes our list of categories and functors: the algebraic K-theory of these categories is (roughly) what I mean by algebraic L-theory. Before establishing the basic relations between them, singling out the important ones and fixing notation, we next give some computations with unitary automorphisms, which will be needed for the proofs.

§2 The elementary unitary group

We begin by recalling those results from the linear case which we wish to imitate, and fixing notation. All modules will be finitely generated projective right R-modules; maps also are written on the right. For M a module, $GL(M)$ is the group of R-automorphisms of M. There are natural injections

$$GL(M) \subset GL(M) \times GL(N) \subset GL(M \oplus N)$$

which we regard as inclusions. We write GL_n for $GL(R^n)$, and GL_∞ for the union of the GL_n, with inclusions defined by

$$R^n \subset R^n \oplus R = R^{n+1} \quad .$$

Similar notations will apply below for other groups defined as functors of M. Since the R^n are cofinal in $\mathcal{P}(R)$, $K_1(R)$ is the commutator quotient group of GL_∞. For any M, we write $SL(M)$ for the kernel of the determinant map so there is an exact sequence

$$1 \to SL(M) \to GL(M) \overset{\det}{\to} K_1(R).$$

Elements of $SL(M)$ are called <u>simple automorphisms</u> of M.

Let $e_1, \ldots, e_n$ denote the standard base of R^n. For

$$r \in R, \quad 1 \leq i, \ j \leq n, \quad i \neq j,$$

let $X_{ij}(r)$ be the automorphism which leaves each $e_k (k \neq i)$ fixed, and takes e_i to $e_i + e_j r$. We call the $X_{ij}(r)$ <u>elementary transvections</u>, and write E_n for the subgroup of GL_n which they generate, and E_∞ for the union of the E_n. The X_{ij} can be expressed as commutators $[x, y] = xyx^{-1}y^{-1}$; in fact, if i, j, k are distinct,

$$[X_{ij}(r), X_{jk}(s)] = X_{ik}(sr).$$

Thus for $n \geq 3$, E_n is contained in the commutator subgroup of GL_n, and a fortiori in SL_n: indeed, it is perfect. Stably, the converse holds.

<u>Lemma 3</u> <u>(Whitehead's lemma)</u>

<u>E_∞ is the commutator subgroup of</u> GL_∞.

<u>Proof</u> We show, in fact, that the commutator subgroup of GL_n lies in E_{2n}. It is convenient to use matrix notation, with blocks of $n \times n$ matrices. Then the matrices of the form

$$\begin{pmatrix} I & A \\ 0 & I \end{pmatrix}$$

form a group, whose product is given by addition of matrices A. If A has only one nonzero element, we have an elementary transvection. Hence all such matrices belong to E_{2n} ; and similarly if the positions of 0 and A are interchanged. Now

$$\begin{pmatrix} A & 0 \\ 0 & A^{-1} \end{pmatrix} = \begin{pmatrix} I & -A \\ 0 & I \end{pmatrix}\begin{pmatrix} I & 0 \\ A^{-1} & I \end{pmatrix}\begin{pmatrix} I & -A \\ 0 & I \end{pmatrix}\begin{pmatrix} I & I \\ 0 & I \end{pmatrix}\begin{pmatrix} I & 0 \\ -I & I \end{pmatrix}\begin{pmatrix} I & I \\ 0 & I \end{pmatrix},$$

and hence belongs to $E(R^{2n})$ and, finally, so does

$$\begin{pmatrix} ABA^{-1}B^{-1} & 0 \\ 0 & I \end{pmatrix} = \begin{pmatrix} AB & 0 \\ 0 & (AB)^{-1} \end{pmatrix}\begin{pmatrix} A^{-1} & 0 \\ 0 & A \end{pmatrix}\begin{pmatrix} B^{-1} & 0 \\ 0 & B \end{pmatrix}$$

since each factor is of the above type.

There is, of course, more to be said, but the above seems the essential basis for understanding the functor $K_1(R)$. We now undertake the corresponding study in the unitary case; we work at a similar depth, but there is more to do: the results are much richer. We supposed fixed an antistructure (R, α, u) in what follows.

For (N, ϕ) a quadratic module - i.e. object of 2 (R, α, u) - write Aut (N, ϕ) for its group of automorphisms in this category. However, we write $U(M)$ for Aut $H(M)$. Write also $GI(M)$ for the subgroup of $U(M)$ of automorphisms leaving the subkernel M of $H(M)$ invariant. The subgroups of $GI(M)$ where the restriction to M of the automorphism belongs to $SL(M)$, $E(M)$ (when $M = R^n$) or is trivial are denoted respectively by

$SI(M)$, $EI(M)$ and $I(M)$. For the corresponding subgroups leaving M^{α} invariant, we use J in place of I ; also if $M = R^n$ we use a suffix n, and have GI_{∞} etc. for the appropriate limits. The hyperbolic functor induces a monomorphism $GL(M) \to GI(M)$; in fact $GI(M)$ is the semidirect product of $GL(M)$ with the normal subgroup $I(M)$; and correspondingly for SI, EI .

The subgroup $EU(M)$ of elementary automorphisms of $H(M)$ is that generated by $I(M)$ and $J(M)$. As in the linear case, we can also give explicit generators. If $e_1, \ldots, e_n$ is (again) the standard base of R^n, we extend by $f_1, \ldots, f_n$ to a base of $H(R^n)$ such that $e_i \cdot f_j = \delta_{ij}$ (the b_θ for a hyperbolic module is denoted by a dot ; we write q for q_θ) and $q(e_i) = q(f_i) = 0$. For $i \neq j$, $1 \leq i, j \leq n$, $r \in R$, we define $E_{ij}(r)$ to be the identity on all basis vectors except

$$e_i \to e_i + f_j r \qquad\qquad e_j \to e_j - f_i r^{\alpha} u$$

and $F_{ij}(r)$ on all save for

$$f_i \to f_i + e_j r \qquad\qquad f_j \to f_j - e_i u^{-1} r^{\alpha} ,$$

so that $E_{ij}(r) \in J_n$, $F_{ij}(r) \in I_n$. Then for i, j, k distinct,

$$[E_{ij}(r), F_{jk}(s)] = H(X_{ik}(-sr)).$$

Since the $X_{ij}(-sr)$ generate E_n , it follows that for $n \geq 3$,

$$EI_n \subset EU_n .$$

We also write

$$\Sigma_{ij} = E_{ij}(1) F_{ij}(u^{-1}) E_{ij}(1) :$$

under Σ_{ij},

$$e_i \to f_j \to - e_i \quad \text{and} \quad e_j \to - f_i u \to - e_j .$$

Next we observe that

$$[H(X_{12}(1)),\ E_{12}(r)]$$

acts as the identity on all basis elements except

$$e_1 \to e_1 + f_1(r - r^{\alpha}u).$$

<u>Lemma 4</u> EU_n <u>is generated (for</u> $n \geq 3$) <u>by the</u> $E_{ij}(r)$ <u>and</u> $F_{ij}(r)$ <u>for</u> $1 \leq i, j \leq n, r \in R$.

<u>Proof</u> It suffices (by symmetry) to show that J_n is contained in the subgroup with these generators. By corollary 2 to lemma 2, each element of J_n is of the form

$$e_i \to e_i + \Sigma_j f_j (b_{ij} - b^{\alpha}_{ji} u),\ f_i \to f_i$$

for some matrix (b_{ij}); Since composition in this group corresponds to matrix addition, it is enough to consider matrices with only one nonzero entry b_{ij}. If $i \neq j$, this is $E_{ij}(b_{ij})$; the case $i = j$ is dealt with by the calculation preceding this lemma.

<u>Corollary</u> <u>For</u> $n \geq 3$, EU_n <u>is perfect</u>.

For, as with E_n, its generators are commutators :

$$E_{ij}(r) = [H(X_{jk}(1)),\ E_{ik}(r)] \ .$$

Our next objective is to show that EU_{∞} is the commutator subgroup of U_{∞} .

<u>Lemma 5</u> <u>If</u> $A \in U(M)$ <u>we can find</u> $A' \in U(M')$ <u>with</u> $A \oplus A' \in EU(M \oplus M')$.

<u>Proof</u> Since $A \oplus A^{-1} \in E(H(M) \oplus H(M))$, it is enough to prove the result under the extra hypothesis that $A \in E(H(M))$. By adding further modules, we may suppose M free of even rank.

Define A' to be the conjugate of A by the (non-unitary) automorphism ι which is 1 on M and -1 on M^{α} : then also $A' \in U(M) \cap E(H(M))$. Now $A \oplus A'$ leaves invariant the subkernel Δ of $H(M) \oplus H(M)$ defined as the graph of ι, and the induced automorphism of Δ belongs to $E(\Delta)$, since $A \in E(H(M))$. If we can find $\mu \in EU(M \oplus M)$ taking $M \oplus M$ isomorphically onto Δ, it will follow that

$$\mu(A \oplus A')\mu^{-1} \in EI(M \oplus M) ,$$

whence $A \oplus A' \in EU(M \oplus M)$, as desired.

It suffices to find μ for $M = R^2$ (we can then take direct sums for other cases. A suitable element of U_4 is, in fact,

$$\mu = \Sigma_{34}\, H(X_{13}(1)\, X_{24}(1)).$$

<u>Theorem 1</u> EU_{∞} <u>is the commutator subgroup of</u> U_{∞}

<u>Proof</u> Since each EU_n is perfect, so is EU_{∞}, so it is contained in the commutator subgroup. Conversely, let $A, B \in U_{\infty}$: say $A, B \in U_m$. By the lemma, there exist (for some r) $A', B' \in U_r$ with $A \oplus A', B \oplus B' \in EU_{m+r}$. Hence EU_{m+2r} contains $A \oplus A' \oplus 1, B \oplus 1 \oplus B'$, hence their commutator $[A,B]$.

It is interesting to note that, as on [17, p. 65] the crux of the proof is the construction of μ (but here we have avoided the matrix identity). Note that μ carries e_1, e_2, e_3, e_4 respectively to $e_1 + e_3, e_2 + e_4, -(f_2 - f_4), (f_1 - f_3)u$ which, by our definitions, is indeed a preferred base of Δ.

Our second basic result is a sort of normal form, analogous to the Bruhat decomposition, for elements of E_{∞}. This result was first mooted in [17, 6.6] and the first formal proof is due to Sharpe [15].

Lemma 6 Let $x \in U(M)$. Then

$x \in SI(M)SJ(M)SI(M)$

if and only if M and Mx have a common based complement in $H(M)$.

Proof If $x = uvw$ is of this form, $M = Mw$ is based complementary to $M^{\alpha}w = M^{\alpha}vw$, and so is $Mx = Muvw = Mvw$. Conversely, if F is the common based complement, we can find $u \in SI(M)$ with $M^{\alpha}u^{-1} = Fx^{-1}$, since $SI(M)$ is transitive on based complements to M (c.f. Lemma 2); and also $w \in SI(M)$ with $M^{\alpha}w = F$. Then $v = u^{-1}xw^{-1}$ preserves the based subkernel M^{α}, so lies in $SJ(M)$.

Remark Our construction of $\Sigma_{12} \in U_2$ was by such a product, and Σ_{12} interchanges $\{e_1, e_2\}$ and the complementary based subkernel $\{f_1u, f_2\}$. We deduce the

Corollary In $H(R^{2n})$, two complementary based subkernels always have a common based complement.

Lemma & (Ranicki) Suppose given based subkernels K_i $(1 \leq i \leq 4)$ in (N, θ) with K_i based complementary to K_{i+1} $(i = 1,2,3)$. Then $K_1 \oplus K_3^{\alpha}$, $K_4 \oplus K_3$ have a common based complement in $(N, \theta) \oplus H(K_3)$.

Proof Up to based isomorphism, we can identify (N, θ) with $H(K_3)$ and K_2^{α} with K_3^{α} (by definition of based complements). To avoid confusion, we then use primes to indicate the second copy of $H(K_3)$. We now claim that the twisted diagonal $\Delta = \Delta(H(K_3))$, defined as in the proof of Lemma 5, is a common based complement.

For by that lemma, it is a based sbukernel. Hence it suffices to show that it is additively a based complement to $K_1 \oplus K_2'$. We change bases by a series of elementary moves. First, change $\Delta = \Delta(K_3) \oplus \Delta(K_2)$ modulo $K_2' \subset K_1 \oplus K_2'$ to obtain $\Delta(K_3) \oplus K_2$. Next change $K_1 \oplus K_2'$ by

$K_2 \subset \Delta(K_3) \oplus K_2$ to obtain $K_3 \oplus K_2'$. Finally, change $\Delta(K_3) \oplus K_2$ by $K_3 \subset K_3 \oplus K_2'$ to obtain $K_3' \oplus K_2$. But by hypothesis, $K_3 \oplus K_2'$ and $K_3' \oplus K_2$ are based complements.

For based subkernels K_1, K_2 of (N, θ) we define $K_1 \sim K_2$ if we can find a based complementary pair (L_1, L_2) such that $K_1 \oplus L_1, K_2 \oplus L_2$ have a common based complement.

<u>Lemma 8</u> $\sim$ <u>is an equivalence relation.</u>

<u>Proof</u> It is clearly reflexive and symmetric. Suppose, then, that $K_1 \sim K_2 \sim K_3$; that $K_1 \oplus L_1$ and $K_2 \oplus L_2$ both have based complement C_1, and that $K_2 \oplus M_2, K_3 \oplus M_3$ have based complement C_2. Applying lemma 7 to $(K_1 \oplus L_1 \oplus M_2, C_1 \oplus M_3, K_2 \oplus L_2 \oplus M_2, C_2 \oplus L_1)$, we find a based complementary pair (N_1, N_2) such that $K_1 \oplus L_1 \oplus M_2 \oplus N_1, C_2 \oplus L_1 \oplus N_2$ have based complement C_3. By the corollary to lemma 6, N_1 and N_2 have a common based complement N_3. Now apply lemma 7 to

$(K_1 \oplus L_1 \oplus M_2 \oplus N_1, C_3, C_2 \oplus L_1 \oplus N_2, K_3 \oplus L_2 \oplus M_3 \oplus N_3)$,

and we obtain the desired conclusion.

<u>Theorem 2</u> <u>For all</u> $x \in EU_n$, <u>we can find</u> $\Sigma \in EU_m$ <u>interchanging the based subkernels</u> $R^m, (R^m)^\alpha$, <u>such that</u>

$$x \oplus \Sigma \in SI_{m+n} \, SJ_{m+n} \, SI_{m+n} .$$

<u>Proof</u> By lemma 6, the conclusion holds if R^{m+n} and $R^n x \oplus (R^m)^\alpha$ have a common based complement; by definition, this holds if $R^n \sim R^n x$. Now EU_n is generated by I_n and J_n. The result holds for $x \in I_n$ since $R^n x = R^n$ and for $x \in J_n$ since R^n and $R^n x$ have common based complement $(R^n)^\alpha$. If it holds for x and for y, then

$R^n \sim R^n x$, so $R^n y \sim R^n xy$, and $R^n \sim R^n y$; so $R^n \sim R^n xy$. The result in general now follows.

Corollary We can improve the conclusion to

$$x \oplus \Sigma \in H(SL_{m+n}) \cdot I_{m+n} \cdot J_{m+n} \cdot I_{m+n} \, .$$

This follows on using the equations

$$SI_r = H(SL_r).I_r = I_r . H(SL_r) \, .$$

§3 K_0 and K_1 of categories of quadratic modules

In §1 we defined the categories $\mathcal{P}(R)$, $\mathcal{Q}(R, \alpha, u)$ and $\mathcal{B}\mathcal{Q}(R, \alpha, u)$. We now obtain some exact sequences relating their algebraic K-groups. In addition to the maps induced by the functors F, H and the forgetful functor $G : \mathcal{B}\mathcal{Q}(R, \alpha, u) \to \mathcal{Q}(R, \alpha, u)$, these involve two further maps: the discriminant map and one which we now define.

Suppose (N, θ, v) a based quadratic module, an object of $\mathcal{B}\mathcal{Q}(R, \alpha, u)$ with class $y \in K_0 \mathcal{B}\mathcal{Q}(R, \alpha, u)$ and α an automorphism of N (as R-module), with determinant $x \in K_1(R)$. Then applying α to a preferred base of M gives another base, whose equivalence class v' depends only on x and v. If (N, θ, v') has class y', we define $\tau(x) = y' - y$. If we replace (N, θ, v) by its direct sum with any (N_2, θ_2, v_2), then (N, θ, v') is affected in the same way, so we obtain the same value for $\tau(x)$. Hence τ is well defined. It is defined for any x, since we can apply an automorphism of M to H(M). Hence we have

$$\tau : K_1(R) \to K_0 \mathcal{B}\mathcal{Q}(R, \alpha, u)$$

<u>Lemma 9</u> <u>The composite</u>

$$K_1(R) \xrightarrow{H_*} K_1 \mathfrak{Q}(R, \alpha, u) \xrightarrow{F_*} K_1(R)$$

<u>is</u> $1 - T$; <u>the composite</u> $\delta \circ \tau = 1 + T$, <u>where</u> T <u>is the involution of</u> $K_1(R)$ <u>induced by</u> (α -) <u>duality</u>.

In matrix terms, T comes from the anti-automorphism of GL_n which sends $A = (a_{ij})$ to $A^* = (a^{\alpha}_{ji})$: note that its square is an inner automorphism, hence induces the identity on $K_1(R)$.

<u>Proof</u> If $x \in K_1(R)$ is represented by the matrix A, $H(x)$ has matrix $\begin{pmatrix} A & 0 \\ 0 & A^{*-1} \end{pmatrix}$, so the first assertion is clear. As to the second, given a (based) quadratic form with matrix B, and change of base with matrix P, the form with its new base has matrix P^*BP, and this result also is immediate.

<u>Note</u> Our description of τ was perhaps vague as to sign: we can take the above as normalising this (unimportant) choice.

<u>Proposition 10</u> <u>The following sequence is exact :</u>

$$0 \to K_1 B\mathfrak{Q}(R, \alpha, u) \xrightarrow{G_*} K_1 \mathfrak{Q}(R, \alpha, u) \xrightarrow{F_*} K_1(R) \xrightarrow{\tau} K_0 B\mathfrak{Q}(R, \alpha, u) \xrightarrow{G_*} K_0 \mathfrak{Q}(R, \alpha, u) .$$

<u>Proof</u> We first show that the sequence has order two. An automorphism in $B\mathfrak{Q}$ must preserve preferred bases by definition, hence is mapped to 0 by F_*. If x is the determinant of an automorphism A of (N, θ), where we may suppose M free since such are cofinal, then we can assign N a preferred base v. Changing this by A, though, gives an isomorphic object of $B\mathfrak{Q}(R, \alpha, u)$, so $\tau(x) = 0$. Finally, if we refer to the definition

$\tau(x) = y' - y$ of τ, we see at once that y, y' have the same image in $K_0\mathcal{Q}(R, \alpha, u)$.

Conversely, let $y \in K_0\mathcal{BQ}(R, \alpha, u)$ be in Ker G_*. Let u be the difference of the classes of (N_1, θ_1, v_1) and $(N_2, \theta_2 v_2)$. Then (N_1, θ_1) and (N_2, θ_2) are stably isomorphic in $\mathcal{Q}(R, \alpha, u)$: since the $H(R^n)$ are cofinal, we can suppose (adding this to each of M_1, M_2) that they are already isomorphic. If A is an isomorphism, and has determinant x with respect to v_1, v_2, it follows from the definition that $\tau(x) = y$.

Next let $\tau(x) = 0$. Stabilising as before, we can suppose (N, θ, v) and (N, θ, v') isomorphic. But then x is the determinant of an automorphism in $\mathcal{Q}(R, \alpha, u)$ of (N, θ). Exactness at $K_1\mathcal{Q}(r, \alpha, u)$ holds by definition of $\mathcal{BQ}$. Finally, G_* is injective, since EU_∞ is the commutator subgroup of U_∞ and, being perfect, also of the subgroup with determinant 0.

Writing $S^{\epsilon}(K_1(R)) = \{x \in K_1(R) : \bar{x} = \epsilon x\}$ for $\epsilon = \pm$, we have

<u>Corollary</u> <u>There is an exact sequence</u>

$$K_1\mathcal{Q}(R, \alpha, u) \xrightarrow{\delta} S^-(K_1(R)) \xrightarrow{\tau} \tilde{K}_0\mathcal{BQ}(R, \alpha, u) \to \tilde{K}_0\mathcal{Q}(R, \alpha, u) \oplus S^+(K_1(R)).$$

This follows at once by diagram-chasing, taking due note of Lemma 9. It is sometimes a more convenient form for calculations.

The above is reasonably straightforward and not unexpected. The following exact sequence, though to some extent it plays a symmetrical rôle below, appears to lie deeper. There is a natural forgetful map

$$K_0\mathcal{BQ}(R, \alpha, u) \to K_0\mathcal{B}(R) \xrightarrow{\epsilon} \mathbb{Z},$$

where ϵ counts the number of elements in a preferred basis. We write $K_0\,\mathcal{B}(R, \alpha, u)$ for the kernel.

<u>Proposition 11</u> The sequence

$$K_0\mathcal{BQ}(R, \alpha, -u) \xrightarrow{\delta} K_1(R) \xrightarrow{H} K_1\mathcal{Q}(R, \alpha, +u)$$

is exact.

Note the change here here u to -u.

Proof We will describe Ker H. Let x be an automorphism of a free module M of even rank, representing $\xi \in K_1(R)$. Then H(x) represents $H(\xi)$, and so does $H(x)\Sigma$, if Σ interchanges the based subkernels M and M^α. Then $H(\xi) = 0$ if and only if this is (stably) in EU_∞, so we can apply the corollary to Theorem 2: replacing x (if necessary) by its direct sum with an identity matrix, we get

$$H(x)\Sigma = H(x_0)uvw$$

with $x_0 \in SL(M)$, u, $w \in I(M)$ and $v \in J(M)$.

By Lemma 1 (c.f. lemma 4), there is a unique $(\alpha, -u)$ - quadratic form θ on M such that for $m \in M$,

$$mv = m + Ab_\theta(m) \quad M \oplus M^\alpha.$$

Since also m = mu, w induces the identity on the submodule M and the quotient module M^α, and muvw M^α, we deduce

$$muvw = Ab_\theta(m).$$

It follows that Ab_θ is an isomorphism, hence θ nonsingular. Next, we see by computing determinants that $\xi = \det(Ab_\theta) = \delta(\theta)$. Thus Ker $H \subset$ Im δ.

We can prove the converse using the same identity as for the Whitehead lemma. Alternatively, if v, Σ are defined as above, Mv is complementary (unbased) to M as well as to $M^\alpha = M\Sigma$ so we can find $w \in I(M)$ with $Mvw = M^\alpha$ and then $x \in J(M)$ such that vwx interchanges M and M^α. Then vwx has the form $H(a)\Sigma$, where

$\det a = \det Ab(\theta) = \delta(M, \theta)$ and $\Sigma \in EU(M)$. Hence $H\delta(M, \theta) = 0$.

We observe various simple corollaries of the last three results - most of which can easily be proved independently.

<u>Proof</u> We will describe Ker H . Let x be an automorphism of a free module M of even rank, representing $\xi \in K_1(R)$. Then $H(x)$ represents $H(\xi)$, and so does $H(x)\,\Sigma$, if Σ interchanges the based subkernels M and M^α. Then $H(\xi) = 0$ if and only if this is (stably) in EU_∞, so we can apply the corollary to Theorem 2 : replacing x (if necessary) by its direct sum with an identity matrix, we get

$$H(x)\,\Sigma = H(x_0)\,uvw$$

with $x_0 \in SL(M)$, $u, w \in I(M)$ and $v \in J(M)$.

By Lemma 1 (c.f. lemma 4), there is a unique $(\alpha, -u)$-quadratic form θ on M such that for $m \in M$,

$$mv = m + Ab_\theta(m) \in M \oplus M^\alpha .$$

Since also $m = mu$, w induces the identity on the submodule M and the quotient module M^α, and $muvw \in M^\alpha$, we deduce

$$muvw = Ab_\theta(m) .$$

It follows that Ab_θ is an isomorphism, hence θ nonsingular. Next, we see by computing determinants that $\xi = \det(Ab_\theta) = \delta\,(\theta)$. Thus Ker H $\subset$ Im δ .

We can prove the converse using the same identity as for the Whitehead lemma. Alternatively, if v, Σ are defined as above, Mv is complementary (unbased) to M as well as to $M^\alpha = M\Sigma$ so we can find $w \in I(M)$ with $Mvw = M^\alpha$ and then $x \in J(M)$ such that vwx interchanges M and M^α. Then vwx has the form $H(a)\,\Sigma$, where $\det a = \det Ab(\theta) = \delta\,(M, \theta)$ and $\Sigma \in EU(M)$. Hence $H\,\delta\,(M, \theta) = 0$.

We observe various simple corollaries of the last three results - most of which can easily be proved independently.

Corollary

$$\mathrm{Im}(1 - T) \subset \mathrm{Ker}\ \tau = \mathrm{Im}\ F_* \subset \mathrm{Ker}\ (1 + T)$$
$$\mathrm{Im}(1 + T) \subset \mathrm{Ker}\ H_* = \mathrm{Im}\ \delta \subseteq \mathrm{Ker}\ (1 - T)\ .$$

To conclude this section, we recall the category $\mathcal{L}(R, \alpha, u)$ of based forms of discriminant 0 . This is a full, cofinal subcategory of $\mathcal{B}\mathcal{Q}(R, \alpha, u)$ - as is clear from the above. Hence we have the easy

Lemma 12
$$K_1\mathcal{L}(R, \alpha, u) = K_1\mathcal{B}\mathcal{Q}(R, \alpha, u)\ .$$
$$\tilde{K}_0\mathcal{L}(R, \alpha, u) = \mathrm{Ker}\ \delta : \tilde{K}_0\mathcal{B}\mathcal{Q}(R, \alpha, u) \to K_1(R).$$

§4 Definitions of the L - groups

We have already drawn attention to the symmetry between Propositions 10 and 11. We now develop a notation to make the most of this. First, write

$$\Lambda_0(R, \alpha, u) = \tilde{K}_0\mathcal{B}\mathcal{Q}(R, \alpha, u)$$
$$\Lambda_1(R, \alpha, u) = K_1\ \mathcal{Q}(R, \alpha, u)$$

and, for any $i \in \mathbb{Z}$,

$$\Lambda_{i+2}(R, \alpha, u) = \Lambda_i(R, \alpha, -u)$$

so that Λ_i is periodic with period 4 in i . Since the transition from u to $-u$ is now dealt with in our suffix, we can write $\Lambda_i(R)$ for the rest of this section without risk of confusion. Now we have exact sequences

$$\Lambda_{i+1}(R) \xrightarrow{\delta_{i+1}} K_1(R) \xrightarrow{\tau_i} \Lambda_i(R)\ ,$$

where δ_{i+1} means F_* or δ, and τ_i is H_* or τ, according to the parity of i, and by Lemma 9,

$$\delta_i \circ \tau_i = 1 + (-1)^i T .$$

Let X be any subgroup of $K_1(R)$ such that $T(X) = X$. Then we define

$$L_i^X(R) = L_i^X(R, \alpha, u) = \delta_i^{-1}(X)/\tau_i(X) .$$

The most obvious (and important) examples are $X = \{0\}$: we will write L_i^S for these. We have

$$L_0^S(R) = \operatorname{Ker} \delta : \tilde{K}_0 \mathcal{B}\mathcal{Q}(R) \to K_1(R) = \tilde{K}_0 \mathcal{L}(R)$$
$$L_1^S(R) = \operatorname{Ker} F_* : K_1 \mathcal{Q}(R) \to K_1(R) = K_1 \mathcal{B}\mathcal{Q}(R) = K_1 \mathcal{L}(R),$$

so these are essentially the K groups of the category $\mathcal{L}(R)$. Next we can take $X = K_1(R)$, and write L_i^K for these groups:

$$L_0^K(R) = \operatorname{Coker} \tau : K_1(R) \to \tilde{K}_0 \mathcal{B}\mathcal{Q}(R) = \tilde{K}_0 \mathcal{Q}(R)$$
$$L_1^K(R) = \operatorname{Coker} H_* : K_1(R) \to K_1 \mathcal{Q}(R) .$$

Although these are the main examples, we will have occasion in other papers to consider : $X = \operatorname{Ker}(K_1(R) \to K_1(S))$ for a ring homomorphism $R \to S$ (of antistructures) and, if R is the integer group ring $\mathbb{Z}\pi$ of a group π also X = the image in $K_1(R)$ of the 1×1 matrices $\pm g$, $g \in \pi$. The latter is the important case for topological applications (c.f. [17]). The idea of defining all the L_i^X was suggested by S. Cappell.

These groups are related by exact sequences. If G is a group with involution T of order 2 - c.g. $K_1(R)$ or X above - we write $H^i(G)$ for the Tate cohomology groups of the action:

$$H^{2i}(G) = \{x \in G : Tx = x\}/\{y + Ty : y \in G\} ,$$
$$H^{2i+1}(G) = \{x \in G : Tx = -x\}/\{y - Ty : y \in G\} .$$

<u>Theorem 3</u> *If* $X \subset Y$ *are* T *-invariant subgroups of* $K_1(R)$, *there is an exact sequence*

$$\dots L_i^X(R) \xrightarrow{j} L_i^Y(R) \xrightarrow{d} H^i(Y/X) \xrightarrow{t} L_{i-1}^X(R) \xrightarrow{j} L_{i-1}^Y(R) \dots$$

<u>Proof</u> It will be convenient to use the temporary notation

$$\Lambda_i^{X,Y} = \delta_i^{-1}(Y)/\tau_i(X).$$

Then δ_i induces a map

$$\delta_i' : \Lambda_i^{X,Y} \to Y/\delta_i\tau_i(X) \to Y/X$$

whose kernel is the set of equivalence classes of elements mapping by δ_i to X, i.e. is $L_i^X(R)$. Similarly, τ_i induces a map

$$\tau_i' : Y/X \to \tau_i(Y)/\tau_i(X) \to \Lambda_i^{X,Y}$$

whose cokernel equals that of $\tau_i : Y \to \delta_i^{-1}(Y)$, i.e. is $L_i^Y(R)$.

Since $\tau_i\ \delta_{i+1} = 0$, $\tau_i'\ \delta_{i+1}' = 0$. Conversely, if $\tau_i'(y + X) = 0$, $\tau_i(y) \in \tau_i(X)$ so for some $x \in X$, $y - x \in \operatorname{Ker} \tau_i = \operatorname{Im}\ \delta_{i+1}$. It follows that $y + X \in \operatorname{Im}\ \delta_{i+1}'$.

We thus have exact sequences

$$0 \to L_{i+1}^X(R) \xrightarrow{j_i} \Lambda_{i+1}^{X,Y} \xrightarrow{\delta_{i+1}'} Y/X \xrightarrow{\tau_i'} \Lambda_i^{X,Y} \xrightarrow{q_i} L_i^Y(R) \to 0 .$$

Also, the relation $\delta_i' \circ \tau_i' = 1 + (-1)^i T$ follows from the corresponding result for $\delta_i \circ \tau_i$. The result thus follows formally from the following elementary lemma, whose proof we leave to the reader.

<u>Lemma 13</u> *Given a sequence of exact sequences*

$$A_{i+1} \xrightarrow{a_{i+1}} B_i \xrightarrow{b_i} A_i$$

write $H_i(B)$ *for the homology of the complex*

$$\dots B_{i+1} \xrightarrow{a_{i+1}b_{i+1}} B_i \xrightarrow{a_i b_i} B_{i-1} \quad \dots .$$

Then there is an exact sequence

$$\ldots \text{Ker } a_{i+1} \to \text{Coker } b_{i+1} \to H_i(B) \to \text{Ker } a_i \to \text{Coker } b_i \ldots$$

<u>Corollary</u> <u>There is an exact sequence</u>

$$L_i^S(R) \to L_i^K(R) \to H^i(K_1(R)) \to L_{i-1}^S(R) \to L_{i-1}^K(R) .$$

A special case of this is due to Rothenberg (see [14]); the general case is also proved in [12].

The above definition makes the L_i^X appear somewhat unnatural. We conclude our discussion by giving a more directly geometrical definition which is, moreover, one which we shall need to refer back to.

Let $K_1(R) \to V$ be a homomorphism with kernel X, equivariant with respect to an action of $\mathbb{Z}/2$ by α. We can, for example, take $V = K_1(R)/X$, but it is sometimes more convenient to let $V = K_1(S)$ with the map induced by a ring map $R \to S$. We use this map to calculate determinants with values in V. Now define the category $\mathcal{B}(R)$ as in §1, but referring to determinant in V. With no further change, we obtain definitions of $\mathcal{B}\mathcal{Q}(R, \alpha, u)$ a forgetful functor $F : \mathcal{B}\mathcal{Q}(R, \alpha, u) \to \mathcal{B}(R)$, and a hyperbolic functor $H : \mathcal{B}_0(R) \to \mathcal{B}\mathcal{Q}(R, \alpha, u)$. The discussion of based subkernels, complementarity, and discriminant at the end of §1 is also unaltered, and we have a category $\mathcal{L}(R, \alpha, u)$. It is sometimes possible to define dual bases and hence H, on free modules of odd rank: for example, if we require (as is often done) $u = \pm 1$, and that the determinant (in V) of $-1 \in R^X$ is zero. As we restrict ourselves to the case when the rank is formally zero, this point is unimportant for us.

It is immediately clear that $L_0^X(R, \alpha, u) = K_0\mathcal{L}(R, \alpha, u)$: forms representing objects in $\mathcal{L}(R, \alpha, u)$ admit free bases: the discriminant in our original sense is restricted to lie in $X \subset K_1(R)$, and the basis is free to change by (the image under τ of) X.

More interesting is the case of L_1^X. We refer to the proof of Theorem 2 (starting with Lemma 6): note that $SI(M)$ now has a new meaning, i.e. automorphisms of $H(M)$ which leave M invariant and induce an automorphism of M with determinant $0 \in V$. To avoid confusion with our earlier notation, let us write $S'I(M)$ for this, $E'U(M)$ for the group generated by $S'I(M)$ and $S'J(M)$ (we make no bones here about listing elementary matrices), $S'U(M)$ for elements of $U(M)$ with determinant $0 \in V$, and conventions as before when $M = R^n$. I claim first that $L_1^X(R, \alpha, u) = S'U_\infty / E'U_\infty$: this is indeed simply a matter of referring back to the definition. We seek, however, a more directly geometric form of the defi[illegible]ion.

[illegible]e (as we see directly) $S'I(M)$ acts transitively on the based complements to M, the proof of Lemma 6 remains valid; so of course does the corollary (the new form is a weaker version than the old). The proof of Lemma 7 remains valid without alteration, and if we define a relation $\sim$ on subkernels as there, we see as before that it is an equivalence relation. Now $S'U_n$ acts (and it clearly acts transitively) on the based subkernels in $H(R^n)$. The given proof of Theorem 2 shows that for $x \in S'U_n$,

$x \in E'U_n$ implies that $R^n \sim R^n x$ and hence for Σ as before

$x \oplus \Sigma \in SI_{m+n}\, SJ_{m+n}\, SI_{m+n}$.

But this in turn implies $x \otimes 1$, $x \otimes \Sigma \in E'U_{m+n}$. The relation $\sim$ between subkernels thus detects neatly the group we want, however, as subkernels are abstractly isomorphic we seek a more intrinsic invariant.

Following Ranicki we define a <u>formation</u> to consist of a triple $(H; F, G)$, where H is a nonsingular based (R, α, u) - quadratic module and F, G are based subkernels in H. A formation is <u>trivial</u> (or split) if F and G are based complements; we define stable equivalence $\approx$ between formations to mean that they can be made isomorphic by adding trivial pairs.

Definition Two formations (H, F, G), (H', F', G') are equivalent ($\sim$) if, after replacing if necessary by stably equivalent formations, we can find a based isomorphism $H \to H'$ taking F to F' and G to G'' with $G'' \sim G'$.

Theorem Equivalence classes of formations form an abelian group under $\oplus$. This group is isomorphic to $L_1^X(R, \alpha, u)$. The isomorphism is induced by taking the class of a based automorphism α of H which takes F to G (as based subkernel).

Proof Any element of $S'I_n$ defines $0 \in L_1^X(R, \alpha, u)$. It follows (transporting by an isomorphism) that so does any automorphism which preserves a based subkernel. Since α is unique up to left and right composition with such automorphisms, its class $\xi \in L_1^X(R, \alpha, u)$ is determined by $(H; F, G)$. Clearly, the sum of two formations has the sum of their invariants. It remains to show that two formations with the same invariant are equivalent.

Up to stable isomorphism, we can identify the formations with $(H(R^n), R^n, R^n x)$ and $(H(R^n), R^n, R^n y)$ where $xy^{-1} \in E'U_\infty$. We seek to show $R^n x \sim R^n y$, or equivalently, $R^n xy^{-1} \sim R^n$. But this was done above.

Note that - absorbing more in the stable equivalence - we can modify $\sim$ to require that G' and G'' have a common complement. Also, $(H, F, G) \sim 0 \Leftrightarrow F \sim G \Leftrightarrow$ stably, F and G have a common complement.

The return from automorphisms to pairs of subkernels brings us closer to the geometry in [17, Chapter 6]. It also now follows that the L-groups of [17] can be described in our present terms as follows. Take $V = Wh\,\pi$, so X is the image in $K_1(\mathbb{Z}\pi)$ of $\{\pm\pi \in (\mathbb{Z}\pi)^\times\}$. Then

$$L_{2k}(\pi) = L_0^X(\mathbb{Z}\pi, \alpha, (-1)^k)$$

$$L_{2k+1}(\pi) = L_1^X(\mathbb{Z}\pi, \alpha, (-1)^k)/\text{class of } \sigma = \begin{pmatrix} 0 & 1 \\ (-1)^k & 0 \end{pmatrix},$$

where α is the anti-involution given by

$$\alpha(g) = w(g)\, g^{-1} \qquad \text{for} \quad g \in \pi$$

(w the orientation homomorphism). These identifications are now immediate on comparing the definitions. Similarly, we obtain the surgery obstruction groups L^h for homotopy equivalence as above, but taking $X = \{0\}$.

§5 Further remarks

Although I regard the above as moreorless in final form it is, in some important respects, incomplete. In this section I discuss desirable generalisations, and compare with the work of other authors.

First, there is the problem of dealing with reflexive bilinear, rather than quadratic forms. The work of Bak [1] [2] has suggested that we should generalise, and consider the concept of 'unitary ring' as formulated in Bass [5]. For (A, α, u) an antistructure, we consider an additive subgroup Λ of A satisfying

(i) $S_{-u}(A) = \{a - a^{\alpha}u : a \in A\} \subseteq \Lambda \subseteq S^{-u}(A) = \{a \in A : a = -a^{\alpha}u\}$

(ii) $a^{\alpha}ra \in \Lambda$ for all $a \in A,\ r \in \Lambda$.

Then a $(-u)$-reflexive form over the unitary ring (A, α, u, Λ) is a $(-u)$-reflexive form ϕ over (A, α, u) with $\phi(x, x) \in \Lambda$ for all x. The module of u-quadratic forms is the quotient of the group of sesquilinear forms by the subgroup of $(-u)$-reflexives. Bass gives generalisations of all our results up to Theorem 1 to quadratic forms in this sense. However, this does not really solve the problem of giving a good account of reflexives. Nor does it seem possible to proceed to analogues of Theorem 2 and its corollaries (which really constitute our main theme), as there is no natural choice of 'dual category' (as we had $\mathcal{Q}(A, \alpha, u)$ and $\mathcal{Q}(A, \alpha, -u)$).

It seems to me that if there is a common generalisation of our two approaches, it should go somewhat as follows. We choose $\Lambda_-(\subset S^{-u}(A))$ and $\Lambda_+(\subset S^u(A))$ independently and then seek (e.g. using some modified version of Witt vectors) a more general notion of form where $\phi(x, x)$ can take any value in Λ_+, and is to be interpreted as $y + y^\alpha u$ where y is defined mod Λ_-. Here, y cannot be assumed to take values in A : we need a larger group. A nontrivial example is Brown's notion of quadratic forms over the field A of 2 elements, taking integers mod 4 as values.

Next, there is the question of higher (and lower!) K (or L) groups. The most suggestive work here has been done by Karoubi, and the ideas can be expressed as follows. We start with the forgetful and hyperbolic functors

$$F : \mathcal{Q}(A, \alpha, u) \to \mathcal{P}(A) \qquad H : \mathcal{P}(A) \to \mathcal{Q}(A, \alpha, u) .$$

Following Quillen and others, from the monoidal category we construct a topological infinite loop space

$$\mathcal{K} = \Omega B|\mathcal{P}(A)| \simeq K_0(A) \times B(GL(A))^{ab} ,$$

$$\Omega B|\mathcal{Q}(A, \alpha, u)| \simeq K_0\mathcal{Q}(A, \alpha, u) \times B(U(A, \alpha, u))^{ab} ;$$

and F, H induce maps between these; in fact, infinite loop maps. Write $\mathcal{U}(A, \alpha, u)$, $\mathcal{V}(A, \alpha, u)$ for the mapping fibres of H, F respectively.

<u>Main conjecture</u> (<u>Karoubi</u>) <u>There is a natural homotopy equivalence</u>

$$\Omega\, \mathcal{U}(A, \alpha, u) \to \mathcal{V}(A, \alpha, -u) .$$

It has been shown by Karoubi [8] that if Quillen's higher K-groups are replaced by those of Karoubi-Villamayor type [9], the corresponding result holds. Also, Sharpe [15] has shown that there is an isomorphism of π_1. We will now describe the periodicity situation which would result from the conjecture.

Write $\mathcal{L}^0 = \Omega B\mathcal{Q}(A, \alpha, u)$, $\mathcal{L}^1 = \mathcal{U}(A, \alpha, u)$,

$\mathcal{L}^2 = \Omega B\mathcal{Q}(A, \alpha, -u)$, $\mathcal{L}^3 = \mathcal{U}(A, \alpha, -u)$,

and regard the p in $\mathcal{L}^p$ as taking values integers mod 4. Then for each p, we have a fibering (up to homotopy)

$$\mathcal{L}^p \to \mathcal{K} \to \mathcal{L}^{p+1} .$$

Define $KU_{p,n} = KU_{p,n}(A, \alpha, u) = \pi_n(\mathcal{L}^{p-n})$. This has period 4 in p, and we have exact sequences

$$\dots KU_{p+1,n} \to K_n \to KU_{p,n} \to KU_{p,n-1} \to K_{n-1} \to \dots$$

Now define $L_{p,n-\frac{1}{2}} = \mathrm{Im}(KU_{p,n} \to KU_{p,n-1})$. This can of course also be defined as a kernel or as a cokernel.

<u>Consequences</u> <u>The composite</u> $K_n \to KU_{p,n} \to K_n$ <u>is</u> $1 + (-1)^p\alpha$.

<u>There are exact sequences</u>

$$\dots L_{p,n+\frac{1}{2}} \to L_{p,n-\frac{1}{2}} \to H^p(K_n) \to L_{p-1,n+\frac{1}{2}} \dots$$

The first should be a simple verification; the second will then follow from Lemma 13. As in §4, we will then also be able to define intermediate L-groups between $L_{p,n-\frac{1}{2}}$ and $L_{p,n+\frac{1}{2}}$ for each α-invariant subgroup X of K_n.

To illustrate this pattern, here are two simple consequences. First, for any (A, α, u), tensor all K and L groups by $\mathbb{Z}[\frac{1}{2}]$. Then $\bar{L}_p = L_{p,n-\frac{1}{2}} \otimes \mathbb{Z}[\frac{1}{2}]$ is independent of n, and we have canonical splittings

$$KU_{p,n} \otimes \mathbb{Z}[\tfrac{1}{2}] = S_{(-1)^p}(K_n \otimes \mathbb{Z}[\tfrac{1}{2}]) \oplus \bar{L}_p .$$

Next suppose A the sum of two anti-isomorphic rings R, S interchanged by α. Then $\mathcal{Q}(A, \alpha, u) \cong \mathcal{P}(R)$, whence

$$KU_{p,n}(A, \alpha, u) \cong K_n(R) \qquad L_{p,n-\frac{1}{2}}(A, \alpha, u) = 0 .$$

Although this development is still conjectural, the description has been justified for low values of n - e.g. the above exact sequence is valid for $n = 1$ (Theorem 3) and $n = 0$ (this and the case $n = 1$ are in Ranicki [12]), thus answering the problems raised in [17, §17D]. Ranicki has also considered the case $n < 0$ where there is a definition analogous to that of Bass [4] for K_n. One would hope here for a spectrum, as Gersten [7] obtains for algebraic K theory.

The above notation illustrates well the difference between what I have described as KU-theory and L-theory. In the former (as studied by the Bass school) the natural spaces are $\mathcal{L}^0$ (and, to lesser degree, other $\mathcal{L}^i$) and the natural sequence of groups is $\pi_n(\mathcal{L}^0)$. In the latter, the natural sequences are the periodic sequences with n fixed, and Ranicki [13] has succeeded in constructing (by simplicial sets) periodic spaces $\mathcal{L}_n$ with $\pi_p(\mathcal{L}_n) = L_{p,n}$ $(n = -\frac{1}{2}, \frac{1}{2}, 1\frac{1}{2})$.

I hope this paper will help explain the viewpoint of L - theory as opposed to KU - theory.

Since we have spaces, relative groups can be defined as homotopy groups of mapping fibres. Algebraic definitions of the relative KU groups in low dimensions are also given by Bass [5]. In general, the relative L groups cannot be very closely related to the relative KU groups: the theory here is clearly susceptible of improvement.

Products have been studied to some extent by Karoubi [8]. It seems, for example, that if A is commutative, $KU_{p,n}(A, \alpha, 1)$ should be a bigraded ring. Again, the complete situation is obscure.

The development likely to be of most value for topological applications would be a definition replacing modules by chain complexes throughout. For the case when 2 is invertible in A, this was achieved by Miscenko [10]. See also the discussion in [17, §17G].

To conclude, we give a dictionary of notations: I will compare others with the systematic notation [S] of this paragraph.

<u>L - theorists</u>

[11] [12] [13]	U_p		V_p		W_p	
[17, §17D]	L^A_p	L^F_p	$L^B_p = L^h_p$	L^C_p	$L^E_p = L^s_p$	L^D_p
This paper			L^K_p	Λ_p	L^S_p	
[S]	$L_{p,-\frac{1}{2}}$	$KU_{p,0}$	$L_{p,\frac{1}{2}}$	$KU_{p,1}$	$L_{p,1\frac{1}{2}}$	$\mathrm{Im}(L_{p,1\frac{1}{2}} \to L_{p,\frac{1}{2}})$

The identifications are not quite precise: the notation of [17] was provisional, but referred to determinants in $Wh(\pi)$, not $K_1(\mathbb{Z}\pi)$; also, the automorphism σ is factored out in the groups of the top two rows.

Karoubi

[8] [9]	${}_1L_n$	${}_{-1}V_{n-1} \cong {}_1U_n$	${}_1W_n$	${}_1L'_n$
[S]	$KU_{n,n}$	$KU_{n+1,n}$	$L_{n,n-\frac{1}{2}}$	$L_{n,n+\frac{1}{2}}$

Changing the prefix from 1 to -1 has the effect of changing the first suffix in the lower row by 2 also. Karoubi also has 'homotopical' versions ${}_1L^{-n}$ etc.

KU - theorists

[3]	$KF_n(A, \lambda, \Lambda)$	$KU_n^\lambda(A)$	$KQ_n^\lambda(A)$	$W_n(A,\lambda,\Lambda)$	$W_n^\lambda(A)$	$WQ_n^\lambda(A)$
[5]	$KU_n^\lambda(A, \Lambda)$	$KU_n^\lambda(A, S^\lambda(A))$	$KU_n^\lambda(A, S_\lambda(A))$	$W_n^\lambda(A, \Lambda)$	$W_n^\lambda(A, S^\lambda(A))$	$W_n^\lambda(A, S_\lambda(A))$
[S]			$KU_{n,n}(A,\alpha,\lambda)$			$L_{n,n-\frac{1}{2}}(A,\alpha,\lambda)$

where $n = 0$ or 1 (usually 0); α is understood.

§6 L - theory of division rings

By way of a simple illustration to the preceding, we now give one calculation. It is not really original: see e.g. [6]. We begin by introducing a new type of elementary matrix.

In $H(R) \oplus (N, \theta)$ we define $\epsilon^1(y, \lambda)$ for $y \in N$, $\lambda \in q_\theta(y)$ by

$$e \mapsto e - f\lambda + y \qquad f \mapsto f$$

and for, $x \in N$, $x \mapsto x - f\, b_\theta(y, x)$.

Then a simple calculation shows that

$$\epsilon^1(y, \lambda)\, \epsilon^1(z, \mu) = \epsilon^1(y + z, \lambda + \mu + b_\theta(y, z)).$$

In the case $(N, \theta) = H(R^n)$, we have

$$\epsilon^1(f_i r, 0) = E_{1i}(r)$$
$$\epsilon^1(e_i r, 0) = \Sigma_{ji}^{-1} E_{1j}(ur) \Sigma_{ji}$$
$$\epsilon^1(0, \mu - \mu^\alpha u) = [\epsilon^1(e_1, 0), \epsilon^1(f_1\mu, 0)]$$

so for $n \geq 2$, all $\epsilon^1(y, \lambda) \in EU_{n+1}$. The same applies, similarly, to $\epsilon^2(y, \lambda)$ defined by

$$e \mapsto e \qquad f \mapsto -e\lambda^\alpha + f + y$$

and for $x \in N$ $\qquad x \mapsto x - eu^{-1}b_\theta(y, x)$

Theorem 5 Let R be a division ring. Then $L_1^K(R, \alpha, u) = 0$ unless R is commutative, α is the identity and $u = 1$, in which case the group has order 2.

Proof We consider a general automorphism of $H(R^n)$, and seek to modify it by elementary and hyperbolic transformations till we obtain a normal form. The argument proceeds by induction on n. Let p be an automorphism of $H(R) \oplus (N, \theta)$, and write

$$ep = ea + fb + x \qquad a, b \in R, \ x \in N .$$

Suppose $a \neq 0$. Then as p is an isometry, $0 = q(e_1) = q(e_1 \, p H(a^{-1})) = \lambda + q(x)$, so $\epsilon^1(-x, ba^\alpha)$ is defined, and $p' = pH(a^{-1})\epsilon^1(-x, ba^\alpha)$ leaves e fixed. Write $fp' = ec + fd + y$. Since p' preserves the inner product $b_\theta(e,f)$, $d = 1$. Then $p'' = p'\epsilon^2(-y, c^{\alpha^{-1}})$ leaves e and f fixed, and thus can be regarded as an automorphism of (N, θ).

Apart from the need for supposing $a \neq 0$, the argument shows by induction that p is the product of elementary and hyperbolic transformations, which is what we are trying to prove. It thus remains to see whether we can always multiply p by an elementary transformation to ensure $a \neq 0$. Now the coefficient of e in $ep\epsilon^2(y, \lambda)$ is minus

$$\lambda^\alpha b + u^{-1}b_\theta(y, x) .$$

If $b = 0$, then $x = ep \neq 0$, so we can choose $b_\theta(y, x) \neq 0$ by nonsingularity. If $b \neq 0$, first try to choose $y = 0$, $\lambda = \mu - \mu^\alpha u \neq 0$. This is possible unless $u = 1$ and $\mu = \mu^\alpha$ for all μ, so α = identity, an antiautomorphism, and R is commutative: we are in the exceptional case. Finally, in this case, $\lambda = q_\theta(y)$ is determined by y. If now $q_\theta(y) b + b_\theta(y, x)$ vanishes for all $y \in N$, the quadratic form q_θ is additive in y, hence

$$0 = q_\theta(y + z) - q_\theta(y) - q_\theta(z) = b_\theta(y, z)$$

for all y, z. Since our form is nonsingular, it follows that $N = 0$.

These results prove that $L_1^K = 0$ save in the exceptional case, and that in that case any nonzero element of L_1^K can be represented by an automorphism of a hyperbolic plane

$$e \mapsto ea + fb$$

$$f \mapsto ec + fd$$

where, moreover, $a = 0$. Since, moreover, we have an isometry of quadratic forms over R it follows that $d = 0$ and $c = b^{-1}$. Multiplying by a hyperbolic automorphism, we reduce to the 'interchange' σ :

$$e\sigma = f \qquad f\sigma = e .$$

It remains to show that σ does not give $0 \in L_1^K(R, 1, 1)$.

If K does not have characteristic 2, this is easy: any elementary or hyperbolic automorphism has determinant (in the naïve sense) + 1, whereas $\det \sigma = -1$. Another proof, which includes the characteristic 2 case, runs as follows. Form the Clifford algebra $C = C_0 \oplus C_1$ of the quadratic form; let Z be the centraliser in C of C_0. Then Z is a quadratic Galois extension of R : either $R \oplus R$ or a field. Any automorphism of the form induces automorphisms of C, C_0 and Z over R. It is now easily shown that any elementary or hyperbolic automorphism induces the identity on Z, whereas σ induces the nontrivial automorphism.

The above theorem follows from those quoted in [18,II] , but this direct proof seems in the spirit of L - theory.

Our argument also yields an unstable result, but since better results are known [2], [5], it does not seem worth pursuing this point. Other L groups for fields were computed in [18,II], and it seems appropriate to quote them here, except for global fields where a better formulation will be given in [18 , V] .

Suppose R a division ring with centre K . If $\alpha|K$ is not the identity, but has fixed field k (type U), our groups L_n have period 2 in n, and vanish for n odd. The exact sequence of Theorem 3, Corollary thus reduces to

$$0 \to H^1(K_1(R)) \to L_0^S(R) \to L_0^K(R) \to H^0(K_1(R)) \to 0 .$$

For R finite, all groups are zero. For R local, the first two are zero; the latter two isomorphic to $k^\times/NK^\times$, hence of order 2 . For $R = \mathbb{C}$, we have

$$0 \to 0 \to 4\mathbb{Z} \to 2\mathbb{Z} \to \{\pm 1\} \to 0 .$$

Next let α be trivial on K , which has characteristic 2 (type SPOT). We suppose R finite (then R = K). Then $L_i^S = L_i^K$ has order 2 and $H^i(K_1(R)) = 0$ for all i .

Finally suppose α trivial on K , of characteristic $\neq 2$. We suppose $L_1(R)$ the commutator quotient of a group which (as algebraic group) is orthogonal (not symplectic); otherwise replace u by -u. We give the table of groups

$$L_3^S \to L_3^K \to H^3(K_1) \to L_2^S \to L_2^K \to H^2(K_1)$$
$$\to L_1^S \to L_1^K \to H^1(K_1) \to L_0^S \to L_0^K \to H^0(K_1)$$

with the convention that 1 denotes a group of order 1, 2 a group of order 2, $G = K^\times/(K^\times)^2$ and $d\mathbb{Z}$ is the subgroup of $\mathbb{Z}$ generated by d .

R finite (G = 2)	$1 - 1 - 2 = 2 - 1 - G$ $G - 2 = 2 - 1 - G = G$	$R = \mathbb{R}$	$1 - 1 - 2 = 2 - 1 - 2$ $2 - 2 = 2 - 4\mathbb{Z} - 2\mathbb{Z} - 2$
R local commutative	$1 - 1 - 2 = 2 - 1 - G$ $G - 2 = 2 - 2 - L_0^K - G$	$R = \mathbb{C}$	$1 - 1 - 2 = 2 - 1 - 1$ $1 - 2 = 2 - 1 - 1 - 1$
R local non-commutative	$1 - 1 - 2 = 2 - 1 - G$ $G - 1 - 2 = 2 - G = G$	$R = \mathbb{H}$	$1 - 1 - 1 - 2\mathbb{Z} = 2\mathbb{Z} - 1$ $1 - 1 - 1 - 1 - 1 - 1$

References

1. A. Bak, 'On modules with quadratic forms', pp.55-66 in 'Algebraic K-theory and its geometric applications', Springer lecture notes no. 108 (1969).

2. A. Bak, 'The stable structure of quadratic modules', preprint, Princeton University, 1970.

3. A, Bak and W. Scharlau, 'Witt groups of orders and finite groups', preprint, Princeton University, 1972.

4. H. Bass, Algebraic K - theory, W.A. Benjamin Inc. 1968.

5. H. Bass, 'Unitary algebraic K - theory', notes, Columbia University, 1972.

6. Connolly,

7. S.M. Gersten, 'On the spectrum of algebraic K - theory', Bull.Amer.Math. Soc. 78 (1972), 216 - 219.

8. M. Karoubi, 'Modules quadratiques et périodicité en K - théorie', preprint, Université de Strasbourg, 1971.

9. M. Karoubi and O. Villamayor, 'K - théorie algébrique et K - théorie topologique', I, Math. Scand. to appear, II, preprint, Université de Strasbourg, 1971.

10. A.S. Miščenko, 'Homotopy invariants of non-simply connected manifolds. I. Rational invariants', Izv. Akad. Nauk, S.S.S.R. ser.mat. 34 (1970) 501 - 514.

11. S.P. Novikov, 'The algebraic construction and properties of Hermitian analogues of K - theory for rings with involution from the point of view of Hamiltonian formalism. Some applications to differential topology and the theory of characteristic classes.' I, Izv. Akad. Nauk. S.S.S.R. ser.Mat. 34 (1970) 253 - 288, II, ibid, 475 - 500.

12. A.A. Ranicki, 'Algebraic L - theory', Proc.London Math. Soc., to appear.

13. A.A. Ranicki, 'Geometric L - theory' preprint, Cambridge University, 1972.

14. J.L. Shaneson, 'Wall's surgery obstruction groups for $G \times \mathbb{Z}$', Ann.of Math. 90 (1969), 296 - 334.

15. R.W. Sharpe, 'On the structure of the unitary Steinberg group', preprint, Yale University, 1971.

16. C.T.C. Wall, 'On the axiomatic foundations of the theory of Hermitian forms', Proc.Cambridge Phil.Soc. 67 (1970) 243 - 250.

17. C.T.C. Wall, Surgery on compact manifolds, Academic Press, 1970.

18. C.T.C. Wall, Classification of Hermitian Forms,

PÉRIODICITÉ DE LA K - THÉORIE HERMITIENNE

MAX KAROUBI

INTRODUCTION

La K-théorie topologique doit son succès aux théorèmes de périodicité de Bott [8] qu'on peut formuler ainsi. Pour tout anneau A soit $GL_n(A)$ le groupe des matrices inversibles d'ordre n à coefficients dans A et soit $GL(A) = \varinjlim GL_n(A)$. On a alors des isomorphismes $\pi_i(GL(\mathbb{R})) \approx \pi_{i+8}(GL(\mathbb{R}))$ et $\pi_i(GL(\mathbb{C})) \approx \pi_{i+2}(GL(\mathbb{C}))$. En fait, ces isomorphismes sont induits par des équivalences d'homotopie explicites $GL(\mathbb{R}) \sim \Omega^8(GL(\mathbb{R}))$ et $GL(\mathbb{C}) \sim \Omega^2(GL(\mathbb{C}))$.

Pour exploiter les théorèmes de périodicité, il est commode de considérer la théorie cohomologique $K^n(X,Y)$ introduite par Atiyah et Hirzebruch. De manière précise, à toute catégorie additive C on peut associer le monoïde abélien (ou semi-groupe) $\Phi(C)$ formé des classes d'isomorphie d'objets de C (la somme dans le monoïde $\Phi(C)$ étant induite par la somme des objets de C). Le groupe abélien K(C) déduit de $\Phi(C)$ par symétrisation est le groupe de Grothendieck de C. Soit maintenant X un espace compact (cette restriction est nécessaire pour la suite) et soit $F = \mathbb{R}$ ou $\mathbb{C}$. La catégorie des F-fibrés vectoriels de rang fini forme une catégorie additive $\xi(X)$ (ou $\xi_F(X)$ si on veut spécifier le corps de base F), la somme dans la catégorie étant induite par la somme de Whitney des fibrés. Le groupe de Grothendieck de $\xi_F(X)$ est noté par abus d'écriture $K_F(X)$, ou simplement K(X) s'il n'y a pas de confusion possible. Par exemple, $K_F(\phi) = 0$ et $K_F(\text{Point}) = \mathbb{Z}$ d'après la théorie de la dimension des espaces vectoriels. Si $f : X \to Y$ est une application continue, f induit un foncteur additif $f^* : \xi_F(Y) \to \xi_F(X)$ (le foncteur "image réciproque"), donc un homomorphisme $K_F(Y) \to K_F(X)$. Ainsi $K_F(X)$ est un foncteur contravariant de X. En particulier, la projection de X sur un point induit un homomorphisme $\mathbb{Z} \approx K_F(\text{Point}) \to K_F(X)$ dont le conoyau est la "K-théorie réduite" $\widetilde{K}_F(X)$. Si X est non vide, on a ainsi une décomposition en somme directe $K_F(X) \approx \mathbb{Z} \oplus \widetilde{K}_F(X)$.

Si Y est un sous-ensemble fermé de X . on pose $K_F^{-n}(X,Y) = \tilde{K}_F(S^n(X/Y))$ pour $n \geq 0$. Comme il est d'usage, X/ϕ désigne l'espace X auquel on a ajouté un point au dehors et $S^n(X/Y)$ désigne la $n^{\text{ième}}$ suspension de X/Y.

Les théorèmes de périodicité de Bott peuvent alors s'exprimer de la maniere suivante : $K_{\mathbb{R}}^{-n}(X,Y) \approx K_{\mathbb{R}}^{-n-8}(X,Y)$ et $K_{\mathbb{C}}^{-n}(X,Y) \approx K_{\mathbb{C}}^{-n-2}(X,Y)$. Cette périodicité des foncteurs K^{-n} permet de définir de manière évidente des foncteurs $K_F^n(X,Y)$ pour $n \in \mathbb{Z}$. On obtient ainsi la théorie cohomologique d'Atiyah et Hirzebruch. Notons au passage que la théorie $K_{\mathbb{R}}^n$ est périodique de période 4 si on néglige la 2 torsion. De manière précise, $K_{\mathbb{R}}^n(X,Y) \underset{\mathbb{Z}}{\otimes} \mathbb{Z}' \approx K_{\mathbb{R}}^{n+4}(X,Y) \underset{\mathbb{Z}}{\otimes} \mathbb{Z}'$ lorsqu'on pose $\mathbb{Z}' = \mathbb{Z}[1/2]$.

Le groupe $K(X)$ peut être défini algébriquement en termes de modules. En effet, désignons par $A = C_F(X)$ l'anneau des fonctions continues sur X à valeurs dans F . Alors, d'après Serre et Swan [4] , la catégorie $\xi_F(X)$ est équivalente à la catégorie $\mathcal{P}(A)$ dont les objets sont les A-modules projectifs de type fini et les morphismes les homomorphismes de A-modules. En particulier, ces deux catégories ont le même groupe de Grothendieck. Par abus d'écriture, on note $K(A)$ le groupe de Grothendieck de la catégorie $\mathcal{P}(A)$ pour tout anneau unitaire A . Avec ces définitions, on a donc $K_F(X) \approx K(A)$ si $A = C_F(X)$. Il convient de remarquer que $K(A)$ est un foncteur covariant de l'anneau A . En effet, un homomorphisme d'anneaux $\theta : A \to B$ induit un foncteur "extension des scalaires" $\mathcal{P}(A) \to \mathcal{P}(B)$ grâce à la formule $M \mapsto M \underset{A}{\otimes} B$, donc un homomorphisme $K(A) \to K(B)$. Par exemple, une application continue $f : X \to Y$ induit un homomorphisme d'anneaux $C_F(Y) \to C_F(X)$, donc un homomorphisme de groupes $K_F(Y) \approx K(C_F(Y)) \to K(C_F(X)) \approx K_F(X)$. On retrouve ainsi le caractère contravariant du foncteur $K(X)$.

Cette interprétation algébrique du foncteur $K(X)$ suggère fortement la possibilité de définir algébriquement les foncteurs $K^n(X,Y)$. De manière précise, pour tout anneau A , on aimerait définir de "manière raisonnable" des foncteurs $K^n(A)$ qui coïncideraient avec les foncteurs $K_F^n(X)$ lorsque $A = C_F(X)$.

Nous nous proposons de voir comment ce programme peut être rempli et quels types d'applications peut en résulter.

En fait, plusieurs définitions des foncteurs $K^n(A)$ ont été proposées par différents auteurs. Nous allons décrire deux d'entre elles seulement (cf les autres exposés de ce colloquium).

Définition de Bass-Milnor-Quillen. Pour distinguer cette définition de la suivante nous la noterons $K_n(A)$ (au lieu de $K^{-n}(A)$) comme il est d'usage.

Soit G un groupe discret quelconque et H un sous-groupe ; H est dit parfait s'il est égal à son sous-groupe des commutateurs $[H , H]$. Nous dirons que G est quasi-parfait si son sous-groupe des commutateurs $H = [G , G]$ est parfait. Par exemple, d'après un théorème bien connu de Bass et Whitehead [4], le groupe $G = GL(A)$ est quasi-parfait. Soit enfin X un CW-complexe tel que $\pi_1(X) = G$ soit un groupe quasi-parfait. Alors, il existe un CW-complexe X^+ et une application continue $i : X \to X^+$ (uniques à homotopie près) qui satisfont aux deux propriétés suivantes

a) $\pi_1(X^+) = G/H$ et l'homomorphisme $\pi_1(i) : \pi_1(X) \to \pi_1(X^+)$ coïncide avec l'homomorphisme quotient $G \to G/H$.

b) Si Λ est un système de coefficients locaux sur X^+, l'application i induit un isomorphisme $H_*(X,i^*\Lambda) \overset{\approx}{\to} H_*(X^+,\Lambda)$.

L'espace X^+ est simplement obtenu en ajoutant à X des cellules de dimension 2 et 3.

Appliquons cette construction à l'espace classifiant $X = B_{GL(A)}$ du groupe $G = GL(A)$. On obtient ainsi un espace $X^+ = B^+_{GL(A)}$ dont les groupes d'homotopie $\pi_n(B^+_{GL(A)})$ sont par définition les groupes $K_n(A)$ pour $n > 0$. On voit aisément que le groupe $K_1(A)$ coïncide avec le groupe de Bass $GL(A)/GL'(A)$ où $GL'(A) = [GL(A),GL(A)]$. De même, le groupe $K_2(A)$ coïncide avec le groupe de Milnor $H_2(GL'(A) ; \mathbf{Z})$.

<u>Définition de Karoubi-Villamayor</u> Cette définition peut être explicitée de plusieurs manières [12][17]. Sans doute la manière la plus simple est de copier la définition des groupes d'homotopie d'un espace topologique. Ainsi, $K^{-n}(A) = \pi^a_{n-1}(GL(A))$ est le $(n-1)^{\text{ième}}$ groupe d'homotopie "algébrique" de $GL(A)$. De manière plus précise, $K^{-1}(A) = \pi^a_0(GL(A))$ est le quotient de $GL(A)$ par la relation d'équivalence suivante : $\alpha_0 \sim \alpha_1 \Leftrightarrow \exists \alpha(x) \in GL(A[x])$ tel que $\alpha(0) = \alpha_0$ et $\alpha(1) = \alpha_1$. Le groupe $K^{-2}(A) = \pi^a_1(GL(A))$ est obtenu en considérant les classes d'homotopie de lacets "algébriques" $\alpha(x)$ tels que $\alpha(0) = \alpha(1) = 1$. Plus généralement, soit Ω^+A le sous-anneau de $A[x]$ formé des polynômes $\alpha(x)$ tels que $\alpha(0) = \alpha(1)$. On a un homomorphisme évident $GL(\Omega^+A) \to GL(A)$ induit par "l'augmentation" $\Omega^+A \to A$ qui associe à un polynôme $\alpha(x)$ sa valeur en 0. Les groupes $K^{-n}(A)$ sont alors définis par récurrence sur n par la formule $K^{-n-1}(A) = \mathrm{Ker}[K^{-n}(\Omega^+A) \to K^{-n}(A)]$, $n > 0$. Gersten a donné une interprétation simpliciale de ces groupes et explicité une suite spectrale dont le terme E^1_{pq} est le groupe $K_q(A[x_1, \ldots, x_p])$ et qui converge vers $K^{-p-q}(A)$. En particulier, si A est un anneau tel que l'inclusion évidente de A dans $A[x_1, \ldots, x_p]$ induise un isomorphisme de $K_q(A)$ sur $K_q(A[x_1, \ldots, x_p])$ pour tout couple d'entiers (p,q), le "edge homomorphisme" de la suite spectrale $K_n(A) \to K^{-n}(A)$ est un isomorphisme. D'après un théorème récent de Quillen , il en est ainsi si A est un anneau noetherien régulier (i.e. si A est un anneau noethérien tel que tout A-module admette une résolution projective finie). Ainsi les théories K_n et K^{-n} coïncident sur une catégorie assez large d'anneaux. La situation est ici formellement analogue à celle des théories homologiques qui satisfont aux axiomes d'Eilenberg-Steenrod : elles coïncident sur une catégorie assez large d'espaces topologiques qui est celle des CW-complexes finis.

Les groupes $K_n(A)$ n'ont été calculés pour toutes les valeurs de n que pour très peu d'anneaux A. Les deux exemples fondamentaux sont les suivants et sont dûs respectivement à Quillen et Borel. Si F_q désigne un corps fini à q éléments, on a $K_n(F_q) = 0$ si n est pair et $K_{2i-1}(F_q) = \mathbb{Z}/(q^i-1)\mathbb{Z}$. Si F est un corps de nombres, on a $K_n(F) \otimes_{\mathbb{Z}} \mathbb{Q} \approx K_{n+4}(F) \otimes_{\mathbb{Z}} \mathbb{Q}$ pour $n > 1$ et le rang de $K_n(F)$ peut être calculé explicitement en fonction du nombre de places complexes et réelles du corps F [7]. Par exemple, $K_n(\mathbb{Q}) \otimes_{\mathbb{Z}} \mathbb{Q} = 0$ si $n \neq 1 \bmod 4$ et $K_n(\mathbb{Q}) \otimes_{\mathbb{Z}} \mathbb{Q} \approx \mathbb{Q}$ si $n = 1 \bmod 4$ et $n > 1$.

Ces résultats suggèrent fortement que les groupes $K_n(A) \otimes_{\mathbb{Z}} \mathbb{Q}$ ou $K^{-n}(A) \otimes_{\mathbb{Z}} \mathbb{Q}$ sont périodiques de période 4 . Il n'en est rien malheureusement comme le montre l'exemple des polynômes laurentiens sur un corps de nombres ou sur un corps fini [16] . D'autre part, si $A = C_F(X)$, l'homomorphisme évident $K^{-n}(A) \to K_F^{-n}(X)$ n'est pas un isomorphisme en général déjà si X est un point et $n = 1$ ou par des arguments plus élaborés si $n > 1$. Ceci montre que la théorie que nous venons de développer est assez loin somme toute de la théorie topologique. Nous allons voir que la situation est sensiblement différente en K-théorie hermitienne.

Alors que la K-théorie ordinaire s'occupe de problèmes de classification de modules, la K-théorie hermitienne a pour objet la classification des modules quadratiques. On peut dire que la K-théorie hermitienne s'intéresse aux groupes orthogonaux, unitaire, symplectique, etc... alors que la K-théorie ordinaire s'intéresse aux phénomènes liés au groupe linéaire. La K-théorie hermitienne apparaît ainsi à priori comme plus complexe que la K-théorie ordinaire. Cependant les problèmes de classification de formes quadratiques (sur un corps seulement jusqu'à une période très récente) ont de tout temps intéressé les mathématiciens. Les travaux récents sur la chirurgie des variétés (en particulier ceux de Wall [29]) conduisent naturellement aux problèmes de classification des formes quadratiques sur des anneaux quelconques.

Soit donc A un anneau muni d'une antiinvolution $a \mapsto \overline{a}$. On a ainsi les identités $\overline{a+b} = \overline{a} + \overline{b}$, $\overline{ab} = \overline{b}\,\overline{a}$ et $\overline{1} = 1$. Pour simplifier, nous supposerons que 2 est inversible dans A (cette restriction est nécessaire aussi pour certains théorèmes). Soit enfin ε un élément du centre de A tel que $\varepsilon\,\overline{\varepsilon} = 1$. Une <u>forme ε-quadratique</u> sur un A-module projectif de type fini M est la donnée d'une application $\mathbb{Z}$-bilinéaire $\theta : M \times M \to A$ qui satisfait aux identités usuelles

$$\theta(x\lambda,y) = \bar{\lambda}\,\theta(x,y)$$

$$\theta(x,y\lambda) = \theta(x,y)\lambda$$

$$\theta(y,x) = \varepsilon\,\overline{\theta(x,y)}\ ,$$

x et y étant des éléments de M et λ étant un élément de A (les A-modules considérés sont des A-modules à droite pour fixer les idées). Soit ${}^tM$ l'ensemble des applications $\mathbb{Z}$-linéaires $f : M \to A$ telles que $f(x\lambda) = \bar{\lambda}f(x)$. Alors ${}^tM$ est un A-module à droite grâce à la formule $(f\lambda)(x) = f(x)\lambda$ et on a ${}^t({}^tM) \approx M$ de manière canonique. Il est clair que la correspondance $M \to {}^tM$ induit un foncteur contravariant de $\mathcal{P}(A)$ dans $\mathcal{P}(A)$ (le foncteur de "dualité"). Une forme ε-quadratique θ définit un homomorphisme $\tilde{\theta} : M \to {}^tM$ par la formule $\tilde{\theta}(y)(x) = \theta(x,y)$.L'homomorphisme "transposé" ${}^t\tilde{\theta}: M \approx {}^t({}^tM) \to {}^tM$ coincide avec $\varepsilon\,\tilde{\theta}$. Réciproquement, un tel homomorphisme $\tilde{\theta}$ définit de manière évidente une forme ε-quadratique θ sur M . La forme θ est dite <u>non dégénérée</u> si $\tilde{\theta}$ est un isomorphisme. Si M et N sont deux A-modules munis de formes ε-quadratiques θ et θ' , une isométrie de M sur N est un isomorphisme A-linéaire $f : M \to N$ tel que $\theta'(f(x),f(y)) = \theta(x,y)$. Les modules ε-quadratiques et les isométries forment ainsi une catégorie qu'on note ${}_\varepsilon Q(A)$. Les classes d'isométrie de modules ε-quadratiques forment un monoïde pour la somme directe des modules. On note ${}_\varepsilon L(A)$ le groupe symétrisé de ce monoïde.

<u>Exemples</u>

1) Soit A le corps $\mathbb{R}$ des nombres réels. Comme il est bien connu, une forme 1-quadratique (ou simplement "forme quadratique" d'après la terminologie classique) non dégénérée est caractérisée par deux entiers p et q. On a donc ${}_1L(\mathbb{R}) = \mathbb{Z} \oplus \mathbb{Z}$. On démontre aisément que ${}_{-1}L(F) = \mathbb{Z}$ pour tout corps F .

2) Plus généralement, soit $A = C_{\mathbb{R}}(X)$,X étant un espace compact. La catégorie des modules ε-quadratiques est alors équivalente à celle des fibrés vectoriels E munis de formes ε-quadratiques en un sens évident. Si $\varepsilon = 1$, il est bien connu qu'on peut scinder E en $E^+ \oplus E^-$, la forme quadratique

étant définie positive sur E^+ et définie négative sur E^-. En outre, cette décomposition est unique à homotopie près. Par conséquent ${}_1L(A) \approx K_{\mathbb{R}}(X) \oplus K_{\mathbb{R}}(X)$. Si $\varepsilon = -1$, il est aussi bien connu qu'on peut munir E d'une structure complexe, cette structure complexe étant caractérisée à homotopie près par l'existence d'un automorphisme J **et d'une** métrique ζ tels que $J^2 = -1$ et $\theta(e,e') = \zeta(Je,e')$, θ étant la forme (-1)-quadratique (i.e. antisymétrique dans ce contexte) sur le fibré E. Par conséquent ${}_{-1}L(A) = K_{\mathbb{C}}(X)$.

3) Soit $A = \mathbb{Q}$ le corps des nombres rationnels. Alors la classification des formes quadratiques sur $\mathbb{Q}$ est bien connue(cf.[26] par exemple). On trouve ${}_1L(\mathbb{Q}) = \mathbb{Z} \oplus \mathbb{Z} \oplus 2$ -torsion, la 2-torsion étant caractérisée par le discriminant et l'invariant de Hasse-Witt qui s'exprime en termes de symboles de Hilbert [26]. Des méthodes analogues s'appliquent aux corps de nombres [23]. Notons que, pour tout corps F, ${}_1L(F)$ est la somme directe d'un groupe libre et d'un groupe de 2-torsion (cf les travaux de Pfister et de Scharlau).

4) Soit enfin B un anneau quelconque et soit B° l'anneau opposé. Soit $A = B \times B^\circ$ avec l'antiinvolution $(x,y) \mapsto (y,x)$. Il est facile de voir que ${}_\varepsilon L(A) \approx K(B)$.

En K-théorie hermitienne, un rôle important est joué par deux foncteurs

$$F : {}_\varepsilon Q(A) \to \overline{\mathcal{P}(A)} \quad \text{et} \quad H : \overline{\mathcal{P}(A)} \to {}_\varepsilon Q(A)$$

où $\overline{\mathcal{P}(A)}$ désigne la catégorie dont les objets sont les objets de $\mathcal{P}(A)$ et dont les morphismes sont les isomorphismes de $\mathcal{P}(A)$. Le foncteur F est simplement le foncteur "oubli" : à un module quadratique on associe le A-module sous-jacent. Le foncteur H est le <u>foncteur hyperbolique</u> explicité dans [5][18][29]. Pour tout module M, on a $H(M) = M \oplus {}^tM$ muni de la forme ε-quadratique θ associée à l'isomorphisme $\tilde{\theta}: (M \oplus {}^tM) \to {}^t(M \oplus {}^tM) \approx M \oplus {}^tM$ défini par la matrice

$$\tilde{\theta} = \begin{pmatrix} 1 & 0 \\ 0 & \varepsilon \end{pmatrix}$$

Si $\alpha : M \to N$ est un isomorphisme, $H(\alpha): M \oplus {}^tM \to N \oplus {}^tN$ est l'isométrie définie par la matrice

$$H(\alpha) = \begin{pmatrix} \alpha & 0 \\ 0 & {}^t\alpha^{-1} \end{pmatrix}$$

le foncteur H induit évidemment un homomorphisme $K(A) \to {}_\varepsilon L(A)$ dont le conoyau est le "groupe de Witt" ${}_\varepsilon W(A)$. Dans l'exemple 2 cité plus haut avec $A = C_{\mathbb{R}}(X)$, l'homomorphisme $K(A) \to {}_\varepsilon L(A)$ coïncide avec l'homomorphisme diagonal $K_{\mathbb{R}}(X) \to K_{\mathbb{R}}(X) \oplus K_{\mathbb{R}}(X)$. Donc ${}_1W(A)$ est isomorphe à $K_{\mathbb{R}}(X)$. Ainsi $K_{\mathbb{R}}(X)$ peut être interprété comme le groupe de Witt de $C_{\mathbb{R}}(X)$; cette interprétation permet de mieux comprendre la périodicité de Bott en K-théorie topologique comme nous allons le voir (interprétation remarquée pour la 1ère fois par Gelfand et Mischenko).

La définition des foncteurs K_n et K^{-n} se transpose aisément en K-théorie hermitienne et permet d'introduire des foncteurs dérivés analogues L_n et L^{-n}. De manière précise, les modules hyperboliques $H(A^n)$ jouent en K-théorie hermitienne le rôle des modules libres A^n en K-théorie ordinaire : tout module quadratique est un facteur direct de $H(A^n)$ pour certain n. Il est donc naturel de considérer le groupe ${}_\varepsilon O_{n,n}(A)$ formé des isométries du module ε-quadratique $H(A^n)$. La limite inductive ${}_\varepsilon O(A) = \varinjlim_n {}_\varepsilon O_{n,n}(A)$ est le groupe ε-orthogonal infini associé à l'anneau A . D'après [18][29] le groupe ${}_\varepsilon O(A)$ est quasi-parfait. En remplaçant $GL(A)$ par ${}_\varepsilon O(A)$ dans la définition des groupes K_n et K^{-n}, on obtient ainsi de nouveaux groupes notés ${}_\varepsilon L_n(A)$ et ${}_\varepsilon L^{-n}(A)$. Ainsi le groupe ${}_\varepsilon L_n(A)$ par exemple est le $n^{\text{ième}}$ groupe d'homotopie de l'espace B_G^+ avec $G = {}_\varepsilon O(A)$. De même, ${}_\varepsilon L^{-n}(A) = \pi_{n-1}^a({}_\varepsilon O(A))$. Les groupes L_n et L^{-n} définis ici sont différents des groupes L_p^s et L_p^h de Wall [29]. En fait, ceux-ci se déduisent de ${}_\varepsilon L_1$ et ${}_\varepsilon L_0$, $\varepsilon = \pm 1$, par des formules simples (cf.§ 4 et 5).

Le foncteur hyperbolique induit un homomorphisme $GL_n(A) \to {}_\varepsilon O_{n,n}(A)$ et, en passant à la limite, un homomorphisme $GL(A) \to {}_\varepsilon O(A)$. Ce dernier induit donc des homomorphismes de groupes abéliens $K_n(A) \to {}_\varepsilon L_n(A)$ et $K^{-n}(A) \to {}_\varepsilon L^{-n}(A)$ dont les conoyaux sont naturellement notés ${}_\varepsilon W_n(A)$ et ${}_\varepsilon W^{-n}(A)$. Un des principaux objectifs de cet article est la démonstration du théorème suivant (on n'en démontrera qu'une partie dans cette rédaction).

0.1. <u>THEOREME</u> <u>Posons</u> ${}_{\varepsilon}\overline{W}_n(A) = {}_{\varepsilon}W_n(A) \underset{\mathbb{Z}}{\otimes} \mathbb{Z}'$ <u>et</u> ${}_{\varepsilon}\overline{W}^{-n}(A) = {}_{\varepsilon}W^{-n}(A) \underset{\mathbb{Z}}{\otimes} \mathbb{Z}'$. <u>Alors l'homomorphisme naturel</u> ${}_{\varepsilon}\overline{W}_n(A) \to {}_{\varepsilon}\overline{W}^{-n}(A)$ <u>est un isomorphisme</u>. <u>En outre, on a des isomorphismes naturels</u> ${}_{\varepsilon}\overline{W}^{-n}(A) \approx {}_{-\varepsilon}\overline{W}^{-n-2}(A) \approx {}_{\varepsilon}\overline{W}^{-n-4}(A) \approx \dots$

On notera l'analogie de ce théorème avec un théorème de périodicité de Bott (sous une forme légèrement plus faible) : $K^n_{\mathbb{R}}(X) \underset{\mathbb{Z}}{\otimes} \mathbb{Z}' \approx K^{n+4}_{\mathbb{R}}(X) \underset{\mathbb{Z}}{\otimes} \mathbb{Z}'$. En fait, une démonstration commune de ces deux théorèmes est donnée dans le paragraphe 4 .

On peut aussi caractériser axiomatiquement la théorie ${}_{\varepsilon}W^{-n}$ indépendamment de toute construction de topologie algébrique. Pour cela, il est commode d'étendre nos définitions aux anneaux A n'ayant pas nécessairement d'élément unité. Dans ce cas, notons A^+ l'ensemble $A \times \mathbb{Z}'$ avec les lois d'addition et de multiplication suivantes qui en fait un anneau avec élément unité :

$$(a,\lambda) + (a', \lambda') = (a + a', \lambda + \lambda')$$

$$(a,\lambda) \cdot (a', \lambda') = (aa' + \lambda' a + \lambda a', \lambda\lambda')$$

L'homomorphisme évident $A^+ \to \mathbb{Z}'$ induit un homomorphisme ${}_{\varepsilon}W(A^+) \to {}_{\varepsilon}W(\mathbb{Z}')$ dont le noyau est désigné par ${}_{\varepsilon}W(A)$. On vérifie aisément que cette définition est cohérente avec la précédente dans le cas où A a déjà un élément unité et qu'elle est fonctorielle vis-à-vis des morphismes quelconques d'anneaux. On pose de même ${}_{\varepsilon}W^{-n}(A) = \mathrm{Ker}[{}_{\varepsilon}W^{-n}(A^+) \to {}_{\varepsilon}W^{-n}(\mathbb{Z}')]$ et ${}_{\varepsilon}\overline{W}^{-n}(A) = {}_{\varepsilon}W^{-n}(A) \underset{\mathbb{Z}}{\otimes} \mathbb{Z}'$

0.2 <u>THEOREME</u> <u>La théorie</u> ${}_{\varepsilon}\overline{W}^{-n}(A)$, $n \geq 0$, <u>est caractérisée par les axiomes suivants</u> (à isomorphisme près) :

1) ${}_{\varepsilon}\overline{W}^{0}(A) = {}_{\varepsilon}\overline{W}(A)$

2) L'homomorphisme évident $A \to A[x]$ induit un isomorphisme

${}_{\varepsilon}\overline{W}^{-n}(A) \approx {}_{\varepsilon}\overline{W}^{-n}(A[x])$.

3) <u>Soit</u> $0 \to A' \to A \to A'' \to 0$

<u>une "suite exacte d'anneaux"</u> (i.e. une suite d'anneaux et d'homomorphismes d'anneaux telle que la suite de groupes abéliens sous-jacents soit exacte). <u>On a alors une "suite exacte de cohomologie"</u>

$$\ldots \quad {}_{\varepsilon}\overline{W}^{-n-1}(A) \to {}_{\varepsilon}\overline{W}^{-n-1}(A'') \to {}_{\varepsilon}\overline{W}^{-n}(A') \to {}_{\varepsilon}\overline{W}^{-n}(A) \to {}_{\varepsilon}\overline{W}^{-n}(A'')$$

0.3. COROLLAIRE Soit

$$\begin{array}{ccc} A & \longrightarrow & A_2 \\ \downarrow & & \downarrow \\ A_1 & \longrightarrow & A' \end{array}$$

un carré cartésien d'anneaux. On a alors la suite exacte (dite "de Mayer-Vietoris")

$${}_{\varepsilon}\overline{W}^{-n-1}(A_1) \oplus {}_{\varepsilon}\overline{W}^{-n-1}(A_2) \to {}_{\varepsilon}\overline{W}^{-n-1}(A') \to {}_{\varepsilon}\overline{W}^{-n}(A) \to {}_{\varepsilon}\overline{W}^{-n}(A_1) \oplus {}_{\varepsilon}\overline{W}^{-n}(A_2) \to {}_{\varepsilon}\overline{W}^{-n}(A')$$

On notera que, d'après Swan , ce corollaire est faux pour les groupes K_n . D'autre part, si A est une algèbre de Banach, on définit de manière évidente des groupes $W_t^{-n}(A)$ en utilisant les groupes d'homotopie usuels au lieu des groupes d'homotopie algébriques.

0.4. THEOREME Soit A une algèbre de Banach involutive quelconque. Alors l'application évidente ${}_{\varepsilon}W^{-n}(A) \to {}_{\varepsilon}W_t^{-n}(A)$ induit un isomorphisme

$${}_{\varepsilon}W^{-n}(A) \otimes_{\mathbb{Z}} \mathbb{Z}' \approx {}_{\varepsilon}W_t^{-n}(A) \otimes_{\mathbb{Z}} \mathbb{Z}'$$

Ce théorème s'applique notamment à l'algèbre de Banach $C_F(X)$ où $F = \mathbb{R}$ ou $\mathbb{C}$. Modulo la 2-torsion, on définit ainsi de manière algébrique les groupes $K_F^{-n}(X)$, ce qui répond en partie à la question posée au début de ce paragraphe.

Dans une certaine mesure, le théorème 0.1 n'est pas entièrement satisfaisant. Il est bien connu par exemple que la 2-torsion est l'élément le plus intéressant dans le groupe de Witt d'un corps. Cette 2-torsion joue aussi un rôle important dans certaines questions de K-théorie topologique réelle. D'autre part, des contre-exemples appropriés montrent que la théorie ${}_{\varepsilon}W^{-n}$ n'est pas périodique en général. Pour comprendre ce qui va suivre, il est bon de revenir à la K-théorie topologique pour un bref moment et tâcher d'interpréter les théorèmes de périodicité de Bott de manière à pouvoir les étendre en K-théorie hermitienne.

Soit donc A une algèbre de Banach involutive. Les foncteurs hyperbolique et oubli induisent des inclusions évidentes $GL_n(A) \subset {}_{\varepsilon}O_{n,n}(A) \subset GL_{2n}(A)$. En passant à la limite, on a $GL(A) \subset {}_{\varepsilon}O(A) \subset GL(A)$

0.5. THEOREME Soit A une algèbre de Banach involutive. On a alors une équivalence d'homotopie entre les composantes des espaces $GL(A)/_{\varepsilon}O(A)$ et $\Omega({}_{-\varepsilon}O(A)/GL(A))$.

Ce théorème est démontré par deux méthodes différentes dans les chapitres 3 et 4 respectivement. Il est facile de voir que ce théorème implique les théorèmes de périodicité de Bott dans toute leur force. En effet, si $A = \mathbb{R}$, $\mathbb{C}$ ou $\mathbb{H}$ avec certaines antiinvolutions, on trouve 6 équivalences d'homotopie $O/U \sim \Omega O$, $\mathbb{Z} \times BO \sim \Omega U/O$, $U/O \sim \Omega Sp/U$, $U/Sp \sim \Omega O/U$, $\mathbb{Z} \times Bsp \sim \Omega U/sp$ $Sp/U \sim \Omega Sp$. Ces équivalences d'homotopie jointes aux équivalences triviales $O \sim \Omega(\mathbb{Z} \times BO)$ et $Sp \sim \Omega(\mathbb{Z} \times Bsp)$ impliquent bien entendu que tous les espaces considérés ont le type d'homotopie de leur huitième espace de lacets itéré. On démontre de même les équivalences d'homotopie $\mathbb{Z} \times BU \sim \Omega U$ et $U \sim (\mathbb{Z} \times BU)$. On notera que cette nouvelle interprétation des théorèmes de périodicité de Bott ne fait pas appel aux algèbres de Clifford contrairement aux tentatives précédentes.

L'analogue des espaces homogènes $GL(A)/_{\varepsilon}O(A)$ et ${}_{\varepsilon}O(A)/GL(A)$ en K-théorie hermitienne est obtenu en considérant les fibres ${}_{\varepsilon}\mathcal{V}(A)$ et ${}_{\varepsilon}\mathcal{U}(A)$ des applications naturelles

$$B^{+}_{{}_{\varepsilon}O(A)} \to B^{+}_{GL(A)} \quad \text{et} \quad B^{+}_{GL(A)} \to B^{+}_{{}_{\varepsilon}O(A)}$$

(ces applications étant transformées en des fibrations de Serre par le procédé standard). Si on pose ${}_{\varepsilon}V_n(A) = \pi_n({}_{\varepsilon}\mathcal{V}(A))$ et ${}_{\varepsilon}U_n(A) = \pi_n({}_{\varepsilon}\mathcal{U}(A))$ $(n \geq 1)$, on a donc les suites exactes

$${}_{\varepsilon}L_{n+1}(A) \to K_{n+1}(A) \to {}_{\varepsilon}V_n(A) \to {}_{\varepsilon}L_n(A) \to K_n(A)$$

$$K_{n+1}(A) \to {}_{\varepsilon}L_{n+1}(A) \to {}_{\varepsilon}U_n(A) \to K_n(A) \to {}_{\varepsilon}L_n(A) \qquad (n \geq 1)$$

En fait, en suivant Bass [4], on peut de même définir ${}_{\varepsilon}V(A) = {}_{\varepsilon}V_0(A)$ et ${}_{\varepsilon}U(A) = {}_{\varepsilon}U_0(A)$ comme les groupes de Grothendieck des foncteurs F et H respectivement. Les suites exactes précédentes sont alors valables au niveau des groupes K_0 et L_0 . Les définitions des groupes ${}_\varepsilon V_n$ et ${}_\varepsilon U_n$ ont été inspirées par le travail de Novikov en K-théorie hermitienne [22] et sont aussi liées aux groupes de Wall L_p^h et L_p^s (p étant défini mod. 4). Cependant, ces définitions ne coincident pas avec les définitions de Novikov.

0.6. CONJECTURE Les composantes connexes des espaces ${}_\varepsilon\mathcal{V}(A)$ et $\Omega({}_{-\varepsilon}\mathcal{U}(A))$ ont naturellement le même type d'homotopie [(1)]

0.7. COROLLAIRE Les théories ${}_\varepsilon V_n(A)$ et ${}_{-\varepsilon}U_{n+1}(A)$ sont naturellement isomorphes.

0.8. COROLLAIRE Soit A un anneau noéthérien régulier. Alors l'application naturelle de A dans A[x] induit des isomorphismes ${}_\varepsilon L_n(A) \approx L_n(A[x])$, ${}_\varepsilon U_n(A[x]) \approx {}_\varepsilon U_n(A[x])$, ${}_\varepsilon V_n(A) \approx {}_\varepsilon V_n(A[x])$.

Le dernier corollaire se démontre par récurrence sur n en utilisant essentiellement le corollaire 0.7 et le théorème de Quillen $K_n(A) \approx K_n(A[x])$. En particulier, on a ainsi ${}_\varepsilon L_1(A[x]) \approx {}_\varepsilon L_1(A)$ pour tout anneau noéthérien régulier. L'auteur ne connait pas de démonstration élémentaire de ce fait. Remarquons aussi que le corollaire 0.7 pour n = 0 a été déjà démontré par R . Sharpe par des méthodes entièrement différentes de celles que nous utilisons. Enfin, si on localise les espaces ${}_\varepsilon\mathcal{V}(A)$ et $\Omega({}_\varepsilon\mathcal{U}(A))$ grâce au système multiplicatif (2^r) , on voit aisément que le corollaire 0.7 est équivalent au théorème 0.1.

Dans la dernière partie de ce paragraphe nous n'avons considéré pour simplifier que les groupes ${}_\varepsilon L_n$, ${}_\varepsilon V_n$, etc. ... définis par la méthode de Quillen (cf. aussi

---

(1) Cette conjecture peut être démontrée si les isomorphismes $K_n(A) \approx K_{n+1}(SA)$ et ${}_\varepsilon L_n(A) \approx {}_\varepsilon L_{n+1}(SA)$ sont définis par des cup-produits. Le premier isomorphisme a été démontré par Gersten semble-t-il mais l'auteur n'a pas encore compris sa démonstration (cf. § 5).

[1][25]). On peut aussi bien considérer des groupes ${}_{\varepsilon}L^{-n}$, ${}_{\varepsilon}V^{-n}$, etc. ... en introduisant des groupes d'homotopie algébriques. Un avantage de cette définition est qu'elle s'étend immédiatement au cas topologique et au cas ultramétrique. On obtient alors des théorèmes du même type ${}_{\varepsilon}V^{n}(A) \approx {}_{-\varepsilon}U^{n-1}(A)$ (les théories étant définies ici pour $n \in \mathbb{Z}$) du moins si A est "K-régulier" (cf. § 1). Les théorèmes de périodicité de Bott apparaissent ainsi comme un cas particulier d'un théorème de nature arithmétique.

Disons maintenant un mot du type d'applications qu'on peut tirer des théorèmes de périodicité. Les méthodes de Borel et Quillen pour calculer les groupes $K_n(A)$ s'appuient essentiellement sur la connaissance de l'homologie du groupe linéaire $GL(A)$. Notre méthode va dans l'autre sens : la périodicité (modulo la 2-torsion) des groupes W nous permet de calculer une partie intéressante de l'homologie (ou de la cohomologie) du groupe orthogonal. Le théorème suivant en est un exemple :

0.9. THEOREME L'homologie à coefficients rationnels du groupe orthogonal infini ${}_{\varepsilon}O(A)$ peut s'écrire comme le produit tensoriel gradué de trois algèbres de Hopf, soit

$$H_*({}_{\varepsilon}O(A)) \approx {}_{\varepsilon}S_A \otimes {}_{\varepsilon}\Lambda_A \otimes {}_{\varepsilon}M_A$$

où :

1) ${}_{\varepsilon}S_A$ est l'algèbre symétrique de l'espace vectoriel gradué

$$\bigoplus_{n>0} {}_{\varepsilon}W^{-2n}(A) \otimes_{\mathbb{Z}} \mathbb{Q}$$

2) ${}_{\varepsilon}\Lambda_A$ est l'algèbre extérieure de l'espace vectoriel gradué

$$\bigoplus_{n>0} {}_{\varepsilon}W^{-2n-1}(A) \otimes_{\mathbb{Z}} \mathbb{Q}$$

3) ${}_{\varepsilon}M_A = \mathrm{Im}[H_*(GL(A)) \to H_*({}_{\varepsilon}O(A))]$ (Cette dernière assertion dépend de la conjecture).

Ce théorème, conjugué avec le théorème 0.1, permet un calcul partiel de l'homologie du groupe orthogonal. Par exemple, si F est un corps commutatif, on a ${}_{-1}W^{-4n}(F) \otimes_{\mathbb{Z}} \mathbb{Q} = {}_{\varepsilon}W^{-4n-1}(F) \otimes_{\mathbb{Z}} \mathbb{Q} = {}_{\varepsilon}W^{-4n-3}(F) \otimes_{\mathbb{Z}} \mathbb{Q} = 0$. Le groupe ${}_{1}W^{-4n}(F) \otimes_{\mathbb{Z}} \mathbb{Q}$

est isomorphe au groupe de Witt $W(F)$ du corps F tensorisé par $\mathbb{Q}$. En particulier, si F est un corps de nombres, le rang du groupe $W(F)$ est égal au nombre de places réelles du corps F . En fait, dans ce cas, le théorème est inclus dans un théorème de Borel analogue à celui cité précédemment. La méthode de Borel (qui est entièrement différente de celle-ci) permet de calculer le facteur ${}_{\varepsilon}M_A$: c'est une algèbre extérieure qui possède r_2 générateurs en chaque degré de la forme $4n + 3$, $2r_2$ représentant le nombre de places complexes du corps F .

Une autre interprétation de la périodicité de Bott en K-théorie a été suggérée initialement par Bass, Heller et Swan [4], puis généralisée par plusieurs auteurs [12][16][28] . On a ainsi les théorèmes suivants par exemple :

0.10. THEOREME Soit A un anneau quelconque et soit $A_z = A[z,z^{-1}]$ l'anneau des polynômes laurentiens à coefficients dans A . On a alors une décomposition naturelle

$$K_{n+1}(A_z) = K_{n+1}(A) \oplus K_n(A) \oplus ?$$

(pour une démonstration voir [16] si $n \leqslant 1$ et [12][28] pour n quelconque).

0.11. THEOREME (Quillen) Soit A un anneau noethérien régulier. On a alors une décomposition naturelle

$$K_{n+1}(A_z) = K_{n+1}(A) \oplus K_n(A)$$

En utilisant le théorème 0.1, on peut démontrer un théorème analogue en "L-théorie" :

0.12. THEOREME Soit A un anneau hermitien (non nécessairement noethérien régulier). On a alors des isomorphismes naturels

$${}_{\varepsilon}\overline{W}^{-n-1}(A_z) = {}_{\varepsilon}\overline{W}^{-n-1}(A) \oplus {}_{\varepsilon}\overline{W}^{-n}(A)$$

Une version plus précise de ce théorème est donnée dans les paragraphes 1 et 5. Bien entendu on a des théorèmes analogues pour les théories ${}_{\varepsilon}W_n$, ${}_{\varepsilon}L_n$, ${}_{\varepsilon}U_n$, etc..

Ce dernier type de résultats suggère fortement la possibilité de retrouver les résultats de Shaneson sur les groupes L_p^h et L_p^s de manière purement algébrique . Une première tentative avait été déjà faite par Novikov (en négligeant la 2-torsion) [22], tentative qui a été reprise par Ranicki avec succès

Notre méthode est sensiblement différente de celle de Novikov et Ranicki. Elle conduit à des résultats plus généraux dans un sens (on considère ici tous les groupes ${}_\varepsilon L_n$ et non seulement ${}_\varepsilon L_1$ et ${}_\varepsilon L_0$) , mais moins généraux dans un autre sens car nous avons été obligés de supposer 2 inversible dans l'anneau A à certains points importants des démonstrations. Il serait évidemment souhaitable d'avoir un théorème du type théorème 0.12 si 2 n'est pas inversible dans A .

Comme il a été précisé dans l'introduction, ce qui précède n'est qu'un résumé des principaux résultats établis dans cet article. Pour des raisons de présentation nous n'avons pas suivi dans le détail le plan esquissé dans cette introduction. Aussi est-il bon de préciser maintenant la substance des paragraphes suivants.

Le premier paragraphe établit un certain nombre de propriétés de base du foncteur ${}_\varepsilon L(A) = {}_\varepsilon L_0(A)$, en particulier son comportement lorsqu'on effectue sur l'anneau A une extension pôlynomiale ou une extension laurentienne. Sa lecture pré-suppose celle de [18].

Le deuxième paragraphe est plus technique et peut être lu rapidement en première lecture. Les foncteurs ${}_\varepsilon U^n$ et ${}_\varepsilon V^n$ y sont définis de manière précise dans l'esprit des groupes d'homotopie algébriques et leurs propriétés "cohomologiques" essentielles établies.

Le troisième paragraphe est assez indépendant des autres. On y étudie en détail le cas topologique (i.e. celui où A est une algèbre de Banach involutive). Contrairement à ce qui se passe dans le cas algébrique, les algèbres de Clifford y jouent un rôle essentiel. En contre-partie, les résultats établis dans ce paragraphe suggèrent fortement des interprétations algébriques qui seront exploitées par la suite. Dans l'ensemble de ce travail, ce paragraphe joue essentiellement un rôle heuristique.

Dans le quatrième paragraphe, on démontre l'analogue du corollaire 0.7.

pour les théories ${}_{\varepsilon}U^n$ et ${}_{\varepsilon}V^n$ ainsi que la périodicité des groupes de Witt ${}_{\varepsilon}\overline{W}^n$. Les méthodes sont fortement inspirées par celles du cas topologique (avec les algèbres de Clifford en moins !).

Enfin, le cinquième paragraphe tâche d'étendre à la théorie de Quillen les résultats établis dans les paragraphes suivants. Sa rédaction est provisoire et sera probablement modifiée de manière essentielle dans une rédaction ultérieure (pour pouvoir démontrer la conjecture 0.6).

En fait, les quatrième et cinquième paragraphes peuvent être généralisés simultanément dans l'esprit des catégories simpliciales introduites par D. Anderson [1]. Ainsi les théories K_n et L_n d'une part et les théories K^{-n} et L^{-n} d'autre part apparaissent comme des cas particuliers d'une même théorie simpliciale. Il est alors sans doute possible de généraliser la conjecture 0.6 dans ce cadre en utilisant les méthodes du cinquième paragraphe.

N.B. Sauf mention expresse du contraire, 2 est inversible dans les anneaux hermitiens A considérés. D'autre part on suppose

a) Soit que A est une algèbre de Banach involutive

b) Soit que $\| \frac{1}{2} \| \leq 1$ et $\|\lambda\| \leq 1$ si $\lambda \in \mathbb{Z}$

(ces hypothèses assurent la convergence de la série $\sqrt{1+\nu}$ si $\nu^n \to 0$ quand $n \to \infty$).

TABLE DES MATIERES

I - INVARIANCE HOMOTOPIQUE DE LA K-THEORIE HERMITIENNE

Soit A un anneau hermitien et soit ${}_{\varepsilon}L'(A)$ le noyau de l'homomorphisme naturel ${}_{\varepsilon}L(A) \to K(A)$[(1)]. Alors ${}_{\varepsilon}L'(A)$ est de manière évidente un foncteur (covariant) de A.

THEOREME 1.1. Le foncteur ${}_{\varepsilon}L'(A)$ est un invariant du type d'homotopie de A. De manière précise, l'inclusion évidente de A dans $A\langle x\rangle$ induit un isomorphisme ${}_{\varepsilon}L'(A) \approx {}_{\varepsilon}L'(A\langle x\rangle)$.

DEMONSTRATION Il est clair que ${}_{\varepsilon}L'(A) \to {}_{\varepsilon}L'(A\langle x\rangle)$ est injective. Montrons donc que l'application est surjective. D'après le théorème 1.4 de [18], tout élément de ${}_{\varepsilon}L'(A\langle x\rangle)$ peut s'écrire $\bar{M}-\bar{N}$ où $\bar{N} = H(A\langle x\rangle)^n$ et où $\bar{M}$ est un $A\langle x\rangle$-module libre de rang $2n$ muni d'une forme quadratique non dégénérée. Posons $\bar{M} = M \otimes_A A\langle x\rangle$ avec $M = A^{2n}$. Puisque 2 est inversible dans A, la donnée d'une forme ε-quadratique non dégénérée s sur $\bar{M}$ est équivalente à la donnée d'un isomorphisme $\alpha(x) : \bar{M} \to {}^t\bar{M}$, $\alpha(x) = \sum_{n=0}^{\infty} \alpha_n x^n$, tel que ${}^t(\alpha(x)) = \varepsilon\alpha(x)$.
En outre, les paires $(M,\alpha(x))$ et $(M, {}^t(\beta(x))\alpha(x)\beta(x))$, où $\beta(x)$ est un automorphisme de $\bar{M}$, donnent le même élément de ${}_{\varepsilon}L'(A\langle x\rangle)$. La donnée du couple $(M,\alpha(x))$ est ainsi équivalente à la donnée suivante : soit M_0 le A-module M muni de la forme ε-quadratique définie par α_0. Si on pose $\alpha'(x) = \alpha_0^{-1}\alpha(x)$, on a $\alpha'(x) = \alpha'(x)^* = 1 + \alpha'_1 x + \ldots + \alpha'_n x^n + \ldots$ et le couple $(M_0,\alpha'(x))$ est équivalent au couple $(M_0, \beta'(x)^*\alpha'(x)\beta'(x))$ où $\beta'(x) = 1 + \sum_{n=1}^{\infty} \beta_n x^n$ est un automorphisme. Pour démontrer le théorème, on voit donc qu'il suffit de trouver un A-module quadratique N_0 et un automorphisme $\beta'(x)$ de $(M_0 \oplus N_0) \otimes_A A\langle x\rangle$ tel que $\beta'(x)^*(\alpha'(x) \oplus 1)\beta'(x) = 1$. Nous allons d'abord nous ramener au cas où $\alpha' = \alpha'(x)$

(1) Voir [18] pour les définitions de base de la K-théorie hermitienne.

est un polynôme en x . En effet, si $\tilde{\alpha}' = \tilde{\alpha}'(x)$ est une somme partielle approchant suffisamment la somme de la série $\alpha'(x)$, on pourra écrire $\tilde{\alpha}' = \tilde{\alpha}'(1+\nu) = (1+\nu^*)\tilde{\alpha}'$ où ν est topologiquement nilpotent. Les séries définissant $\sqrt{1+\nu}$ et $\sqrt{1+\nu^*}$ convergent et on peut écrire $\alpha' = \sqrt{1+\nu^*}\ \tilde{\alpha}'\ \sqrt{1+\nu}$. Supposons maintenant que α' soit un polynôme en x et montrons, par récurrence sur le degré, qu'on peut se ramener au cas où α' est de degré un . En effet, si $\alpha'(x)$ est de degré p , soit $\alpha'(x) = 1+\alpha'_1 x + \ldots + \alpha'_p x^p$, soit M'_0 le A-module M muni de la forme quadratique $(-\alpha_0)$ et soit $N_0 = M_0 \oplus M'_0$. Considérons alors l'automorphisme $\beta'(x)$ de $\overline{(M_0 \oplus N_0)} = (M_0 \oplus M_0 \oplus M'_0) \underset{A}{\otimes} A[x]$ défini par la matrice

$$\beta'(x) = \begin{pmatrix} 1 & 0 & 0 \\ u & 1 & 0 \\ v & 0 & 1 \end{pmatrix}$$

où on pose

$$u = \frac{(1-\alpha_p)}{2} x^{p/2} \quad \text{et} \quad v = \frac{(1+\alpha_p)}{2} x^{p/2} \quad \text{si } p \text{ est pair} \geq 2$$

$$u = \frac{(1-\alpha_p x)x^{\frac{p-1}{2}}}{2} \quad \text{et} \quad v = \frac{(1+\alpha_p x)x^{\frac{p-1}{2}}}{2} \quad \text{si } p \text{ est impair} > 1 .$$

Alors la forme quadratique définie par $\alpha''(x) = \beta'(x)^*(\alpha'(x) \oplus Id_{N_0} \otimes A(x))\beta'(x)$ s'exprime par la matrice

$$\begin{pmatrix} \alpha + u^2 - v^2 & u & -v \\ u & 1 & 0 \\ v & 0 & 1 \end{pmatrix}$$

qui est bien de degré $\leq p-1$. Enfin, si $\alpha'(x) = 1+\nu x$ est de degré un , ν est topologiquement nilpotent. On utilise encore l'identité $\alpha'(x) = (\sqrt{1+\nu x}))^2$ avec ν auto-adjoint. Ceci achève la démonstration du théorème 1.1.

Remarque (Villamayor). Le théorème précédent est en défaut lorsque A est le corps $\mathbb{Z}/2$. En effet, considérons sur A^2 les formes quadratiques non dégénérées définies par les matrices

$$\begin{pmatrix} 1 & 1 \\ 0 & 1 \end{pmatrix} \quad \text{et} \quad \begin{pmatrix} 0 & 1 \\ 0 & 0 \end{pmatrix}.$$

La matrice

$$\begin{pmatrix} t & 1 \\ 0 & t \end{pmatrix}$$

définit une forme quadratique non dégénérée sur $(A[t])^2$ qui réalise une homotopie entre les deux formes quadratiques précédentes. Cependant, si x est l'élément central non trivial de la partie homogène de degré zéro de l'algèbre de Clifford des deux formes quadratiques, on trouve dans le premier cas l'équation $x^2 + x + 1 = 0$ et dans le second l'équation $x^2 + x = 0$. En d'autres termes, les invariants de Arf des deux formes sont différents. Donc les deux formes quadratiques ne sont pas isomorphes.

Le théorème précédent, susceptible de nombreuses applications, va nous permettre de construire de manière plus explicite que dans [18] les suites exactes de la K-théorie hermitienne. En suivant [16], posons $A_{n,p} = A\langle x_1, \ldots, x_n, t_1, \ldots, t_p, t_1^{-1}, \ldots, t_p^{-1}\rangle$ pour tout anneau de Banach A. Nous dirons que A est <u>(n,p)-régulier</u> si l'inclusion canonique de A dans $A_{n,p}$ induit un isomorphisme $K(A) \approx K(A_{n,p})$. Nous dirons que A est <u>n-régulier</u> (resp. <u>K-régulier</u>) si A est $(n,0)$-régulier (resp. (n,p)-régulier pour tout couple (n,p)).

<u>THEOREME 1.2.</u> <u>Soit A un anneau n-régulier. Alors l'inclusion de A dans $A_{n,0}$ induit un isomorphisme</u> ${}_{\varepsilon}L(A) \approx {}_{\varepsilon}L(A_{n,0})$.

<u>DEMONSTRATION</u> Considérons le diagramme

$$\begin{array}{ccccccc} 0 & \longrightarrow & {}_{\varepsilon}L'(A) & \longrightarrow & {}_{\varepsilon}L(A) & \longrightarrow & K(A) \\ & & \uparrow \alpha & & \uparrow \beta & & \uparrow \gamma \\ 0 & \longrightarrow & {}_{\varepsilon}L'(A_{n,0}) & \longrightarrow & {}_{\varepsilon}L(A_{n,0}) & \longrightarrow & K(A_{n,0}) \end{array}$$

Puisque α et γ sont injectifs, il en est de même de β. Donc β est un isomorphisme.

Si $A \xrightarrow{\psi} A''$ est un homomorphisme surjectif d'anneaux de Banach (resp. d'anneaux hermitiens), Ψ est une GL-fibration d'ordre n(resp. ${}_\varepsilon O$-fibration d'ordre n) si, pour tout élément $\alpha'' = \alpha''(x_1, \ldots, x_n)$ de $GL(A''_{n,0})$(resp. ${}_\varepsilon O(A_{n,0})$) tel que $\alpha''(0, \ldots, 0) = 1$, il existe un élément α de $GL(A_{n,0})$ (resp. ${}_\varepsilon O(A_{n,0})$) tel que $\Psi(\alpha) = \alpha''$ en un sens évident. L'homomorphisme Ψ est une GL-fibration (resp. une ${}_\varepsilon O$-fibration) si c'est une GL-fibration (resp. une ${}_\varepsilon O$-fibration) d'ordre n pour tout n.

PROPOSITION 1.3. *Soit A" un anneau hermitien 1-régulier. Alors* ${}_\varepsilon L_1(SA'') \approx {}_\varepsilon L^{-1}(SA'') \approx {}_\varepsilon L^0(A'')$. *Si A" est n-régulier et si* $f : A \to A''$ *est un homomorphisme surjectif, l'homomorphisme* $Sf : SA \to SA''$ *est une* ${}_\varepsilon O$*-fibration d'ordre* n.

DEMONSTRATION : L'identité ${}_\varepsilon L_1(SA'') \approx {}_\varepsilon L(A'')$ est vraie pour tout anneau hermitien A"[18]. D'autre part ${}_\varepsilon L^{-1}(SA'')$ est le conoyau de la double flèche ${}_\varepsilon L_1((SA'')\langle x\rangle) \rightrightarrows {}_\varepsilon L_1(SA'')$, soit ${}_\varepsilon L(A''\langle x\rangle) \rightrightarrows {}_\varepsilon L(A'')$. D'autre part, on a aussi ${}_\varepsilon L_1(SA''\langle x_1, \ldots, x_n\rangle) \approx L(A''\langle x_1, \ldots, x_n\rangle) \approx {}_\varepsilon L(A'') \approx {}_\varepsilon L_1(SA'')$, ce qui démontre évidemment la deuxième assertion en vertu du théorème 2.6. de [18].

THEOREME 1.4. *Soit*

$$0 \to A' \to A \to A'' \to 0$$

une suite exacte d'anneaux hermitiens (A' étant muni de la norme induite et A" de la norme quotient). *Supposons* A', A *et* A" *(n+1)-réguliers. On a alors la suite exacte*

$${}_\varepsilon L^{-n}(A') \to {}_\varepsilon L^{-n}(A) \to {}_\varepsilon L^{-n}(A'') \to {}_\varepsilon L^{-n+1}(A') \to {}_\varepsilon L^{-n+1}(A) \to \ldots$$

DEMONSTRATION Puisque $\Omega^p B$ est (n-p+1)-régulier si B est n+1-régulier (c'est la même démonstration que celle de la proposition 3.7. de [16]), on peut écrire les isomorphismes ${}_\varepsilon L^{-p-1}(SB) \approx {}_\varepsilon L^{-1}(\Omega^p SB) \approx {}_\varepsilon L^{-1}(S\Omega^p B) \approx {}_\varepsilon L^0(\Omega^p B) \approx L^{-p}(B)$ pour $p \leq n$ avec B = A', A et A". Il suffit d'écrire alors la suite exacte de la K-théorie hermitienne associée à la ${}_\varepsilon O$-fibration d'ordre n

$$0 \to SA' \to SA \to SA'' \to 0$$

Remarque 1 Si $n < 0$, la suite précédente est exacte sans restriction sur A', A ou A''.

Remarque 2 Le théorème est une généralisation du théorème suivant de K-théorie :

Soit

$$0 \to A' \to A \to A'' \to 0$$

une suite exacte d'anneaux de Banach n+1-réguliers. On a alors la suite exacte

$$K^{-n}(A') \to K^{-n}(A) \to K^{-n}(A'') \to K^{-n+1}(A') \to K^{-n+1}(A) \to \dots$$

Bien entendu, ce théorème peut aussi se démontrer directement en suivant la même méthode.

Les méthodes de [16] se généralisent sans peine en K-théorie hermitienne. De manière précise, soit A un anneau hermitien quelconque (2 n'étant pas nécessairement inversible dans A) et soit $A < t,t^{-1}> = A_{0,1}$ l'anneau de Banach des séries convergentes $\sum_{n=-\infty}^{+\infty} a_n t^n$ muni de l'antiinvolution $\sum_{n=-\infty}^{+\infty} a_n t^n \to \sum_{n=\infty}^{+\infty} \bar{a}_n t^{-n}$, $x \to \bar{x}$ représentant l'antiinvolution sur A.

THEOREME 1.5. *Soit $\beta : {}_{\varepsilon}L(A) \to {}_{\varepsilon}L_1(A < t,t^{-1}>)$ l'homomorphisme défini par $\beta(E) = d(E \otimes_A A < t,t^{-1}>, 1 \otimes t)$. Alors β est une injection de ${}_{\varepsilon}L(A)$ sur un facteur direct dans ${}_{\varepsilon}L_1(A<t,t^{-1}>)$. En particulier, on a les décompositions*

$${}_{\varepsilon}L_1(A < t,t^{-1}>) \approx {}_{\varepsilon}L_1(A) \oplus {}_{\varepsilon}L(A) \oplus ?$$

$${}_{\varepsilon}L^n(A < t,t^{-1}>) \approx {}_{\varepsilon}L^n(A) \oplus {}_{\varepsilon}L^{n+1}(A) \oplus ? \quad \text{pour } n > 0$$

$$L^n(A < t,t^{-1}>) \approx {}_{\varepsilon}L^n(A) \oplus {}_{\varepsilon}L^{n+1}(A) \oplus ? \quad \text{pour } n < 0 \text{ si } 2$$

est inversible dans A *et si* A *est (-n)-régulier.*

La démonstration de ce théorème est en tout point analogue à la démonstration du théorème analogue en K-théorie (cf. la démonstration du lemme 2.4 de [16]). Contrairement à ce qui se passe en K-théorie, le facteur ? est réellement non connu, la démonstration du théorème 2.2 de [16] ne pouvant s'étendre au cas hermitien . (cf. § 5).

THEOREME 1.6. Soit A un anneau hermitien K-régulier et soit ${}_{\varepsilon}L^{p,q}(A)$ le *groupe* ${}_{\varepsilon}L(S^p\Omega^q A)$. *On a alors* ${}_{\varepsilon}L^{p,q}(A) \approx {}_{\varepsilon}L^{p-q}(A)$.

DEMONSTRATION Il suffit de prouver que ${}_{\varepsilon}L(S\Omega A) \approx {}_{\varepsilon}L(A)$ d'après les considérations du § 3 de [16]. En effet, des ${}_{\varepsilon}O$-fibrations

$$0 \to \widetilde{A} \to CA \to SA \to 0$$

$$0 \to S\Omega A \to SEA \to SA \to 0 ,$$

on déduit les isomorphismes ${}_{\varepsilon}L^{-1}(SA) \approx {}_{\varepsilon}L(S\Omega A)$ et ${}_{\varepsilon}L^{-1}(SA) \approx {}_{\varepsilon}L(\widetilde{A}) \approx {}_{\varepsilon}L(A)$.

Les structures multiplicatives décrites dans [16] § 4 s'étendent sans peine à la K-théorie hermitienne, l'homomorphisme fondamental

$${}_{\varepsilon}L(A) \times {}_{\eta}L(B) \to {}_{\varepsilon\eta}L(A \underset{\mathbb{Z}'}{\otimes} B)$$

étant défini par le cup-produit des modules quadratiques (plus précisément de modules munis de formes hermitiennes puisque 2 est inversible dans les anneaux que nous considérons). Il convient de remplacer aussi, partout où cela se présente, l'anneau de base $\mathbb{Z}$ par $\mathbb{Z}' = \mathbb{Z}[\frac{1}{2}]$. Le théorème 1.6 nous permet en particulier de construire un élément u de ${}_{1}L(S\Omega\mathbb{Z}')$, image de 1 par l'isomorphisme ${}_{1}L(\mathbb{Z}') \approx {}_{1}L(S\Omega\mathbb{Z}')$. L'isomorphisme de ${}_{\varepsilon}L^{p,q}(A)$ sur ${}_{\varepsilon}L^{p+1,q+1}(A)$ est alors défini par le cup-produit par u .

Remarque Le théorème 1.6 (et son interprétation grâce aux cup-produits) nous sera très utile par la suite. Son intérêt provient évidemment du fait qu'un théorème général sur ${}_{\varepsilon}L^{n_0}$, n_0 fixé, implique un théorème général pour tous les ${}_{\varepsilon}L^n$. Notons aussi que l'hypothèse "K-régulier" n'est pas trop restrictive. Les anneaux noethériens réguliers discrets, les corps complets à valuation discrète, les algèbres de Banach sur $\mathbb{R}$ ou $\mathbb{C}$ sont des exemples d'anneaux de Banach K-réguliers.

II - LES THEORIES ${}_{\varepsilon}V^n$ ET ${}_{\varepsilon}U^n$

Le but de ce paragraphe est de construire des théories intermédiaires ${}_{\varepsilon}V^*$ et ${}_{\varepsilon}U^*$ entre la K-théorie hermitienne et la K-théorie ordinaire.

Ces théories joueront un rôle fondamental par la suite.

Soit donc $\mathcal{C}$ la catégorie hermitienne $\mathcal{P}(A)$ et soient

$$H = {}_{\varepsilon}H : \bar{\mathcal{C}} \longrightarrow {}_{\varepsilon}Q(\mathcal{C}) \quad \text{et} \quad J = {}_{\varepsilon}J : {}_{\varepsilon}Q(\mathcal{C}) \longrightarrow \bar{\mathcal{C}} \quad ^{(1)}$$

les foncteurs "hyperbolique" et "oubli" respectivement décrits dans [18] § 1 .

DEFINITION 2.1. Soit A un anneau hermitien. On désigne par ${}_{\varepsilon}U(A)$ (resp. ${}_{\varepsilon}V(A)$) le "groupe de Grothendieck" du foncteur ${}_{\varepsilon}H$ (resp. ${}_{\varepsilon}J$).

Cette définition mérite d'être précisée dans l'esprit de [4], [13]. On considère pour cela l'ensemble des triples (E,F,α) où E et F sont des objets de $\mathcal{C}$ (resp. ${}_{\varepsilon}Q(\mathcal{C})$) et où $\alpha : {}_{\varepsilon}H(E) \to H(F)$ est une isométrie (resp. $\alpha : E \to F$ est un isomorphisme des modules sous-jacents). Alors ${}_{\varepsilon}U(A)$ (resp. ${}_{\varepsilon}V(A)$) est le quotient du groupe libre engendré par cet ensemble par le sous-groupe engendré par les relations

$$(E,F,\alpha) + (E',F',\alpha') = (E \oplus E', F \oplus F', \alpha \oplus \alpha')$$

$(E,F,\alpha) = 0$ si le triple (E,F,α) est homotope à un triple de la forme (G,G,Id).

Dans le cas de la définition de ${}_{\varepsilon}V(A)$, il est facile de voir, compte-tenu du théorème 1.1 , qu'on peut remplacer la deuxième relation par la relation plus faible $(E,E,\alpha) = 0$ si α est un isomorphisme des modules sous-jacents homotope à l'identité.

En suivant le schéma décrit dans [4], [13], on en déduit la suite exacte

$${}_{\varepsilon}L^{-1}(A) \to K^{-1}(A) \to {}_{\varepsilon}V(A) \to {}_{\varepsilon}L(A) \to K(A)$$

Si A est 1-régulier, on a aussi la suite exacte

$$K^{-1}(A) \to {}_{\varepsilon}L^{-1}(A) \to {}_{\varepsilon}U(A) \to K(A) \to {}_{\varepsilon}L(A)$$

En suivant de nouveau [13], on peut donner une autre interprétation des groupes ${}_{\varepsilon}V(A)$ et ${}_{\varepsilon}U(A)$. Considérons l'ensemble des triples (E, g_1, g_2) où E est un objet de $\mathcal{C}$ et où $g_j : E \to {}^{t}E$, j = 1,2 , est un isomorphisme tel que ${}^{t}g_j = \varepsilon g_j$.

(1) Par la suite, on écrira simplement ${}_{\varepsilon}Q(A)$ au lieu de ${}_{\varepsilon}Q(\mathcal{C})$

Le groupe ${}_{\varepsilon}V(A)$ est alors le quotient du groupe libre engendré par cet ensemble par le sous-groupe engendré par les relations suivantes

$$(1) \quad (E,g_1,g_2) + (E',g_1',g_2') = (E \oplus E', g_1 \oplus g_1', g_2 \oplus g_2')$$

$$(2) \quad (E,g_1,g_2) = 0 \text{ si } g_1 \text{ est homotope à } g_2 .$$

A la paire (E,g_1,g_2) on peut en effet associer la paire (E_1,E_2,α) où E_j est le A-module E muni de la forme quadratique g_j et où α est l'identité sur les A-modules sous-jacents.

De même, considérons l'ensemble des triples (F,η_1,η_2) où F est un objet de ${}_{\varepsilon}Q(\mathcal{C})$ et où η_j , $j = 1,2$, est une "graduation gauche" de F , c'est-à-dire un automorphisme du module sous-jacent tel que $\eta_j^* = -\eta_j$ et $(\eta_j)^2 = 1$. La donnée d'une telle graduation η_j équivaut à écrire $F = E_j \oplus {}^tE_j$ où $E_j = \mathrm{Ker}(1-\eta_j)$, et ${}^tE_j = \mathrm{Ker}(1+\eta_j)$, $E_j \oplus {}^tE_j$ étant muni de la forme ε-hyperbolique canonique. Le groupe ${}_{\varepsilon}U(A)$ est alors le quotient du groupe libre engendré par cet ensemble par le sous-groupe engendré par les relations suivantes

$$(1) \quad (F,\eta_1,\eta_2) + (F',\eta_1',\eta_2') = (F \oplus F', \eta_1 \oplus \eta_1', \eta_2 \oplus \eta_2')$$

$$(2) \quad (F,\eta_1,\eta_2) = 0 \text{ si } \eta_1 \text{ est homotope à } \eta_2 .$$

Au triple (F,η_1,η_2) on peut en effet associer le triple (E_1,E_2,α) où $E_j = \mathrm{Ker}(1-\eta_j)$ et où $\alpha : H(E_1) \to H(E_2)$ est le composé des isomorphismes $H(E_1) \approx F \approx H(E_2)$. On notera que dans cette définition de ${}_{\varepsilon}V(A)$ et de ${}_{\varepsilon}U(A)$, deux triples isomorphes donnent le même élément des groupes en question [(1)]. En effet, raisonnons dans le cas du groupe ${}_{\varepsilon}U(A)$ par exemple. On a alors l'identité $(F,\eta_1,\eta_2) + (F,\eta_2,\eta_3) + (F,\eta_3,\eta_1) = (F \oplus F \oplus F, \eta_1 \oplus \eta_2 \oplus \eta_3 , \eta_2 \oplus \eta_3 \oplus \eta_1)$. Posons $\eta_1' = \eta_1 \oplus \eta_2 \oplus \eta_3$ et $\eta_2' = \eta_2 \oplus \eta_3 \oplus \eta_1$. Alors $\eta_2' = \alpha\eta_1'\alpha^{-1}$ où $\alpha = \sigma.z.\sigma^{-1}.z^{-1}$ avec

---

[(1)] En d'autres termes, la relation d'isomorphie est une conséquence des relations (1) et (2) écrites ci-dessus.

$$\sigma = \begin{pmatrix} 0 & 1 & 0 \\ 1 & 0 & 0 \\ 0 & 0 & 1 \end{pmatrix} z = \begin{pmatrix} 0 & 0 & 1 \\ 0 & 1 & 0 \\ 1 & 0 & 0 \end{pmatrix}$$

D'après le lemme 2.12 de [18] , il en résulte que α est homotope à 1 . Donc η_2' est homotope à η_1' . En particulier, on a $(F,\eta_1,\eta_2) = -(F,\eta_2,\eta_1)$ dans le groupe ${}_{\varepsilon}U(A)$. Soient maintenant (F,η_1,η_2) et (F',η_1',η_2') deux triples isomorphes grâce à un isomorphisme h . Considérons la somme $(F,\eta_1,\eta_2) + (F',\eta_2',\eta_1') + (F,\eta_1,\eta_1) = (F \oplus F' \oplus F, \eta_1 \oplus \eta_2' \oplus \eta_1, \eta_2 \oplus \eta_1' + \eta_1)$. On a alors l'identité

$$\begin{pmatrix} 0 & 0 & 1 \\ h & 0 & 0 \\ 0 & h^{-1} & 0 \end{pmatrix} \begin{pmatrix} \eta_1 & 0 & 0 \\ 0 & \eta_2' & 0 \\ 0 & 0 & \eta_1 \end{pmatrix} \begin{pmatrix} 0 & h^{-1} & 0 \\ 0 & 0 & h \\ 1 & 0 & 0 \end{pmatrix}$$

$$= \begin{pmatrix} \eta_1 & 0 & 0 \\ 0 & \eta_1' & 0 \\ 0 & 0 & \eta_2 \end{pmatrix}$$

La première matrice étant un commutateur par une identité analogue à l'identité précédente, on a bien l'assertion annoncée.

Si A n'a pas nécessairement d'élément unité, posons ${}_{\varepsilon}V(A) = \mathrm{Ker}({}_{\varepsilon}V(A^+) \to {}_{\varepsilon}V(\mathbb{Z}'))$, où A^+ désigne l'algèbre augmentée sur $\mathbb{Z}' = \mathbb{Z}[\frac{1}{2}]$ associée à A . Il est clair qu'on a encore la suite exacte

$${}_{\varepsilon}L^{-1}(A) \longrightarrow K^{-1}(A) \longrightarrow {}_{\varepsilon}V(A) \longrightarrow {}_{\varepsilon}L(A) \longrightarrow K(A)$$

D'autre part, le théorème 1.1 (vrai aussi si A n'a pas d'élément unité) montre que ${}_{\varepsilon}V(A)$ est un invariant du type d'homotopie de A .En particulier, ${}_{\varepsilon}V(EA) = 0$ pour tout anneau hermitien A . Cette propriété nous conduit à définir ${}_{\varepsilon}V^{-n}(A)$ comme étant le groupe ${}_{\varepsilon}V(\Omega^n A)$.

<u>THEOREME 2.1.</u> <u>Pour tout anneau hermitien</u> A <u>on a la suite exacte</u>

$${}_{\varepsilon}L^{-n-1}(A) \to K^{-n-1}(A) \to {}_{\varepsilon}V^{-n}(A) \to {}_{\varepsilon}L^{-n}(A) \to K^{-n}(A)$$

DEMONSTRATION : Considérons la suite exacte

$$ {}_{\varepsilon}L^{-1}(\Omega A) \to K^{-1}(\Omega A) \to {}_{\varepsilon}V(\Omega A) \to {}_{\varepsilon}L(\Omega A) \to K(\Omega A) $$

Puisque ${}_{\varepsilon}L^{-1}(A) \approx \mathrm{Ker}({}_{\varepsilon}L(\Omega A) \to {}_{\varepsilon}L(EA))$ et $K^{-1}(A) \approx \mathrm{Ker}[K(\Omega A) \to K(EA)]$ et que ${}_{\varepsilon}V(EA) = 0$, on a aussi la suite exacte

$$ {}_{\varepsilon}L^{-2}(A) \to K^{-2}(A) \to {}_{\varepsilon}V^{-1}(A) \to {}_{\varepsilon}L^{-1}(A) \to K^{-1}(A) $$

Pour $n > 1$, la suite du théorème se déduit de celle-ci en remplaçant A par $\Omega^{n-1}A$.

Pour $n > 0$, définissons de même ${}_{\varepsilon}V^{n}(A)$ comme étant ${}_{\varepsilon}V(S^{n}A)$.

On a alors le théorème suivant

THEOREME 2.2. *Soit A un anneau hermitien (1,n-1)-régulier* (par exemple K-régulier). *On a alors la suite exacte*

$$ {}_{\varepsilon}L^{n-1}(A) \to K^{n-1}(A) \to {}_{\varepsilon}V^{n}(A) \to {}_{\varepsilon}L^{n}(A) \to K^{n}(A) $$

DEMONSTRATION La méthode décrite dans [16] permet de montrer que $S^{n-1}A$ est 1-régulier. Par suite, $K^{-1}(S^{n}A) \approx K^{-1}(S\, S^{n-1}A) \approx K(S^{n-1}A) = K^{n-1}(A)$. Il suffit de considérer alors la suite exacte

$$ {}_{\varepsilon}L^{-1}(S^{n}A) \to K^{-1}(S^{n}A) \to {}_{\varepsilon}V(S^{n}A) \to {}_{\varepsilon}L(S^{n}A) \to K(S^{n}A) $$

On peut montrer que la théorie ${}_{\varepsilon}V^{n}$ est une "théorie de la cohomologie" en un sens raisonnable sur la catégorie des anneaux hermitiens. Un premier pas dans cette direction est la proposition suivante

PROPOSITION 2.3. *Soit* $0 \to A' \to A \xrightarrow{\psi_*} A'' \to 0$ *une* GL-${}_{\varepsilon}$O-*fibration d'ordre un d'anneaux hermitiens. On a alors la suite exacte*

$$ {}_{\varepsilon}V(A') \to {}_{\varepsilon}V(A) \xrightarrow{\psi_*} {}_{\varepsilon}V(A'') $$

DEMONSTRATION Supposons d'abord que A et A'' aient des éléments unités et que Ψ respecte les éléments unités. Soit $u = d(E,g_1,g_2)$(1) un élément de ${}_{\varepsilon}V(A)$ tel que $\Psi_*(u) = 0$. Sans restreindre la généralité on peut supposer que $E = {}_{\varepsilon}H(A^{n})$ et que g_1 est la forme ε-quadratique

---

(1) Comme il est d'usage, on note d la classe d'un triple dans K,L,V,U, etc ...

canonique sur E . D'après le théorème 1.1. , ceci implique qu'après stabilisation éventuelle, il existe un automorphisme $\alpha''(x)$ de $\Psi_*(E)$ tel que $\alpha''(0) = 1$ et que $\Psi_*(g_2) = {}^t\alpha''(1)\Psi_*(g_1)\alpha''(1)$. Puisque Ψ est une GL-fibration, il existe un automorphisme $\alpha(x)$ de E tel que $\alpha(0) = 1$ et $\Psi_*(\alpha(x)) = \alpha''(x)$ Alors $d(E,g_1,g_2) = d(E,g_1,g_2')$ avec $g_2' = {}^t\alpha(1)^{-1}g_2\alpha(1)$. Puisque $\Psi_*(g_2') = \Psi_*(g_1)$ on voit que u est l'image d'un élément de ${}_\varepsilon V(A^+)$ par l'homomorphisme obtenu en appliquant le foncteur ${}_\varepsilon V$ à la première flèche horizontale du diagramme

$$\begin{array}{ccc} (A')^+ & \longrightarrow & A \\ \downarrow & & \downarrow \\ \mathbb{Z}' & \longrightarrow & A'' \end{array}$$

il en résulte évidemment l'exactitude de la suite dans ce cas. Si maintenant A , A'' et Ψ sont quelconques, on considère le diagramme

$$\begin{array}{ccccc} 0 & & 0 & & 0 \\ \downarrow & & \downarrow & & \downarrow \\ {}_\varepsilon V(A') & \longrightarrow & {}_\varepsilon V(A) & \longrightarrow & {}_\varepsilon V(A'') \\ \downarrow & & \downarrow & & \downarrow \\ {}_\varepsilon V(A') & \longrightarrow & {}_\varepsilon V(A^+) & \longrightarrow & {}_\varepsilon V(A''^+) \\ \downarrow & & \downarrow & & \downarrow \\ 0 & & {}_\varepsilon V(\mathbb{Z}) & \longrightarrow & {}_\varepsilon V(\mathbb{Z}) \\ & & \downarrow & & \downarrow \\ & & 0 & & 0 \end{array}$$

<u>LEMME 2.4.</u> <u>Soit</u>

$$0 \to A' \to A \to A'' \to 0$$

<u>une suite exacte d'anneaux hermitiens qui est à la fois une</u> GL-<u>fibration et une</u> ${}_\varepsilon$0-<u>fibration d'ordre un</u> . Si A <u>est contractile, le groupe</u>

${}_\varepsilon L'(A') = \operatorname{Ker}[{}_\varepsilon L(A') \to K(A')]$ s'identifie naturellement au groupe

${}_\varepsilon L'^{-1}(A'') = \operatorname{Ker}[{}_\varepsilon L^{-1}(A'') \to K^{-1}(A'')]$.

<u>DEMONSTRATION</u> Il suffit de considérer le diagramme naturel

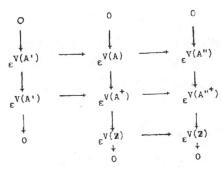

$$\begin{array}{ccccccc}
 & & 0 & & 0 & & 0\\
 & & \downarrow & & \downarrow & & \downarrow\\
0 & \longrightarrow & {}_{\varepsilon}L'^{-1}(A'') & \longrightarrow & {}_{\varepsilon}L'(A') & \longrightarrow & {}_{\varepsilon}L'(A)\\
 & & \downarrow & & \downarrow & & \downarrow\\
0 & \longrightarrow & {}_{\varepsilon}L^{-1}(A'') & \longrightarrow & {}_{\varepsilon}L(A') & \longrightarrow & {}_{\varepsilon}L(A)\\
 & & \downarrow & & \downarrow & & \downarrow\\
0 & \longrightarrow & K^{-1}(A'') & \longrightarrow & K(A') & \longrightarrow & K(A)
\end{array}$$

où ${}_{\varepsilon}L'(A) = 0$ puisque ${}_{\varepsilon}L'(A)$ est un invariant du type d'homotopie de A (théorème 1.1).

COROLLAIRE 2.5. *Considérons le diagramme commutatif*

$$\begin{array}{ccccccccc}
0 & \longrightarrow & A' & \longrightarrow & A & \longrightarrow & A'' & \longrightarrow & 0\\
 & & \downarrow & & \downarrow & & \downarrow & & \\
0 & \longrightarrow & B' & \longrightarrow & B & \longrightarrow & B'' & \longrightarrow & 0
\end{array}$$

où A *et* B *sont contractiles et où* $A'' \to B''$ *induit un isomorphisme sur les groupes* K^{-i} et ${}_{\varepsilon}L^{-i}$, $i = 1,2$. *Si les suites horizontales sont des* ${}_{\varepsilon}O$-GL-*fibrations d'ordre 2* , *l'homomorphisme* $A' \to B'$ *induit un isomorphisme* ${}_{\varepsilon}V(A') \to {}_{\varepsilon}V(B')$.

DEMONSTRATION On a le diagramme commutatif

$$\begin{array}{ccccccccc}
{}_{\varepsilon}L^{-1}(A') & \to & K^{-1}(A') & \to & {}_{\varepsilon}V(A') & \to & {}_{\varepsilon}L'(A') & \to & 0\\
\downarrow & & \downarrow & & \downarrow & & \downarrow & & \\
{}_{\varepsilon}L^{-1}(B') & \to & K^{-1}(B') & \to & {}_{\varepsilon}V(B') & \to & {}_{\varepsilon}L'(B') & \to & 0
\end{array}$$

Les isomorphismes naturels ${}_{\varepsilon}L^{-1}(A') \approx {}_{\varepsilon}L^{-2}(A'')$, $K^{-1}(A') \approx K^{-2}(A'')$, etc. permettent de conclure.

Soit

$$0 \to A' \to A \to A'' \to 0$$

une ${}_{\varepsilon}O$-GL-fibration d'ordre 2 . On peut comme dans [17] lui associer le diagramme

$$\begin{array}{ccccccccc}
0 & \longrightarrow & A' & \longrightarrow & A & \longrightarrow & A'' & \longrightarrow & 0 \\
 & & \uparrow & & \uparrow & & \| & & \\
0 & \longrightarrow & D & \longrightarrow & EA & \longrightarrow & A'' & \longrightarrow & 0 \\
 & & \downarrow & & \downarrow & & \| & & \\
0 & \longrightarrow & \Omega A' & \longrightarrow & EA'' & \longrightarrow & A'' & \longrightarrow & 0
\end{array} .$$

D'après le corollaire précédent, on a ${}_{\varepsilon}V(D) \approx {}_{\varepsilon}V(\Omega A'') \approx {}_{\varepsilon}V^{-1}(A'')$. L'homomorphisme $D \to A'$ permet donc de définir un homomorphisme de connexion $\partial^{-1} : {}_{\varepsilon}V^{-1}(A'') \to V(A')$.

<u>PROPOSITION 2.6.</u> <u>Soit</u>

$$0 \to A' \to A \to A'' \to 0$$

<u>une</u> ${}_{\varepsilon}$<u>0-GL-fibration d'ordre 2</u> . <u>On a alors la suite exacte</u>

$${}_{\varepsilon}V^{-1}(A) \to {}_{\varepsilon}V^{-1}(A'') \xrightarrow{\partial^{-1}} {}_{\varepsilon}V(A') \to {}_{\varepsilon}V(A)$$

<u>DEMONSTRATION</u> La GL-fibration évidente

$$0 \to \Omega A \to D \to A' \to 0$$

nous permet de voir que la suite

$$V^{-1}(A) \to V^{-1}(A'') \xrightarrow{\partial^{-1}} V(A')$$

est exacte d'après la proposition 2.3. Soit maintenant R l'anneau produit fibré dans le diagramme

$$\begin{array}{ccc}
R & \longrightarrow & EA'' \\
\downarrow & & \downarrow \\
A & \longrightarrow & A''
\end{array}$$

On a alors une ${}_{\varepsilon}$0-GL-fibration d'ordre 2

$$0 \to A' \to R \to EA'' \to 0$$

D'après le lemme des cinq et le théorème 1.1. on voit aisément que l'homomorphisme $A' \to R$ induit un isomorphisme entre les groupes ${}_{\varepsilon}V(A')$ et ${}_{\varepsilon}V(R)$ puisque $K^{-1}(A') \approx K^{-1}(R)$, ${}_{\varepsilon}L^{-1}(A') \approx {}_{\varepsilon}L^{-1}(R)$ et que ${}_{\varepsilon}L'(A') \approx {}_{\varepsilon}L'(R)$.

D'autre part, on a aussi la ${}_{\varepsilon}$0-GL-fibration

$$0 \to \Omega A'' \to R \to A \to 0$$

ce qui permet de démontrer une suite exacte du type

$$ {}_{\varepsilon}V^{-1}(A'') \xrightarrow{\sigma^{-1}} {}_{\varepsilon}V(A') \to {}_{\varepsilon}V(A) $$

où ${}_{\varepsilon}V(A') \longrightarrow {}_{\varepsilon}V(A)$ est l'homomorphisme naturel et où σ^{-1} est un nouvel opérateur de connexion. En appliquant ce résultat à la ${}_{\varepsilon}$O-GL-fibration précédente, on a aussi la suite exacte

$$ {}_{\varepsilon}V^{-1}(A) \to {}_{\varepsilon}V^{-1}(A'') \xrightarrow{\sigma^{-1}} {}_{\varepsilon}V(A') , $$

l'homomorphisme ${}_{\varepsilon}V^{-1}(A) \to {}_{\varepsilon}V^{-1}(A'')$ étant inconnu. Il résulte de cette discussion que σ^{-1} et ∂^{-1} réalisent tous deux, en tant qu'opérateurs de connexion, un isomorphisme de ${}_{\varepsilon}V^{-1}(A'')$ sur ${}_{\varepsilon}V(\Omega A'')$. Un argument classique nous permet de conclure à l'existence d'un automorphisme γ de $V^{-1}(A'')$ tel que $\sigma^{-1} = \partial^{-1}\gamma$. Il suffit en effet de considérer le diagramme

$$
\begin{array}{ccc}
V^{-1}(A'') & \xrightarrow{\sigma^{-1} \text{ ou } \partial^{-1}} & V(A') \\
\| & & \uparrow \\
V^{-1}(A'') & \xrightarrow{\sigma^{-1} \text{ ou } \partial^{-1}} & V(D) \\
\| & & \uparrow \approx \\
V^{-1}(A'') & \xrightarrow[\approx]{\sigma^{-1} \text{ ou } \partial^{-1}} & V(\Omega A'')
\end{array}
$$

La proposition est démontrée.

<u>COROLLAIRE 2.7.</u> <u>Soit</u>

$$ 0 \to A' \to A \to A'' \to 0 $$

<u>une</u> ${}_{\varepsilon}$O-GL-<u>fibration d'ordre</u> $n+1$. <u>On a alors la suite exacte</u>

$$ {}_{\varepsilon}V^{-n}(A') \to {}_{\varepsilon}V^{-n}(A) \to {}_{\varepsilon}V^{-n}(A'') \to {}_{\varepsilon}V^{-n+1}(A') \to {}_{\varepsilon}V^{-n+1}(A) $$

<u>DEMONSTRATION</u> On applique la proposition précédente à la suite

$$ 0 \to \Omega^{n-1} A' \to \Omega^{n-1} A \to \Omega^{n-1} A'' \to 0 $$

Pour pouvoir appliquer le corollaire précédent, rappelons que si B est un anneau de Banach $(n+1)$-régulier, le cup-produit par un générateur de $K^{-1}(S\mathbb{Z}) \approx K^{-1}(S\mathbb{Z}') \approx \mathbb{Z}$ induit un isomorphisme de $K^{-n}(B)$ sur $K^{-n+1}(SB)$. De même, le cup-produit par l'élément u explicité à la fin du premier paragraphe induit un isomorphisme de ${}_{\varepsilon}L^{-n}(B)$ sur ${}_{\varepsilon}L^{-n-1}(SB)$ compatible avec le précédent.

Le lemme des cinq appliqué aux suites exactes

$$\begin{array}{ccccccccc}
{}_{\varepsilon}L^{-n}(B) & \to & K^{-n}(B) & \to & {}_{\varepsilon}V^{-n+1}(B) & \to & {}_{\varepsilon}L^{-n+1}(B) & \to & K^{-n+1}(B) \\
\downarrow & & \downarrow & & \downarrow & & \downarrow & & \downarrow \\
{}_{\varepsilon}L^{-n-1}(SB) & \to & K^{-n-1}(SB) & \to & {}_{\varepsilon}V^{-n}(SB) & \to & {}_{\varepsilon}L^{-n}(SB) & \to & K^{-n}(SB)
\end{array}$$

permet donc de voir que ${}_{\varepsilon}V^{-n+1}(B) \approx {}_{\varepsilon}V^{-n}(SB)$ (on regarde la théorie ${}_{\varepsilon}V$ comme "module sur la théorie ${}_{1}L$"). On en déduit le théorème suivant :

<u>THEOREME 2.8.</u> <u>Soit</u>

$$0 \to A' \to A \to A \to 0$$

<u>une suite exacte d'anneaux hermitiens</u> (n+2)-<u>réguliers</u>. <u>On a alors la suite exacte</u>

$${}_{\varepsilon}V^{-n}(A') \to {}_{\varepsilon}V^{-n}(A) \to {}_{\varepsilon}V^{-n}(A'') \to {}_{\varepsilon}V^{-n+1}(A') \to {}_{\varepsilon}V^{-n+1}(A) \to {}_{\varepsilon}V^{-n+1}(A'')$$

La démonstration de ce théorème est analogue à celle du théorème 1.4.

Nous nous proposons de même de prolonger la suite exacte du théorème 2.8. vers la droite. Supposons pour cela que A', A et A" soièent 2-réguliers. On a alors ${}_{\varepsilon}V^{-1}(SB) \approx {}_{\varepsilon}V(B)$ pour $B = A'$, A et A" et la suite

$$0 \to SA' \to SA \to SA'' \to 0$$

est une ${}_{\varepsilon}$O-GL-fibration d'ordre 2 . Le corollaire 2.7. nous permet donc d'écrire la suite exacte

$${}_{\varepsilon}V^{-1}(SA') \to {}_{\varepsilon}V^{-1}(SA) \to {}_{\varepsilon}V^{-1}(SA'') \to {}_{\varepsilon}V(SA') \to {}_{\varepsilon}V(SA) \to {}_{\varepsilon}V(SA'')$$

soit

$${}_{\varepsilon}V(A') \to {}_{\varepsilon}V(A) \to {}_{\varepsilon}V(A'') \to {}_{\varepsilon}V^{1}(A') \to {}_{\varepsilon}V^{1}(A) \to {}_{\varepsilon}V^{1}(A'')$$

En appliquant le foncteur S^n , on en déduit le théorème suivant :

<u>THEOREME 2.9.</u> <u>Soit</u>

$$0 \to A' \to A \to A'' \to 0$$

<u>une suite exacte d'anneaux hermitiens</u> (2,n)-<u>réguliers</u> (par exemple K-réguliers).

<u>On a alors la suite exacte</u>

$${}_{\varepsilon}V^{n}(A') \to {}_{\varepsilon}V^{n}(A) \to {}_{\varepsilon}V^{n}(A'') \to {}_{\varepsilon}V^{n+1}(A') \to {}_{\varepsilon}V^{n+1}(A) \to {}_{\varepsilon}V^{n+1}(A'')$$

Il est possible de caractériser axiomatiquement la théorie $_{\varepsilon}V^n$, $n \in \mathbb{Z}$, sur une catégorie d'anneaux suffisamment réguliers en suivant le schéma décrit dans [17]. Les détails sont laissés au lecteur.

On peut de même construire une théorie $_{\varepsilon}U^n$, $n \in \mathbb{Z}$, avec un peu plus d'hypothèses sur les anneaux hermitiens. Le premier pas dans cette direction est la proposition suivante

PROPOSITION 2.10. Soit

$$0 \to A' \to A \xrightarrow{\Psi} A'' \to 0$$

une $_{\varepsilon}O$*-fibration telle que* A'' *soit* 1-*régulier. On a alors la suite exacte*

$$_{\varepsilon}U(A') \to {}_{\varepsilon}U(A) \xrightarrow{\Psi_*} {}_{\varepsilon}U(A'')$$

DEMONSTRATION Supposons d'abord que A et A'' soient des anneaux unitaires et que Ψ respecte les éléments unités. Soit $d(F,\eta_1,\eta_2)$ un élément de $_{\varepsilon}U(A)$ tel que $\Psi_*(u) = 0$. Sans restreindre la généralité, on peut supposer que $F = H(A^n)$, η_1 représentant la graduation gauche canonique. En stabilisant éventuellement, on a donc $\Psi_*(\eta_2)$ homotope à $\Psi_*(\eta_1)$ parmi les graduations gauches de $\Psi_*(F)$. Puisque A'' est 1-régulier, $K(A''\langle x\rangle) \approx K(A'')$ et, en stabilisant de nouveau, on voit qu'il existe un automorphisme orthogonal $\alpha''(x)$ de $\Psi_*(F)$ tel que $\alpha''(0) = \mathrm{Id}$ et que $\Psi_*(\eta_2) = \alpha''(1)\Psi_*(\eta_1)\alpha''(1)^{-1}$. Puisque Ψ est une $_{\varepsilon}O$-fibration, il existe un automorphisme orthogonal $\alpha(x)$ de F tel que $\alpha(0) = 1$ et que $\Psi(\alpha(x)) = \alpha''(x)$. Soit η'_2 la graduation gauche définie par $\eta'_2 = \alpha(1)^{-1}\eta_2\alpha(1)$. Il est clair que les triples (F,η_1,η_2) et (F,η_1,η'_2) sont homotopes, donc $d(F,\eta_1,\eta_2) = d(F,\eta_1,\eta'_2)$. Par ailleurs $d(F,\eta_1,\eta'_2)$ provient clairement de $_{\varepsilon}U(A'^+)$ par l'homomorphisme induit par la première flèche du diagramme

$$\begin{array}{ccc} A'^+ & \longrightarrow & A \\ \downarrow & & \downarrow \\ Z' & \longrightarrow & A'' \end{array}$$

puisque $\Psi_*(\eta_1) = \Psi_*(\eta'_2)$. Il en résulte que la suite de la proposition est exacte dans le cas envisagé. Si A et A" sont quelconques, la démonstration s'achève comme celle de la proposition 2.3.

LEMME 2.11. Considérons le diagramme commutatif

$$\begin{array}{ccccccccc} 0 & \longrightarrow & A' & \longrightarrow & A & \longrightarrow & A'' & \longrightarrow & 0 \\ & & \downarrow & & \downarrow & & \downarrow & & \\ 0 & \longrightarrow & B' & \longrightarrow & B & \longrightarrow & B'' & \longrightarrow & 0 \end{array}$$

où A et B sont des anneaux contractiles tels que $K(A) = K(B) = {}_{\varepsilon}L(A) = {}_{\varepsilon}L(B)=0$ et où $A'' \to B''$ induit un isomorphisme sur les groupes K^{-i} et ${}_{\varepsilon}L^{-i}$, $i = 1,2$. Si les suites horizontales sont des ${}_{\varepsilon}$O-GL-fibrations d'ordre 2, l'homomorphisme $A' \to B'$ induit un isomorphisme ${}_{\varepsilon}U(A') \xrightarrow{\approx} {}_{\varepsilon}U(B')$.

DEMONSTRATION On a le diagramme commutatif

$$\begin{array}{ccccccccc} K^{-1}(A') & \longrightarrow & {}_{\varepsilon}L^{-1}(A') & \longrightarrow & {}_{\varepsilon}U(A') & \longrightarrow & K(A') & \longrightarrow & {}_{\varepsilon}L(A') \\ \downarrow & & \downarrow & & \downarrow & & \downarrow & & \downarrow \\ K^{-1}(B') & \longrightarrow & {}_{\varepsilon}L^{-1}(B') & \longrightarrow & {}_{\varepsilon}U(B') & \longrightarrow & K(B') & \longrightarrow & {}_{\varepsilon}L(B') \end{array}$$

et les hypothèses assurent que les quatre flèches extrêmes du diagramme sont des isomorphismes. Le lemme des cinq permet de conclure.

PROPOSITION 2.12. Soit

$$0 \to A' \to A \to A'' \to 0$$

une ${}_{\varepsilon}$O-GL-fibration d'ordre 2. Si A', A et A" sont 1-réguliers, on a la suite exacte

$${}_{\varepsilon}U^{-1}(A) \to {}_{\varepsilon}U^{-1}(A'') \to {}_{\varepsilon}U(A') \to {}_{\varepsilon}U(A) \to {}_{\varepsilon}U(A'')$$

La démonstration de cette proposition est analogue à celle de la proposition 2.6.

COROLLAIRE 2.13. Soit

$$0 \to A' \to A \to A'' \to 0$$

une ${}_{\varepsilon}$O-GL-fibration d'ordre n+1. Si A', A et A" sont n-réguliers, on a la suite exacte

$$ {}_{\varepsilon}U^{-n}(A) \to {}_{\varepsilon}U^{-n}(A'') \to {}_{\varepsilon}U^{-n+1}(A') \to {}_{\varepsilon}U^{-n+1}(A) \to {}_{\varepsilon}U^{-n+1}(A'') $$ où l'on

pose ${}_{\varepsilon}U^{-p}(B) = {}_{\varepsilon}U(\Omega^p B)$.

Si A est un anneau hermitien quelconque, on a le diagramme commutatif

$$
\begin{array}{ccccccccc}
K^{-2}(SA) & \to & {}_{\varepsilon}L^{-2}(SA) & \to & {}_{\varepsilon}U^{-1}(SA) & \to & K^{-1}(SA) & \to & {}_{\varepsilon}L^{-1}(SA) \\
\uparrow & & \uparrow & & \uparrow & & \uparrow & & \uparrow \\
K^{-1}(A) & \to & {}_{\varepsilon}L^{-1}(A) & \to & {}_{\varepsilon}U(A) & \to & K(A) & \to & {}_{\varepsilon}L(A)
\end{array}
$$

où l'homomorphisme ${}_{\varepsilon}U(A) \to {}_{\varepsilon}U^{-1}(SA)$ est défini par le cup-produit par l'élément u de ${}_{1}L^{-1}(S\mathbb{Z}')$ image de l'élément unité de l'anneau ${}_{1}L(\mathbb{Z}')$ par l'isomorphisme canonique ${}_{1}L(\mathbb{Z}') \approx {}_{1}L^{-1}(S\mathbb{Z}')$. Si A est $(2,1)$-régulier, le lemme des cinq permet donc d'affirmer que l'homomorphisme ${}_{\varepsilon}U(A) \to {}_{\varepsilon}U^{-1}(SA)$ est un isomorphisme. On en déduit le théorème suivant

<u>THEOREME 2.14</u> <u>Soit</u>

$$ 0 \longrightarrow A' \longrightarrow A \longrightarrow A'' \longrightarrow 0 $$

<u>une suite exacte d'anneaux hermitiens telle que</u> A' , A <u>et</u> A'' <u>soient</u> $(n+2,1)$-<u>réguliers</u> (par exemple K-réguliers). <u>On a alors la suite exacte</u>

$$ {}_{\varepsilon}U^{-n}(A') \to {}_{\varepsilon}U^{-n}(A) \to {}_{\varepsilon}U^{-n}(A'') \to {}_{\varepsilon}U^{-n+1}(A') \to {}_{\varepsilon}U^{-n+1}(A) $$

Posons de même ${}_{\varepsilon}U^{p}(B) = {}_{\varepsilon}U(S^p B)$, $p > 0$. Le théorème suivant est analogue au théorème 2.9. et se démontre de la même manière.

<u>THEOREME 2.15.</u> <u>Soit</u>

$$ 0 \longrightarrow A' \longrightarrow A \longrightarrow A'' \longrightarrow 0 $$

<u>une suite exacte d'anneaux hermitiens</u> $(2,n+1)$-<u>réguliers</u> (par exemple K-réguliers). <u>On a alors la suite exacte</u>

$$ {}_{\varepsilon}U^{n}(A') \to {}_{\varepsilon}U^{n}(A) \to {}_{\varepsilon}U^{n}(A'') \to {}_{\varepsilon}U^{n+1}(A') \to {}_{\varepsilon}U^{n+1}(A) \to {}_{\varepsilon}U^{n+1}(A'') $$

On peut aussi comparer la théorie ${}_{\varepsilon}U^{n}$ aux théories ${}_{\varepsilon}L^{n}$ et K^{n} . La suite exacte fondamentale (A étant 1-régulier)

$$ K^{-1}(A) \to {}_{\varepsilon}L^{-1}(A) \to {}_{\varepsilon}U(A) \to K(A) \to {}_{\varepsilon}L(A) $$

permet de démontrer les deux théorèmes suivants :

THEOREME 2.16. Soit A un anneau hermitien n-régulier. On a alors la suite exacte

$$K^{-n}(A) \to {}_{\varepsilon}L^{-n}(A) \to {}_{\varepsilon}U^{-n+1}(A) \to K^{-n+1}(A) \to {}_{\varepsilon}L^{-n+1}(A) \ , \ n \geq 1$$

THEOREME 2.17. Soit A un anneau hermitien (1,n)-régulier (par exemple K-régulier). On a alors la suite exacte

$$K^{n-1}(A) \to {}_{\varepsilon}L^{n-1}(A) \to {}_{\varepsilon}U^{n}(A) \to {}_{\varepsilon}K^{n}(A) \to {}_{\varepsilon}L^{n}(A) \quad , \quad n \geq 1 .$$

Remarque générale sur le § 2 .

Nous avons introduit les foncteurs ${}_{\varepsilon}U^{n}(A)$ et ${}_{\varepsilon}V^{n}(A)$, A étant un anneau hermitien. Il est facile de voir que nos méthodes permettent de définir de même des foncteurs ${}_{\varepsilon}U^{n}(\mathcal{C})$ et ${}_{\varepsilon}V^{n}(\mathcal{C})$ pour une catégorie hermitienne $\mathcal{C}$ ayant des propriétés analogues. Si $\mathcal{C} = P(A)$, on retrouve bien entendu ${}_{\varepsilon}U^{n}(A)$ et ${}_{\varepsilon}V^{n}(A)$.

Terminons ce paragraphe par quelques mots sur la théorie "relative" qui est une généralisation mineure (mais techniquement utile) de la théorie que nous venons de développer. Si $\Psi : A \to A''$ est un homomorphisme surjectif d'anneaux hermitiens unitaires, on définit des groupes relatifs ${}_{\varepsilon}V(\Psi)$ et ${}_{\varepsilon}U(\Psi)$ en suivant le même schéma que dans [13]. Définissons de manière précise le groupe ${}_{\varepsilon}U(\Psi)$ par exemple.

On considère pour cela l'ensemble des triples (F,η_1,η_2) où F est un objet de ${}_{\varepsilon}Q(\mathcal{C})$, $\mathcal{C} = P(A)$, et où η_1 et η_2 sont des graduations gauches de F telles que $\Psi_*(\eta_1) = \Psi_*(\eta_2)$, $\Psi_* : \mathcal{P}(A) \to \mathcal{P}(A'')$ étant le foncteur extension des scalaires. Le groupe ${}_{\varepsilon}U(\Psi)$ est alors le quotient du groupe libre engendré par cet ensemble par le sous-groupe engendré par les relations suivantes :

$$(F,\eta_1,\eta_2) + (F',\eta_1',\eta_2') = (F \oplus F',\eta_1 \oplus \eta_1',\eta_2 \oplus \eta_2')$$

$(F,\eta_1,\eta_2) = 0$ si η_1 est homotope à η_2 parmi les graduations gauches η de F telles que $\Psi_*(\eta) = \Psi_*(\eta_1) = \Psi_*(\eta_2)$.

Le raisonnement fait dans [18] montre que ${}_{\varepsilon}U(\Psi) \approx {}_{\varepsilon}U(A')$ où $A' = \operatorname{Ker} \Psi$ (théorème d'excision). Des définitions et des propriétés analogues sont aussi vraies en ${}_{\varepsilon}V$-théorie. Si $n \in \mathbb{Z}$, on définit ${}_{\varepsilon}U^n(\Psi)$ (resp. ${}_{\varepsilon}V^n(\Psi)$) comme le groupe ${}_{\varepsilon}U(S^n\Psi)$ ou ${}_{\varepsilon}U(\Omega^{-n}\Psi)$ (resp. ${}_{\varepsilon}V(S^n\Psi)$ ou ${}_{\varepsilon}V(\Omega^{-n}\Psi)$) suivant le signe de n. Bien entendu, on a aussi ${}_{\varepsilon}U^n(\Psi) \approx {}_{\varepsilon}U^n(A')$ et ${}_{\varepsilon}V^n(\Psi) \approx {}_{\varepsilon}V^n(A')$.

III - LA THEORIE TOPOLOGIQUE

Dans ce paragraphe nous allons restreindre provisoirement notre étude au cas où A (ou plus généralement $\mathcal{C}$) est une algèbre de Banach réelle munie d'une antiinvolution bornée $\sigma: A \to A$ (resp. une catégorie de Banach dans le sens de [13] qui est de surcroît hermitienne). Nous allons voir que, dans ce cadre, nous pourrons obtenir des résultats théoriques satisfaisants grâce à l'introduction des algèbres de Clifford. Les méthodes utilisées sont pour l'essentiel inspirées de [13][14] . Mis à part son intérêt propre, la théorie topologique nous fournira aussi des indications pour l'étude du cas général dans le paragraphe suivant.

Soit donc $\mathcal{C}$ une catégorie de Banach hermitienne et soit E un objet de ${}_{\varepsilon}Q(\mathcal{C})$. En suivant [13] , on désigne par $C^{p,q}$ l'algèbre de Clifford de $\mathbb{R}^{p+q}$ muni de la forme quadratique $-x_1^2 - \ldots - x_p^2 + x_{p+1}^2 + \ldots + x_{p+q}^2$. Une structure de $C^{p,q}$-module (resp. $C^{p,q}$-module gauche) sur E est la donnée de $p+q$ automorphismes e_i , ε_j , $i = 1, \ldots, p$, $j = 1, \ldots, q$, satisfaisant aux relations suivantes

$$(e_i)^2 = -(\varepsilon_j)^2 = -1$$

$$e_r e_{r'} + e_{r'} e_r = \varepsilon_s \varepsilon_{s'} + \varepsilon_{s'} \varepsilon_s = 0 \quad \text{pour } r \neq r' \text{ et } s \neq s'$$

$$e_i e_i^* = \varepsilon_j \varepsilon_j^* = 1 \text{ (resp. } e_i e_i^* = \varepsilon_j \varepsilon_j^* = -1)$$

On note ${}_{\varepsilon}\mathcal{C}_L^{p,q}$ (resp. ${}_{\varepsilon}\mathcal{C}_U^{p,q}$) la catégorie évidente dont les objets sont les objets de ${}_{\varepsilon}Q(\mathcal{C})$ munis d'une structure de $C^{p,q}$-module (resp. de $C^{p,q}$-module gauche).

DEFINITION 3.1. Soit $\mathcal{C}$ une catégorie de Banach hermitienne. On désigne par ${}_\varepsilon L^{p,q}(\mathcal{C})$ (resp. ${}_\varepsilon U^{p,q}(\mathcal{C})$) le groupe de Grothendieck du foncteur restriction des scalaires.

$$ {}_\varepsilon\mathcal{C}_L^{p,q+1} \longrightarrow {}_\varepsilon\mathcal{C}_L^{p,q} \quad (\text{resp. } {}_\varepsilon\mathcal{C}_U^{p,q+1} \longrightarrow {}_\varepsilon\mathcal{C}_U^{p,q}) $$

Si $\mathcal{C}$ est la catégorie $\mathcal{P}(A)$, on écrit simplement ${}_\varepsilon L^{p,q}(A)$ (resp. ${}_\varepsilon U^{p,q}(A)$) le groupe ${}_\varepsilon L^{p,q}(\mathcal{C})$ (resp. ${}_\varepsilon U^{p,q}(\mathcal{C})$).

Remarque : Dans cette définition, le "groupe de Grothendieck" est à entendre dans le sens de [13]. La topologie intervient pour identifier deux triples homotopes.

PROPOSITION 3.1 bis Le groupe ${}_\varepsilon U^{0,0}(\mathcal{C})$ s'identifie au groupe ${}_\varepsilon U(\mathcal{C})$ défini dans le paragraphe précédent. En particulier, ${}_\varepsilon U^{0,0}(A) \approx {}_\varepsilon U(A)$.

DEMONSTRATION Soit E un objet de $\mathcal{C}$ et soit $H(E) = E \oplus {}^tE$ l'objet ε-hyperbolique associé. On peut alors munir $H(E)$ d'une structure de $C^{0,1}$-module gauche en posant

$$ \eta_1 = \begin{pmatrix} 1 & 0 \\ 0 & -1 \end{pmatrix} $$

Si donc $\alpha : H(E) \to H(F)$ est un isomorphisme, on voit que le triple $(H(E),H(F),\alpha)$ définit bien un élément de ${}_\varepsilon U^{0,0}(\mathcal{C})$. Réciproquement, si $d(G,H,\alpha)$ est un élément de ${}_\varepsilon U^{0,0}(\mathcal{C})$, l'action de ε_1 sur G et H équivaut à un scindage $G = E \oplus {}^tE$, $H = F \oplus {}^tF$ (à isomorphisme canonique près). Enfin, les homomorphismes réciproques l'un de l'autre ${}_\varepsilon U^{0,0}(\mathcal{C}) \longrightarrow {}_\varepsilon U(\mathcal{C})$ et ${}_\varepsilon U(\mathcal{C}) \to {}_\varepsilon U^{0,0}(\mathcal{C})$ sont bien définis, l'homotopie "banachique" étant clairement équivalente à l'homotopie topologique.

PROPOSITION 3.2. Le groupe ${}_\varepsilon U^{0,1}(\mathcal{C})$ s'identifie au groupe ${}_{-\varepsilon}V(\mathcal{C})$ défini dans le paragraphe précédent. En particulier ${}_\varepsilon U^{0,1}(A) \approx {}_{-\varepsilon}V(A)$.

Avant de commencer la démonstration de la proposition proprement dite, il est bon de donner une description du groupe ${}_{\varepsilon}L^{p,q}(\mathcal{C})$ (resp. ${}_{\varepsilon}U^{p,q}(\mathcal{C})$) en termes de graduations qui est analogue à celle proposée dans [13] . De manière précise, considérons l'ensemble des triples (E,η_1,η_2) où E est un objet de ${}_{\varepsilon}\mathcal{C}_L^{p,q}$ (resp. ${}_{\varepsilon}\mathcal{C}_U^{p,q}$) et où η_1 et η_2 sont des "graduations" (resp."graduations gauches") de E . Par graduation (resp. graduation gauche) on entend un automorphisme involutif de l'objet de $\mathcal{C}$ sous-jacent qui anticommute aux générateurs de l'algèbre de Clifford et tel que $\eta^* = \eta$ (resp. $\eta^* = -\eta$). Le groupe ${}_{\varepsilon}L^{p,q}(\mathcal{C})$ (resp. ${}_{\varepsilon}U^{p,q}(\mathcal{C})$) est le quotient du groupe libre engendré par cet ensemble par le sous-groupe engendré par les relations suivantes

$$(E,\eta_1,\eta_2) + (E',\eta_1',\eta_2') = (E \oplus E',\eta_1 \oplus \eta_1',\eta_2 \oplus \eta_2')$$

$$(E,\eta_1,\eta_2) = 0 \text{ si } \eta_2 \text{ est homotope à } \eta_1 .$$

DEMONSTRATION DE LA PROPOSITION 3.2. Compte-tenu de la traduction précédente, un élément de ${}_{\varepsilon}U^{0,1}(\mathcal{C})$ s'écrit comme la classe d'un triple (E,η_1,η_2) où E est un $C^{0,1}$ module (resp. $C^{0,1}$-module gauche). L'automorphisme ε_1 permet de scinder canoniquement E en $F \oplus {}^tF$, ε_1 étant représenté par la matrice

$$\varepsilon_1 = \begin{pmatrix} 1 & 0 \\ 0 & -1 \end{pmatrix}$$

Dans ces conditions η_j , $j = 1,2$ doit s'écrire sous la forme

$$\eta_j = \begin{pmatrix} 0 & g_j^{-1} \\ g_j & 0 \end{pmatrix}$$

où $g_j : F \to {}^tF$ est un isomorphisme tel que ${}^tg_j = -\varepsilon g_j$. La correspondance $(E,\eta_1,\eta_2) \longmapsto (F,g_1,g_2)$ définit l'isomorphisme cherché entre les groupes ${}_{\varepsilon}U^{0,1}(\mathcal{C})$ et ${}_{-\varepsilon}V(\mathcal{C})$.

PROPOSITION 3.3. Les groupes ${}_{\varepsilon}L^{p,q}(\mathcal{C})$ (resp. ${}_{\varepsilon}U^{p,q}(\mathcal{C})$) ne dépendent que de la différence p-q à isomorphisme près.

Si on pose ${}_\varepsilon L^n(\mathcal{C}) = {}_\varepsilon L^{p,q}(\mathcal{C})$ (resp. ${}_\varepsilon U^n(\mathcal{C}) = {}_\varepsilon U^{p,q}(\mathcal{C})$) pour $p-q = n$, on a un isomorphisme de périodicité ${}_\varepsilon L^n(\mathcal{C}) \approx {}_\varepsilon L^{n+8}(\mathcal{C})$ (resp. ${}_\varepsilon U^n(\mathcal{C}) \approx {}_\varepsilon U^{n+8}(\mathcal{C})$) . Supposons en outre que la catégorie $\mathcal{C}$ soit "hermitienne complexe" c'est-à-dire qu'il existe un automorphisme I de $\mathcal{C}$ tel que $I^2 = -1 = -I^*I$ et qui soit l'identité sur les objets. On a alors ${}_\varepsilon L^n(\mathcal{C}) \approx {}_\varepsilon L^{n+2}(\mathcal{C})$ (resp. ${}_\varepsilon U^n(\mathcal{C}) \approx {}_\varepsilon U^{n+2}(\mathcal{C})$).

La démonstration de cette proposition est analogue à la démonstration standard en K-théorie [13].

Il est possible d'interpréter "concrètement" les différents groupes ${}_\varepsilon L^n$ et ${}_\varepsilon U^n$. Il est commode de supposer que $\mathcal{C} = \mathcal{P}(A)$, auquel cas on notera ${}_\varepsilon U^n(A)$ et ${}_\varepsilon L^n(A)$ les groupes ${}_\varepsilon L^n(\mathcal{C})$ et ${}_\varepsilon U^n(\mathcal{C})$. Ce faisant, nous anticipons sur le théorème fondamental de ce § affirmant précisément que cette définition de ${}_\varepsilon L^n(A)$ (resp. ${}_\varepsilon U^n(A)$) coïncide avec celle décrite dans [18] (resp. dans le paragraphe précédent). Notons $\mathbb{C}$ (resp. $\mathbb{C}'$) le corps des nombres complexes muni de l'involution changeant i en $-i$ (resp. de l'involution triviale). Notons de même $\mathbb{H}$ (resp. $\mathbb{H}'$) le corps des quaternions muni de l'antiinvolution $i \to -i$, $j \to -j$, $k \to -k$ (resp. $i \to i$, $j \to j$, $k \to -k$). Le groupe ${}_\varepsilon L^n(A)$ est alors le groupe de Grothendieck du foncteur suivant.

Pour	$n = 0$	${}_\varepsilon Q(A)$	$\to$	$*$ (foncteur nul)(1)
Pour	$n = 1$	${}_\varepsilon Q(A)$	$\to$	${}_\varepsilon Q(A \otimes \mathbb{C})$ (extension des scalaires)
Pour	$n = 2$	${}_\varepsilon Q(A \otimes \mathbb{C})$	$\to$	${}_\varepsilon Q(A \otimes \mathbb{H})$ (extension des scalaires)
Pour	$n = 3$	${}_\varepsilon Q(A \otimes \mathbb{H})$	$\to$	${}_\varepsilon Q(A \otimes \mathbb{H}) \times {}_\varepsilon Q(A \otimes \mathbb{H})$ (foncteur diagonal)
Pour	$n = 4$	${}_\varepsilon Q(A \otimes \mathbb{H})$	$\to$	$*$ (foncteur nul)
Pour	$n = 5$	${}_\varepsilon Q(A \otimes \mathbb{H})$	$\to$	${}_\varepsilon Q(A \otimes \mathbb{C})$ (restriction des scalaires)
Pour	$n = 6$	${}_\varepsilon Q(A \otimes \mathbb{C})$	$\to$	${}_\varepsilon Q(A)$ (restriction des scalaires)
Pour	$n = 7$	${}_\varepsilon Q(A)$	$\to$	${}_\varepsilon Q(A) \times {}_\varepsilon Q(A)$ (foncteur diagonal)

Ces assertions proviennent du théorème de classification des algèbres de Clifford explicité dans [13] . On montre que ${}_\varepsilon U^n(A)$ est le groupe de

---

(1) On écrit simplement ${}_\varepsilon Q(A)$ au lieu de ${}_\varepsilon Q(\mathcal{P}(A))$.

Grothendieck du foncteur suivant.

Pour $n = 0$ $\quad {}_{\varepsilon}P(A) \to {}_{\varepsilon}Q(A)$ (foncteur hyperbolique)

$n = 1$ $\quad {}_{\varepsilon}Q(A) \to {}_{\varepsilon}Q(A \otimes \mathbb{C}')$ (extension des scalaires)

$n = 2$ $\quad {}_{\varepsilon}Q(A \otimes \mathbb{C}') \to {}_{\varepsilon}Q(A \otimes \mathbb{H}')$ (extension des scalaires)

$n = 3$ $\quad {}_{\varepsilon}Q(A \otimes \mathbb{H}')) \to \mathbb{P}(A \otimes \mathbb{H}')$ (foncteur "oubli")

$n = 4$ $\quad \mathbb{P}(A \otimes \mathbb{H}') \to {}_{-\varepsilon}Q(A \otimes \mathbb{H}')$ (foncteur hyperbolique)

$n = 5$ $\quad {}_{\varepsilon}Q(A \otimes \mathbb{H}') \to {}_{-\varepsilon}Q(A \otimes \mathbb{C}')$ (restriction des scalaires)

$n = 6$ $\quad {}_{\varepsilon}Q(A \otimes \mathbb{C}') \to {}_{-\varepsilon}Q(A)$ (restriction des scalaires)

$n = 7$ $\quad {}_{\varepsilon}Q(A) \to \mathbb{P}(A)$ (foncteur oubli)

En suivant de nouveau [13] , on définit des groupes ${}_{\varepsilon}L^{p,q}(X,Y;\mathcal{C})$ pour X compact et Y fermé dans X . Par exemple, le groupe ${}_{\varepsilon}U^{p,q}(X;\mathcal{C}) = {}_{\varepsilon}U^{p,q}(X,\emptyset;\mathcal{C})$ est le groupe ${}_{\varepsilon}U^{p,q}$ de la catégorie hermitienne $\mathcal{C}(X)$ des $\mathcal{C}$-fibrés sur X (cf.[13] où on montre que $\mathcal{C}(X)$ est une catégorie de Banach ; le fait que $\mathcal{C}(X)$ soit hermitienne résulte évidemment des considérations de la fin du § 1). De manière générale, les sorites développés dans [13] et [14] sur les groupes $K^{p,q}(X,Y ; \mathcal{C})$ se transposent sans peine aux groupes plus généraux ${}_{\varepsilon}L^{p,q}(X,Y ; \mathcal{C})$ et ${}_{\varepsilon}U^{p,q}(X,Y ; \mathcal{C})$. Si $\mathcal{C} = \mathbb{P}(A)$, on note simplement ${}_{\varepsilon}L^{p,q}(X,Y ; A)$ et ${}_{\varepsilon}U^{p,q}(X,Y ; A)$ les groupes ${}_{\varepsilon}L^{p,q}(X,Y ; \mathbb{P}(A))$ et ${}_{\varepsilon}U^{p,q}(X,Y ; \mathbb{P}(A))$. Il est facile alors d'expliciter des espaces classifiants pour les foncteurs $X \mapsto {}_{\varepsilon}L^{p,q}(X,\emptyset ; A)$ et $X \mapsto {}_{\varepsilon}U^{p,q}(X,\emptyset ; A)$ (ce qui suffit en raison de la propriété d'excision à laquelle satisfont les groupes ${}_{\varepsilon}L^{p,q}$ et ${}_{\varepsilon}U^{p,q}$ comme les groupes $K^{p,q}$). En suivant [15] , on trouve que la composante neutre des espaces classifiants pour ${}_{\varepsilon}U^{p,q}$ par exemple , $p-q = n$, est la composante neutre des espaces suivants

Pour $n = 0$ $\quad {}_{\varepsilon}O(A)/GL(A)$

$n = 1$ $\quad {}_{\varepsilon}O(A \otimes \mathbb{C}')/{}_{\varepsilon}O(A)$

$n = 2$ $\quad {}_{\varepsilon}O(A \otimes \mathbb{H}')/{}_{\varepsilon}O(A \otimes \mathbb{C}')$

$n = 3$ $\quad GL(A \otimes \mathbb{H}')/{}_{\varepsilon}O(A \otimes \mathbb{H}')$

$n = 4$ $\quad {}_{-\varepsilon}O(A \otimes \mathbb{H}')/GL(A \otimes \mathbb{H}')$

$n = 5$ $\quad {}_{-\varepsilon}O(A \otimes \mathbb{C}')/{}_{-\varepsilon}O(A \otimes \mathbb{H}')$

$n = 6 \qquad {}_{-\varepsilon}O(A)/{}_{-\varepsilon}O(A \otimes \mathbb{C}')$

$n = 7 \qquad GL(A)/{}_{-\varepsilon}O(A)$

En suivant encore [13], on définit deux homomorphismes fondamentaux

$$t_L : {}_{\varepsilon}L^{p,q+1}(X,Y ; \mathcal{C}) \to {}_{\varepsilon}L^{p,q}((X,Y)\wedge(D^1,S^0) ; \mathcal{C})$$

$$t_U : {}_{\varepsilon}U^{p,q+1}(X,Y ; \mathcal{C}) \to {}_{\varepsilon}U^{p,q}((X,Y)\wedge(D^1,S^0) ; \mathcal{C})$$

On va se contenter d'expliciter t_U, t_L se laissant définir de manière analogue. Un élément de ${}_{\varepsilon}U^{p,q+1}(X,Y ; \mathcal{C})$ s'écrit $d(E,\eta_1,\eta_2)$ où E est un $\mathcal{C}$-fibré sur X muni d'une forme ε-quadratique non dégénérée et d'une structure de $C^{p,q+1}$-module gauche, η_1 et η_2 étant des graduations gauches de E telles que $\eta_1|_Y = \eta_2|_Y$. Représentons D^1 comme le demi-cercle formé des points $e^{i\theta}$, $0 \leq \theta \leq \pi$. Alors $t_U(d(E,\eta_1,\eta_2)) = d(\pi^*E_1,\xi_1,\xi_2)$ où E_1 est le $C^{p,q}$-module sous-jacent à E, $\pi : X \times D^1 \to X, \xi_j = \xi_j(x,\theta)$ étant la graduation gauche de π^*E_1 définie par la formule

$$\xi_j(x,\theta) = \eta_j(x)\sin\theta + \varepsilon_{q+1}\cos\theta ,$$

ε_{q+1} étant le dernier générateur de l'algèbre $C^{p,q+1}$

<u>THEOREME 3.4.</u> (théorème fondamental de la K-théorie hermitienne topologique)

<u>Les homomorphismes</u> t_L <u>et</u> t_U <u>sont des isomorphismes.</u>

Ce théorème est malheureusement de démonstration plus délicate que celle du théorème analogue en K-théorie dont il est la généralisation [13] . Sa démonstration va utiliser à peu près toutes les techniques connues pour démontrer la périodicité de Bott usuelle (à l'exception de la théorie de Morse)! La démonstration qui va suivre ne pourra donc être lue avec profit que par un lecteur expert déjà familier avec [2] [3] [13] [14] [16] . En voici les principales étapes :

1) En introduisant des "opérateurs de Fredholm" convenables, nous définissons des groupes ${}_{\varepsilon}\bar{L}^{p,q}(X ; \mathcal{C})$ pour X compact (plus généralement localement compact), des isomorphismes $j : {}_{\varepsilon}\bar{L}^{p,q}(X \times \mathbb{R} ; \mathcal{C}) \approx {}_{\varepsilon}L^{p,q+1}(X ; \mathcal{C})$ et un homomorphisme $\bar{t}_L : {}_{\varepsilon}\bar{L}^{p,q+1}(X ; \mathcal{C}) \to {}_{\varepsilon}\bar{L}^{p,q}(X \times \mathbb{R} ; \mathcal{C})$ tel que le diagramme suivant commute

$$\begin{array}{ccc} {}_{\varepsilon}\bar{L}^{p+1,q+1}(X \times \mathbb{R} \,;\, \mathcal{C}) & \xrightarrow{j} & {}_{\varepsilon}L^{p,q+1}(X \,;\, \mathcal{C}) \\ \Big\downarrow \bar{t}_L & & \Big\downarrow t_L \\ {}_{\varepsilon}\bar{L}^{p+1,q}(X \times \mathbb{R}^2;\, \mathcal{C}) & \xrightarrow{j} & {}_{\varepsilon}L^{p,q}(X \times \mathbb{R} \,;\, \mathcal{C}) \approx {}_{\varepsilon}L^{p,q}(X \times D^1,\, X \times S^0;\, \mathcal{C}) \end{array}$$

Les mêmes constructions s'appliquent à la théorie ${}_{\varepsilon}U^{p,q}$

2) Nous introduisons l'analogue de la théorie KR[2][13] pour les théories ${}_{\varepsilon}\bar{U}^{p,q}$ et ${}_{\varepsilon}\bar{L}^{p,q}$. Si $\mathbb{R}^{m,n}$ représente $\mathbb{R}^{m+n}$ muni de l'involution $(x,y) \mapsto (-x,y)$, on généralise $\bar{t}_L$ et $\bar{t}_U$ en des homomorphismes

$$\bar{t}_L : {}_{\varepsilon}\bar{L}^{p+m,q+n}(X \,;\, \mathcal{C}) \to {}_{\varepsilon}\bar{L}^{p,q}(X \times \mathbb{R}^{m,n};\, \mathcal{C})$$

$$\bar{t}_U : {}_{\varepsilon}\bar{U}^{p+m,q+n}(X,\mathcal{C}) \to {}_{\varepsilon}\bar{U}^{p,q}(X \times \mathbb{R}^{m,n};\, \mathcal{C})$$

3) En particulier, pour $p = q = 0$ et $m = n = 1$, on montre que $\bar{t}_L$ est essentiellement l'expression de l'isomorphisme ${}_{\varepsilon}L(\mathcal{C}) \approx {}_{\varepsilon}L^{-1}(S\mathcal{C}) \approx \mathrm{Coker}$ $[{}_{\varepsilon}L^{-1}(\mathcal{C}) \to {}_{\varepsilon}L^{-1}(\mathcal{C}\langle z,z^{-1}\rangle)]$.

4) Des structures multiplicatives adéquates nous permettent d'interpréter les homomorphismes composés

$${}_{\varepsilon}\bar{U}^{p,q}(X,\mathcal{C}) \approx {}_{\varepsilon}\bar{U}^{p+1,q+1}(X;\mathcal{C}) \to {}_{\varepsilon}\bar{U}^{p,q}(X \times \mathbb{R}^{1,1};\, \mathcal{C})$$

$${}_{\varepsilon}\bar{L}^{p,q}(X,\mathcal{C}) \approx {}_{\varepsilon}\bar{L}^{p+1,q+1}(X;\mathcal{C}) \to {}_{\varepsilon}\bar{L}^{p,q}(X \times \mathbb{R}^{1,1};\, \mathcal{C})$$

comme le cup-produit par un élément remarquable $u \in {}_{1}L^{0,0}(\mathbb{R}^{1,1};\mathcal{E})$, $\mathcal{E}$ étant la catégorie des espaces vectoriels réels de dimension finie.

5) La démonstration se termine alors comme dans [14], exposé V , en appliquant le lemme des cinq et en interprétant les groupes ${}_{\varepsilon}\bar{L}^{p,q}$ et ${}_{\varepsilon}\bar{U}^{p,q}$ comme des groupes de Grothendieck de foncteurs hermitiens entre des catégories hermitiennes convenables.

Décrivons maintenant de manière un peu plus détaillée les différentes étapes.

1) <u>Introduction des opérateurs de Fredholm</u>

Soit $C\mathcal{C}$ le cône de la catégorie hermitienne $\mathcal{C}$ [18] . Si E et F sont deux objets de $C\mathcal{C}$, un morphisme $f : E \to F$ est dit un "opérateur de Fredholm" si la classe $\breve{f}$ de f dans la catégorie suspension $S\mathcal{C}$ est un isomorphisme (cf. [1] pour une justification de cette terminologie). Soient

E un $C\mathcal{C}$-fibré sur X muni d'une forme ε-quadratique non dégénérée et d'une structure de $C^{p,q+1}$-module gauche. On le suppose trivial.

. $D : E \to E$ une famille continue d'opérateurs de Fredholm telle que $D^* = -D$ et que $D^2 = \lambda(x)+k$ où $\lambda(x)$ est une fonction positive adéquate (i.e. $\lambda(x) \geq C > 0$) et où k est une famille continue d'opérateurs complètement continus dans la terminologie de [14] exp. IV. Comme condition algébrique, on suppose en outre que D anti-commute aux générateurs de l'algèbre $C^{p,q+1}$. Enfin, si X n'est pas compact, on doit supposer aussi que $D^2 \sim \mu(x)$ pour $x \to \infty$ où $\mu(x)$ est une deuxième fonction adéquate sur X .

On désigne par ${}_\varepsilon\bar{U}^{p,q}(X ; \mathcal{C})$ le groupe de Grothendieck de l'ensemble des classes d'homotopie de telles paires (E,D) . En fait, on voit aisément (en suivant [14] exp. III) que ce groupe n'est autre que ${}_\varepsilon U^{p,q+1}(X ; S\mathcal{C})$, les fonctions adéquates se laissant facilement déformer en 1 . On définit de même le groupe ${}_\varepsilon\bar{L}^{p,q}(X ; \mathcal{C}) \approx {}_\varepsilon L^{p,q+1}(X ; S\mathcal{C})$.

Du fait que les groupes ${}_\varepsilon L^{p,q}$, ${}_\varepsilon U^{p,q}$ sont les groupes de Grothendieck de foncteurs entre catégories hermitiennes, les théorèmes généraux sur les catégories filtrées [14] [18] permettent de démontrer des isomorphismes

$${}_\varepsilon\bar{U}^{p,q}(X \times \mathbb{R} ; \mathcal{C}) \approx {}_\varepsilon U^{p,q+1}(X \times \mathbb{R} ; S\,\mathcal{C}) \approx {}_\varepsilon U^{p,q+1}(X ; \mathcal{C})$$

$${}_\varepsilon\bar{L}^{p,q}(X \times \mathbb{R} ; \mathcal{C}) \approx {}_\varepsilon L^{p,q+1}(X \times \mathbb{R} ; S\,\mathcal{C}) \approx {}_\varepsilon L^{p,q+1}(X ; \mathcal{C}) ,$$

les groupes ${}_\varepsilon L^{p,q}$ et ${}_\varepsilon U^{p,q}$ d'un espace localement compact se laissant définir comme dans [14] exp. III. Essayons alors de traduire comme dans [14] l'homomorphisme $t_U : {}_\varepsilon U^{p,q+1}(X ; S\,\mathcal{C}) \to {}_\varepsilon U^{p,q}(X \times \mathbb{R} ; S\,\mathcal{C})$ en termes de groupes $\bar{U}^{p,q}$. On trouve alors un homomorphisme

$$\bar{t}_U : {}_{\varepsilon}\bar{U}^{p,q+1}(X ; \mathcal{C}) \to {}_{\varepsilon}\bar{U}^{p,q}(X \times \mathbb{R} ; \mathcal{C})$$

qui s'explicite par la formule $\bar{t}_U(\sigma(E,D)) = \sigma(\pi^* E_1, \Delta)$ où $\pi : X \times \mathbb{R} \to X$, où E_1 est le $C^{p,q}$-module sous-jacent à E et où Δ est l'opérateur de Fredholm défini au-dessus du point (x,λ) par $D(x) + \varepsilon_{q+1}\lambda$.
Bien entendu, ces considérations sont aussi valables en théorie L .

2) Introduction des analogues de la théorie KR

Supposons que X soit un espace muni d'une involution σ . Il convient de modifier alors la définition des groupes ${}_{\varepsilon}\bar{U}^{p,q}(X)$ et ${}_{\varepsilon}\bar{L}^{p,q}(X)$ comme suit. On considère toujours des paires (E,D) comme ci-dessus, la différence étant que D est un morphisme $E \otimes C \to E \otimes \mathcal{C}$ tel que $\overline{D(x)} = D(\sigma(x))$ en un sens évident. Si l'involution est triviale, on retrouve la définition précédente. L'homomorphisme généralisé

$$\bar{t}_U : {}_{\varepsilon}\bar{U}^{p+m,q+n}(X ; \mathcal{C}) \to {}_{\varepsilon}\bar{U}^{p,q}(X \times \mathbb{R}^{m,n} ; \mathcal{C})$$

s'explicite par la formule $t_U(\sigma(E,D)) = \sigma(\pi^* E_1, \Delta)$, $\pi : X \times \mathbb{R}^{m,n} \to X$, où E_1 est le $C^{p,q}$-fibré sous-jacent à E et où Δ est l'opérateur de Fredholm défini au-dessus du point (x,λ,μ) de $X \times \mathbb{R}^{m,n}$ par $\Delta(x,\lambda,\mu) = \overline{D(x)} + i\rho(\lambda) + \rho(\mu)$ où $\rho : \mathbb{R}^{m+n} \to \mathrm{End}\,(E)$ est induit par la structure de $C^{m,n}$-module sous-jacente. On définit de même l'homomorphisme généralisé $\bar{t}_L$.

3) Description de t_L et t_U pour $p = q = 0$ et $m = n = 1$. Il est clair d'après les formules précédentes que l'on peut définir de même les théories ${}_{\varepsilon}U^{p,q}(X ; \mathcal{C})$ et ${}_{\varepsilon}L^{p,q}(X ; \mathcal{C})$ pour un espace X avec involution. On n'est donc pas obligé de supposer que $\mathcal{C}$ est une suspension. Pour $p = q = 0$ et $m = n = 1$, on a ainsi un diagramme

$$\begin{array}{ccc} {}_{\varepsilon}L^{0,0}(X ; \mathcal{C}) \approx {}_{\varepsilon}L^{1,1}(X ; \mathcal{C}) & \longrightarrow & {}_{\varepsilon}L^{0,0}(X \times D^{1,1}, X \times S^{1,1} ; \mathcal{C}) \\ \wr & & \wr \\ {}_{\varepsilon}L(X ; \mathcal{C}) & \xrightarrow{\beta} & {}_{\varepsilon}L(X \times D^{1,1}, X \times S^{1,1} ; \mathcal{C}) \end{array}$$

qui s'explicite de la même manière que l'homomorphisme analogue en K-théorie (cf. [14] exp.V). De manière plus précise, l'homomorphisme β associe à un objet E de ${}_{\varepsilon}\mathcal{Q}(\mathcal{C}(X))$ l'élément $d(\pi^{*}E,\pi^{*}E,\alpha)$ où $\pi : X \times D^{1,1} \to X$ et où $\alpha : E \otimes \mathbb{C}_{|X \times S^{1,1}} \to E \otimes \mathbb{C}_{|X \times S^{1,1}}$ est l'automorphisme défini au-dessus du point $e^{i\theta} \in S^{1,1}$ par la multiplication par $\cos\theta + i\sin\theta$. En suivant [16], on peut encore interpréter cet homomorphisme d'une autre manière. En effet, la théorie des séries de Fourier montre que ${}_{\varepsilon}L(X \times D^{1,1}, X \times S^{1,1} ; \mathcal{C}) \approx {}_{\varepsilon}\tilde{L}^{-1}(\mathcal{C}'\langle z,z^{-1}\rangle)$ où $\tilde{L}^{-1}(\mathcal{C}'\langle z,z^{-1}\rangle)$ désigne le conoyau de l'application naturelle ${}_{\varepsilon}L^{-1}(\mathcal{C}') \to {}_{\varepsilon}L^{-1}(\mathcal{C}'\langle z,z^{-1}\rangle)$ avec $\mathcal{C}' = \mathcal{C}(X)$ (poser $z = e^{i\theta}$). Par ailleurs, on dispose d'un homomorphisme naturel ${}_{\varepsilon}L^{-1}(\mathcal{C}'\langle z,z^{-1}\rangle) \to {}_{\varepsilon}L^{-1}(S\mathcal{C}')$ induit par le foncteur $\mathcal{C}'\langle z,z^{-1}\rangle \to S\mathcal{C}'$ introduit dans [16]. Puisque ${}_{\varepsilon}L^{-1}(S\mathcal{C}') \approx {}_{\varepsilon}L^{0}(\mathcal{C}')$, on en déduit finalement un homomorphisme $\beta' : {}_{\varepsilon}L(X \times \mathbb{R}^{1,1}; \mathcal{C}) \to {}_{\varepsilon}L(X ; \mathcal{C})$ compatible avec les structures de ${}_{1}L(? ;\mathcal{C})$-module et qui est inverse à gauche de l'homomorphisme $\beta : {}_{\varepsilon}L(X ; \mathcal{C}) \to {}_{\varepsilon}L(X \times \mathbb{R}^{1,1}; \mathcal{C})$. Un argument topologique formel dû à Atiyah [3] et utilisant essentiellement le fait que la bijection compacte de $\mathbb{R}^2$ dans $\mathbb{R}^2$ définie par $(x,y) \to (y,x)$ est homotope (parmi les applications compactes) à la bijection $(x,y) \to (-x,y)$ permet de montrer de même que $\beta\beta'$ est l'identité. Donc β est un isomorphisme pour toute catégorie hermitienne $\mathcal{C}$.

4) Interprétation des homomorphismes composés

$$\beta_U : {}_{\varepsilon}\bar{U}^{p,q}(X ; \mathcal{C}) \xrightarrow{\approx} {}_{\varepsilon}\bar{U}^{p+1,q+1}(X ; \mathcal{C}) \xrightarrow{\bar{t}_L} \bar{U}^{p,q}(X \times \mathbb{R}^{1,1} ; \mathcal{C}) \quad \text{et}$$

$$\beta_L : {}_{\varepsilon}\bar{L}^{p,q}(X ; \mathcal{C}) \xrightarrow{\approx} {}_{\varepsilon}\bar{L}^{p+1,q+1}(X ; \mathcal{C}) \xrightarrow{\bar{t}_L} {}_{\varepsilon}\bar{L}^{p,q}(X \times \mathbb{R}^{1,1} ; \mathcal{C})$$

Raisonnons dans le cadre de la théorie ${}_{\varepsilon}\bar{U}^{p,q}$ pour fixer les idées. L'homomorphisme précédent s'explicite alors comme suit. A la paire (E,D) on associe la paire (F,Δ) où $F = \pi^{*}(E \oplus E')$, $\pi : X \times \mathbb{R}^{1,1} \to X$, E' étant E muni de la structure de $C^{p,q+1}$-module conjuguée et de la forme quadratique opposée, et où

Δ est l'opérateur de Fredholm défini au-dessus du point (x,λ,μ) de $X \times \mathbb{R}^{1,1}$ par la formule

$$\Delta(x,\lambda,\mu) = \begin{pmatrix} D(x) & \mu + i\lambda \\ \mu - i\lambda & -D(x) \end{pmatrix}$$

Il s'agit de montrer que β_U s'interprète toujours comme le cup-produit par un élément u de ${}_1L(\mathbb{R}^{1,1};\mathcal{E})$ qui est l'image de 1 par l'isomorphisme ${}_1L(\mathcal{E}) \to {}_1L(\mathbb{R}^{1,1};\mathcal{E})$ explicité dans le point 3 . Pour cela une digression s'impose au sujet de la définition de $L(X) = {}_1L(X;\mathcal{E})$ lorsque X est localement compact. De manière précise, considérons l'ensemble des couples (E,α) où E et α s'explicitent comme suit :

. E est un fibré "réel" (dans la terminologie de [2]) muni d'une forme 1-quadratique non dégénérée, qui est facteur direct d'un fibré trivial et qui est $\mathbb{Z}_2$-gradué c'est-à-dire muni d'une involution auto-adjointe.

. α est un endomorphisme de fibrés réels de degré un , symétrique gauche $(\alpha^*=-\alpha)$ tel que $\alpha^2 \sim \lambda(x)$ pour $x \to \infty$, $\lambda(x)$ étant une fonction adéquate.

La paire (E,α) est dite acyclique si α^2 est une fonction adéquate sur tout X . Deux paires (E,α) et (E',α') sont homotopes s'il existe une isométrie $f : E \to E'$ telle que α soit homotope à $f^{-1}.\alpha'.f$ parmi l'ensemble des endomorphismes qui satisfont aux propriétés ci-dessus. Enfin, deux paires σ et σ' sont équivalentes s'il existe deux paires acycliques τ et τ' telles que $\sigma \oplus \tau$ et $\sigma' \oplus \tau'$ soient homotopes. Les classes d'équivalence forment un groupe (pour la somme des paires) qu'on notera $\bar{L}(X)$.

<u>Lemme 3.5.</u> <u>L'homomorphisme de</u> $\bar{L}(X)$ <u>dans</u> $L(X) = {}_1L(X;\mathcal{E})$ <u>qui associe à la paire</u> (E,α) <u>la différence</u> $E^0 - E^1$, E^0 <u>et</u> E^1 <u>étant les parties homogènes de degré zéro et un de</u> E , <u>est un isomorphisme lorsque</u> X <u>est un espace compact à involution</u> .

La démonstration de ce lemme est évidente. On a maintenant le lemme plus général suivant :

LEMME 3.6. Soit X un espace localement compact à involution. Alors $\bar{L}(X) \approx \mathrm{Ker}(L(X^+) \to L(\mathrm{Point}))$, X^+ désignant le compactifié d'Alexandroff de X.

Démonstration Soit (E,α) une paire définissant un élément de $\bar{L}(X)$. Quitte à ajouter une paire acyclique, on peut supposer que E^0 est la restriction à X d'un fibré quadratique sur X^+ que nous noterons $\tilde{E}^0$. Soit alors $\tilde{E}^1$ le fibré réel sur X^+ obtenu par recollement de $E^0 \otimes \mathbb{C}|_{X^+-K}$ et de $E'^1 \otimes \mathbb{C}$ au moyen de l'isométrie induite par α au-dessus de $X-K$, K compact convenable. L'homomorphisme de $\bar{L}(X)$ dans $\mathrm{Ker}[L(X^+) \to L(\mathrm{Point})]$ associe alors à la paire (E,α) la différence $\tilde{E}^0-\tilde{E}^1$. En sens inverse, si G et H sont deux fibrés quadratiques sur X^+ et si $\beta : G|_{\mathrm{Point}} \to H|_{\mathrm{Point}}$ est une isométrie, posons $E = G \oplus H'|_X$ et $\alpha = \alpha'|_X$ où $\alpha' : (G \oplus H') \otimes \mathbb{C} \to (G \oplus H') \otimes \mathbb{C}$ est un prolongement symétrique gauche à X^+ de l'endomorphisme de degré un défini par la matrice

$$\begin{pmatrix} 0 & \beta^{-1} \\ \beta & 0 \end{pmatrix}$$

Il est clair que les homomorphismes ainsi définis sont inverses l'un de l'autre.

Si X et Y sont deux espaces localement compacts à involution définissons un cup-produit

$${}_{\varepsilon}\bar{U}^{p,q}(X\ ;\ \mathcal{C}) \times \bar{L}(Y) \longrightarrow {}_{\varepsilon}\bar{U}^{p,q}(X \times Y\ ;\ \mathcal{C})$$

par la formule de [14] exp. III :

$$\sigma(E,D) \smile \sigma(F,\alpha) = \sigma(E \otimes F, D \otimes \eta + 1 \otimes \alpha)\ ,$$

η étant la graduation de F et la structure de $C^{p,q}$-module de $E \otimes F$ étant induite par les automorphismes $e_j \otimes \eta$ et $\varepsilon_k \otimes \eta$.

LEMME 3.7. Le diagramme suivant est commutatif

$$\begin{array}{ccc} {}_{\varepsilon}\bar{U}^{p,q}(X\ ;\ \mathcal{C}) \times \bar{L}(Y) & \longrightarrow & {}_{\varepsilon}\bar{U}^{p,q}(X \times Y\ ;\ \mathcal{C}) \\ \downarrow\uparrow & & \downarrow\uparrow \\ {}_{\varepsilon}\bar{U}^{p,q}(X\ ;\ \mathcal{C}) \times L(Y^+) & \longrightarrow & {}_{\varepsilon}\bar{U}^{p,q}(X \times Y^+;\ \mathcal{C}) \end{array}$$

La démonstration de ce lemme est évidente.

Dans le contexte précédent, l'élément u de $\bar{L}(\mathbb{R}^{1,1})$ est la classe de la paire (E,α) où $E = \pi^* \mathbb{R}^2$, $\pi : \mathbb{R}^{1,1} \to$ Point et où $\alpha : E \otimes \mathbb{C} \to E \otimes \mathbb{C}$ est l'endomorphisme symétrique gauche de degré un défini par la matrice

$$\begin{pmatrix} 0 & \mu + i\lambda \\ \mu - i\lambda & 0 \end{pmatrix}$$

Il suffit en effet d'expliciter $\bar{t}_L : {}_1L^{1,1}(\text{Point};\mathcal{C}) \to {}_1\bar{L}^{0,0}(\mathbb{R}^{1,1};\mathcal{E})$. D'après le lemme 3.7., il est clair que l'homomorphisme $\beta_U : {}_\varepsilon\bar{U}^{p,q}(X ; \mathcal{C}) \to \bar{U}^{p,q}(X \times \mathbb{R}^{1,1}; \mathcal{C})$ est induit par le cup-produit par l'élément $u \in \bar{L}(\mathbb{R}^{1,1}) \approx {}_1L(\mathbb{R}^{1,1} : \mathcal{E})$ pour toutes les définitions du cup-produit envisagées. On peut faire le même raisonnement pour l'homomorphisme β_L.

5) <u>Les homomorphismes β_U et β_L sont des isomorphismes.</u> Raisonnons encore dans le cas de la théorie U pour fixer les idées. Si $\mathcal{C}$ est une catégorie hermitienne et si A est un anneau hermitien, on définit de manière évidente la catégorie hermitienne $\mathcal{C} \otimes A$ (catégorie des A-modules objets de $\mathcal{C}$). En particulier, munissons l'algèbre de Clifford $C^{p,q}$ de l'antiinvolution induite par $e_i \to e_i$ et $\varepsilon_j \to -\varepsilon_j$ (on notera $C_U^{p,q}$ l'anneau hermitien ainsi obtenu). Il est clair que le groupe ${}_\varepsilon U^{p,q}(\mathcal{C})$ est le groupe ${}_\varepsilon L(\varphi)$ associé au foncteur hermitien $\varphi : \mathcal{C} \otimes C_U^{p,q+1} \to \mathcal{C} \otimes C_U^{p,q}$. Considérons le diagramme

$$\begin{array}{ccccccccc}
{}_\varepsilon L^{-1}(\mathcal{C}\otimes C_U^{p,q+1}) & \to & {}_\varepsilon L^{-1}(\mathcal{C}\otimes C_U^{p,q}) & \to & {}_\varepsilon U^{p,q}(\mathcal{C}) & \to & {}_\varepsilon L(\mathcal{C}\otimes C_U^{p,q+1}) & \to & {}_\varepsilon L(\mathcal{C}\otimes C_U^{p,q}) \\
\downarrow & & \downarrow & & \downarrow & & \downarrow & & \downarrow \\
{}_\varepsilon L^{-1}(\mathbb{R}^{1,1};\mathcal{C}\otimes C_U^{p,q+1}) & \to & {}_\varepsilon L^{-1}(\mathbb{R}^{1,1};\mathcal{C}\otimes C_U^{p,q}) & \to & {}_\varepsilon U^{p,q}(\mathbb{R}^{1,1};\mathcal{C}) & \to & {}_\varepsilon L(\mathbb{R}^{1,1};\mathcal{C}\otimes C_U^{p,q+1}) & \to & {}_\varepsilon L(\mathbb{R}^{1,1};\mathcal{C}\otimes C_U^{p,q})
\end{array}$$

L'argument utilisé pour la démonstration du théorème de Thom en K-théorie équivariante ([14] exp. V) permet de démontrer de même que le cup-produit par u induit un isomorphisme ${}_\varepsilon U^{p,q}(\mathcal{C}) \approx {}_\varepsilon U^{p,q}(\mathbb{R}^{1,1}; \mathcal{C})$ pour toute catégorie hermitienne $\mathcal{C}$. Donc on a aussi un isomorphisme ${}_\varepsilon U^{p,q}(X ; \mathcal{C}) \approx {}_\varepsilon U^{p,q}(X \times \mathbb{R}^{1,1}; \mathcal{C})$ pour X compact ou plus généralement localement compact d'après la propriété d'excision. En remplaçant $\mathcal{C}$ par $S\mathcal{C}$, l'argument du point 1 permet de montrer de même les isomorphismes ${}_{\varepsilon}U^{p,q}(X ; \mathcal{C}) \approx {}_{\varepsilon}U^{p+1,q+1}(X ; \mathcal{C}) \xrightarrow{t_U} {}_{\varepsilon}U^{p,q}(X \times \mathbb{R}^{1,1}; \mathcal{C})$

6) Fin de la démonstration du théorème 3.4.

La transitivité évidente des homomorphismes $\bar{t}_U$ et la suite

$${}_{\varepsilon}\bar{U}^{p+1,q+1}(X;\mathcal{C}) \xrightarrow{t_U} {}_{\varepsilon}\bar{U}^{p,q+1}(X \times \mathbb{R}^{1,0}; \mathcal{C}) \xrightarrow{t_U} {}_{\varepsilon}\bar{U}^{p,q}(X \times \mathbb{R}^{1,1};\mathcal{C}) \xrightarrow{t_U} {}_{\varepsilon}\bar{U}^{p-1,q}(X \times \mathbb{R}^{1,1}; \mathcal{C})$$

permet de montrer comme dans [14] que t_U est un isomorphisme. On en déduit que $t_U : {}_{\varepsilon}U^{p+m,q+n}(X;\mathcal{D}) \to {}_{\varepsilon}U^{p,q}(X \times \mathbb{R}^{m,n};\mathcal{D})$ est un isomorphisme lorsque $\mathcal{D}$ est une catégorie de la forme $S\mathcal{C}$ d'après le point 1. En remplaçant X par $X \times \mathbb{R}^{0,1}$, on en déduit le théorème 3.4. (sous une forme plus générale).

Le raisonnement déjà fait dans [13][15] montre que les foncteurs $(X,Y) \mapsto {}_{\varepsilon}L^{p,q}(X,Y ; A)$ et $(X,Y) \mapsto {}_{\varepsilon}U^{p,q}(X,Y ; A)$, $p-q = n$, constituent les éléments d'une théorie de la cohomologie périodique de période 8 (2 dans le cas hermitien complexe). On en déduit en particulier le théorème suivant :

THEOREME 3.8. Soit A une algèbre de Banach munie d'une antiinvolution bornée. Alors les groupes ${}_{\varepsilon}L^{n}(A)$ définis dans [18] et les groupes ${}_{\varepsilon}U^{n}(A)$ définis dans le § précédent coïncident respectivement avec les groupes ${}_{\varepsilon}L^{p,q}(A)$ et ${}_{\varepsilon}U^{p,q}(A)$ pour $p-q = n$. En particulier, les premiers groupes sont périodiques par rapport à n de période 8 (ou 2 si A est "hermitienne complexe" c'est-à-dire s'il existe un élément i du centre de A tel que $i^2 = -1$ et $\sigma(i) = -i$).

COROLLAIRE 3.9. Soit A une algèbre de Banach munie d'une antiinvolution bornée(1) et soit ${}_{\varepsilon}O(A) = \varinjlim {}_{\varepsilon}O_{n,n}(A)$. Alors $\pi_i({}_{\varepsilon}O(A)) \approx {}_{\varepsilon}L^{-i-1}(A)$ est un groupe périodique de période 8 par rapport à i. Si A est hermitienne complexe, la période peut être réduite à 2.

En fait, le corollaire 3.9. est lui-même une conséquence d'un résultat plus

(1) On dira simplement algèbre de Banach involutive par la suite

général qui exprime des équivalences d'homotopie entre certains espaces homogènes : il suffit de considérer les espaces classifiants associés aux foncteurs $X \mapsto {}_{\varepsilon}L^n(X\ ;\ A)$. En examinant de même les espaces classifiants associés aux foncteurs $X \to {}_{\varepsilon}U^n(X\ ;\ A)$, on obtient le théorème suivant :

THEOREME 3.10. Soit A une algèbre de Banach involutive. On a alors des équivalences d'homotopie entre les composantes neutres des espaces suivants :

$$GL(A)/{}_{-\varepsilon}O(A) \quad \text{et } \Omega({}_{\varepsilon}O(A)/GL(A))$$
$${}_{\varepsilon}O(A)/GL(A) \quad \text{et } \Omega({}_{\varepsilon}O(A \otimes \mathbb{C}')/{}_{\varepsilon}O(A))$$
$$O(A \otimes \mathbb{C}')/{}_{\varepsilon}O(A) \quad \text{et } \Omega({}_{\varepsilon}O(A \otimes \mathbb{H}')/{}_{\varepsilon}O(A \otimes \mathbb{C}'))$$
$${}_{\varepsilon}O(A \otimes \mathbb{H}')/{}_{\varepsilon}O(A \otimes \mathbb{C}') \quad \text{et } \Omega(GL(A \otimes \mathbb{H}')/{}_{\varepsilon}O(A \otimes \mathbb{H}'))$$
$$GL(A \otimes \mathbb{H}')/{}_{\varepsilon}O(A \otimes \mathbb{H}') \quad \text{et } \Omega({}_{-\varepsilon}O(A \otimes \mathbb{H}')/GL(A \otimes \mathbb{H}'))$$
$${}_{-\varepsilon}O(A \otimes \mathbb{H}')/GL(A \otimes \mathbb{H}') \quad \text{et } \Omega({}_{-\varepsilon}O(A \otimes \mathbb{C}')/{}_{-\varepsilon}O(A \otimes \mathbb{H}'))$$
$${}_{-\varepsilon}O(A \otimes \mathbb{C}')/{}_{-\varepsilon}O(A \otimes \mathbb{H}') \quad \text{et } \Omega({}_{-\varepsilon}O(A)/{}_{-\varepsilon}O(A \otimes \mathbb{C}'))$$
$${}_{-\varepsilon}O(A)/{}_{-\varepsilon}O(A \otimes \mathbb{C}') \quad \text{et } \Omega(GL(A)/{}_{-\varepsilon}O(A))$$

Si A est hermitienne complexe, ces équivalences d'homotopie se réduisent aux deux suivantes

$$GL(A)/O(A) \quad \text{et} \quad \Omega(O(A)/GL(A))$$
$$O(A)/GL(A) \quad \text{et} \quad \Omega(GL(A)/O(A))$$

avec $\quad O(A) = {}_{1}O(A) \approx {}_{-1}O(A)$

Nous allons consacrer le reste du paragraphe à démontrer des suites exactes permettant de comparer les théories K^n, ${}_{\varepsilon}L^n$ et ${}_{\varepsilon}U^n$. D'après le paragraphe précédent, nous avons la suite exacte

$$K^{-1}(\mathcal{C}) \to {}_{\varepsilon}L^{-1}(\mathcal{C}) \to {}_{\varepsilon}U(\mathcal{C}) \to K(\mathcal{C}) \to {}_{\varepsilon}L(\mathcal{C})$$

En appliquant les foncteurs S et Ω, on en déduit la suite exacte

$$K^{n-1}(\mathcal{C}) \to {}_{\varepsilon}L^{n-1}(\mathcal{C}) \to {}_{\varepsilon}U^n(\mathcal{C}) \to K^n(\mathcal{C}) \to {}_{\varepsilon}L^n(\mathcal{C})$$

Nous allons tâcher de reconnaître de manière algébrique les homomorphismes définis dans cette suite (cf. [13]). En d'autres termes, nous allons définir une suite du

$$K^{p,q+1}(\mathcal{C}) \to {}_{\varepsilon}L^{p,q+1}(\mathcal{C}) \overset{\partial^{p,q+1}}{\to} {}_{\varepsilon}U^{p,q}(\mathcal{C}) \to K^{p,q}(\mathcal{C}) \to {}_{\varepsilon}L^{p,q}(\mathcal{C})$$

L'homomorphisme de $K^{p,q}$ dans ${}_{\varepsilon}L^{p,q}$ est clair ;c'est celui induit par le foncteur hyperbolique. L'homomorphisme de ${}_{\varepsilon}U^{p,q}$ dans $K^{p,q}$ est le morphisme "oubli" de structures hermitiennes. Il nous reste à définir $\partial^{p,q+1} : {}_{\varepsilon}L^{p,q+1}(\mathcal{C}) \longrightarrow {}_{\varepsilon}U^{p,q}(\mathcal{C})$ Soit donc $d(E,\eta_1,\eta_2)$ un élément de ${}_{\varepsilon}L^{p,q+1}(\mathcal{C})$. Le $C^{p,q+1}$-module quadratique E se scinde en tant que module quadratique sous la forme d'une somme $F \oplus G$, le générateur ε_{q+1} opérant grâce à l'involution définie par la matrice

$$\begin{pmatrix} 1 & 0 \\ 0 & -1 \end{pmatrix}$$

Les autres générateurs de l'algèbre de Clifford ainsi que η_1 et η_2 opèrent par des automorphismes orthogonaux s'écrivant sous la forme

$$\begin{pmatrix} 0 & \gamma \\ \gamma' & 0 \end{pmatrix}$$

Soit alors $\tilde{E} = F \oplus G'$, G' désignant G muni de la forme quadratique opposée. Si on oublie l'action de ε_{q+1} , il est clair que le triple $(\tilde{E},\eta_1,\eta_2)$ définit un élément de ${}_{\varepsilon}U^{p,q}(\mathcal{C})$, l'opération consistant à passer de E à $\tilde{E}$ échangeant les caractères auto-adjoint et symétrique gauche. Si $p = q = 0$, on voit par inspection que $\partial^{0,1}$ coïncide avec l'homomorphisme ${}_{\varepsilon}L^{-1}(\mathcal{C}) \to {}_{\varepsilon}U(\mathcal{C})$. Puisque $\partial^{p,q+1}$ est compatible avec les isomorphismes t_U et t_L , on en déduit le théorème suivant :

<u>THEOREME 3.11.</u> <u>Soit $\mathcal{C}$ une catégorie de Banach hermitienne. La suite naturelle</u>

$$K^{p,q+1}(\mathcal{C}) \to {}_{\varepsilon}L^{p,q+1}(\mathcal{C}) \xrightarrow{\partial^{p,q+1}} {}_{\varepsilon}U^{p,q}(\mathcal{C}) \to K^{p,q}(\mathcal{C}) \to {}_{\varepsilon}L^{p,q}(\mathcal{C})$$

<u>où $\partial^{p,q+1}$ est défini ci-dessus, est une suite exacte.</u>

De la même manière, considérons la suite exacte du paragraphe précédent

$${}_{-\varepsilon}L^{-1}(\mathcal{C}) \to K^{-1}(\mathcal{C}) \to {}_{-\varepsilon}V(\mathcal{C}) \to {}_{-\varepsilon}L(\mathcal{C}) \to K(\mathcal{C})$$

Puisque $_{-\varepsilon}V(\mathcal{C}) \approx {}_{\varepsilon}U^{0,1}(\mathcal{C})$ (proposition 3.2), il est naturel de chercher à démontrer une suite exacte du type

$$_{-\varepsilon}L^{p,q+1}(\mathcal{C}) \to K^{p,q+1}(\mathcal{C}) \to {}_{\varepsilon}U^{p,q+1}(\mathcal{C}) \xrightarrow{\delta^{p,q+1}} {}_{-\varepsilon}L^{p,q}(\mathcal{C}) \to K^{p,q}(\mathcal{C})$$

Dans cette suite, l'homomorphisme de $_{-\varepsilon}L^{p,q}$ dans $K^{p,q}$ est évidemment induit par le foncteur oubli tandis que l'homomorphisme de $K^{p,q}$ dans $_{\varepsilon}U^{p,q}$ est induit par le foncteur hyperbolique. Il reste à expliciter $\delta^{p,q+1}$. Soit donc $d(E,\eta_1,\eta_2)$ un élément de $_{\varepsilon}U^{p,q+1}(\mathcal{C})$. L'automorphisme involutif symétrique gauche ε_{q+1} permet de scinder E en $F \oplus {}^tF$, ε_{q+1} s'écrivant sous la forme

$$\varepsilon_{q+1} = \begin{pmatrix} 1 & 0 \\ 0 & -1 \end{pmatrix}$$

Si on considère maintenant $F \oplus {}^tF$ comme un module $(-\varepsilon)$-hyperbolique $\tilde{E}$, il est clair que les automorphismes e_i, ε_j, $i = 1,\ldots,p, j = 1,\ldots,q$ de l'algèbre de Clifford opèrent sur $\tilde{E}$ de manière unitaire. On en déduit évidemment l'homomorphisme $\delta^{p,q+1}$ ainsi que le théorème attendu :

<u>THEOREME 3.12.</u> <u>Soit $\mathcal{C}$ une catégorie de Banach hermitienne</u>

$$_{-\varepsilon}L^{p,q+1}(\mathcal{C}) \to K^{p,q+1}(\mathcal{C}) \to {}_{\varepsilon}U^{p,q+1}(\mathcal{C}) \xrightarrow{\delta^{p,q+1}} {}_{-\varepsilon}L^{p,q}(\mathcal{C}) \to K^{p,q}(\mathcal{C})$$

<u>est une suite exacte.</u>

<u>Exemples.</u>

La théorie topologique que nous venons de développer est en fait bien connue dans le cas des algèbres de Banach de fonctions continues. En voici quelques exemples :

1) $A = \mathbb{C}(X)$, algèbre des fonctions continues à valeurs complexes, munie de l'involution définie par la conjugaison complexe. Alors

$_1L(A) \approx {}_{-1}L(A) \approx KU(X) \oplus KU(X)$, $K(A) \approx KU(X)$, $_1U(A) \approx {}_{-1}U(A) \approx KU^{-1}(X)$

2) $A = \mathbb{C}(X)$ munie de l'involution triviale. Alors

$_1L(X) \approx KO(X)$, $_{-1}L(X) \approx Ksp(X)$, $_1U(A) \approx KO^6(X)$, $_{-1}U(A) \approx KO^2(X)$

3) $A = \mathbb{C}(X)$ munie de l'involution induite par une involution de X.

Alors ${}_1L(A) \approx KR(X)$.

4) $A = \mathbb{R}(X)$, algèbre des fonctions continues à valeurs réelles, munie de l'involution triviale. Alors ${}_1L(A) \approx KO(X) \oplus KO(X)$, ${}_{-1}L(A) \approx KU(X)$, ${}_1U(A) \approx KO^{-1}(X)$ et ${}_{-1}U(A) \approx KO^1(X)$. Bien entendu, $K(A) \approx KO(X)$.

IV - LES THEOREMES FONDAMENTAUX

Soit A est une algèbre de Banach involutive.

Nous avons vu dans le paragraphe précédent que les groupes ${}_\varepsilon V^n(A)$ et ${}_{-\varepsilon}U^{n-1}(A)$ sont isomorphes. Le théorème fondamental de ce paragraphe est l'isomorphisme ${}_\varepsilon V^n(A) \approx {}_{-\varepsilon}U^{n-1}(A)$ pour tout anneau hermitien A K-régulier satisfaisant aux hypothèses signalées à la fin de l'introduction. Nous démontrerons également le théorème de périodicité sur les groupes ${}_\varepsilon \widetilde{W}^n$. Les méthodes vont s'inspirer de l'étude topologique détaillée précédemment. En effet, pour commencer, essayons d'expliciter l'homomorphisme $t_U : {}_{-\varepsilon}U^{0,1}(A) \to {}_{-\varepsilon}U^{0,0}(D^1, S^0; A)$

défini dans le paragraphe précédent (cas topologique). Tout élément de ${}_{-\varepsilon}U^{0,1}(A) \approx {}_\varepsilon V(A)$ peut s'écrire sous la forme $d(E \oplus {}^tE, \eta_1, \eta_2)$ où $E \oplus {}^tE$ est muni de la forme ε-hyperbolique canonique et où η_j, $j = 1,2$ se représente par la matrice

$$\eta_j = \begin{pmatrix} 0 & g_j^{-1} \\ g_j & 0 \end{pmatrix}$$

avec ${}^tg_j = \varepsilon g_j$. La graduation gauche de $E \oplus {}^tE$ est évidemment définie par la matrice

$$\eta_0 = \begin{pmatrix} 1 & 0 \\ 0 & -1 \end{pmatrix}$$

D'après le paragraphe précédent, on a $t_U(d(E \oplus {}^tE,\eta_1,\eta_2)) = d(P^*(E \oplus {}^tE)$, $\eta_0 \cos\theta + \eta_1 \sin\theta, \eta_0 \cos\theta + \eta_2 \sin\theta)$, $0 \leq \theta \leq \pi$, $P : D^1 \to$ Point. La graduation gauche $\eta_0 \cos\theta + \eta_j \sin\theta$ s'écrit aussi $(\cos\theta/2 - \eta_0\eta_j \sin\theta/2)\eta_0(\cos\theta/2 + \eta_0\eta_j \sin\theta/2)$. L'application $\theta \mapsto \cos\theta/2 - \eta_0\eta_j \sin\theta/2$ est un chemin dans le groupe des automorphismes orthogonaux de $E \oplus {}^tE$ reliant l'identité à l'automorphisme

$$\begin{pmatrix} 0 & -g_j^{-1} \\ +g_j & 0 \end{pmatrix}$$

Pour définir un homomorphisme analogue dans le cas général, nous devons remplacer ce chemin par un chemin homotope qui est algébrique. De manière précise, considérons le chemin $x \mapsto h_j(x)$, $0 \leq x \leq 1$, où

$$h_j(x) = \begin{pmatrix} 1 & -g_j^{-1}x \\ 0 & 1 \end{pmatrix} \begin{pmatrix} 1 & 0 \\ g_jx & 1 \end{pmatrix} \begin{pmatrix} 1 & -g_j^{-1}x \\ 0 & 1 \end{pmatrix}$$

$$= \begin{pmatrix} 1-x^2 & g_j^{-1}(x^3-2x) \\ g_jx & 1-x^2 \end{pmatrix}$$

Si on fait le changement de paramètre $x = \sqrt{2}\sin\theta/4$ $0 \leq \theta \leq \pi$, on voit que l'homotopie entre les deux chemins est donnée par la matrice

$$\sigma_j(\theta,t) = \begin{pmatrix} a(\theta,t) & b(\theta,t)g_j^{-1} \\ c(\theta,t)g_j & a(\theta,t) \end{pmatrix}$$

avec $a(\theta,t) = \cos\theta/2$

$c(\theta,t) = (t\sqrt{2} + 2(1-t)\cos\theta/4)\sin\theta/4$

$b(\theta,t) = \frac{a^2-1}{c} = \frac{-4\sin\theta/4\cos\theta/4}{t\sqrt{2}+2(1-t)\cos\theta/4}$

En résumé, nous voyons par ce calcul que l'homomorphisme t_U s'interprête dans le cadre des anneaux hermitiens en un homomorphisme.

$$\tau_V : {}_\varepsilon V(A) \to {}_{-\varepsilon}U(\Psi) \approx {}_{-\varepsilon}U(\Omega A) \quad , \Psi : A\langle x\rangle \to A \times A$$

défini par la formule

$\tau_V(d(E,g_1,g_2)) = d(F,\eta_1,\eta_2)$ où $F = E \oplus {}^tE$ est muni de la forme $(-\varepsilon)$-hyperbolique canonique et où $\eta_j = \eta_j(x) = h_j(x)\,\eta_0 h_j(x)^{-1}$, η_0 et $h_j(x)$ étant définis par les formules ci-dessus. Nous montrerons plus loin (th.4.8) que τ_V est un isomorphisme si A est K-régulier.

Par des techniques analogues, on peut aussi définir un homomorphisme

$$\tau_V^1 : {}_\varepsilon V^1(A) \to {}_{-\varepsilon}U(A) \approx {}_{-\varepsilon}U(\tilde{A}) \approx U(\varphi) \ , \ \varphi : CA \to SA$$

De manière précise, un élément de ${}_\varepsilon V(SA)$ peut toujours s'écrire $d(\check{E},g_1,g_2)$ où E est un CA-module et où $\check{E} = E \otimes_{CA} SA$. Soit alors $F = E \oplus {}^tE$ le CA-module $(-\varepsilon)$-hyperbolique canonique et soit r_j, $j = 1,2$, l'automorphisme orthogonal de F défini par

$$r_j = \begin{pmatrix} 0 & -g_j^{-1} \\ g_j & 0 \end{pmatrix} = \begin{pmatrix} 1 & -g_j^{-1} \\ 0 & 1 \end{pmatrix} \begin{pmatrix} 1 & 0 \\ g_j & 0 \end{pmatrix} \begin{pmatrix} 1 & -g_j^{-1} \\ 0 & 1 \end{pmatrix}$$

Soient h_j et h'_j des endomorphismes de E tels que $\check{h}_j = -g_j^{-1}$, $\check{h}'_j = g_j$ et tels que h_j et h'_j appartiennent à l'image de l'opérateur "d'antisymétrisation" $1-T\varepsilon$. Posons

$$s_j = \begin{pmatrix} 1 & h_j \\ 0 & 1 \end{pmatrix} \begin{pmatrix} 1 & 0 \\ h'_j & 1 \end{pmatrix} \begin{pmatrix} 1 & h_j \\ 0 & 1 \end{pmatrix} \text{ et } \eta_j = s_j \eta_0 s_j^{-1}$$

avec

$$\eta_0 = \begin{pmatrix} 1 & 0 \\ 0 & -1 \end{pmatrix}$$

On pose alors $\tau_V^1(d(E,g_1,g_2)) = d(F,\eta_1,\eta_2)$. On vérifie aisément que τ_V^1 est ainsi bien défini. Si A est (2-2)-régulier (par exemple K-régulier) , l'homomorphisme composé $_{\varepsilon}V(SA) \to {}_{-\varepsilon}U(A) \to {}_{-\varepsilon}U(\Omega SA)$ n'est autre que l'homomorphisme τ_V appliqué à l'anneau SA .

Pour démontrer que τ_V(ou τ_V^1) est un isomorphisme, nous allons définir un homomorphisme en sens inverse

$$\tau_U : {}_{-\varepsilon}U(A) \to {}_{\varepsilon}V(SA)$$

En fait, τ_U va être l'homomorphisme composé de

$$\tau'_U : {}_{-\varepsilon}U(A) \to {}_{\varepsilon}V(A<z,z^{-1}>)$$

et de l'homomorphisme de $_{\varepsilon}V(A<z,z^{-1}>)$ dans $_{\varepsilon}V(SA)$ induit par l'homomorphisme d'anneaux $A<z,z^{-1}> \to SA$. De manière précise, tout élément de $_{-\varepsilon}U(A)$ peut s'écrire $d(F,\eta_1,\eta_2)$ où $F = E \oplus {}^tE$ est muni de la forme $(-\varepsilon)$-hyperbolique canonique et où η_j est la graduation gauche définie par $\eta_j = 2p_j - 1$ avec

$$p_j = \begin{pmatrix} a_j & b_j \\ c_j & d_j \end{pmatrix}$$

Identifions F et ^tF au moyen de la forme 1-hyperbolique canonique. On pose alors $\tau'_U(d(F,\eta_1,\eta_2)) = d(F,g_1,g_2)$ où $g_j : F \to {}^tF \approx F$ est défini par la matrice

$$g_j = g_j(z) = \begin{pmatrix} a_j+(1-a_j)z & -b_j(z+z^{-1}-2) \\ -\varepsilon c_j & \varepsilon(d_jz^{-1}+(1-d_j)) \end{pmatrix}$$

Puisque $p_j^* = 1-p_j$, on a $^tg_j = \varepsilon g_j$. Puisque p_j est un projecteur, g_j est inversible et son inverse est défini par la matrice

$$g_j^{-1} = \begin{pmatrix} a_j+(1-a_j)z^{-1} & \varepsilon b_j(z+z^{-1}-2) \\ c_j & \varepsilon(d_jz+1-d_j) \end{pmatrix}$$

D'autre part, cette définition est indépendante du scindage de F choisi. En effet, si $E' \oplus {}^tE'$ est un autre scindage et si on désigne par g'_j les formes quadratiques qui s'en déduisent par la construction précédente, on a l'identité

$d(F,g_1,g_2) + d(F,g'_2,g'_1) = d(F \oplus F \oplus F,h_1,h_2)$ où h_1 est la forme ε-quadratique déduite de la graduation gauche

$$\xi_1 = \begin{pmatrix} \eta_1 & 0 & 0 \\ 0 & \eta_2 & 0 \\ 0 & 0 & \eta_1 \end{pmatrix}$$

et où h_2 est la forme ε-quadratique déduite de la graduation gauche $\alpha\,\xi_1\alpha^{-1}$ où α est la matrice de permutation

$$\alpha = \begin{pmatrix} 0 & 1 & 0 \\ 0 & 0 & 1 \\ 1 & 0 & 0 \end{pmatrix}$$

Puisque α est un commutateur, α est stablement homotope à l'identité, ce qui démontre l'assertion. De ces vérifications il résulte que τ'_U, donc τ_U est bien défini.

Remarque Si $\Psi : A \to A''$ est un homomorphisme surjectif d'anneaux hermitiens, les mêmes considérations permettent de définir un homomorphisme $\tau'_U : {}_{-\varepsilon}U(\Psi) \to {}_{\varepsilon}V(\Psi\langle z,z^{-1}\rangle)$ et $\tau_U : {}_{-\varepsilon}U(\Psi) \to {}_{\varepsilon}V(S\Psi)$. Ces homomorphismes correspondent aux homomorphismes ${}_{-\varepsilon}U(A') \to {}_{\varepsilon}V(A'\langle z,z^{-1}\rangle)$ et ${}_{-\varepsilon}U(A') \to {}_{\varepsilon}V(SA')$ avec $A' = \mathrm{Ker}\Psi$ grâce aux isomorphismes d'excision.

Il convient d'interprèter τ'_U lorsque A est une algèbre de Banach involutive. Nous allons voir que, dans ce cas, τ'_U n'est autre qu'une reformulation de l'homomorphisme défini lors de la démonstration du théorème 3.4.

$$t_U : {}_{-\varepsilon}U(A) \approx {}_{-\varepsilon}U^{1,1}(A) \to {}_{-\varepsilon}U^{0,1}(D^{1,0},S^{1,0};A)$$

En effet, soit $x' = d(F,\eta_1,\eta_2)$ un élément de ${}_{-\varepsilon}U(A)$. L'élément de ${}_{-\varepsilon}U^{1,1}(A)$ qui lui correspond est $x = d(F \oplus F',\gamma_1,\gamma_2)$ qui s'explicite ainsi :

• E' est le module $(-\varepsilon)$-quadratique E muni de la forme quadratique opposée et $E \oplus E'$ est muni de la structure de $C^{1,1}$-module gauche induite par les automorphismes

$$e_1 = \begin{pmatrix} 0 & 1 \\ -1 & 0 \end{pmatrix} , \quad \varepsilon_1 = \begin{pmatrix} 0 & 1 \\ 1 & 0 \end{pmatrix}$$

• γ_1 et γ_2 sont les graduations gauches définies par les formules

$$\gamma_j = \begin{pmatrix} \eta_j & 0 \\ 0 & -\eta_j \end{pmatrix}$$

Représentons $D^{1,0}$ comme l'ensemble des points de $\mathbb{C}$ de coordonnée $\exp(i\theta)$, $-\pi/2 \leq \theta \leq +\pi/2$. Avec ces notations, on a $t_U(x) = d(F \oplus F', \gamma_1(\theta), \gamma_2(\theta))$ où

$$\gamma_j(\theta) = \begin{pmatrix} \eta_j \cos\theta & +i\sin\theta \\ -i\sin\theta & -\eta_j\cos\theta \end{pmatrix}$$

et où $F \oplus F'$ est regardé comme $C^{0,1}$-module (par restriction des scalaires).

Par ailleurs, désignons par $h : F \to {}^tF$ l'isomorphisme définissant la forme $(-\varepsilon)$-hermitienne sur F. Si on munit $F \oplus {}^tF$ de la forme $(-\varepsilon)$-hyperbolique canonique, on voit que $F \oplus {}^tF$ et $F \oplus F'$ sont isomorphes grâce à l'isomorphisme défini par la matrice

$$\begin{pmatrix} 1 & h^{-1}/2 \\ 1 & -h^{-1}/2 \end{pmatrix}$$

On en déduit que $t_U(x) = d(F \oplus {}^tF, \xi_1(\theta), \xi_2(\theta))$ où $F \oplus {}^tF$ est muni de la structure de $C^{0,1}$-module gauche définie par la matrice

$$\begin{pmatrix} 1 & 0 \\ 0 & -1 \end{pmatrix}$$

et où $\xi_j(\theta)$ est la graduation gauche définie par la matrice

$$\xi_j(\theta) = \begin{pmatrix} 0 & (\eta_j \cos\theta - i\sin\theta)h^{-1} \\ h(\eta_j \cos\theta + i\sin\theta) & 0 \end{pmatrix}$$

Compte tenu de l'isomorphisme $_{-\varepsilon}U^{0,1} \approx {}_{\varepsilon}V^0$, on déduit donc du théorème 3.4., ou plutôt de sa généralisation explicitée dans le § 3, le corollaire suivant :

PROPOSITION 4.1. Soit $\tilde{\tau}_U : {}_{-\varepsilon}U(A) \to {}_{\varepsilon}V(D^{1,0},S^{1,0};A)$ l'homomorphisme défini par $\tilde{\tau}_U(d(F,\eta_1,\eta_2)) = d(\pi^*F,\tilde{g}_1,\tilde{g}_2)$ où $\pi : D^{1,0} \to$ Point et où $g_j : F \otimes C \to {}^tF \otimes \bar{C}$ est l'isomorphisme défini au-dessus du point $e^{i\theta}$ de $D^{1,0}(-\pi/2 \leq \theta \leq +\pi/2)$ par la formule $\tilde{g}_j(\theta) = h(\eta_j \cos\theta + i\sin\theta)$. Alors $\tilde{\tau}_U$ coïncide, modulo les isomorphismes explicités plus haut, avec l'homomorphisme t_U . En particulier, $\tilde{\tau}_U$ est un isomorphisme pour toute algèbre de Banach involutive.

Pour obtenir un homomorphisme plus proche de τ_U il convient d'identifier le groupe $_{\varepsilon}V(D^{1,0},S^{1,0};A) \approx {}_{\varepsilon}\tilde{V}(D^{1,0}/S^{1,0};A)$ à $_{\varepsilon}\tilde{V}(A\langle z,z^{-1}\rangle)$ (1) en posant $z = -\exp(-2i\theta)$. On va voir alors que $\tilde{\tau}_U$ coïncide, modulo ces légères modifications, avec l'homomorphisme général τ'_U défini précédemment. En effet, soit $d(F,\eta_1,\eta_2)$ un élément de $_{-\varepsilon}U(A)$ avec $F = E \oplus {}^tE$ et

$$\eta_j = \begin{pmatrix} u_j & v_j \\ w_j & y_j \end{pmatrix}$$

Si on pose $z = -e^{-2i\theta}$, $\tau'_U(d(F,\eta_1,\eta_2)) = d(E \oplus {}^tE,g_1,g_2)$ avec

$$g_j = \begin{pmatrix} \frac{1+u_j}{2} - \frac{1-u_j}{2}e^{-2i\theta} & 2v_j\cos^2\theta \\ -\frac{\varepsilon w_j}{2} & \varepsilon(\frac{1-y_j}{2}-\frac{1+y_j}{2}e^{2i\theta}) \end{pmatrix}$$

avec $\pi/2 \leq \theta \leq +\frac{\pi}{2}$.

---

(1) On pose $_{\varepsilon}\tilde{V}(A\langle z,z^{-1}\rangle) = \mathrm{coker}[{}_{\varepsilon}V(A) \to {}_{\varepsilon}V(A\langle z,z^{-1}\rangle)]$.

mais non nécessairement $g_j'(-\pi/2) = g_j'(\pi/2)$ (on a "coupé" en deux le cercle trigonométrique, opération qui n'est pas algébrique dans notre contexte). Dans ces conditions, il est clair que $g_j = g_j(\theta)$ est homotope à la matrice

$$g_j'(\theta) = \begin{pmatrix} \frac{1+u_j}{2} - \frac{1-u_j}{2}e^{-2i\theta} & v_j \cos\theta \\ -\varepsilon\, w_j \cos\theta & \varepsilon\left(\frac{1-y_j}{2} - \frac{1+y_j}{2}e^{2i\theta}\right) \end{pmatrix}$$

Enfin le triple (F, g_1', g_2') est isomorphe (donc homotope stablement) au triple (F, g_1'', g_2'') où g_j'' est défini par le produit de matrices

$$g_j'' = \begin{pmatrix} e^{i\frac{\theta}{2}} & 0 \\ 0 & e^{-i\frac{\theta}{2}} \end{pmatrix} \begin{pmatrix} \frac{1+u_j}{2} - \frac{1-u_j}{2}e^{-2i\theta} & v_j \cos\theta \\ -\varepsilon w_j \cos\theta & \varepsilon\left(\frac{1-y_j}{2} - \frac{1+y_j}{2}e^{2i\theta}\right) \end{pmatrix} \begin{pmatrix} e^{i\frac{\theta}{2}} & 0 \\ 0 & e^{-i\frac{\theta}{2}} \end{pmatrix}$$

$= h(\eta_j \cos\theta + i \sin\theta)$ avec

$$h = \begin{pmatrix} 1 & 0 \\ 0 & -\varepsilon \end{pmatrix}$$

représentant la forme $(-\varepsilon)$-hyperbolique canonique sur $F = E \oplus {}^tE$. On a ainsi démontré de manière topologique la proposition suivante :

PROPOSITION 4.2. L'homomorphisme composé

$$\tau_U : {}_{-\varepsilon}U(A) \to {}_{\varepsilon}\tilde{V}(A\langle z, z^{-1}\rangle) \to {}_{\varepsilon}V(SA)$$

est un isomorphisme pour toute algèbre de Banach involutive A .

Remarque. L'isomorphisme ${}_{\varepsilon}\tilde{V}(A\langle z,z^{-1}\rangle) \approx {}_{\varepsilon}V(SA)$ résulte évidemment du théorème 3.4. (convenablement généralisé dans le cadre de la KR-théorie). J'ignore dans quelle mesure l'application ${}_{\varepsilon}\tilde{V}(A\langle z,z^{-1}\rangle) \to {}_{\varepsilon}V(SA)$ est injective pour un anneau hermitien quelconque A .

Pour démontrer les théorèmes fondamentaux, nous aurons besoin de quelques théorèmes bien connus sur la classification des formes quadratiques sur $\mathbb{Q}$ et sur $\mathbb{Z}' = \mathbb{Z}[\frac{1}{2}]$. Pour commencer, démontrons le lemme suivant :

LEMME 4.3. *L'homomorphisme ${}_1L(\mathbb{Z}') \to {}_1L(\mathbb{Q})$ est injectif.*

DEMONSTRATION : Considérons le diagramme commutatif

$$\begin{array}{ccccccc} K(\mathbb{Z}') & \longrightarrow & {}_1L(\mathbb{Z}') & \longrightarrow & {}_1W(\mathbb{Z}') & \longrightarrow & 0 \\ \downarrow & & \downarrow & & \downarrow & & \\ K(\mathbb{Q}) & \longrightarrow & {}_1L(\mathbb{Q}) & \longrightarrow & {}_1W(\mathbb{Q}) & \longrightarrow & 0 \end{array}$$

où on pose de manière générale ${}_\varepsilon W(A) = \mathrm{Coker}(K(A) \to {}_\varepsilon L(A))$ (${}_\varepsilon W(A)$ est le "groupe de Witt" de A). Ce diagramme montre qu'il suffit de démontrer que l'homomorphisme ${}_1W(\mathbb{Z}') \to {}_1W(\mathbb{Q})$ est injectif. Soit donc E un module quadratique sur $\mathbb{Z}'$ tel que $E \otimes \mathbb{Q}$ soit hyperbolique. Si q désigne la forme quadratique sur E, il existe un sous-module L de E tel que $q|_L = 0$. En outre, on peut choisir L maximal puisque $\mathbb{Z}'$ est noéthérien. Dans ce cas, il est clair que E/L est sans torsion donc projectif. Par suite, L est facteur direct dans E et l'application transposée de l'inclusion, soit ${}^tE \longrightarrow {}^tL$, est surjective. Il en est donc de même de l'application $E/L \longrightarrow {}^tL$ déduite de ${}^tE \longrightarrow {}^tL$ par passage au quotient. D'autre part, puisque $E \otimes \mathbb{Q}$ est hyperbolique, $E/L \otimes \mathbb{Q} \longrightarrow {}^tL \otimes \mathbb{Q} \approx {}^t(L \otimes \mathbb{Q})$ est bijective et il en est de même de $E/L \longrightarrow {}^tL$ d'après ce qui précède. Si on choisit une section quelconque de $E \approx {}^tE \longrightarrow {}^tL$, on identifie ainsi E à $L \oplus {}^tL$ en tant que module quadratique.

Remarque Cette démonstration vaut aussi si on remplace $\mathbb{Z}'$ par un anneau principal A et $\mathbb{Q}$ par le corps de fractions de A.

PROPOSITION 4.4. *Soient q et q' deux formes 1-quadratiques non dégénérées de rang n et d'indice j (nombre de carrés négatifs) sur $\mathbb{Z}'$. Pour que q et q' soient stablement isomorphes, il faut et il suffit qu'elles aient même discriminant. Par conséquent ${}_1L(\mathbb{Z}') \approx \mathbb{Z} \oplus \mathbb{Z} \oplus (\mathbb{Z}/2)^2$.*

DEMONSTRATION Remarquons tout d'abord que q et q' sont isomorphes sur le corps fini $\mathbb{Z}/p\mathbb{Z}$, $p \neq 2$, les formes quadratiques sur ce corps étant précisément caractérisées par n et le discriminant. Par des arguments classiques d'approximation, on en déduit qu'elles sont isomorphes sur l'anneau des entiers p-adiques $\mathbb{Z}_p$: en effet, si $\alpha : (E,q) \to (E',q')$ est un isomorphisme non nécessairement unitaire tel que $\alpha^*\alpha = 1 + \nu$, ν topologiquement nilpotent, on remplace α par $\alpha' = \alpha(\sqrt{1+\nu})^{-1}$.

Par conséquent, q et q' sont isomorphes sur $\mathbb{Q}_p$ et les invariants locaux $\varepsilon_p(q)$ et $\varepsilon_p(q')$, $p \neq 2$, sont égaux [26] . D'après la formule du produit, on a aussi $\varepsilon_2(q) = \varepsilon_2(q')$. En appliquant le théorème de classification des formes quadratiques sur les rationnels [26] , on en déduit que q et q' sont isomorphes sur les rationnels. La proposition résulte alors du lemme précédent.

THEOREME 4.5. Soit A un anneau hermitien (2,1)-régulier (par exemple K-régulier) et soit $\varphi : A\langle x\rangle \to A \times A$. Alors l'homomorphisme composé

$$ {}_{\varepsilon}V(A) \xrightarrow{\tau_V} {}_{-\varepsilon}U(\varphi) \xrightarrow{\tau_U} {}_{\varepsilon}V(S\varphi) \approx {}_{\varepsilon}V(S\Omega A) \approx {}_{\varepsilon}V(A) $$

est un isomorphisme.

DEMONSTRATION Considérons l'homomorphisme composé

$$ {}_{\varepsilon}V(A) \xrightarrow{\tau_V} {}_{-\varepsilon}U(\varphi) \xrightarrow{\tau_U} {}_{\varepsilon}V(\psi) $$

où $\psi: A\langle x,z,z^{-1}\rangle \longrightarrow A\langle z,z^{-1}\rangle \times A\langle z,z^{-1}\rangle$. Cet homomorphisme associe à l'élément $d(E,g_1,g_2)$ la classe du triple $(E \oplus {}^tE, \sigma_1, \sigma_2)$ où $\sigma_j = \sigma_j(x,z)$ s'explicite par la formule

$$ \sigma_j(x,z) = \begin{pmatrix} \frac{1+u_j}{2} + \frac{1-u_j}{2}z & -\frac{v_j}{2}(z+z^{-1}-2) \\ -\varepsilon\frac{w_j}{2} & \varepsilon\left(\frac{1-y_j}{2} + \frac{1+y_j}{2}z^{-1}\right) \end{pmatrix} $$

avec

$$ u_j = 2x^4 - 4x^2+1 = -y_j $$
$$ v_j = 2x(x^4-3x^2+2)g_j^{-1} $$
$$ w_j = 2x(1-x^2)g_j $$

En particulier, on a

$$\sigma_j(0,z) = \begin{pmatrix} 1 & 0 \\ 0 & \varepsilon \end{pmatrix}, \quad \sigma_j(1,z) = \begin{pmatrix} z & 0 \\ 0 & \varepsilon z^{-1} \end{pmatrix}$$

Soit P le module sur $\Omega_+(\mathbb{Z}'\langle z,z^{-1}\rangle)$ obtenu par "recollement" (cf. [21]') à partir du $\mathbb{Z}'\langle x,z,z^{-1}\rangle$-trivial de rang 2 , soit M , grâce à l'isomorphisme de M/xM sur $M/(x-1)M$ défini par la matrice

$$\begin{pmatrix} 1 & 0 \\ 0 & z \end{pmatrix}$$

En suivant Milnor, on voit que P s'identifie, en tant que $\mathbb{Z}'$-module, au sous-ensemble de $M = (\mathbb{Z}'\langle x,z,z^{-1}\rangle)^2$ formé des couples $(\gamma_1(x,z),\gamma_2(x,z))$ tels que $\gamma_1(1,z) = \gamma_1(0,z)$ et $\gamma_2(1,z) = z\,\gamma_2(0,z)$. Considérons maintenant la matrice

$$h(x,z) = \varepsilon \begin{pmatrix} a(x,z) & b(x,z) \\ c(x,z) & d(x,z) \end{pmatrix} = \varepsilon \begin{pmatrix} \frac{1+u}{2} + \frac{1-u}{2}z & -\frac{v}{2}(z+z^{-1}-2) \\ \frac{-w}{2} & \frac{1-y}{2}+\frac{1+y}{2}z^{-1} \end{pmatrix}$$

avec

$$u = 2x^4 - 4x^2+1 = -y$$
$$v = 2x(x^4-3x^2+2)$$
$$w = 2x(1-x^2)$$

La matrice h induit une forme 1-hermitienne sur P . De manière explicite, le produit scalaire entre les couples (γ_1,γ_2) et (γ'_1,γ'_2) est l'expression $\varepsilon(\bar{\gamma}_1 c\gamma'_1 + \bar{\gamma}_1 d\gamma'_2 + \bar{\gamma}_2 a\gamma'_1 + \bar{\gamma}_2 b\gamma'_2)$.

D'autre part, rappelons que chaque fois qu'est donné un "bimorphisme" d'anneaux de Banach (cf. [16]) $A \times B \to C$ compatible avec les antiinvolutions, le produit tensoriel des formes permet de définir un "cup-produit" ,

$${}_{\varepsilon}V(A) \times {}_{\eta}L(B) \longrightarrow {}_{\varepsilon\eta}V(C)$$

De manière plus précise, si $d(E,g_1,g_2)$ est un élément de ${}_{\varepsilon}V(A)$ et si P est un module η-quadratique, $h : P \to {}^tP$ étant la forme η-hermitienne non dégénérée correspondante, leur cup-produit est la classe $d(E \otimes P, g_1 \otimes h, g_2 \otimes h)$. En particulier, le cup-produit par le module 1-quadratique P sur $\Omega_+(\mathbf{Z}'\langle z, z^{-1}\rangle)$ considéré ci-dessus définit un homomorphisme

$$\beta_V : {}_{\varepsilon}V(A) \longrightarrow {}_{\varepsilon}V(\Omega_+(A\langle z, z^{-1}\rangle))$$

que nous nous proposons de comparer avec $\tau'_U . \tau_V$. En fait, nous allons raisonner dans le conoyau de l'homomorphisme naturel ${}_{\varepsilon}V(\Omega_+ A) \to {}_{\varepsilon}V(\Omega_+(A\langle z, z^{-1}\rangle))$. Le module $E \otimes P \subset E \otimes_{\mathbf{Z}'} (\mathbf{Z}'\langle x, z, z^{-1}\rangle)^2$ peut s'identifier au sous-module formé des couples $(\gamma'_1(x,z), \gamma'_2(x,z))$ tels que $\gamma'_1(1,z) = \gamma'_1(0,z)$ et $\gamma'_2(1,z) = z\gamma'_2(0,z)$. Le diagramme commutatif

$$\begin{array}{ccc}
E \otimes P \subset E^2 & \xrightarrow{\begin{pmatrix} 1 & 0 \\ 0 & g_j \end{pmatrix}} & E \oplus {}^tE \supset G \\
\Big\downarrow g_j \otimes h & & \Big\downarrow s_j \\
{}^tE \otimes {}^tP \subset {}^t(E^2) & \xrightarrow{\begin{pmatrix} 1 & 0 \\ 0 & \varepsilon g_j^{-1} \end{pmatrix}} & {}^tE \oplus E \supset {}^tG
\end{array}$$

avec $h = \varepsilon \begin{pmatrix} c & d \\ a & b \end{pmatrix}$ et $s_j = \begin{pmatrix} \varepsilon g_j c(x,z) & \varepsilon d(x,z) \\ a(x,z) & g_j^{-1} b(x,z) \end{pmatrix}$

montre que, modulo un élément de ${}_{\varepsilon}V(\Omega_+ A)$, l'élément $d(E \otimes P, g_1 \otimes h, g_2 \otimes h)$ peut aussi s'écrire $d(G, s_1, s_2)$ où G est le sous-ensemble de $E \otimes_{\mathbf{Z}'} \mathbf{Z}'\langle x, z, z^{-1}\rangle \oplus ({}^tE \otimes_{\mathbf{Z}'} \mathbf{Z}'\langle x, z, z^{-1}\rangle)$ formé des couples $(\gamma''_1(x,z), \gamma''_2(x,z))$ tels que $\gamma''_1(1,z) = \gamma''_1(0,z)$, $\gamma''_2(1,z) = z\gamma''_2(0,z)$. Grâce à ces formules, on voit ainsi que le triple (G, s_1, s_2) définit en fait un élément du groupe relatif ${}_{\varepsilon}V(\chi)$ où $\chi : \Omega_+(A\langle z, z^{-1}\rangle) \to A\langle z, z^{-1}\rangle$. Par extension des scalaires dans le diagramme

$$\begin{array}{ccc} \Omega_+(A\langle z,z^{-1}\rangle) & \longrightarrow & A\langle x,z,z^{-1}\rangle \\ \downarrow & & \downarrow \\ A\langle z,z^{-1}\rangle & \longrightarrow & A\langle z,z^{-1}\rangle \times A\langle z,z^{-1}\rangle) \end{array}$$

on obtient bien l'élément dont nous sommes partis. Puisque l'image de ${}_{\varepsilon}V(P)$ dans ${}_{\varepsilon}V(SB)$ par l'homomorphisme composé $B \to B\langle z,z^{-1}\rangle \to SB$ est égale à zéro, on voit ainsi que l'homomorphisme ${}_{\varepsilon}V(A) \to {}_{-\varepsilon}V(\Omega A) \to {}_{\varepsilon}V(S\Omega A)$ est induit par le cup-produit par un certain élément u de ${}_1L(S\Omega\mathbb{Z}') \approx {}_1L(\mathbb{Z}')$ indépendant de l'anneau hermitien A. Puisque A est (2,1)-régulier, il suffit de démontrer que u est une unité de l'anneau ${}_1L(\mathbb{Z}')$ pour achever la démonstration. Remarquons alors que l'image de u dans $K(\mathbb{Z}')$ est par construction l'élément unité de $K(\mathbb{Z}')$ et que son image dans ${}_1L(\mathbb{R}) \approx \mathbb{Z} \times \mathbb{Z}$ est un couple (λ,μ) avec $\lambda - \mu = \pm 1$ car on sait que l'homomorphisme $\tau_U\tau_V : {}_{\varepsilon}V(\mathbb{R}) \to {}_{\varepsilon}V(S\Omega\mathbb{R})$ est un isomorphisme ($\mathbb{R}$ étant muni de sa topologie usuelle). D'après la proposition 4.4, u représente donc la forme quadratique αx^2 avec $\alpha = \pm 1$ ou ± 2. Puisque α est une unité de $\mathbb{Z}'$, ceci achève la démonstration du théorème 4.5.

Avant de démontrer que $\tau_V\tau_U$ est un isomorphisme, il nous faut vérifier la compatibilité de τ_V et τ_U avec l'homomorphisme "discriminant" $\Delta : {}_{\varepsilon}V(A) \to K^{-1}(A)$. Cet homomorphisme associe au triple (E,g_1,g_2) la classe de $(E,g_1^{-1}g_2)$ dans le groupe $K^{-1}(A)$.

PROPOSITION 4.6. Les diagrammes

$$\begin{array}{ccc} {}_{-\varepsilon}U(A) & \longrightarrow & K(A) \\ \tau_U\downarrow & & \downarrow\approx \\ {}_{\varepsilon}V(SA) & \xrightarrow{\Delta} & K^{-1}(SA) \end{array} \qquad \begin{array}{ccc} {}_{\varepsilon}V(A) & \xrightarrow{\Delta} & K^{-1}(A) \\ \downarrow\tau_V & & \wr\wr \\ {}_{-\varepsilon}U(\Omega A) & \longrightarrow & K(\Omega A) \end{array}$$

sont commutatifs.

DEMONSTRATION Pour le premier diagramme, il est plus commode de vérifier la commutativité du diagramme analogue

$$\begin{array}{ccc} {}_{-\varepsilon}U(A) & \longrightarrow & K(A) \\ \tau'_U\downarrow & & \downarrow \\ {}_{\varepsilon}V(A\langle z,z^{-1}\rangle) & \longrightarrow & K^{-1}(A\langle z,z^{-1}\rangle) \end{array}$$

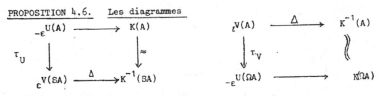

Soit donc $d(E,\eta_1,\eta_2)$ un élément de ${}_{-\varepsilon}U(A)$ avec $F = E \oplus {}^tE$ muni de la forme $(-\varepsilon)$-hyperbolique canonique. L'homomorphisme à travers ${}_{\varepsilon}V(A<z,z^{-1}>)$ associe à cet élément la classe de $(F,g_1^{-1}g_2)$ avec

$$g_j = \begin{pmatrix} a_j + (1-a_j)z & -b_j(z+z^{-1}-2) \\ -\varepsilon c_j & \varepsilon(d_j z^{-1}+1-d_j) \end{pmatrix}$$

Notons que le fait que g_j soit inversible dépend uniquement du fait que η_j est une graduation (gauche ou non). On voit donc que g_j dépend seulement du choix d'une graduation particulière sur F. Considérons alors $F' = E \oplus {}^tE \oplus {}^tE$ où le deuxième facteur ${}^tE$ est de degré zéro. Il est clair qu'on peut déformer $\eta_j \oplus 1$ en une graduation qui vaut 1 sur le premier facteur ${}^tE$. L'automorphisme g_j est donc homotope (stablement) à l'automorphisme g_j' obtenu en considérant tous les automorphismes de degré zéro. On obtient ainsi $g_j' = \frac{1+\eta_j}{2} + \frac{1-\eta_j}{2}z$ à un facteur εz^{-1} près. On retrouve (stablement) l'homomorphisme canonique de $K(A) \approx K^0(A)$ dans $K^{-1}(A<z,z^{-1}>)$. Considérons maintenant le deuxième diagramme ainsi qu'un élément $d(E,g_1,g_2)$ de ${}_{\varepsilon}V(A)$. Alors $\tau_V(d(E,g_1,g_2)) = d(F,h_1(x)\eta_0 h_1(x)^{-1}, h_2(x)\eta_0 h_2(x)^{-1})$ où $F = E \oplus {}^tE$ et où η_0 et $h_j(x)$ sont définis par les matrices

$$\eta_0 = \begin{pmatrix} 1 & 0 \\ 0 & -1 \end{pmatrix}, \quad h_j(x) = \begin{pmatrix} 1 & -g_j^{-1}x \\ 0 & 1 \end{pmatrix}\begin{pmatrix} 1 & 0 \\ g_j x & 1 \end{pmatrix}\begin{pmatrix} 1 & -g_j^{-1}x \\ 0 & 1 \end{pmatrix}$$

En fait, cet élément est dans le groupe ${}_{-\varepsilon}U(\Psi)$ avec $\Psi : A<x> \to A \times A$ et peut s'écrire aussi $d(F,\eta_0,k(x)\eta_0 k(x)^{-1})$ avec $k(x) = h_1(x)^{-1}h_2(x)$. D'après [17] (théorème d'excision), l'élément de $K^{-1}(A)$ qui lui correspond est la classe du couple $(E,k(1)|_E) = (E,g_1^{-1}g_2) = \Delta(d(E,g_1,g_2))$.

THEOREME 4.7. *L'homomorphisme composé*

$${}_{-\varepsilon}U(A) \xrightarrow{\tau_U} {}_{\varepsilon}V(SA) \xrightarrow{\tau_V} {}_{-\varepsilon}U(\Omega SA) \approx {}_{-\varepsilon}U(A)$$

est un isomorphisme si A *est (2,2)-régulier.*

Démonstration   Pour tout anneau hermitien B et pour tout bimorphisme $A \times B \to C$ (compatible avec les structures hermitiennes), on a le diagramme commutatif

$$\begin{array}{ccccc}
{}_{-\varepsilon}U(A) \times {}_{\eta}L(B) & \xrightarrow{\tau_U \times Id} & {}_{\varepsilon}V(SA) \times {}_{\eta}L(B) & \xrightarrow{\tau_V \times Id} & {}_{-\varepsilon}U(\Omega SA) \times {}_{\eta}L(B) \\
\downarrow & & \downarrow & & \downarrow \\
{}_{-\varepsilon\eta}U(C) & \xrightarrow{\tau_U} & {}_{\varepsilon\eta}V(SC) & \xrightarrow{\tau_V} & {}_{-\varepsilon\eta}U(\Omega SC)
\end{array}$$

Considérons alors le cas particulier où $B = S\Omega\mathbb{Z}'$, $C = \Omega SA$, le bimorphisme étant évident. Le cup-produit par l'élément unité de ${}_1L(S\Omega\mathbb{Z}') \approx {}_1L(\mathbb{Z}')$ définit des isomorphismes ${}_{-\varepsilon}U(A) \approx {}_{-\varepsilon}U(S\Omega A)$, etc. compatibles avec les homomorphismes τ_U et τ_V. Sans restreindre la généralité, on peut donc supposer d'emblée que A est une suspension. En changeant A en SA, on est ramené à démontrer que l'homomorphisme composé

$${}_{-\varepsilon}U(SA) \xrightarrow{\tau_U} {}_{\varepsilon}V(S^2A) \xrightarrow{\tau_V} {}_{-\varepsilon}U(\Omega S^2A) \approx {}_{-\varepsilon}U(SA)$$

est un isomorphisme. Considérons le diagramme commutatif

$$\begin{array}{ccc}
{}_{-\varepsilon}L(A) \times {}_1U(S\mathbb{Z}') & \longrightarrow & {}_{-\varepsilon}U(SA) \\
\Big\downarrow{\scriptstyle Id \times \tau_U} & & \Big\downarrow{\scriptstyle \tau_U} \\
{}_{-\varepsilon}L(A) \times {}_{-1}V(S^2\mathbb{Z}') & \longrightarrow & {}_{\varepsilon}V(S^2A) \\
\Big\downarrow{\scriptstyle Id \times \tau_V} & & \Big\downarrow{\scriptstyle \tau_V} \\
{}_{-\varepsilon}L(A) \times {}_1U(\Omega S^2\mathbb{Z}') & \longrightarrow & {}_{-\varepsilon}U(\Omega S^2A) \approx {}_{-\varepsilon}U(SA)
\end{array}$$

On a alors ${}_1U(\Omega S^2\mathbb{Z}') \approx {}_1U^1(\mathbb{Z}') \approx \mathrm{Coker}(K(\mathbb{Z}') \to {}_1L(\mathbb{Z}')) \approx \mathbb{Z} \oplus (\mathbb{Z}/2)^2$ car $K^1(\mathbb{Z}') = 0$. En outre, le cup-produit par l'élément unité de ${}_1U^1(\mathbb{Z}')$ définit l'opérateur de connexion

$${}_{-\varepsilon}L(A) \longrightarrow {}_{-\varepsilon}U^1(A) \approx {}_{-\varepsilon}U(SA)$$

Puisque les opérateurs τ_U et τ_V sont les restrictions à $S\mathbb{Z}'$ des opérateurs topologiques t_U sur $S\mathbb{R}$, on en déduit que $\tau_V\tau_U$ est un isomorphisme sur $\mathrm{Im}\,\partial$ d'après la proposition 3.4. D'autre part, pour tout anneau hermitien B(1,1)-régulier, on a le diagramme commutatif (proposition 4.6.)

$$\begin{array}{ccc} {}_{-\varepsilon}U(B) & \longrightarrow & K^0(B) \\ \downarrow \tau_U & & \downarrow \\ {}_{\varepsilon}V(SB) & \longrightarrow & K^{-1}(SB) \\ \downarrow \tau_V & & \downarrow \\ {}_{-\varepsilon}U(\Omega SB) & \longrightarrow & K^0(\Omega SB) \approx K^0(B) \end{array}$$

où l'application composée $K^0(B) \to K^0(\Omega SB) \to K^0(B)$ est l'identité. En particulier, si on pose $B = SA$, on a donc le diagramme commutatif

$$\begin{array}{ccccccccc} 0 & \longrightarrow & \operatorname{Im} \partial & \longrightarrow & {}_{-\varepsilon}U(SA) & \longrightarrow & K^1(A) & \longrightarrow & L^1(A) \\ & & \downarrow r & & \downarrow \tau_V\tau_U & & \downarrow \mathrm{Id} & & \downarrow \mathrm{Id} \\ 0 & \longrightarrow & \operatorname{Im} \partial & \longrightarrow & {}_{-\varepsilon}U(SA) & \longrightarrow & K^1(A) & \longrightarrow & L^1(A) \end{array}$$

où r est un isomorphisme. Donc $\tau_V\tau_U$ est un isomorphisme.

Les homomorphismes τ_U et τ_V se "suspendent" et se "désuspendent" en appliquant les foncteurs Ω et S à l'anneau hermitien. A. On obtient ainsi des homomorphismes entre ${}_{\varepsilon}V^n(A)$ et ${}_{-\varepsilon}U^{n-1}(A)$ que nous noterons encore τ_U et τ_V. Le théorème suivant est alors une reformulation un peu moins précise des théorèmes 4.5. et 4.7.

THEOREME 4.8. ("théorème fondamental"). *Soit A un anneau hermitien K-régulier. Alors τ_U et τ_V définissent des isomorphismes naturels entre les groupes ${}_{\varepsilon}V^n(A)$ et ${}_{-\varepsilon}U^{n-1}(A)$. De manière plus précise, $\tau_V\tau_U$ et $\tau_U\tau_V$ sont induits par le cup-produit par un élément inversible de l'anneau ${}_1L(\mathbb{Z}')$.*

Supposons maintenant que les $K^i(A)$ soient nuls pour $i = n-2$, $n-1$ et n. Le diagramme commutatif

$$\begin{array}{ccccccc} & & K^{n-1}(A) & \longrightarrow & {}_{\varepsilon}V^n(A) & \longrightarrow & {}_{\varepsilon}L^n(A) \longrightarrow K^n(A) \\ & & & & \uparrow \tau_U & & \\ K^{n-2}(A) & \longrightarrow & {}_{-\varepsilon}L^{n-2}(A) & \longrightarrow & {}_{-\varepsilon}U^{n-1}(A) & \longrightarrow & K^{n-1}(A) \end{array}$$

montre que dans ce cas ${}_{-\varepsilon}L^{n-2}(A) \approx {}_{\varepsilon}L^n(A)$. Posons de manière générale

${}_{\varepsilon}W^p(A) = \mathrm{Coker}(K^p(A) \to {}_{\varepsilon}L^p(A))$. Alors, si $K^i(A) = 0$ pour $i = n-1$ et n seulement, on a ${}_{-\varepsilon}W^{n-2}(A) \approx {}_{\varepsilon}L^n(A) \approx {}_{\varepsilon}W^n(A)$. Ceci s'applique notamment lorsque $n > 2$ et lorsque A est noéthérien régulier discret (par ex. $A = \mathbb{Z}'$). Notons u_2 l'image de 1 par l'homomorphisme ${}_1L(\mathbb{Z}') \to {}_{-1}L^2(\mathbb{Z}')$. En suspendant ou en désuspendant de manière convenable, on voit que le morphisme de ${}_{-\varepsilon}L^{n-2}(A)$ dans ${}_{\varepsilon}L^n(A)$ déduit du diagramme précédent est induit par le cup-produit par u_2 (ceci sans hypothèse sur $K^i(A)$). On en déduit le théorème suivant :

THEOREME 4.9 ("premier théorème de périodicité"). *Soit A un anneau hermitien K-régulier tel que $K^i(A) = 0$ pour $i = n-2, n-1$ et n (resp. $i=n-1$ et n). Alors le cup-produit par $u_2 \in {}_1L^2(\mathbb{Z}')$ induit un isomorphisme ${}_{-\varepsilon}L^{n-2}(A) \approx {}_{\varepsilon}L^n(A)$ (resp. ${}_{-\varepsilon}W^{n-2}(A) \approx {}_{\varepsilon}W^n(A)$). En particulier, si A est un anneau noethérien régulier discret, on a des isomorphismes*

$$ {}_{\varepsilon}W(A) = {}_{\varepsilon}W^0(A) \approx {}_{-\varepsilon}L^2(A) \approx {}_{\varepsilon}L^4(A) \approx {}_{-\varepsilon}L^6(A) \approx \dots $$

Ce théorème prouve évidemment la non trivialité des foncteurs ${}_{\varepsilon}L^n$ pour $n > 0$ sur la catégorie des anneaux noéthériens réguliers discrets. Ce résultat contraste avec la trivialité des foncteurs K^n dans le même cas prouvée par Bass [4] . D'autre part, il est intéressant et utile de connaître l'image de u_2 dans le groupe ${}_{-1}L^2(\mathbb{R})$ par l'homomorphisme naturel ${}_1L^2(\mathbb{Z}') \to {}_{-1}L^2(\mathbb{R}) \approx KU^2(\mathrm{Point}) \approx \mathbb{Z}$ ($\mathbb{R}$ est bien entendu muni de sa topologie usuelle et nous appliquons la périodicité de Bott topologique). Dans ce cas, le théorème 4.9 s'applique de nouveau (car $KO^i(\mathrm{Point}) = 0$ pour $i = 1,2$) pour montrer que cette image est le générateur topologique usuel.

Nous allons maintenant nous intéresser au comportement des foncteurs ${}_{\varepsilon}L^n$ et ${}_{\varepsilon}W^n$ pour $n < 0$. Un premier pas dans cette direction va consister à construire un élément remarquable u_{-2} de ${}_{-1}L^{-2}(\mathbb{Z}')$ dont l'image dans ${}_{-1}L^{-2}(\mathbb{R}) \approx KU^{-2}(\mathrm{Point}) \approx \mathbb{Z}$ est aussi le générateur topologique. En effet, pour A anneau hermitien K-régulier, on a les suites exactes

$$ {}_1L^{-1}(A) \longrightarrow K^{-1}(A) \longrightarrow {}_1V(A) \longrightarrow {}_1L(A) \longrightarrow K(A) $$
$$ \hphantom{{}_1L^{-1}(A) \longrightarrow K^{-1}(A) \longrightarrow {}_1}\wr $$
$$ K^{-2}(A) \longrightarrow {}_{-1}L^{-2}(A) \xrightarrow{\partial} {}_{-1}U^{-1}(A) \to K^{-1}(A) \longrightarrow {}_{-1}L^{-1}(A) $$

Si $A = \mathbb{R}$, on retrouve des suites exactes bien connues (X=point).

$$\begin{array}{ccccccccc} KO^{-1}(X) \oplus KO^{-1}(X) & \to & KO^{-1}(X) & \to & KO(X) & \to & KO(X) \oplus KO(X) & \to & KO(X) \\ & & & & \wr & & & & \\ KO^{-2}(X) & \to & KU^{-2}(X) & \to & KO(X) & \to & KO^{-1}(X) & \to & KU^{-1}(X) \end{array}$$

Celles-ci sont aussi des corollaires évidents des théorèmes 3.11 et 3.12 où tous les homomorphismes sont explicités à l'aide des algèbres de Clifford[1]. Considérons maintenant l'élément a de $_1V(\mathbb{Z}')$ défini par le triple $(\mathbb{Z}',1,-1)$. Puisque $2\Delta(a)^{-1}$ est l'élément nul de $K^{-1}(\mathbb{Z}')$, il existe un élément u_{-2} de $_{-1}L^{-2}(\mathbb{Z}')$ tel que $\partial\tau_U(u_{-2}) = 2a$. Cet élément est déterminé à l'image de $K^{-2}(\mathbb{Z}') \approx \mathbb{Z}/2$ près. Le diagramme topologique ci-dessus nous permet de voir que l'image de u_{-2} dans $_{-1}L^{-2}(\mathbb{R}) \approx KU^{-2}(\text{Point}) \approx \mathbb{Z}$ est le générateur topologique (au signe près).

Considérons enfin le cup-produit $u_2 \cup u_{-2}$ dans le groupe $_1L(\mathbb{Z}')$ et cherchons son image dans le groupe $_1L(\mathbb{R}) \approx \mathbb{Z} \times \mathbb{Z}$. Si A (resp. A') est l'algèbre des fonctions continues à valeurs réelles sur un espace compact X(resp. X') , le cup-produit

$$_{-1}L(A) \times {_{-1}L(A')} \longrightarrow {_1L(A'')}$$

où A" est l'anneau des fonctions continues à valeurs réelles sur $X \times X'$ s'interprète comme l'homomorphisme bilinéaire

$$KU(X) \times KU(X') \xrightarrow{\theta} KO(X) \oplus KO(X)$$

défini par $\theta(E,F) = (\varepsilon_0(E \otimes F), \varepsilon_0(E \otimes \overline{F}))$, $\varepsilon_0 : KU(X) \to KO(X)$ étant l'homomorphisme de "réalification". Il résulte de ces considérations topologiques que l'image de $u_2 \cup u_{-2}$ dans $_1L(\mathbb{R}) \approx \mathbb{Z} \times \mathbb{Z}$ est le couple (2,-2). Par conséquent, d'après la proposition 4.4., l'image de $u_2 \cup u_{-2}$ dans l'anneau de Witt $_1W(\mathbb{Z}')$ est quatre fois un élément inversible de cet anneau. On en déduit le théorème suivant

<u>THEOREME 4.10.</u> (Deuxième théorème de périodicité) <u>Soit</u> A <u>un anneau hermitien</u>

---

[1] Comparer avec la thèse de D.W. ANDERSON (non publiée).

<u>K-régulier et soit</u> $\beta : {}_{\varepsilon}W^n(A) \to {}_{-\varepsilon}W^{n+2}(A)$ <u>(resp.</u> $\beta' : {}_{-\varepsilon}W^{n+2}(A) \to {}_{\varepsilon}W^n(A)$) <u>l'homomorphisme induit par le cup-produit par</u> u_2 (resp. u_{-2}). <u>Alors les homomorphismes composés</u> $\beta'\beta$ <u>et</u> $\beta\beta'$ <u>sont (à isomorphisme près) la multiplication par</u> 4 . <u>En particulier,</u> ${}_{\varepsilon}W^n(A) \otimes_{\mathbf{Z}} \mathbf{Z}' \approx {}_{-\varepsilon}W^{n+2}(A) \otimes_{\mathbf{Z}} \mathbf{Z}' \approx {}_{\varepsilon}W^{n+4}(A) \otimes_{\mathbf{Z}} \mathbf{Z}'$ (1)

Cette "périodicité" modulo 2 des foncteurs ${}_{\varepsilon}W^n$ peut être traduite sur les foncteurs ${}_{\varepsilon}L^n$ de la manière suivante. Posons ${}_{\varepsilon}\bar{L}^n(A) = {}_{\varepsilon}L^n(A) \otimes_{\mathbf{Z}} \mathbf{Z}'$ et ${}_{\varepsilon}\bar{W}^n(A) = {}_{\varepsilon}W^n(A) \otimes_{\mathbf{Z}} \mathbf{Z}'$. Alors ${}_{\varepsilon}\bar{L}^0(A) \approx {}_{\varepsilon}\bar{W}^0(A) \oplus {}_{\varepsilon}\Gamma^0(A)$ où ${}_{\varepsilon}\Gamma^0(A)$ est l'image de $K^0(A) \otimes_{\mathbf{Z}} \mathbf{Z}'$ par l'homomomorphisme induit par le foncteur hyperbolique et où ${}_{\varepsilon}\bar{W}^0(A)$ est contenu dans ${}_{\varepsilon}\bar{L}^0(A)$ grâce à l'homomorphisme qui associe au module quadratique (E,Q) la différence $(E,Q) - (E,-Q)$. En suspendant ou en désuspendant, on en déduit une décomposition analogue ${}_{\varepsilon}\bar{L}^n(A) \approx {}_{\varepsilon}\bar{W}^n(A) \oplus {}_{\varepsilon}\Gamma^n(A)$ de ${}_{\varepsilon}\bar{L}^n(A)$. Ainsi ${}_{\varepsilon}\bar{L}^n(A)$ contient comme facteur direct un groupe périodique non trivial en général.

(1) On remarquera que ce théorème implique l'isomorphisme bien connu $KO^n(X) \otimes_{\mathbf{Z}} \mathbf{Z}' \approx KO^{n+4}(X) \otimes_{\mathbf{Z}} \mathbf{Z}'$ dans le cas où A est l'algèbre des fonctions continues réelles sur X .

Remarque importante Un examen attentif de la construction de $_\varepsilon\overline{W}^n(A) \approx \mathrm{Ker}(_\varepsilon\overline{L}^n(A) \to \overline{K}^n(A))$ montre que dans les théorèmes où intervient $_\varepsilon\overline{W}^n$ l'hypothèse de K-régularité peut être supprimée En effet, cette hypothèse intervient lorsqu'on veut montrer que $_\varepsilon\overline{W}^0(S^p\Omega^q A)$ ne dépend que de la différence p-q. Mais l'invariance homotopique et la semi-exactitude de la théorie $_\varepsilon\overline{W}^0 \approx {}_\varepsilon L'^0$ permet de démontrer ce fait directement. Nous y reviendrons dans le § suivant.

Pour exploiter pleinement le théorème 4.8., notamment pour la 2-torsion, il est commode d'introduire une suite exacte à 12 termes très voisine de la "suite exacte de chirurgie" due à Rothenberg et Shaneson. Pour cela, rappelons que si A est un anneau hermitien, $\mathbb{Z}_2$ opère sur $GL_n(A)$ par l'involution $\alpha \mapsto ({}^t\bar{\alpha})^{-1}$. Cette involution induit une involution $x \mapsto \bar{x}$ sur les groupes $K^n(A)$ compatible avec les isomorphismes $K^{n-1}(A) \approx K^n(\Omega A)$ et $K^{n+1}(A) \approx K^n(SA)$. On note $k^n(A)$ (resp. $k'^n(A)$) le groupe de cohomologie de Tate $\hat{H}^{pair}(\mathbb{Z}_2 ; K^n(A))$ (resp. $\hat{H}^{impair}(\mathbb{Z}_2 ; K^n(A))$. De manière plus précise, $k^n(A)$ est le quotient du sous-groupe de $K^n(A)$ formé des éléments x tels que $x = \bar{x}$ par le sous-groupe formé des éléments qui s'écrivent $y + \bar{y}$. De même, $k'^n(A)$ est le quotient du sous-groupe de $K^n(A)$ formé des éléments x tels que $\bar{x} = -x$ par le sous-groupe formé des éléments qui s'écrivent $y - \bar{y}$. Les groupes $k^n(A)$ et $k'^n(A)$ sont en fait des espaces vectoriels sur $\mathbb{Z}_2$.

THEOREME 4.11. (suite exacte des 12). Soit A un anneau hermitien K-régulier. On a alors une suite exacte à 12 termes :

$$\begin{array}{ccccccccccc}
k^n(A) & \xrightarrow{\alpha} & {}_\varepsilon W^{n-1}(A) & \xrightarrow{\beta} & {}_{-\varepsilon}L'^{n+1}(A) & \xrightarrow{d} & k'^n(A) & \xrightarrow{\gamma} & {}_\varepsilon L'^n(A) & \xrightarrow{c} & {}_\varepsilon W^n(A) \\
\uparrow r & & & & & & & & & & \downarrow r \\
{}_{-\varepsilon}W^n(A) & \xleftarrow{c} & {}_{-\varepsilon}L'^n(A) & \xleftarrow{\gamma} & k'^n(A) & \xleftarrow{d} & {}_\varepsilon L'^{n+1}(A) & \xleftarrow{\beta} & {}_{-\varepsilon}W^{n-1}(A) & \xleftarrow{\alpha} & k^n(A)
\end{array}$$

DEMONSTRATION Compte-tenu des isomorphismes $K^{n-1}(A) \approx K^n(\Omega A)$, $K^{n+1}(A) \approx K^n(SA)$, etc..., il suffit de faire la démonstration pour $n = -1$.

a) Définition de d. Rappelons qu'on a un homomorphisme "discriminant"

$\Delta : {}_{\varepsilon}V^{0}(A) \longrightarrow K^{-1}(A)$ en associant à la classe d'un triple (E, g_1, g_2) la classe de l'automorphisme $g_1^{-1} g_2$ de E.

Cet homomorphisme rend commutatif le diagramme

$$\begin{array}{ccccccccc}
{}_{\varepsilon}L^{-1}(A) & \xrightarrow{f} & K^{-1}(A) & \xrightarrow{\partial} & {}_{\varepsilon}V^{0}(A) & \xrightarrow{h} & {}_{\varepsilon}L'(A) & \longrightarrow & 0 \\
 & & & & \downarrow{\tau_V} & \searrow^{\Delta} & & & \\
 & & {}_{-\varepsilon}L^{-2}(A) & \xrightarrow{\delta} & {}_{-\varepsilon}U^{-1}(A) & \xrightarrow{s} & K^{-1}(A) & \longrightarrow & {}_{-\varepsilon}L^{-1}(A)
\end{array}$$

Si $x \in K^{-1}(A)$, on a $(\Delta\partial)(x) = x - \bar{x}$. En relevant les éléments de ${}_{\varepsilon}L'(A)$ dans ${}_{\varepsilon}V^{0}(A)$ puis en leur appliquant l'homomorphisme Δ, on obtient un homomorphisme bien défini d de ${}_{\varepsilon}L'(A)$ dans $k^{-1}(A)$ que nous appellerons encore le discriminant (si A est un corps, c'est le discriminant classique).

b) <u>Définition de β</u> : C'est l'homomorphisme zigzag $h.\tau_V^{-1}.\delta$ induit sur W^{-2}.

c) <u>Définition de α</u> : Soit H le sous-groupe de $K^{-1}(A)$ formé des éléments x tels que $\bar{x} = +x$. Alors l'homomorphisme $s\tau_V\partial$ induit 0 sur H, donc définit en fait un homomorphisme de H dans $\operatorname{Im} \delta = {}_{-\varepsilon}W^{-2}(A)$. Soit H' le sous-groupe de H formé des éléments qui s'écrivent $y + \bar{y}$. Alors H' est contenu dans $\operatorname{Im} f$ et l'homomorphisme précédent s'annule sur H' puisque la suite

$${}_{\varepsilon}L^{-1}(A) \longrightarrow K^{-1}(A) \longrightarrow V^{0}(A)$$

est exacte. Ceci définit bien un homomorphisme α de $k^{-1}(A) = H/H'$ dans ${}_{-\varepsilon}W^{-2}(A)$.

d) <u>Définition de r</u>. L'homomorphisme "rang" ${}_{\varepsilon}L^{n}(A) \longrightarrow K^{n}(A)$ induit de manière évidente un autre homomorphisme r (appelé aussi rang) de ${}_{\varepsilon}W^{n}(A)$ dans $k^{n}(A)$.

e) <u>Définition de c</u>. C'est la restriction à ${}_{\varepsilon}L'^{n}(A)$ de l'homomorphisme quotient ${}_{\varepsilon}L^{n}(A) \longrightarrow {}_{\varepsilon}W^{n}(A)$.

f) <u>Définition de γ</u>. C'est l'homomorphisme induit par le foncteur hyperbolique sur les groupes K^{-1} et L^{-1}.

Avec ces définitions, la démonstration de l'exactitude de la suite des 12 est une simple "chasse au diagramme" laissée en exercice au lecteur.

Une autre façon d'écrire la suite exacte des 12 est de poser les définitions suivantes (suggérées par Wall).

$$L_h^{n,0}(A) = {}_1L'^n(A) \qquad\qquad L_s^{n,0}(A) = {}_{-1}W^{n-2}(A)$$

$$L_h^{n,2}(A) = {}_{-1}L'^n(A) \qquad\qquad L_s^{n,2}(A) = {}_1W^{n-2}(A)$$

$$L_h^{n,1}(A) = {}_1W^{n-1}(A) \qquad\qquad L_s^{n,1}(A) = {}_1L'^{n-1}(A)$$

$$L_h^{n,3}(A) = {}_{-1}W^{n-1}(A) \qquad\qquad L_s^{n,3}(A) = {}_{-1}L'^{n-1}(A)$$

$$k^n(A) = k^{n,0}(A) = k^{n,2}(A)$$

$$k'^n(A) = k^{n,1}(A) = k^{n,3}(A)$$

On a alors la suite exacte

$$\begin{array}{ccccccccccc}
k^{n-1,0} & \longrightarrow & L_s^{n,0} & \longrightarrow & L_h^{n,0} & \longrightarrow & k^{n-1,1} & \longrightarrow & L_s^{n,1} & \longrightarrow & L_h^{n,1} \\
\uparrow & & & & & & & & & & \downarrow \\
L_h^{n,3} & \longleftarrow & L_s^{n,3} & \longleftarrow & k^{n-1,3} & \longleftarrow & L_h^{n,2} & \longleftarrow & L_s^{n,2} & \longleftarrow & k^{n-1,2}
\end{array}$$

qui, pour $n =$, est analogue à celle démontrée par Rothenberg et Shaneson dans un contexte différent. On notera que $L_s^{n,p}$ et $L_h^{n,p}$ sont définies pour $n \in \mathbb{Z}$ et $p = 0 \bmod 4$ et qu'on a les isomorphismes $L_s^{n,p} = L_h^{n-1,p-1}$.

Pour calculer de manière explicite les homomorphismes de la suite des 12 nous aurons besoin de quelques lemmes.

<u>LEMME 4.12.</u> <u>Soit</u> A <u>un anneau hermitien K-régulier et soit</u> σ <u>l'homomorphisme composé</u>

$$K(A) \xrightarrow{\ \partial\ } {}_{+\epsilon}V^1(A) \xrightarrow{\ \tau_U^{-1}\ } {}_{-\epsilon}U(A)$$

<u>Alors l'homomorphisme</u> σ <u>associe à la classe du module</u> E <u>la classe du triple</u> $(E \oplus {}^tE\ ,\ \eta\ ,\ -\eta)$ avec

$$\eta = \begin{pmatrix} 1 & 0 \\ 0 & -1 \end{pmatrix}$$

DEMONSTRATION Soit σ' l'homomorphisme associé à la correspondance $E \longmapsto x = (E \oplus {}^tE, \eta, -\eta)$. L'homomorphisme τ_U (explicité p. 56) se factorise à travers ${}_{+\varepsilon}V(A\langle z,z^{-1}\rangle)$ et l'élément de ${}_{+\varepsilon}V(A\langle z,z^{-1}\rangle)$ qui correspond à x est la classe du triple $(E \oplus {}^tE, g_0, g_1)$ où

$$g_0 = \begin{pmatrix} 1 & 0 \\ 0 & \varepsilon \end{pmatrix} \qquad g_1 = \begin{pmatrix} z & 0 \\ 0 & \varepsilon z^{-1} \end{pmatrix}$$

D'autre part, l'homomorphisme $K(A) \xrightarrow{\partial} V^1(A)$ se calcule comme l'homomorphisme composé $K(A) \xrightarrow{\beta_K} K^{-1}(A\langle z,z^{-1}\rangle) \xrightarrow{\partial'} {}_{\varepsilon}V(A\langle z,z^{-1}\rangle) \longrightarrow {}_{\varepsilon}V(SA)$. Or β_K associe à la classe de E la classe du couple (E,z) et ∂' associe à la classe du couple (E,z) la classe du triple $(E \oplus {}^tE, g_0, g_1 = \alpha g_0 {}^t\alpha)$ avec

$$\alpha = \begin{pmatrix} z & 0 \\ 0 & 1 \end{pmatrix}$$

LEMME 4.13 L'homomorphisme composé

$$k(A) \xrightarrow{\alpha} {}_{\varepsilon}W^{-1}(A) \xrightarrow{r} k^{-1}(A)$$

associe à la classe d'un module η-quadratique E la classe du couple $(E, -\varepsilon\eta)$ (en d'autres termes, cet homomorphisme induit le cup-produit par $(Z', -\varepsilon\eta) \in {}_1L(Z')$ sur l'image de l'homomorphisme ${}_{\eta}L(A) \longrightarrow K(A)$).

DEMONSTRATION Il suffit de suivre pas à pas les définitions. En effet, un élément de $k(A)$ est la classe d'un module E tel qu'il existe un isomorphisme $\alpha : {}^tE \longrightarrow E$.

D'après le lemme précédent, $\sigma([E])$ est la classe du triple $(E \oplus {}^tE, \eta, -\eta)$. Mais η peut s'écrire

$$\eta = \begin{pmatrix} 0 & \varepsilon\alpha \\ {}^t\alpha^{-1} & 0 \end{pmatrix} \begin{pmatrix} 1 & 0 \\ 0 & -1 \end{pmatrix} \begin{pmatrix} 0 & {}^t\alpha \\ \varepsilon\alpha^{-1} & 0 \end{pmatrix}$$

où $\begin{pmatrix} 0 & \varepsilon\alpha \\ {}^t\alpha^{-1} & 0 \end{pmatrix}$ est la matrice d'un automorphisme ε-orthogonal de

$E \oplus {}^tE$. L'élément de $_\varepsilon W^{-1}(A)$ qui est associé à E par l'homomorphisme composé est précisément la classe de cet automorphisme.

Supposons maintenant que E soit le module sous-jacent à un module η-quadratique. On peut alors choisir α en sorte que ${}^t\alpha = \eta\alpha$

Donc $\begin{pmatrix} 0 & \varepsilon\alpha \\ {}^t\alpha^{-1} & 0 \end{pmatrix} = \begin{pmatrix} 0 & \varepsilon\alpha \\ \eta\alpha^{-1} & 0 \end{pmatrix} \sim \begin{pmatrix} \varepsilon\eta & 0 \\ 0 & 1 \end{pmatrix}$

LEMME 4.14. L'homomorphisme composé

$$k'(A) \longrightarrow {}_\varepsilon L'(A) \longrightarrow k'^{-1}(A)$$

est défini par le cup-produit par -1 sur les éléments de $k'(A)$ qui sont des classes de modules ε-quadratiques.

DEMONSTRATION Un élément de $k'(A)$ est dit une classe de module ε-quadratique s'il peut s'écrire $[E] - [F]$ où E et F sont des modules ε-quadratiques tels que $E \oplus {}^tE \approx F \oplus {}^tF$ en tant que A-modules sous-jacents. Soit q(resp. r) la forme quadratique sur E (resp. F). L'image de $[E] - [F]$ dans $_\varepsilon L'(A)$ par l'homomorphisme $k'(A) \longrightarrow {}_\varepsilon L'(A)$ est $(E,q) + (E,-q) - (F,r) - (F,-r) = (E,q) - (E,-q) - (F,r) + (F,-r) \bmod. 2L'(A)$. Or cet élément est l'image de $d(E,q,-q) - d(F,r,-r)$ par l'homomorphisme $_\varepsilon V(A) \longrightarrow {}_\varepsilon L'(A)$. Son discriminant dans $K^{-1}(A)$ est précisément la classe de $(E \oplus F\ ,\ -1 \oplus -1)$, d'où le résultat annoncé.

LEMME 4.15. On a un diagramme commutatif

$$\begin{array}{ccc} k'^{-1}(A) & \xrightarrow{h'} & {}_\varepsilon L^{-1}(A) \\ d \uparrow & & \uparrow d' \\ {}_\varepsilon L'(A) & \longrightarrow & {}_\varepsilon L(A) \end{array}$$

où d' associe à la classe E du module ε-quadratique E, la classe du couple $(E,-1)$ et où h' est induit par la foncteur hyperbolique.

DEMONSTRATION Soit x un élément de ${}_{\varepsilon}L'(A)$ c'est-à-dire un élément de l'image de l'homomorphisme ${}_{\varepsilon}V(A) \xrightarrow{h} {}_{\varepsilon}L(A)$. Alors $d(x)$ est la classe du couple $(E, g_2^{-1}g_1)$ lorsque $h(d(E,g_1,g_2)) = x$. Donc $h'(d(x))$ est la classe du couple $(E \oplus {}^tE, \alpha)$ où

$$\alpha = \begin{pmatrix} g_2^{-1}g_1 & 0 \\ 0 & {}^tg_2\,{}^tg_1^{-1} \end{pmatrix} = \begin{pmatrix} g_2^{-1}g_1 & 0 \\ 0 & g_2g_1^{-1} \end{pmatrix}$$

$$= \begin{pmatrix} 0 & -\frac{g_2^{-1}}{2} \\ -2g_2 & 0 \end{pmatrix}\begin{pmatrix} 0 & \frac{g_1^{-1}}{2} \\ -2g_1 & 0 \end{pmatrix} = \alpha_{g_2} \cdot \alpha_{g_1} \quad \text{où on pose}$$

$$\alpha_g = \begin{pmatrix} 0 & -\frac{g^{-1}}{2} \\ -2g & 0 \end{pmatrix}$$

D'une manière générale, si E est un module muni d'une forme ε-quadratique $g : E \longrightarrow {}^tE$ et si E' désigne le même module muni de la forme $-g$, on a un isomorphisme $f_g : E \oplus E' \longrightarrow E \oplus {}^tE$ défini par la matrice

$$f_g = \begin{pmatrix} 1/2 & 1/2 \\ g & -g \end{pmatrix}$$

Donc $f_g^{-1}.\alpha_g.f_g$

$$= \begin{pmatrix} 1 & g^{-1}/2 \\ 1 & -g^{-1}/2 \end{pmatrix}\begin{pmatrix} 0 & -g^{-1}/2 \\ -2g & 0 \end{pmatrix}\begin{pmatrix} 1/2 & 1/2 \\ g & -g \end{pmatrix}$$

$$= \begin{pmatrix} -1 & 0 \\ 0 & 1 \end{pmatrix}$$

Par conséquent $d(E \oplus {}^tE,\alpha) = d(E \oplus {}^tE,\alpha_{g_2}.\alpha_{g_1}) = d(E \oplus {}^tE,\alpha_{g_2}) + d(E \oplus {}^tE,\alpha_{g_1}) =$ $d(E_1 , -1) + d(E_2 , -1) = d(E_1 , -1) - d(E_2 , -1)$ où E_i désigne E muni de la forme g_i.

Nous allons appliquer la suite exacte des 12 au cas d'un corps commutatif F muni de l'involution triviale. Dans ce cas, $k^{-1}(F) = \pm 1$ et $k'^{-1}(F) = F^*/(F^*)^2$. Puisque tout élément de $K^{-2}(F)$ peut s'écrire $\Sigma\, x_i \cup y_i$ où x_i et y_i sont des éléments de $K^{-1}(F)$, l'involution $z \mapsto \bar{z}$ est l'identité sur le groupe $K^{-2}(F)$. Par conséquent $k^{-2}(F) = K^{-2}(F)/2K^{-2}(F)$ et $k'^{-2}(F)$ est le sous-groupe formé de la 2-torsion de $K^{-2}(F)$.

Une partie de la suite exacte des 12 s'écrit donc ainsi

$${}_1W^{-1}(F) \xrightarrow{r} \pm 1 \xrightarrow{\alpha} {}_{-1}W^{-2}(F) \xrightarrow{\beta} {}_1L'^0(F) \xrightarrow{d} F^*/(F^*)^2$$

Il est clair que les homomorphismes ${}_1W^{-1}(F) \longrightarrow \pm 1$ et ${}_1L'^0(F) \longrightarrow F^*/(F^*)^2$ sont surjectifs et se scindent. On a donc ainsi la suite exacte scindée

(1) $0 \longrightarrow {}_{-1}W^{-2}(F) \xrightarrow{\hat{\beta}} {}_1L'^0(F) \xrightarrow{d} F^*/(F^*)^2 \longrightarrow 0$ ce qui permet de calculer ${}_{-1}W^{-2}(F)$ en fonction de ${}_1L'^0(F)$.

Considérons maintenant l'homomorphisme ${}_{-1}L^{-2}(F) \longrightarrow K^{-2}(F)$ induit par le foncteur oubli. On voit aisément que l'homomorphisme $\pi_1(SL_2(F)) \longrightarrow K^{-2}(F)$ est surjectif. Puisque $SL_2(F) = Sp_2(F) \approx {}_{-1}O_{1,1}(F)$, ${}_{-1}L^{-2}(F) \longrightarrow K^{-2}(F)$ est épi, d'où la suite exacte

(2) $0 \longrightarrow {}_{-1}L'^{-2}(F) \longrightarrow {}_{-1}L^{-2}(F) \longrightarrow K^{-2}(F) \longrightarrow 0$

Pour calculer ${}_{-1}L'^{-2}(F)$, utilisons un autre bout de la suite exacte des 12, soit

(3) $F^*/(F^*)^2 \xrightarrow{d} k'^{-2}(F) \xrightarrow{\gamma} {}_{-1}L'^{-2}(F) \xrightarrow{c} {}_{-1}W^{-2}(F) \xrightarrow{r} k^{-2}(F) \longrightarrow 0$

Tous les homomorphismes de cette suite sont clairs à l'exception de d. Pour le calculer nous aurons besoin du lemme suivant

<u>LEMME 4.16</u> <u>L'homomorphisme</u> $K^{-2}(F) \longrightarrow {}_1L^{-2}(F)$ <u>est injectif.</u>

DEMONSTRATION Il suffit de voir que ${}_1V^{-2}(F) = 0$. Or , d'après le théorème fondamental, ${}_1V^{-2}(F) \approx {}_{-1}U^{-1}(F)$ où ${}_{-1}U^{-1}(F)$ s'insère dans la suite exacte

$${}_{-1}L^{-2}(F) \longrightarrow K^{-2}(F) \longrightarrow {}_{-1}U^{-1}(F) \longrightarrow {}_{-1}L^{-1}(F) \longrightarrow K^{-1}(F)$$

Puisque ${}_{-1}L^{-2}(F) \longrightarrow K^{-2}(F)$ est épi et que ${}_{-1}L^{-1}(F)$ est égal à 0 (car c'est un quotient de ${}_1L_1(F) = Sp(F)/[Sp(F),Sp(F)] = 0)$, on a bien le lemme annoncé.

Calculons maintenant l'homomorphisme d de la suite exacte 3 . Le lemme précédent montre qu'il suffit de calculer l'homomorphisme $F^*(F^*)^2 \longrightarrow k'^{-2}(F) \hookrightarrow {}_1L^{-2}(F)$. On est alors dans les conditions d'application du lemme 4.14 (avec des notations légèrement différentes). Celui-ci implique que d est simplement défini par $d(a) = (a , -1)$ (soit le cup-produit de a par -1).

En effet, il suffit de contempler le diagramme

$$\begin{array}{ccc} & k'^{-2}(F) & \longrightarrow & {}_1L^{-2}(F) \\ \nearrow & & \nearrow & \\ K^{-1}(F) & \longrightarrow & {}_1L'^{-1}(F) & \end{array}$$

où les flèches obliques sont définies par le cup-produit par $-1 \in K^{-1}(F)$ ou ${}_1L^{-1}(F)$ et où les flèches horizontales sont induites par le foncteur hyperbolique.

De la même manière on peut essayer de calculer ${}_{-1}W^{-2}(F)$ et ${}_1L'^{-2}(F)$. Une partie de la suite exacte des 12 s'écrit

$${}_{-1}W^{-1}(F) \longrightarrow \pm 1 \longrightarrow {}_1W^{-2}(F) \longrightarrow {}_{-1}L'^{0}(F)$$

Puisque ${}_{-1}W^{-1}(F) = {}_{-1}L'^{0}(F) = 0$, on a donc ${}_1W^{-2}(F) \approx \mathbb{Z}_2$. Compte-tenu du lemme 4.15 , on a ainsi la suite exacte

$$(4) \qquad 0 \longrightarrow K^{-2}(F) \longrightarrow {}_1L^{-2}(F) \longrightarrow \mathbb{Z}_2 \longrightarrow 0$$

D'autre part, ${}_1L'^{-2}(F)$ se calcule à partir de la suite exacte

$$0 \longrightarrow k'^{-2}(F) \longrightarrow {}_1L'^{-2}(F) \longrightarrow {}_1W^{-2}(F) \longrightarrow k^{-2}(F)$$

Le générateur de ${}_1W^{-2}(F)$ provient du générateur de $k^{-1}(F) = \pm 1$ grâce à l'homomorphisme α . D'après le lemme 4.13. , l'image de ce générateur dans le groupe $k^{-2}(F)$ est simplement la classe du symbole $(-1 , -1)$. D'où la proposition suivante:

PROPOSITION 4.17. Si le symbole $(-1, -1) \neq 0$ dans $k^{-2}(F)$, on a ${}_1L'^{-2}(F) \approx k'^{-2}(F)$. Sinon on a la suite exacte

$$(5) \quad 0 \longrightarrow k'^{-2}(F) \longrightarrow {}_1L'^{-2}(F) \longrightarrow \mathbb{Z}_2 \longrightarrow 0$$

Pour terminer notre étude des groupes K^{-2} et L^{-2}, il convient de chercher le noyau de l'homomorphisme $K^{-2}(F) \longrightarrow {}_{-1}L^{-2}(F)$ induit par le foncteur hyperbolique. Pour cela considérons le diagramme

$$\begin{array}{ccccccccc} {}_1L^{-2}(F) & \longrightarrow & K^{-2}(F) & \longrightarrow & {}_1V^{-1}(F) & \longrightarrow & {}_1L^{-1}(F) & \longrightarrow & K^{-1}(F) \\ & & & & \wr & & & & \\ K^{-3}(F) & \longrightarrow & {}_{-1}L^{-3}(F) & \longrightarrow & {}_{-1}U^{-2}(F) & \longrightarrow & K^{-2}(F) & \longrightarrow & {}_{-1}L^{-2}(F) \end{array}$$

L'homomorphisme zigzag $K^{-2}(F) \longrightarrow {}_1V^{-1}(F) \approx {}_{-1}U^{-2}(F) \longrightarrow K^{-2}(F)$ associe à x l'élément $x - \bar{x} = 0$ car tout élément de $K^{-2}(F)$ est symétrique. D'autre part, ${}_1L'^{-1}(F) \approx F^*/(F^*)^2$ et le noyau de $K^{-2}(F) \longrightarrow {}_{-1}L^{-2}(F)$ s'identifie à l'image de l'homomorphisme $F^*/(F^*)^2 \xrightarrow{d} K^{-2}(F)$, soit $a \longmapsto (a, -1)$ d'après ce qui précède. On obtient donc ainsi la suite exacte [1]

$$(6) \quad F^*/(F^*)^2 \xrightarrow{d} K^{-2}(F) \longrightarrow {}_{-1}L^{-2}(F) \longrightarrow {}_{-1}W^{-2}(F) \longrightarrow 0$$

qui est analogue à la suite (3).

Nous allons maintenant calculer ${}_1W^{-3}(F)$ et ${}_{-1}W^{-3}(F)$.

Dans le premier cas, on a la suite exacte

$${}_{-1}L'^{-2}(F) \longrightarrow {}_{-1}W^{-2}(F) \longrightarrow k^{-2}(F) \longrightarrow {}_1W^{-3}(F) \longrightarrow {}_{-1}L'^{-1}(F)$$

Nous avons déjà démontré que ${}_{-1}L'^{-1}(F) = 0$ et que ${}_{-1}W^{-2}(F) \longrightarrow k^{-2}(F)$ est épi. Donc ${}_1W^{-3}(F) = 0$. Dans le deuxième cas on a la suite exacte

$${}_1L'^{-2}(F) \longrightarrow {}_1W^{-2}(F) \longrightarrow k^{-2}(F) \longrightarrow {}_{-1}W^{-3}(F) \longrightarrow {}_1L'^{-1}(F) \xrightarrow{d} k'^{-2}(F)$$

Nous avons déjà vu que ${}_1W^{-2}(F) \approx \mathbb{Z}_2$ et que l'image de l'élément non trivial de ${}_1W^{-2}(F)$ dans $k^{-2}(F)$ est la classe du symbole $(-1, -1)$. D'autre part, ${}_1L'^{-1}(F) \approx F^*/(F^*)^2$ et d est défini par $a \longmapsto (a, -1)$. En rassemblant tous ces résultats, on obtient finalement la proposition suivante :

---

[1] Une suite exacte analogue a été démontrée par Bak.

PROPOSITION 4.18 On a une suite exacte

$$0 \longrightarrow k^{-2}(F)/(-1\ ,\ -1) \longrightarrow {}_{-1}W^{-3}(F) \longrightarrow F^*/(F^*)^2 \longrightarrow k'^{-2}(F) \quad \text{où} \quad d(a) = (a, -1).$$

Pour illustrer ces différentes propositions, considérons le cas où $F = \mathbb{Q}$ le corps des nombres rationnels. Dans ce cas, Ker $({}_1L'^0(F) \longrightarrow F^*/(F^*)^2)$ s'identifie à l'ensemble des classes de formes quadratiques de rang et de discriminant zéro. Donc ce noyau est la somme directe de $\mathbb{Z}$ et d'une somme infinie de groupes $\mathbb{Z}_2$ (autant que l'ensemble S des nombres premiers). D'après Bass et Tate on a aussi $k^{-2}(\mathbb{Q}) \approx \bigoplus_{p \in S} \mathbb{Z}/2$ et l'homomorphisme surjectif ${}_{-1}W^{-2}(\mathbb{Q}) \longrightarrow k^{-2}(\mathbb{Q})$ induit un isomorphisme sur la composante 2-primaire. Il serait intéressant de calculer le noyau de ${}_{-1}W^{-2}(F) \longrightarrow k^{-2}(F)$ en général (comparer avec [21]). D'autre part, la suite (3) permet de calculer ${}_{-1}L'^{-2}(\mathbb{Q})$. En effet, en plongeant $\mathbb{Q}$ dans $\mathbb{Q}_p$ on voit aisément que Coker d $= \bigoplus_{p \in S'} \mathbb{Z}_2$ où S' est l'ensemble des nombres premiers s'écrivant $4n + 1$. On obtient donc ainsi la suite exacte

$$0 \longrightarrow \bigoplus_{p \in S'} \mathbb{Z}_2 \longrightarrow {}_{-1}L'^{-2}(\mathbb{Q}) \longrightarrow \mathbb{Z} \longrightarrow 0$$

De même, la proposition 4.18 permet d'écrire la suite exacte

$$0 \longrightarrow \bigoplus_{p \in S} \mathbb{Z}_2 \longrightarrow {}_{-1}W^{-3}(\mathbb{Q}) \longrightarrow \bigoplus_{p \in S'} \mathbb{Z}_2 \longrightarrow 0$$

Voici une autre application de la suite exacte des 12 qui est de nature différente. D'après le travail de Quillen, on a une application injective $\mathbb{Z}_{24} \approx \pi_3^s \longrightarrow K_3(\mathbb{Z}) \approx K^{-3}(\mathbb{Z})$. Nous allons montrer que cette application n'est pas surjective au niveau de la composante 2-primaire (ce qui contredit une conjecture récente de Lichtenbaum). Tout d'abord, si on localise $\mathbb{Z}$ par rapport au système multiplicatif (2^p), on obtient une suite exacte de localisation (due à Quillen)

$$K^{-3}(F_2) \longrightarrow K^{-3}(\mathbb{Z}) \longrightarrow K^{-3}(\mathbb{Z}') \longrightarrow K^{-2}(F_2)$$

où F_2 désigne le corps à deux éléments. Puisque $K^{-2}(F_2) = 0$ et que $K^{-3}(F_2) \approx \mathbb{Z}/3$, la composante 2-primaire de $K^{-3}(\mathbb{Z})$ s'applique bijectivement sur la composante 2-primaire de $K^{-3}(\mathbb{Z}')$. Ceci étant dit, l'homomorphisme

$\pi_3^s \longrightarrow K^{-3}(\mathbb{Z}')$ se factorise à travers ${}_1L'^{-3}(\mathbb{Z}')$ car il est induit par l'homomorphisme de groupes simpliciaux $f : \Sigma_r \longrightarrow GL_{2r}(\mathbb{Z}')$ défini par

$$f(\sigma) = \alpha \begin{pmatrix} \sigma & 0 \\ 0 & 1 \end{pmatrix} \alpha$$

avec $\alpha = \begin{pmatrix} 1/2 & 1/2 \\ 1 & -1 \end{pmatrix}$ [(1)]. En effet cet homomorphisme se factorise de manière évidente à travers ${}_1O_{r,r}(\mathbb{Z}')$

Considérons maintenant le groupe ${}_1W^{-3}(\mathbb{Z}')$. Il s'insère dans la suite exacte

$$ {}_{-1}W^{-2}(\mathbb{Z}') \longrightarrow k^{-2}(\mathbb{Z}') \longrightarrow {}_1W^{-3}(\mathbb{Z}') \longrightarrow {}_{-1}L'^{-1}(\mathbb{Z}')$$

Comme il est bien connu, on a $Sp(\mathbb{Z}') = [Sp(\mathbb{Z}'), Sp(\mathbb{Z}')]$ (ceci résulte du fait que $\mathbb{Z}'$ est un anneau euclidien et que toute matrice ε-élémentaire dans le sens de [18]est un produit de commutateurs). Donc ${}_{-1}L^{-1}(\mathbb{Z}') = {}_{-1}L'^{-1}(\mathbb{Z}') = 0$. D'autre part, la suite exacte de la localisation $\mathbb{Z} \longrightarrow \mathbb{Z}'$ s'écrit au niveau K^{-2} sous la forme

$$K^{-2}(F_2) \longrightarrow K^{-2}(\mathbb{Z}) \longrightarrow K^{-2}(\mathbb{Z}') \longrightarrow K^{-1}(F_2)$$

Donc $K^{-2}(\mathbb{Z}') \approx K^{-2}(\mathbb{Z}) \approx \mathbb{Z}/2$, le générateur étant la classe du lacet algébrique $x \longmapsto h(x)^4$ où

$$h(x) = \begin{pmatrix} 1 & x \\ 0 & 1 \end{pmatrix} \begin{pmatrix} 1 & 0 \\ -x & 1 \end{pmatrix} \begin{pmatrix} 1 & x \\ 0 & 1 \end{pmatrix}$$

Puisque $Sp_2(\mathbb{Z}') = SL_2(\mathbb{Z}')$, on voit ainsi que l'homomorphisme naturel ${}_{-1}L^{-2}(\mathbb{Z}') \longrightarrow K^{-2}(\mathbb{Z}')$ est surjectif, ce qui implique la surjectivité du morphisme ${}_{-1}W^{-2}(\mathbb{Z}') \longrightarrow k^{-2}(\mathbb{Z}')$ qui en est déduit par passage au quotient.

Il résulte des calculs précédents que ${}_1W^{-3}(\mathbb{Z}') = 0$. Considérons maintenant le diagramme

---

(1) Nous utilisons ici l'interprétation simpliciale des groupes K^{-n} due à Gersten [12]. Cette interprétation vaut aussi pour les groupes ${}_\varepsilon L^{-n}$.

$$\begin{array}{ccc} & & L^{-3}(\mathbb{Z}') \\ & \overset{\alpha'}{\nearrow} & F \downarrow \quad \uparrow H \\ \pi_3^s & \underset{\alpha}{\searrow} & \\ & & K^{-3}(\mathbb{Z}') \end{array}$$

où F (resp. H) est induit par le foncteur oubli (resp. hyperbolique) et où $\alpha = F.\alpha'$ est l'homomorphisme explicité par Quillen. Soit x l'élément d'ordre 8 dans π_3^s et soit y un élément de $K^{-3}(\mathbb{Z}')$ tel que $H(y) = \alpha'(x)$. Alors y n'est pas dans l'image de α. En effet, dans le cas contraire, on aurait $(F.H)(y) = y + \bar{y} = 2y = \alpha(x)$. Si on pose $y = \alpha(x')$, on a ainsi $\alpha(x) = 2\alpha(x')$, soit $x = 2x'$ puisque α est injective. Ceci est absurde puisque $\mathbb{Z}_8$ est la composante primaire de π_3^s. Ainsi le groupe $K_3(\mathbb{Z}) \approx K^{-3}(\mathbb{Z})$ contient au moins un sous-groupe de 48 éléments. Est-il un facteur direct dans $K_3(\mathbb{Z})$?

Parallèlement au groupe ${}_1W^{-3}(\mathbb{Z}')$ dont nous venons de montrer l'intérêt, il est instructif de calculer de même le groupe ${}_{-1}W^{-3}(\mathbb{Z}')$. Celui-ci s'insère dans la suite exacte

$${}_1L'^{-2}(\mathbb{Z}') \longrightarrow {}_1W^{-2}(\mathbb{Z}') \longrightarrow k^{-2}(\mathbb{Z}') \longrightarrow {}_{-1}W^{-3}(\mathbb{Z}') \longrightarrow {}_1L'^{-1}(\mathbb{Z}') \longrightarrow k'^{-2}(\mathbb{Z}')$$

Dans cette suite on a $k'^{-2}(\mathbb{Z}') \approx k^{-2}(\mathbb{Z}') \approx \mathbb{Z}/2$ qui est engendré par le symbole $(-1\ ,\ -1)$. L'homomorphisme ${}_1W^*(\mathbb{Z}') \longrightarrow k^*(\mathbb{Z}')$ étant de manière évidente un homomorphisme d'anneaux gradués et -1 étant dans l'image de ${}_1W^{-1}(\mathbb{Z}') \longrightarrow k^{-1}(\mathbb{Z}')$, l'homomorphisme ${}_1W^{-2}(\mathbb{Z}') \longrightarrow k^{-2}(\mathbb{Z}')$ est épi. D'autre part, le calcul fait dans [18] montre que ${}_1L'^{-1}(\mathbb{Z}') \approx \mathbb{Z}'^*/(\mathbb{Z}'^*)^2 \approx \mathbb{Z}/2 \times \mathbb{Z}/2$. En appliquant le lemme 4.14 on voit aussitôt que le noyau de ${}_1L'^{-1}(\mathbb{Z}') \rightarrow k'^{-2}(\mathbb{Z}')$ est engendré par la classe de 2 . Donc ${}_{-1}W^{-3}(\mathbb{Z}') \approx \mathbb{Z}/2$.

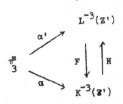

V - THEORIE DE QUILLEN ET HOMOLOGIE DU GROUPE ORTHOGONAL (rédaction provisoire)

Nous allons appliquer les idées précédentes au calcul partiel de l'homologie du groupe orthogonal ${}_{\varepsilon}O(A) = \varinjlim {}_{\varepsilon}O_{n,n}(A)$ lorsque A est un anneau discret (par "homologie", on entend l'homologie de l'espace classifiant $B_{{}_{\varepsilon}O(A)}$). Cependant, la K-théorie la plus proche de l'homologie n'est pas celle que nous venons de développer dans les § précédents mais celle définie récemment par Quillen [24]. Il nous faut donc transposer un certain nombre de considérations dans ce cadre.

Rappelons d'abord les définitions de Quillen (pour un exposé plus systématique voir [12] [28] par ex.). Soit X un espace topologique ayant le type d'homotopie d'un CW-complexe et soit $G \subset \pi_1(X)$ un sous-groupe normal parfait. Alors il existe un espace X^+ et une application continue $i : X \to X^+$ caractérisés à homotopie près par les propriétés suivantes :

a) $\pi_1(X^+) = \pi_1(X)/G$

b) Pour tout $\pi_1(X^+)$-module M, l'application $X \to X^+$ induit un isomorphisme $H_*(X ; M) \to H_*(X^+; M)$.

Cette construction s'applique notamment à la situation suivante. D'après [29] (voir aussi [18] lemmes 2.12. et 2.13), le groupe orthogonal infini contient son sous-groupe des commutateurs ${}_\varepsilon O'(A) = [{}_\varepsilon O(A), {}_\varepsilon O(A)]$ comme sous-groupe normal parfait. Posons $X = B_{\varepsilon O(A)}$ et $X^+ = B^+_{\varepsilon O(A)}$, le "+" étant pris par rapport au sous-groupe $G = {}_\varepsilon O'(A)$. La somme directe des matrices induit alors une structure de H-espace sur $B^+_{\varepsilon O(A)}$. On définit de même $B^+_{GL(A)} = B^+_{\varepsilon O(A \times A^0)}$

<u>DEFINITION 5.1.</u> <u>Soit</u> A <u>un anneau hermitien discret</u>. <u>Pour</u> $n > 0$, <u>on pose</u> ${}_\varepsilon L_n(A) = \pi_n(B^+_{\varepsilon O(A)})$, $K_n(A) = \pi_n(B^+_{GL(A)})$. <u>Pour</u> $n \leq 0$, <u>on pose</u> ${}_\varepsilon L_n(A) = {}_\varepsilon L^{-n}(A)$ $K_n(A) = K^{-n}(A)$. <u>Enfin, pour</u> $n \in \mathbb{Z}$, <u>on pose</u> ${}_\varepsilon W_n(A) = \mathrm{Coker}[K_n(A) \to {}_\varepsilon L_n(A)]$.

On notera qu'avec cette définition on a ${}_\varepsilon L_1(A) \approx {}_\varepsilon O(A)/{}_\varepsilon O'(A)$ et ${}_\varepsilon L_2(A) = H_2({}_\varepsilon O'(A))$ comme il est d'usage. D'autre part, si A est contenu dans une algèbre de Banach A', l'application continue $B_{\varepsilon O(A)} \to B_{\varepsilon O(A')}$ ($O(A')$ étant muni de la topologie induite par celle de A') induit un homomorphisme ${}_\varepsilon L_n(A) \to \pi_n(B_{\varepsilon O(A')}) \approx {}_\varepsilon L^{-n}(A')$. Enfin le raisonnement fait en [12] [28] montre que ${}_\varepsilon L_n(S^n A) \approx {}_\varepsilon L(A)$.

<u>LEMME 5.2.</u> <u>Soit</u> $A = \mathbb{Z}'$ <u>et</u> $A' = \mathbb{R}$ <u>comme ci-dessus. Alors l'homomorphisme de</u> ${}_{-1}L_2(A)$ <u>dans</u> ${}_{-1}L^{-2}(A') \approx \mathbb{Z}$ <u>est surjectif.</u>

<u>DEMONSTRATION</u> Soient G un groupe quelconque, H un sous-groupe distingué tel que $K = G/H$ soit parfait. Démontrons d'abord la suite exacte

$$H_2(K) \xrightarrow{\partial} H/[G,H] \longrightarrow G/[G,G]$$

En effet, la suite exacte

$$1 \longrightarrow H/[G,H] \longrightarrow G/[G,H] \longrightarrow K \longrightarrow 1$$

définit une extension centrale de K par $H/[G,H]$, d'où un homomorphisme canonique de $H_2(K)$ dans $H/[G,H]$ grâce à l'extension universelle [6]

$$1 \to H_2(K) \to \tilde{K} \to K \to 1$$

L'homomorphisme composé $H_2(K) \to H/[G,H] \to G/[G,G]$ est bien zéro en raison du diagramme commutatif

$$\begin{array}{ccccccc} 1 \to & G/[G,G] & \to & G/[G,G] \times K & \to & K \to 1 \\ & \uparrow & & \uparrow & & \uparrow \\ 1 \to & H_2(K) & \to & \tilde{K} & \text{---} & K \to 1 \end{array}$$

D'autre part, soit $\bar{h}$ un élément de $H/[G;H]$ et soit h un représentant de $\bar{h}$ dans H tel que $h = \prod_i [\alpha_i,\beta_i]$ où α_i et β_i sont des éléments de G . Soient α_i' et β_i' leurs images dans K . Il est clair que $\bar{h}$ ne dépend que de α_i' et de β_i' ; en effet, des relèvements arbitraires $\overline{\alpha_i}$ et $\overline{\beta_i}$ de α_i' et β_i' dans $G/[G,H]$ permettent de retrouver $\bar{h} = \pi\overline{[\alpha_i,\beta_i]}$. Il en résulte évidemment que $\bar{h} = \partial(\alpha)$ où α est un élément de $H_2(K)$. Enfin, il est clair que l'homomorphisme composé $H_2(G) \to H_2(K) \to H/[G,H]$ est nul. Appliquons ces considérations au cas où $G = {}_{-1}O'(\mathbb{Z}'[x])$, $K = {}_{-1}O'(\mathbb{Z}') \times {}_{-1}O'(\mathbb{Z}')$ et $H = \mathrm{Ker}\ {}_{-1}O'(\mathbb{Z}'[x]) \to {}_{-1}O'(\mathbb{Z}') \times {}_{-1}O'(\mathbb{Z}')$(resp. $G_1 = {}_{-1}O'(\mathbb{R}[x])$, $K_1 = {}_{-1}O'(\mathbb{R}) \times {}_{-1}O'(\mathbb{R})$ et $H_1 = \mathrm{Ker}({}_{-1}O'(\mathbb{R}[x]) \to {}_{-1}O'(\mathbb{R}) \times {}_{-1}O'(\mathbb{R}))$

$$\begin{array}{ccccc} H_2(K) & \text{———} & H_2(K') & \xrightarrow{u} & \pi_1({}_{-1}O'(\mathbb{R})) \\ \downarrow & & \downarrow & & \| \\ H/[G,H] & \longrightarrow & H_1/[G_1,H_1] & \longrightarrow & \pi_1({}_{-1}O'(\mathbb{R})) \\ \downarrow & & \downarrow & & | \\ {}_{-1}L^{-2}(\mathbb{Z}) & \longrightarrow & {}_{-1}L^{-2}(\mathbb{R}_d) & \longrightarrow & {}_{-1}L^{-2}(\mathbb{R}) \end{array}$$

Dans ce diagramme $\mathbb{R}_d$ désigne le corps des réels muni de la topologie discrète et u représente l'homomorphisme explicité par Milnor [21]' dans un contexte légèrement différent (de manière précise, on associe à tout élément de $H_2(K') \subset K'$ l'élément de $\pi_1({}_{-1}O'(\mathbb{R}))$ obtenu grâce à l'homomorphisme canonique de $\tilde{K}'$ dans le recouvrement universel de ${}_{-1}O'(\mathbb{R})$). Considérons maintenant le lacet $(h(x))^4$ où

$$h(x) = \begin{pmatrix} 1 & x \\ 0 & 1 \end{pmatrix} \begin{pmatrix} 1 & 0 \\ -x & 1 \end{pmatrix} \begin{pmatrix} 1 & x \\ 0 & 1 \end{pmatrix}$$

dans $\mathbb{Z}'[x] \oplus {}^t\mathbb{Z}'[x]$. L'argument utilisé au début du quatrième paragraphe montre que $(h(x))^4$ est un lacet algébrique dont l'image dans $\pi_1({}_{-1}O'(\mathbb{R})) \approx \mathbb{Z}$ est le générateur topologique usuel. Ce lacet est un élément de H dont la classe dans $G/[G,G]$ est nulle car $h(x)$ est un produit de matrices orthogonales élémentaires (cf. [18] lemme 2.13). De la suite exacte

$${}_{-1}L_2(\mathbb{Z}') \longrightarrow H/[G,H] \longrightarrow G/[G,G]$$

avec ${}_{-1}L_2(\mathbb{Z}') = H_2({}_{-1}O(\mathbb{Z}'))$, on déduit évidemment l'assertion annoncée.

<u>Remarque</u> Nous avons admis implicitement que l'homomorphisme $H_2(K_1) \overset{\psi}{\to} \pi_1({}_{-1}O'(\mathbb{R}))$ correspondait modulo les diverses identifications à l'homomorphisme $\pi_2(B^+_{K_1}) \to \pi_2(B_{{}_{-1}O'(\mathbb{R})})$. Pour le démontrer, considérons le diagramme

$$\begin{array}{ccc}
0 & \longrightarrow & 0 \\
\downarrow & & \downarrow \\
H_2(K_1) & \longrightarrow & \pi_1({}_{-1}O'(\mathbb{R})) \\
\downarrow & & \downarrow \\
\tilde{K}_1 & \longrightarrow & {}_{-1}\tilde{O}'(\mathbb{R}) \\
\downarrow & & \downarrow \\
K_1 & \longrightarrow & {}_{-1}O'(\mathbb{R}) \\
\downarrow & & \downarrow \\
0 & & 0
\end{array}$$

où ${}_{-1}\tilde{O}'(\mathbb{R})$ désigne le revêtement universel de ${}_{-1}O'(\mathbb{R})$. Si on applique le foncteur B^+ à ce diagramme on trouve des fibrations (cf. [28] par exemple). On en déduit une identification canonique de $\pi_2(B^+_{K'_1})$ avec $\pi_1(B^+_{H_2(K'_1)}) \approx H_2(K'_1)$, etc...

Soit maintenant

$$\varphi : A \times A' \longrightarrow A''$$

un bimorphisme d'anneaux hermitiens discrets. Soit $p : A^{2n} \to A^{2n}$ un projecteur auto-adjoint dans $A^{2n} \approx {}_{\varepsilon}H(A^n)$ et soit α un élément de ${}_{\eta}O_{m,n}(A')$. Nous pouvons associer au couple (p,α) l'élément β de ${}_{\varepsilon\eta}O_{2nm,2nm}(A'')$ défini par la formule $\alpha \otimes p + 1 \otimes (1-p)$ à condition d'identifier ${}_{\varepsilon}H(A''^n) \otimes {}_{\eta}H(A''^m)$ à ${}_{\varepsilon\eta}H(A''^{2nm})$. L'entier n étant fixé ainsi que le projecteur p, on définit ainsi un homomorphisme du groupe ${}_{\eta}O(A')$ dans le groupe ${}_{\varepsilon\eta}O(A'')$, donc une application continue de $B^{+}_{{}_{\varepsilon}O(A')}$ dans $B^{+}_{{}_{\varepsilon\eta}O(A'')}$. On notera $\theta_{\varphi}(p)$ l'application continue ainsi obtenue. Puisqu'un automorphisme intérieur de ${}_{\varepsilon\eta}O(A'')$ induit une application $B^{+}_{{}_{\varepsilon\eta}O(A'')}$ dans lui-même homotope à l'identité, on voit que $\theta_{\varphi}(p)$ ne dépend en fait que de la classe de Imp dans l'ensemble ${}_{\varepsilon}\Phi(A)$ des classes d'isomorphie de A-modules ε-quadratiques. On en déduit une application ${}_{\varepsilon}\Phi(A) \times \pi_i(B^{+}_{{}_{\eta}O(A')}) \to \pi_i(B^{+}_{{}_{\varepsilon\eta}O(A'')})$ qui est bilinéaire en un sens évident, donc un homomorphisme bilinéaire

$$ {}_{\varepsilon}L(A) \times {}_{\eta}L_i(A') \longrightarrow {}_{\varepsilon\eta}L_i(A'') $$

(appelé "cup-produit").

<u>LEMME 5.3.</u> <u>Pour tout projecteur auto-adjoint</u> p <u>dans</u> ${}_{\varepsilon}H(A^n)$ <u>on a le diagramme commutatif où les suites verticales sont des fibrations</u>[1]

$$
\begin{array}{ccc}
B^{+}_{{}_{\varepsilon}O(\tilde{A}')} & \xrightarrow{\theta_{\varphi}(p)} & B^{+}_{{}_{\varepsilon\eta}O(\tilde{A}'')} \\
\downarrow & & \downarrow \\
B^{+}_{{}_{\varepsilon}O(CA')} & \xrightarrow{\theta_{C\varphi}(p)} & B^{+}_{{}_{\varepsilon\eta}O(CA'')} \\
\downarrow & & \downarrow \\
B^{+}_{{}_{\varepsilon}O'(SA')} & \xrightarrow{\theta_{S\varphi}(p)} & B^{+}_{{}_{\varepsilon\eta}O'(SA'')}
\end{array}
$$

<u>et on pose</u> ${}_{\varepsilon}O'(B) = [{}_{\varepsilon}O(B), {}_{\varepsilon}O(B)]$ <u>de manière générale</u>.

<u>Remarque</u> Dans ce qui précède on a défini le cup-produit à gauche par L. On définirait de même le cup-produit à droite.

---

(1) $\tilde{A}'$ et $\tilde{A}''$ ne sont pas des anneaux unitaires. Cependant, on peut construire l'espace B^{+} qui leur est attaché grâce à la méthode décrite dans [28] par exemple.

DEMONSTRATION   On a le diagramme commutatif

$$\begin{array}{ccc} B_{\eta}O(\tilde{A}') & \longrightarrow & B_{\varepsilon\eta}O(\tilde{A}'') \\ \downarrow & & \downarrow \\ B_{\eta}O(CA') & \longrightarrow & B_{\varepsilon\eta}O(CA'') \\ \downarrow & & \downarrow \\ B_{\eta}O'(SA') & \longrightarrow & B_{\varepsilon\eta}O'(SA'') \end{array}$$

d'après la construction fonctorielle de l'espace classifiant. La construction de B^+ peut aussi être rendue fonctorielle grâce aux méthodes décrites dans [12] ou dans [28] . Le lemme en résulte.

COROLLAIRE 5.4. *Soit s_n l'élément de ${}_1L_n(S^n\mathbb{Z}')$ correspondant à l'élément unité de ${}_1L(\mathbb{Z}')$ grâce à l'isomorphisme ${}_1L(\mathbb{Z}') \approx {}_1L_n(S^n\mathbb{Z}')$. On a alors le diagramme commutatif*

$$\begin{array}{ccc} {}_{\varepsilon}L(A) & \xrightarrow{\ \Psi_n\ } & {}_{\varepsilon}L_n(S^nA) \\ \| & & \wr\wr \\ {}_{\varepsilon}L(A) & \xrightarrow{\ \Psi_{n+1}\ } & {}_{\varepsilon}L_{n+1}(S^{n+1}A) \end{array}$$

où Ψ_j désigne le cup-produit par s_j .

En effet, les suites verticales du lemme 5.3. sont des fibrations dont l'espace total est contractile d'après l'argument utilisé dans [12] [28] . Pour $n \geq 1$ le corollaire en résulte donc bien (faire $A' = S^n\mathbb{Z}'$). Pour $n = 0$, la démonstration est analogue à la démonstration standard en K-théorie [16] .

COROLLAIRE 5.5. *Le cup-produit par s_n induit un isomorphisme ${}_{\varepsilon}L(A) \approx {}_{\varepsilon}L_n(S^nA)$.*

La démonstration se fait par récurrence sur n .

Si n et m sont des entiers fixés, on peut aussi définir un homomorphisme de groupes

$${}_{\varepsilon}O_{n,n}(A') \times {}_{\eta}O_{m,n}(A') \longrightarrow {}_{\varepsilon\eta}O_{2nm,2nm}(A'')$$

grâce au produit tensoriel des matrices. Soit ${}_{\varepsilon}E_{n,n}(B)$ le sous-groupe normal de ${}_{\varepsilon}O_{n,n}(B)$ engendré par les matrices orthogonales ε-élémentaires. D'après

[29] (voir aussi [18] lemme 2.13), ce sous-groupe est parfait si $n \geq 3$. On en déduit une application continue

$$\gamma^+ : B^+_{\varepsilon O_{n,n}(A)} \times B^+_{\eta O_{m,m}(A')} \longrightarrow B^+_{\varepsilon\eta O_{2nm,2nm}(A'')}$$

Soit alors $A=A'=A''=\mathbb{Z}'$ et $n=m=3$. L'élément v_2 de ${}_{-1}L_2(\mathbb{Z}')$ est représenté par une application continue $S^2 \to B^+_{-1O_{3,3}(\mathbb{Z}')} \to B^+_{-1O(\mathbb{Z}')}$ et l'application γ^+ permet de définir une application continue

$$\sigma : S^2 \times S^2 \to B^+_{+1O_{18,18}(\mathbb{Z}')} \to B^+_{1O(\mathbb{Z}')}$$

On peut modifier σ en $\tilde{\sigma}$ grâce à la formule

$\tilde{\sigma}(x,y) = \sigma(x,y) - \sigma(x,a) - \sigma(a,x) + \sigma(a,a)$. Dans cette formule a est le point de base de S^2 et le "-" est compris au sens de la structure d'espace de lacets de $B^+_{1O(\mathbb{Z}')}$. On obtient ainsi de manière explicite un élément v_4 de ${}_1L_4(\mathbb{Z}')$ dont l'image dans ${}_1L_4(\mathbb{R}) = {}_1L^{-4}(\mathbb{R}) \approx \mathbb{Z} \oplus \mathbb{Z}$ est le couple $(2, -2)$. En procédant ainsi de proche en proche, on construit des éléments $v_{4n} \in {}_1L_{4n}(\mathbb{Z}')$ dont l'image dans ${}_1L_{4n}(\mathbb{R}) = {}_1L^{-4n}(\mathbb{R}) \approx \mathbb{Z} \oplus \mathbb{Z}$ est le couple $(2^{2n-1}, 2^{2n-1})$.

On construit de même un élément v_{4n+2} de ${}_{-1}L_{4n+2}(\mathbb{Z}')$ dont l'image dans ${}_{-1}L_{4n+2}(\mathbb{R}) = {}_{-1}L^{-4n-2}(\mathbb{R}) \approx \mathbb{Z}$ est 4^n fois le générateur. Désignons par v'_{2n} la classe de v_{2n} dans ${}_\varepsilon W_{2n}(\mathbb{Z}')$ avec $\varepsilon = (-1)^n$. Il est clair que les structures multiplicatives précédentes passent aux groupes W, ce qui permet de définir deux homomorphismes fondamentaux

$$\beta : {}_\eta W_0(A) \longrightarrow {}_{\eta\varepsilon}W_{2n}(A)$$

$$\beta' : {}_{\eta\varepsilon}W_{2n}(A) \longrightarrow {}_\eta W_{2n}(S^{2n}A) \approx {}_\eta W_0(A) \quad (\varepsilon = (-1)^n)$$

en considérant le cup-produit par v'_{2n} et par l'élément $v'_{-2n} \in {}_\varepsilon W_{-2n}(\mathbb{Z}') = {}_\varepsilon W^{2n}(\mathbb{Z}')$ construit dans le paragraphe précédent.

<u>THEOREME 5.6.</u> <u>L'homomorphisme composé</u> ${}_\eta W_0(A) \to {}_{\varepsilon\eta}W_{2n}(A) \to {}_\eta W_0(A)$ <u>coïncide avec la multiplication par</u> 4^n. En particulier, ${}_{\varepsilon\eta}W_{2n}(A) \otimes_{\mathbb{Z}} \mathbb{Z}'$ <u>contient</u> ${}_\eta W_0(A) \otimes_{\mathbb{Z}} \mathbb{Z}'$ <u>comme facteur direct</u> (avec $\varepsilon = (-1)^n$).

<u>DEMONSTRATION</u> Ceci résulte du diagramme commutatif (associativité du

cup-produit)

$$\begin{array}{ccc} {}_{\eta}W_0(A) \times {}_{\varepsilon}W_{2n}(\mathbb{Z}') \times {}_{\varepsilon}W_0(S^{2n}\mathbb{Z}') & \longrightarrow & {}_{\eta}W_0(A) \times {}_{1}W_{2n}(S^{2n}\mathbb{Z}') \\ \downarrow & & \downarrow \\ {}_{\eta\varepsilon}W_{2n}(A) \times {}_{\varepsilon}W_0(S^{2n}\mathbb{Z}') & \longrightarrow & {}_{\eta}W_{2n}(S^{2n}A) \end{array}$$

et du corollaire 5.5.

Pour démontrer des théorèmes plus forts, nous allons avoir besoin d'interpréter les cup-produits de manière plus formelle. En effet, il résulte du travail d'Anderson et Segal [1] [25] qu'on peut définir des cup-produits

$${}_{\varepsilon}L_n(A) \times {}_{\eta}L_p(A') \longrightarrow {}_{\varepsilon\eta}L_{n+p}(A'')$$

à partir du foncteur

$${}_{\varepsilon}Q(A) \times {}_{\eta}Q(A') \to {}_{\varepsilon\eta}Q(A'')$$

induit par le bimorphisme $A \times A' \to A''$ et par le produit tensoriel des modules quadratiques. D'autre part, on a un diagramme commutatif (à isomorphisme canonique près)

$$\begin{array}{ccc} \overline{\mathbb{P}(A)} \times {}_{\eta}Q(A') & \longrightarrow & \overline{\mathbb{P}(A'')} \\ \downarrow & & \downarrow \\ {}_{\varepsilon}Q(A) \times {}_{\eta}Q(A') & \longrightarrow & {}_{\varepsilon\eta}Q(A'') \end{array}$$

ainsi qu'un diagramme analogue en permutant les rôles de A et A'. On en déduit que l'image des K_n dans les L_n forme un "idéal" pour la multiplication, ce qui permet de définir un cup-produit

$${}_{\varepsilon}W_n(A) \times {}_{\eta}W_p(A') \longrightarrow {}_{\varepsilon\eta}W_{n+p}(A'')$$

En particulier, on peut définir un homomorphisme

$$\beta : {}_{\varepsilon}W_n(A) \longrightarrow {}_{-\varepsilon}W_{n+2}(A)$$

en considérant le cup-produit par la classe de v_2 dans ${}_{-1}W_2(\mathbb{Z}')$ (le bimorphisme $A \times \mathbb{Z}' \to A$ étant évident). De même, on peut définir un homomorphisme

$$\beta' : {}_{-\varepsilon}W_{n+2}(A) \longrightarrow {}_{\varepsilon}W_{n+2}(S^2A) \approx {}_{\varepsilon}W_n(A)$$

en considérant le cup-produit par la classe de v_{-2} dans ${}_{-1}W_{-2}(\mathbb{Z}')$. Le théorème suivant généralise alors le théorème 5.6. :

"THEOREME 5.7." Les homomorphismes $\beta'\beta$ et $\beta\beta'$ coïncident avec la multiplication par 4 (modulo la généralisation du théorème de Gersten citée plus bas).

DEMONSTRATION En raison de l'associativité du cup-produit, les homomorphismes $\beta\beta'$ et $\beta'\beta$ coïncident avec l'homomorphisme β'' : ${}_\varepsilon W_p(A) \to {}_\varepsilon W_{p+2}(S^2A)$, $p = n$ ou $n+2$, qui est défini par le cup-produit par $4v$ où v est l'élément de ${}_1W_2(S^2\mathbb{Z}')$ correspondant à l'élément unité de ${}_1W_0(\mathbb{Z}') \approx \mathbb{Z} \oplus (\mathbb{Z}/2)^2$ grâce à l'isomorphisme ${}_1W_0(\mathbb{Z}') \approx {}_1W_2(S^2\mathbb{Z}')$. Il résulte du travail de Gersten [12] que l'isomorphisme ${}_\varepsilon L_p(B) \approx {}_\varepsilon L_{p+2}(S^2B)$ (donc ${}_\varepsilon W_p(B) \approx {}_\varepsilon W_{p+2}(S^2B)$) est précisément induit par le cup-produit par v. Par conséquent, la situation pour la théorie de Quillen est la même que celle envisagée dans le quatrième paragraphe (théorème 4.10). Le théorème est démontré.

Remarque (écrite après la rédaction de ce chapitre) En fait le résultat de Gersten cité plus haut semble pour l'instant hypothétique. Il convient donc de remplacer ${}_\varepsilon W_n(A)$ par $\varinjlim_p {}_\varepsilon W_{n+p}(S^pA)$ pour assurer ce théorème et ceux qui vont suivre. On notera que cette limite inductive est facteur direct de $W_n(A)$.

COROLLAIRE 5.8. Soit ${}_\varepsilon\overline{W}_n(A) = {}_\varepsilon W_n(A) \otimes_{\mathbb{Z}} \mathbb{Z}'$ et soit

$$\begin{array}{ccc} A & \longrightarrow & A_1 \\ \downarrow & & \downarrow \\ A_2 & \longrightarrow & A' \end{array}$$

un diagramme cartésien d'anneaux hermitiens unitaires discrets. On a alors la suite exacte (Mayer-Victoris) :

$${}_\varepsilon\overline{W}_{n+2}(A') \to {}_\varepsilon\overline{W}_{n+1}(A) \to {}_\varepsilon\overline{W}_{n+1}(A_1) \oplus {}_\varepsilon\overline{W}_{n+1}(A_2) \to {}_\varepsilon\overline{W}_{n+1}(A') \to {}_\varepsilon\overline{W}_n(A) \to \quad \ldots (n \in \mathbb{Z}).$$

Ce corollaire peut encore être formulé de manière plus suggestive sous la forme suivante : posons ${}_\varepsilon\overline{W}_n(A) = \mathrm{Ker}[{}_\varepsilon\overline{W}_n(A^+) \to {}_\varepsilon\overline{W}_n(\mathbb{Z}')]$ pour tout anneau hermitien A unitaire ou pas.

COROLLAIRE 5.9. Soit

$$0 \to A' \to A \to A'' \to 0$$

une suite exacte d'anneaux hermitiens discrets. On a alors la suite exacte

$$ {}_\varepsilon\overline{W}_n(A) \to {}_\varepsilon\overline{W}_n(A'') \to {}_\varepsilon\overline{W}_{n-1}(A') \to {}_\varepsilon\overline{W}_{n-1}(A) \to {}_\varepsilon\overline{W}_{n-1}(A''), \quad n \in \mathbb{Z} $$

DEMONSTRATION DES DEUX COROLLAIRES Soit $d : {}_\varepsilon L_0(B) \to {}_\varepsilon L_0(B)$ l'endomorphisme associant au module quadratique (M,q) la différence $(M,q)-(M,-q)$. Il est clair que d s'annule sur l'image de $K(B)$ dans $L_0(B)$, donc définit un homomorphisme de groupe noté encore $d : {}_\varepsilon W_0(B) \to {}_\varepsilon L_0(B)$. Si on désigne par $\lambda : {}_\varepsilon L_0(B) \to {}_\varepsilon W_0(B)$ l'homomorphisme quotient, λd coïncide avec la multiplication par 2. En remplaçant éventuellement B par une suspension, on en déduit que la théorie ${}_\varepsilon\overline{W}_n(B)$ est facteur direct naturel de la théorie ${}_\varepsilon\overline{L}_n(B) = {}_\varepsilon L_n(B) \otimes_{\mathbb{Z}} \mathbb{Z}[\frac{1}{2}]$. Si

$$ 0 \to A' \to A \to A'' \to 0 $$

est une suite exacte d'anneaux hermitiens, il résulte des considérations de [8] (théorème 3.1.) que l'on a la suite exacte

$$ {}_\varepsilon L_n(A) \to {}_\varepsilon L_n(A'') \to {}_\varepsilon L_{n-1}(A') \to {}_\varepsilon L_{n-1}(A) \to {}_\varepsilon L_{n-1}(A'') $$

pour $n \leq 0$. Le corollaire 5.9. en résulte de manière évidente pour $n \leq 0$. Pour $n > 0$, on applique le théorème de périodicité pour se ramener au cas précédent. Le corollaire 5.8. s'en déduit ou se démontre de la même manière

LEMME 5.10. *Posons ${}_\varepsilon L'_0(B) = \mathrm{Ker}[{}_\varepsilon L_0(B) \to K_0(B)]$ et ${}_\varepsilon \overline{L}'_0(B) = {}_\varepsilon L'_0(B) \otimes_{\mathbb{Z}} \mathbb{Z}[\frac{1}{2}]$. Alors l'homomorphisme évident $\lambda' : {}_\varepsilon L'_0(B) \to {}_\varepsilon W_0(B)$ induit un isomorphisme ${}_\varepsilon \overline{L}'_0(B) \xrightarrow{\approx} {}_\varepsilon W_0(B)$.*

DEMONSTRATION L'homomorphisme $d : {}_\varepsilon W_0(B) \to {}_\varepsilon L_0(B)$ explicité plus haut se factorise en fait à travers $L'_0(B)$, soit $d' : {}_\varepsilon W_0(B) \to {}_\varepsilon L'_0(B)$. Il résulte des considérations ci-dessus que $\lambda' \cdot d' = 2$ avec $\lambda' = \lambda|_{{}_\varepsilon L'_0(B)}$. D'autre part, tout élément de ${}_\varepsilon L'_0(B)$ peut s'écrire $(M,q)-(M,q_0)$ d'après [18] (théorème 1.4.). Si on applique l'opérateur $d'\lambda'$, on trouve la classe de $(M,q)-(M,-q)-(M,q_0)+(M,-q_0)$, soit la classe de $2(M,q)-2(M,-q_0)$ toujours d'après [18] (loc.cit). Il en résulte que $d'\lambda'$ et $\lambda' d'$ coïncident avec la multiplication par 2.

COROLLAIRE 5.11. *Le groupe ${}_\varepsilon\overline{W}_n(A)$ est un invariant homotopique de A. En d'autres termes ${}_\varepsilon\overline{W}_n(A) \approx {}_\varepsilon\overline{W}_n(A[x])$.*

En effet $_\varepsilon\overline{W}_n(A) \approx {}_\varepsilon\overline{W}_{-p}(A)$ pour un certain $p = n+4k > 0$. D'après le lemme 5.10 et le théorème 1.1., $_\varepsilon\overline{W}_{-p}(A) \approx {}_\varepsilon\overline{W}_0(S^pA) \approx {}_\varepsilon\overline{W}_0((S^pA[x]) \approx {}_\varepsilon\overline{W}_0(S^p(A[x]))$ $\approx {}_\varepsilon W_n(A[x])$.

THEOREME 5.12. Soit A un anneau hermitien (non nécessairement K-régulier). Alors les groupes $_\varepsilon\overline{W}_n(A) : {}_\varepsilon\overline{W}_n(A) \otimes_{\mathbb{Z}} \mathbb{Z}'$ et $_\varepsilon\overline{W}^{-n}(A) = {}_\varepsilon\overline{W}^{-n}(A) \otimes_{\mathbb{Z}} \mathbb{Z}'$ sont naturellement isomorphes et sont périodiques par rapport à n de période 4 . De manière précise

$$_\varepsilon\overline{W}_n(A) \approx {}_\varepsilon\overline{W}^{-n}(A) \approx {}_{-\varepsilon}\overline{W}_{n+2}(A) \approx {}_{-\varepsilon}\overline{W}^{-n-2}(A)$$

DEMONSTRATION Par définition, les groupes $_\varepsilon W_n(A)$ et $_\varepsilon W^{-n}(A)$ coïncident pour $n \leq 0$. Il en est de même de $_\varepsilon\overline{W}_n(A)$ et de $_\varepsilon\overline{W}^{-n}(A)$. Raisonnons maintenant par récurrence sur n . La $_\varepsilon O$-fibration standard

$$0 \to \Omega A \to EA \to A \to 0$$

induit un isomorphisme $_\varepsilon\overline{W}^{-n-1}(A) \approx {}_\varepsilon\overline{W}^{-n}(\Omega A)$ et de même $_\varepsilon\overline{W}_{n+1}(A) \approx {}_\varepsilon\overline{W}_n(\Omega A)$ d'après les corollaires 5.9. et 5.11 , l'anneau hermitien EA étant contractile [17] [18] . L'isomorphisme entre les théories $_\varepsilon\overline{W}_n$ et $_\varepsilon\overline{W}^{-n}$ implique donc bien un isomorphisme entre les théories $_\varepsilon\overline{W}_{n+1}$ et $_\varepsilon\overline{W}^{-n-1}$. La dernière partie du théorème 5.12. résulte du théorème 5.7.

Remarque Ce théorème améliore le théorème 4.10 dans le cas discret, l'hypothèse de K-régularité ayant été supprimée.

Nous allons tâcher d'interpréter le théorème précédent sous une forme plus "homotopique". Considérons l'automorphisme $\alpha \mapsto \alpha'$ de $_\varepsilon O_{n,n}(A)$ défini par

$$\alpha = \begin{pmatrix} a & b \\ c & d \end{pmatrix} \qquad \alpha' = \begin{pmatrix} d & c \\ b & a \end{pmatrix}$$

Cet automorphisme correspond essentiellement à changer la forme quadratique canonique de $_\varepsilon H(A^n)$ en son opposée. Par conséquent $\alpha \oplus \alpha'$ est conjuguée de $_\varepsilon H(\alpha)$ dans le groupe $_\varepsilon O_{2n,2n}(A)$ et ceci de manière canonique. Soit

$$\gamma : B^+_\varepsilon O(A) \longrightarrow B^+_\varepsilon O(A)$$

l'application continue induite par l'homomorphisme $\alpha \mapsto \alpha \oplus \alpha'$. Il est clair que $\gamma^2 = 2\gamma$ pour la structure de H-espace de $B^+_{\varepsilon O(A)}$. Soit $\tilde{B}^+_{\varepsilon O(A)}$ le H-espace obtenu à partir de $B^+_{\varepsilon O(A)}$ en localisant par rapport au système multiplicatif (2^n). Si on pose $\tilde{\gamma} = \gamma/2$, $\tilde{\gamma}$ est une application canonique continue de $B^+_{\varepsilon O(A)}$ dans lui-même telle que γ^2 soit homotope à γ et qui est en outre compatible avec la structure de H-espace de $B^+_{\varepsilon O(A)}$. Pour tout CW-complexe X , l'application $\tilde{\gamma}$ induit un scindage du foncteur semi-exact $F(X) = [X, \tilde{B}^+_{\varepsilon O(A)}]$ en $F(X) = F'(X) \oplus F''(X)$. Si $X = S^n$, on a $F'(X) \approx {}_\varepsilon W_n(A) \otimes_{\mathbb{Z}} \mathbb{Z}[\frac{1}{2}]$. Le théorème de représentabilité des foncteurs semi-exacts nous permet donc d'énoncer le théorème suivant :

THEOREME 5.13. Soit $\tilde{B}^+_{\varepsilon O(A)}$ le H-espace obtenu en localisant $B^+_{\varepsilon O(A)}$ par rapport au système multiplicatif (2^p). Au type d'homotopie près, on a alors une décomposition de $\tilde{B}^+_{\varepsilon O(A)}$ en le produit de deux H-espaces connexes $B^{++}_{\varepsilon O(A)} \times Y$ tels que $\pi_n(B^{++}_{\varepsilon O(A)}) \approx {}_\varepsilon W_n(A) \otimes_{\mathbb{Z}} \mathbb{Z}[\frac{1}{2}] = {}_\varepsilon \overline{W}_n(A)$. En particulier, on a $\pi_{n+4}(B^{++}_{\varepsilon O(A)}) \approx \pi_n(B^{++}_{\varepsilon O(A)})$, les premiers groupes d'homotopie étant respectivement :

$$\pi_1(B^{++}_{\varepsilon O(A)}) \approx {}_\varepsilon \overline{W}_1(A) \approx {}_\varepsilon \overline{W}^{-1}(A)$$

$$\pi_2(B^{++}_{\varepsilon O(A)}) \approx {}_{-\varepsilon} \overline{W}_0(A)$$

$$\pi_3(B^{++}_{\varepsilon O(A)}) \approx {}_{-\varepsilon} \overline{W}_1(A) \approx {}_{-\varepsilon} \overline{W}^{-1}(A)$$

$$\pi_4(B^{++}_{\varepsilon O(A)}) \approx {}_\varepsilon \overline{W}_0(A)$$

COROLLAIRE 5.14. L'algèbre d'homologie $H_*({}_\varepsilon O(A); \mathbb{Q})$ du groupe orthogonal infini à coefficients rationnels peut s'écrire comme le produit tensoriel gradué de trois algèbres de Hopf, soit

$$H_*({}_\varepsilon O(A); \mathbb{Q}) \approx {}_\varepsilon S_*(A) \hat{\otimes} {}_\varepsilon \Lambda_*(A) \hat{\otimes} {}_\varepsilon M_*(A) \quad \text{où } {}_\varepsilon M_0 = \mathbb{Q} \text{ et où}$$

${}_\varepsilon S_*(A)$ et ${}_\varepsilon \Lambda_*(A)$ s'explicitent comme suit :

1) ${}_\varepsilon S_*(A)$ est l'algèbre symétrique de l'espace vectoriel gradué $({}_\varepsilon W_2(A) \oplus {}_\varepsilon W_4(A) \oplus \ldots \oplus {}_\varepsilon W_{2n}(A) \oplus \ldots) \otimes_{\mathbb{Z}} \mathbb{Q}$

2) ${}_{\varepsilon}\Lambda_{*}(A)$ <u>est l'algèbre extérieure de l'espace vectoriel gradué</u>

$({}_{\varepsilon}W_1(A) \oplus {}_{\varepsilon}W_3(A) \oplus \ldots \oplus {}_{\varepsilon}W_{2n+1}(A) \oplus \ldots) \otimes_{\mathbb{Z}} \mathbb{Q}$.

<u>Exemples</u> : 1) Si A est un corps commutatif F muni de l'involution triviale, on a ${}_1L_1(F) \approx \mathbb{Z}/2 \times F^*/(F^*)^2$, ${}_{-1}L_1(F) = {}_{-1}W_0(F) = 0$ [18] . L'algèbre ${}_{\varepsilon}S_{*}(A) \otimes {}_{\varepsilon}\Lambda_{*}(A)$ se réduit donc à l'algèbre symétrique de l'espace vectoriel $({}_1W_4(F) \oplus {}_1W_8(F) \oplus \ldots \oplus {}_1W_{4n}(F) \oplus \ldots) \otimes_{\mathbb{Z}} \mathbb{Q}$ si $\varepsilon = 1$ où à celle de $({}_{-1}W_2(F) \oplus {}_{-1}W_6(F) \oplus \ldots \oplus {}_{-1}W_{4n+2}(F) \oplus \ldots) \otimes_{\mathbb{Z}} \mathbb{Q}$ si $\varepsilon = -1$. On notera que ${}_{-1}W_{4n+2}(F) \otimes_{\mathbb{Z}} \mathbb{Q} \approx {}_1W_{4n}(F) \otimes_{\mathbb{Z}} \mathbb{Q} \approx W(F) \otimes_{\mathbb{Z}} \mathbb{Q}$ où $W(F) = {}_1W_0(F)$ est l'anneau de Witt de F .

2) Si A est un corps de nombres F muni de l'involution triviale, l'homologie $H_{*}({}_{\varepsilon}\mathcal{O}(F); \mathbb{Q})$ a été entièrement déterminée par Borel [7] . Si $\varepsilon = 1$, ${}_{\varepsilon}\mathcal{O}_{n,n}(F)$ est désigné classiquement par $O_{n,n}(F)$ et $\varinjlim H_{*}(O_{n,n}(F); \mathbb{Q}) = \bigotimes_{v \in V} I_v$ où V est l'ensemble formé des places réelles et d'un représentant de chaque paire de places complexes conjuguées et où

$$I_v = \mathbb{Q}[x_4, x_8, \ldots] \quad \text{si } v \text{ est réelle}$$

$$I_v = \Lambda[x_3, x_7, \ldots] \quad \text{si } v \text{ est complexe}$$

Si $\varepsilon = -1$, ${}_{-1}O_{n,n}(F)$ est désigné classiquement par $Sp_{2n}(F)$ et on a aussi $\varinjlim H_{*}(Sp_{2n}(F); \mathbb{Q}) = \bigotimes_{v \in V} I'_v$ où $I'_v = I_v$ si v est complexe et où $I'_v = \mathbb{Q}[x_2, x_6, \ldots]$ si v est réelle. On voit donc que les places réelles "se conservent" dans le groupe ${}_{\varepsilon}S_{*}(A) \hat{\otimes} {}_{\varepsilon}\Lambda_{*}(A)$. Il n'en est pas de même des places complexes qui proviennent donc de l'homologie du groupe linéaire, homologie que les techniques développées dans cet article ne permettent pas de calculer, semble-t-il.

Nous allons essayer maintenant de reformuler le théorème fondamental du § 4 dans le cadre de la théorie de Quillen. Soit ${}_{\varepsilon}\mathcal{V}_A$ (resp. ${}_{\varepsilon}\mathcal{U}_A$) la fibre au sens homotopique de l'application

$B^+_{{}_\varepsilon O(A)} \to B^+_{GL(A)}$ (resp. $B^+_{GL(A)} \to B^+_{{}_\varepsilon O(A)}$ induite par l'homomorphisme naturel ${}_\varepsilon O(A) \to GL(A)$ (resp. $GL(A) \to {}_\varepsilon O(A)$) associé au foncteur oubli (resp. le foncteur hyperbolique). Pour $n > 0$, on définit alors le groupe ${}_\varepsilon V_n(A)$ (resp. ${}_\varepsilon U_n(A)$) comme le groupe d'homotopie $\pi_n({}_\varepsilon \mathcal{V}_A)$ (resp. $\pi_n({}_\varepsilon \mathcal{U}_A)$). On a alors des isomorphismes naturels ${}_\varepsilon V_n(A) \approx {}_\varepsilon V_{n+1}(SA)$ et ${}_\varepsilon U_n(A) \approx {}_\varepsilon U_{n+1}(SA)$ induits par le cup-produit par $u \in {}_1L_1(SZ')$. Ceci nous permet de définir ${}_\varepsilon V_n(A)$ et ${}_\varepsilon U_n(A)$) pour $n \leq 0$ grâce aux formules de récurrence ${}_\varepsilon V_n(A) = {}_\varepsilon V_{n+1}(SA)$ et ${}_\varepsilon U_n(A) = {}_\varepsilon U_{n+1}(SA)$. Avec ces définitions, on a donc les suites exactes

$$K_{n+1}(A) \to {}_\varepsilon L_{n+1}(A) \to {}_\varepsilon U_n(A) \to K_n(A) \to {}_\varepsilon L_n(A)$$

$${}_\varepsilon L_{n+1}(A) \to K_{n+1}(A) \to {}_\varepsilon V_n(A) \to {}_\varepsilon L_n(A) \to K_n(A)$$

CONJECTURE 5.15. Les théories ${}_\varepsilon U_{n+1}(A)$ et ${}_{-\varepsilon} V_n(A)$ sont isomorphes.

Comme nous l'avons montré formellement dans le paragraphe précédent, cette conjecture impliquerait, compte-tenu des suites exactes citées plus haut, les théorèmes démontrés jusqu'à maintenant dans ce paragraphe. D'autre part il est facile de voir que cette conjecture rejoint le théorème 4.8. pour $n > -1$ si A est K-régulier. Enfin, la conjecture 5.15. est essentiellement équivalente à la remarque après le théorème 5.7. lorsqu'on tensorise tous les groupes par $\mathbb{Z}' = \mathbb{Z}[\frac{1}{2}]$ Dans le reste de ce paragraphe nous allons voir comment s'exprime (et se démontre partiellement) cette conjecture pour de petites valeurs de n.

Soit $\varphi : G \to K$ un homomorphisme entre groupes discrets et soit $\tilde{K}$ le revêtement universel de $K' = [K,K]$ au sens de [6] p.1 (K' étant supposé parfait). On a donc le diagramme commutatif

$$\begin{array}{ccc} H & \xrightarrow{\beta} & \tilde{K} \\ \downarrow{\scriptstyle \gamma} & & \downarrow{\scriptstyle \theta_K} \\ G & \xrightarrow{\varphi} & K \end{array}$$

où H désigne le produit fibré et où l'homomorphisme θ_K est l'homomorphisme composé $\tilde{K} \to [K,K] \to K$. Soit $\tilde{G}$ le revêtement universel de $G' = [G,G]$ (supposé parfait également) et soit $\alpha : \tilde{G} \to H$ l'homomorphisme déduit de la fonctorialité de la construction des revêtements universels. Soit enfin H_0 l'image de $\tilde{G}$ dans H par l'homomorphisme α.

<u>LEMME 5.16.</u> <u>Le groupe H_0 est un sous-groupe distingué de H et on a la suite exacte</u>

$$H_2(K') \xrightarrow{\partial} H/H_0 \xrightarrow{n} G/[G,G] \xrightarrow{m} K/[K,K]$$

<u>Si φ est un homomorphisme injectif, on a la suite exacte plus complète</u>

$$H_2(G') \xrightarrow{m_1} H_2(K') \xrightarrow{\partial} H/H_0 \xrightarrow{n} G/[G,G] \xrightarrow{m} K/[K,K]$$

<u>DEMONSTRATION</u> Soit $h = (g,\tilde{k})$ un élément de $H \subset G \times \tilde{K}$ et soit $\tilde{g}$ un élément de $\tilde{G}$. Puisque les automorphismes intérieurs de G induisent des automorphismes de $\tilde{G}$ de manière naturelle, on a $h\alpha(\tilde{g})h^{-1}=\alpha(\tilde{g}_1)$ où $\tilde{g}_1$ désigne le transformé de $\tilde{g}$ par l'automorphisme de $\tilde{G}$ induit par l'automorphisme $u \mapsto gug^{-1}$ de G. Donc H_0 est un sous-groupe distingué de H. Démontrons maintenant l'exactitude de la suite

a) <u>Exactitude en $G/[G,G]$</u>

- Si $h = (g,\tilde{k})$ est un élément de H de classe $\bar{h}$ dans H/H_0, $(m.n)(\bar{h})$ est la classe de $\theta_K(\tilde{k})$ dans $K/[K,K]$, classe qui est nulle puisque θ_K se factorise à travers $[K,K]$.
- Réciproquement, soit g un élément de G tel que $\varphi(g)$ soit un commutateur. Alors $g = \gamma(h)$ avec $h = (g,\tilde{k})$, $\tilde{k}$ étant un relèvement quelconque de $\varphi(g)$ dans $\tilde{K}$.

b) <u>Exactitude en H/H_0</u>

- La définition de ∂ s'explicite ainsi : si x est un élément de $H_2(K') = \operatorname{Ker} \theta_K$, $\partial(x)$ est la classe de $h = (g,\tilde{k})$ où $g = 1$ et $\tilde{k} = x$. Dans ces conditions, il est clair que $(m.\partial)(x) = 0$. Réciproquement, désignons par θ_G l'homomorphisme canonique de $\tilde{G}$ dans G. Si $h = (g,\tilde{k})$ est un élément de H tel que g soit un commutateur, i.e. $g = \theta_G(\tilde{g})$, on a $h\alpha(\tilde{g})^{-1} = (1,x)$ avec $\theta_K(x) = 1$.

Donc $\bar{h} = \partial(x)$.

c) Exactitude en $H_2(K')$ (si φ est injectif).

- Soit y un élément de $H_2(G) = \mathrm{Ker}\ \theta_G$. On a alors $(m_1.\partial)(y) = \bar{h}$ avec $h = (\theta_G(y),(\beta\alpha)(y)) = \alpha(y) \in H_0$. Réciproquement, soit x un élément de $H_2(K') = \mathrm{Ker}\ \theta_K$ tel que $x = (\beta\alpha)(\tilde{g})$. Puisque φ est injectif, on a $\theta_G(\tilde{g}) = 1$ et $x = m_1(y)$ avec $y = \tilde{g}$.

PROPOSITION 5.17. *On suppose que $\varphi : G \to K$ est un homomorphisme injectif. Alors H/H_0 est canoniquement isomorphe à $\pi_1(F)$, F étant la fibre au sens homotopique de l'application $B_G^+ \to B_K^+$ ($G' = [G,G]$ et $K' = [K,K]$ étant des sous-groupes parfaits de G et de K respectivement). En particulier, si $G = GL(A)$ et $K = {}_\varepsilon O(A)$ et si φ est l'homomorphisme induit par le foncteur hyperbolique, le groupe H/H_0 s'identifie à ${}_\varepsilon U_1(A)$.*

DEMONSTRATION : On a la suite exacte

$$\begin{array}{ccccccccc}
\pi_2(B_G^+) & \to & \pi_2(B_K^+) & \to & \pi_1(F) & \to & \pi_1(B_G^+) & \to & \pi_1(B_K^+) \\
\wr & & \wr & & & & & & \\
\pi_2(B_{G'}^+) & \to & \pi_2(B_{K'}^+) & & & & & & \\
\wr & & \wr & & & & & & \\
H_2(G') & \to & H_2(K') & & & & & &
\end{array}$$

Considérons maintenant le diagramme commutatif suivant où F_1 désigne le produit fibré.

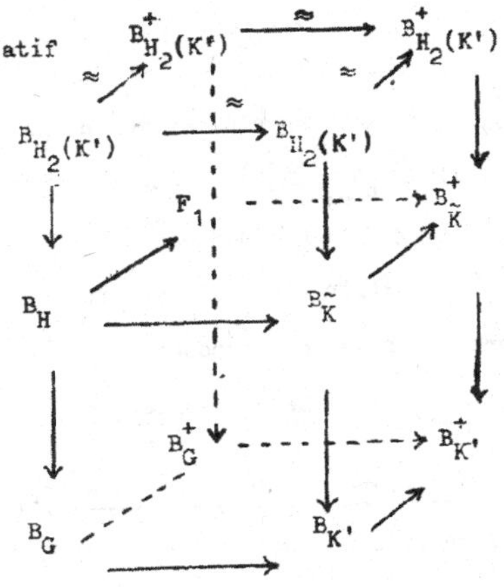

On notera que la suite $B_{H_2(K')} \to B^+_{\tilde K} \to B^+_{K'}$ est une fibration (au sens homotopique) d'après [28], lemme 3.1. par exemple. Puisque $\pi_1(B^+_{\tilde K}) = \pi_2(B^+_{\tilde K}) = 0$, on a $\pi_1(F) \approx \pi_1(F_1)$ ainsi que la suite exacte

$$\pi_2(B^+_{G'}) \to \pi_2(B^+_{K'}) \to \pi_1(F_1) \to \pi_1(B^+_G) \to \pi_1(B^+_K)$$

L'application $B_H \to F_1$ définit un homomorphisme $H \approx \pi_1(B_H) \to \pi_1(F_1)$. Cet homomorphisme s'annule sur $H_0 = \mathrm{Im}(\pi_1(B_{\tilde G}) \to \pi_1(B_H))$ car $\tilde G = [\tilde G, \tilde G]$. On voit par inspection que le diagramme suivant est commutatif

$$\begin{array}{ccccccccc}
\pi_2(B^+_{G'}) & \longrightarrow & \pi_2(B^+_{K'}) & \longrightarrow & \pi_1(F_1) & \longrightarrow & \pi_1(B^+_G) & \longrightarrow & \pi_1(B^+_K) \\
\wr & & \wr & & \uparrow & & \wr & & \wr \\
H_2(G') & \longrightarrow & H_2(K') & \longrightarrow & H/H_0 & \longrightarrow & G/[G,G] & \longrightarrow & K/[K,K]
\end{array}$$

Donc $H/H_0 \approx \pi_1(F_1) \approx \pi_1(F)$ d'après le lemme des cinq.

Compte-tenu de l'identification $H/H_0 \approx {}_{\varepsilon}U_1(A)$ avec les notations de la proposition 5.17, nous pouvons définir un homomorphisme ${}_{\varepsilon}U_1(A) \to {}_{\varepsilon}U^{-1}(A) \approx {}_{\varepsilon}U(\Psi)$ où $\Psi : A[x] \to A \times A$ de la manière suivante. Il existe un seul homomorphisme $s : {}_{\varepsilon}\widetilde{O(A)} \to {}_{\varepsilon}O(EA)/{}_{\varepsilon}O^0(\Omega A)$, où ${}_{\varepsilon}O^0(\Omega A)$ désigne le groupe des lacets dans ${}_{\varepsilon}O(A)$ qui sont homotopes à l'identité, qui rende commutatif le diagramme

$$\begin{array}{ccc}
{}_{\varepsilon}\widetilde{O(A)} & \xrightarrow{\ s\ } & {}_{\varepsilon}O(EA)/{}_{\varepsilon}O^0(\Omega A) \\
\downarrow & \swarrow_{p_1} & \\
{}_{\varepsilon}O(A) & &
\end{array}$$

Si $h = (g, \tilde k)$ est un élément de H, choisissons un représentant $k(x)$ de $s(k)$. Donc $k(x) \in {}_{\varepsilon}O_{n,n}(A[x])$ pour un certain n, $k(0) = 1$ et $k(1) = \varphi(g)$. Si η_0 désigne la graduation gauche canonique de $A^{2n} = H(A^n)$, le triple $(H(A^n), \eta_0, k(x)\eta_0 k(x)^{-1})$ définit l'élément de ${}_{\varepsilon}U(\Psi)$ cherché. On vérifie aisément que cette correspondance induit bien un homomorphisme de ${}_{\varepsilon}U_1(A)$ dans ${}_{\varepsilon}U(\Psi) \approx {}_{\varepsilon}U^{-1}(A)$ que nous noterons r.

PROPOSITION 5.18. Supposons que $K_1(A) \approx K^{-1}(A)$ (ce qui est vrai si A est noethérien régulier par exemple). Supposons 6 inversible dans A. Il existe

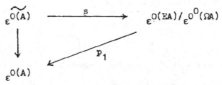

<u>alors un homomorphisme naturel</u> $\tilde{\tau}_V : {}_{-\varepsilon}V(A) \longrightarrow {}_{\varepsilon}U_1(A)$ <u>tel que le diagramme suivant commute</u>

$$\begin{array}{ccc} & {}_{-\varepsilon}V(A) & \\ \tilde{\tau}_V \swarrow & & \searrow \tau_V \\ {}_{\varepsilon}U_1(A) & \xrightarrow{\ r\ } & {}_{\varepsilon}U^{-1}(A) \end{array}$$

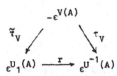

<u>l'homomorphisme</u> τ_V <u>étant celui défini dans le paragraphe précédent.</u>

<u>DEMONSTRATION</u> Nous allons baser notre démonstration sur le diagramme déjà écrit plus haut

$$\begin{array}{ccc} {}_{\varepsilon}\widetilde{O(A)} & \xrightarrow{\ \cdot s\ } & {}_{\varepsilon}O(EA)/{}_{\varepsilon}O^{0}(\Omega A) \\ \theta_G \searrow & & \swarrow P_1 \\ & {}_{\varepsilon}O(A) & \end{array}$$

La matrice

$$\begin{pmatrix} 1 & -x \\ 0 & 1 \end{pmatrix}$$

peut s'écrire $\alpha_1(x)a_1(\alpha_1(x))^{-1}a_1^{-1}$ avec

$$a_1 = \begin{pmatrix} 2 & 0 \\ 0 & 1/2 \end{pmatrix}, \quad \alpha_1(x) = \begin{pmatrix} 1 & +x/3 \\ 0 & 1 \end{pmatrix}$$

De même la matrice

$$\begin{pmatrix} 1 & 0 \\ x & 1 \end{pmatrix}$$

peut s'écrire $\alpha_2(x)a_2(\alpha_2(x))^{-1}a_2^{-1}$ avec

$$a_2 = \begin{pmatrix} 1/2 & 0 \\ 0 & 2 \end{pmatrix}, \quad \alpha_2(x) = \begin{pmatrix} 1 & 0 \\ -x/3 & 1 \end{pmatrix}$$

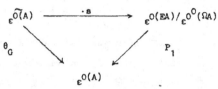

Si g est une matrice $n \times n$ qui est $(-\varepsilon)$-symétrique, les matrices

$$\begin{pmatrix} 1 & -g^{-1}x \\ 0 & 1 \end{pmatrix} \quad \text{et} \quad \begin{pmatrix} 1 & 0 \\ gx & 1 \end{pmatrix}$$

appartiennent à ${}_{+\varepsilon}O_{n,n}(A[X])$ et peuvent de même s'écrire
$\alpha_1^g(x)a_1(\alpha_1^g(x))^{-1}a_1^{-1}$ et $\alpha_2^g(x)a_2(\alpha_2^g(x))^{-1}a_2^{-1}$ avec

$$\alpha_1^g(x) = \begin{pmatrix} 1 & g^{-1}x/3 \\ 0 & 1 \end{pmatrix} \quad \text{et} \quad \alpha_2^g(x) = \begin{pmatrix} 1 & 0 \\ -gx/3 & 1 \end{pmatrix}$$

Soit f_{a_i} l'automorphisme de ${}_{+\varepsilon}\widetilde{O(A)}$ induit par l'automorphisme intérieur de ${}_{\varepsilon}O(A)$ associé à a_i , $i = 1,2$. Soit de même $\tilde{\alpha}_i^g$ un relèvement quelconque de $\alpha_i^g(1)$ dans ${}_{\varepsilon}\widetilde{O(A)}$. Soit β_i^g l'élément $\tilde{\alpha}_i^g f_{a_i}((\tilde{\alpha}_i^g)^{-1})$ de ${}_{\varepsilon}\widetilde{O(A)}$. Alors $s(\beta_1^g\beta_2^g(\beta_1^g)^{-1}))$ à la même classe que le produit des matrices

$$\begin{pmatrix} 1 & -g^{-1}x \\ 0 & 1 \end{pmatrix}\begin{pmatrix} 1 & 0 \\ gx & 1 \end{pmatrix}\begin{pmatrix} 1 & -g^{-1}x \\ 0 & 1 \end{pmatrix}$$

Puisque β_i^g est indépendant du choix du relèvement $\tilde{\alpha}_i^g$, l'élément $\beta(g) = \beta_1^g\beta_2^g(\beta_1^g)^{-1}$ est parfaitement défini. Après ces préliminaires un peu techniques, nous sommes prêts à définir τ_V . Remarquons tout d'abord que le groupe ${}_{-\varepsilon}V(A)$ peut être décrit comme le quotient du groupe libre engendré par les triples de la forme $A^{2n} = {}_{-\varepsilon}H(A^n)$ et où $g_i^* = -\varepsilon g_i$ par le sous-groupe engendré par les relations suivantes :

$$(A^{2n}, g_1, g_2) + (A^{2n'}, g_1', g_2') = (A^{2n+2n'}, g_1 \oplus g_1' , g_2 \oplus g_2')$$

$$(A^{2n}, g_1, g_2) + (A^{2n}, g_2, g_3) = (A^{2n}, g_1, g_3)$$

A un tel triple (A^{2n}, g_1, g_2) , on va associer un élément de ${}_{\varepsilon}U_1(A)$ comme suit on considère le couple $(g,\underset{\sim}{k})$ dans le produit fibré

$$\begin{array}{ccc} H & \longrightarrow & {}_{\varepsilon}\widetilde{O(A)} \\ \downarrow & & \downarrow \theta_K \\ GL(A) & \xrightarrow{\varphi} & {}_{\varepsilon}O(A) \end{array}$$

où $g = g_1^{-1}g_2$ est un élément de $GL(A)$ et où $\tilde{k} = \beta(g_1)\beta(g_2)^{-1}$. Il est clair que $\tilde{\tau}_V : {}_{-\varepsilon}V(A) \to {}_{\varepsilon}U_1(A) \approx H/H_0$ est ainsi bien défini et que le diagramme

$$\begin{array}{ccc} & {}_{-\varepsilon}V(A) & \\ {}^{\tilde{\tau}_V}\swarrow & & \searrow{}^{\tau_V} \\ {}_{\varepsilon}U_1(A) & \xrightarrow{r} & {}_{\varepsilon}\dot{U}^{-1}(A) \end{array}$$

est commutatif par construction.

THEOREME 5.19. *Soit* $\tilde{\tau}_U : {}_{\varepsilon}U_1(A) \to {}_{-\varepsilon}V(A)$ *l'homomorphisme composé* $\tau_U \cdot r$ *Alors, si* $K_1(A) \approx K^{-1}(A)$, *si* A *est* (2,1)*-régulier et si* 6 *est inversible dans* A , *l'homomorphisme* $\tilde{\tau}_U\tilde{\tau}_V$ *est un automorphisme de* ${}_{-\varepsilon}V(A)$. *En particulier* ${}_{-\varepsilon}V(A)$ *est un facteur direct dans* ${}_{\varepsilon}U_1(A)$.

Ce théorème est une conséquence immédiate du théorème 4.8. et des considérations qui précèdent.

COROLLAIRE 5.20. *Sous les hypothèses précédentes, le groupe* ${}_{\varepsilon}W_2(A) =$ $\mathrm{Coker}\left[H_2(GL'(A) ; \mathbb{Z}) \to H_2({}_{\varepsilon}O'(A) ; \mathbb{Z})\right]$ *contient* ${}_{\varepsilon}W^{-2}(A)$ *comme facteur direct* (cf. la proposition 4.11. et le corollaire 4.12. pour le calcul de ${}_{\varepsilon}W^{-2}(A)$).

Dans le cas où A est un corps commutatif quelconque F de caractéristique différente de 2 muni d'une involution triviale, Matsumoto a démontré récemment que ${}_{\varepsilon}W_2(F) \approx {}_{-\varepsilon}W^{-2}(F)$ avec nos notations. Notons que la conjecture 5.15. implique l'isomorphisme ${}_{\varepsilon}W_2(A) \approx {}_{\varepsilon}W^{-2}(A)$ pour tout anneau hermitien K-régulier A tel que $K_1(A) \approx K^{-1}(A)$.

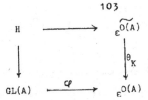

APPENDICES

1. Sur la conjecture 5.15.

Rick Sharpe a démontré cette conjecture pour $n = 0$ grâce à des techniques fines sur les cocycles de Steinberg. En particulier, si 2 est inversible dans A, on a $_{\varepsilon}W^{-2}(A) \approx {}_{\varepsilon}W_2(A)$, ce qui améliore le corollaire précédent. Notons aussi la conséquence intéressante suivante :

THEOREME 5.21. Soit A un anneau hermitien discret tel que 2 soit inversible dans A et tel que $K(A[x]) \approx K(A)$ (resp. $K_1(A[x]) \approx K_1(A)$). Alors $_{\varepsilon}W_1(A[x]) \approx {}_{\varepsilon}W_1(A)$ (resp. $_{\varepsilon}W_2(A[x]) \approx {}_{\varepsilon}W_2(A)$). En particulier, si on suppose en outre que $K_1(A[x]) \approx K_1(A)$ (resp. $K_2(A[x]) \approx K_2(A)$) , on a $_{\varepsilon}L_1(A[x]) \approx {}_{\varepsilon}L_1(A)$ (resp. $_{\varepsilon}L_2(A[x]) \approx {}_{\varepsilon}L_2(A)$).

DEMONSTRATION On écrit les suites exactes :

$$K_{n+1}(B) \longrightarrow {}_{\varepsilon}L_{n+1}(B) \longrightarrow {}_{\varepsilon}U_n(B) \longrightarrow K_n(B) \longrightarrow {}_{\varepsilon}L_n(B)$$

$$\wr\wr$$

$$_{-\varepsilon}L_n(B) \longrightarrow K_n(B) \longrightarrow {}_{-\varepsilon}V_{n-1}(B) \longrightarrow {}_{-\varepsilon}L_{n-1}(B) \longrightarrow K_{n-1}(B)$$

pour $n = 0$, pour $B = A$, $A[x]$, SA et $SA[x]$. Puis on applique le théorème 1.1. et le lemme des cinq un certain nombre de fois.

Remarque Ce théorème s'applique notamment lorsque A est un anneau noethérien régulier (resp. un corps) d'après le travail de Bass (resp. Sylvester et Quillen).

2. Sur le calcul de $\overline{W}^{-n}(A)$ lorsque A est une algèbre de Banach involutive munie de la topologie discrète

L'objet de ce deuxième appendice est de démontrer le théorème suivant annoncé sous une forme plus faible dans l'introduction :

THEOREME Soit $_{\varepsilon}W_t^{-n}(A)$ (resp $_{\varepsilon}W_d^{-n}(A)$) le groupe $_{\varepsilon}W^{-n}$ de l'algèbre de Banach A muni de sa topologie usuelle (resp. discrète). Alors l'application évidente

${}_{\varepsilon}W_d^{-n}(A) \longrightarrow {}_{\varepsilon}W_t^{-n}(A)$ induit un isomorphisme ${}_{\varepsilon}\overline{W}_d^{-n}(A) \approx {}_{\varepsilon}\overline{W}_t^{-n}(A)$

DEMONSTRATION En raison des théorèmes de périodicité il suffit de faire la démonstration pour $n = 0$ et 1. Pour $n = 0$, il n'y a rien à démontrer, le groupe ${}_{\varepsilon}W_t^0(A)$ étant défini de manière algébrique. Pour $n = 1$, l'application

$$ {}_{\varepsilon}W_d^{-1}(A) \longrightarrow {}_{\varepsilon}W_t^{-1}(A) $$

est évidemment surjective. Pour démontrer l'injectivité il suffit de démontrer le fait suivant : soit

$$ M = \begin{pmatrix} 1+a & b \\ c & 1+d \end{pmatrix} $$

une matrice $n \times n$ qui est ε-orthogonale et où a, b, c et d sont "suffisamment petits". Alors M s'écrit comme le produit de matrices ε-élémentaires et de matrices de la forme

$$ H(\alpha) = \begin{pmatrix} \alpha & 0 \\ 0 & {}^t\alpha^{-1} \end{pmatrix} $$

En effet, choisissons précisément $\alpha = 1 + a$. En multipliant M par $H(\alpha)^{-1}$, on voit qu'on peut supposer $a = 0$. Dans ce cas ${}^tb = -\varepsilon b$. En multipliant M à droite par la matrice ε-élémentaire

$$ \begin{pmatrix} 1 & -b \\ 0 & 1 \end{pmatrix} $$

on peut de même supposer $b = 0$. Dans ce cas, la matrice

$$ \begin{pmatrix} 1 & 0 \\ c & 1+d \end{pmatrix} $$

est ε-orthogonale si et seulement si $d = 0$ et ${}^tc = -\varepsilon c$. Ceci achève la démonstration du théorème.

Remarque Il est faux que l'homomorphisme ${}_{\varepsilon}W_d^{-n}(A) \longrightarrow {}_{\varepsilon}W_t^{-n}(A)$ soit un isomorphisme en général.

En effet, si $n = 2$ et si $A = \mathbb{C}$ muni de l'involution définie par la conjugaison complexe, la suite exacte des 12 permet de voir que l'homomorphisme ${}_1W_d^{-2}(\mathbb{C}) \longrightarrow {}_1W_t^{-2}(\mathbb{C})$ n'est pas surjectif.

3. L-théorie d'une extension laurentienne

Il s'agit ici de calculer ${}_\varepsilon L^{-n}(A) \langle z,z^{-1}\rangle$) où A est un anneau hermitien K-régulier. Ceci est donc un complément au chapitre I de cet article. En remplaçant A par une suspension ou une désuspension convenable, on se ramène au cas où $n = -1$.

En suivant Novikov, on posera $A_z = A\langle z,z^{-1}\rangle$, $A_{z,t} = A\langle z,t,z^{-1},t^{-1}\rangle$, etc..
On posera aussi ${}_\varepsilon\tilde{L}^{-1}(A_z) = \mathrm{Coker}\left[{}_\varepsilon L^{-1}(A) \longrightarrow {}_\varepsilon L^{-1}(A_z)\right]$ et
${}_\varepsilon\tilde{\tilde{L}}^{-1}(A_{z,t}) = \mathrm{Coker}\left[{}_\varepsilon L^{-1}(A_z) \oplus {}_\varepsilon L^{-1}(A_t) \longrightarrow {}_\varepsilon L^{-1}(A_{z,t})\right]$.

Soit $\beta_A = \beta$: ${}_\varepsilon L(A) \longrightarrow {}_\varepsilon\tilde{L}^{-1}(A_z)$ l'homomorphisme défini par le cup-produit par la classe du couple $(\mathbb{Z}',z)$ dans le groupe ${}_1L^{-1}(\mathbb{Z}'_z)$.

<u>THEOREME</u> <u>Si A est un anneau hermitien K-régulier, l'homomorphisme β est un isomorphisme. Par conséquent, on a</u> ${}_\varepsilon L^n(A_z) \approx {}_\varepsilon L^n(A) \oplus {}_\varepsilon L^{n+1}(A)$

<u>DEMONSTRATION</u> Celle-ci va s'appuyer sur une astuce dûe à Atiyah [3] qui a été déjà utilisée dans le § 3 . De manière précise, on définit un homomorphisme $\sigma_A = \sigma$: ${}_\varepsilon\tilde{L}^{-1}(A_z) \longrightarrow {}_\varepsilon L(A)$ inverse à gauche de β simplement comme l'homomorphisme obtenu par composition des homomorphismes évidents ${}_\varepsilon\tilde{L}^{-1}(A_z) \to {}_\varepsilon L^{-1}(SA) \approx {}_\varepsilon L(A)$. Il s'agit maintenant de montrer que $\beta.\ \sigma = \mathrm{Id}$.

Soit Γ le foncteur de la catégorie des anneaux (sans éléments unités) dans elle même qui associe à un anneau A l'anneau ΓA formé des séries laurentiennes $\sum_{n=-\infty}^{+\infty} a_n z^n$ telles que $\sum_{n=-\infty}^{+\infty} a_n = 0$ .Soit $\Delta A = (\Omega\Gamma)(A)$. Alors β peut s'interpréter comme un homomorphisme

$${}_\varepsilon L(A) \quad \to {}_\varepsilon L(\Delta A)$$

défini par le cup-produit par l'image de u dans ${}_1L(\Delta\,\mathbb{Z}')$. De même, on peut considérer σ comme un homomorphisme de ${}_\varepsilon L(\Delta\,A)$ dans ${}_\varepsilon L(A)$. Enfin, si on a un bimorphisme $A \times B \longrightarrow C$,on en déduit un autre bimorphisme

$$\Delta A \times B \longrightarrow \Delta C$$ et

un diagramme commutatif

$$\begin{array}{ccc} {}_{\varepsilon}L(A) \times {}_{\eta}L(B) & \longrightarrow & {}_{\varepsilon\eta}L(C) \\ \sigma \times \uparrow Id & & \uparrow \sigma \\ {}_{\varepsilon}L(\Delta A) \times {}_{\eta}L(B) & \longrightarrow & {}_{\varepsilon\eta}L(\Delta C) \end{array}$$

compatibilité de σ avec les cup-produits). Considérons maintenant la suite

$$ {}_{\varepsilon}L(A) \xrightarrow{\beta} {}_{\varepsilon}L(\Delta_1 A) \xrightarrow{\beta_{\Delta_1 A} = \beta'} {}_{\varepsilon}L(\Delta_2\Delta_1 A)$$

où $\Delta_1 A \approx \Delta A$ est un sous-anneau de $A \langle x_1, z_1, z_1^{-1} \rangle$ et où $\Delta_2\Delta_1 A$ est un sous-anneau de $A \langle x_1, x_2, z_1, z_2, z_1^{-1} z_2^{-1} \rangle$. Soit $\sigma_1 : L(\Delta_2\Delta_1 A) \xrightarrow{\sigma(\Delta_1 A)} {}_{\varepsilon}L(\Delta_2 A) \approx {}_{\varepsilon}L(\Delta_1 A)$. D'après la compatibilité de σ avec les cup-produits, on a $(\beta\,\sigma)(x) = (\sigma_1 \beta')(x)$ $x \in {}_{\varepsilon}L(\Delta_1 A)$. Pour achever la démonstration, il suffit donc de démontrer que la permutation des variables x_1, x_2 et z_1, z_2 induit la transformation identique sur le groupe ${}_{\varepsilon}L(\Delta_2\Delta_1 A)$. Or la même démonstration que celle déjà faite en K-théorie [16] montre que la permutation des variables x_1, x_2 induit la transformation -Id sur les groupes ${}_{\varepsilon}L$. Pour terminer la démonstration, il suffit donc de démontrer le lemme suivant:

<u>LEMME</u> <u>Soit</u> $\alpha(z,t)$ <u>un élément de</u> ${}_{\varepsilon}O(A_z, t)$. <u>Alors l'élément</u> $\alpha(z,t).\alpha(t,z)$ <u>est homotope à un produit d'éléments de</u> ${}_{\varepsilon}O(A_z)$ <u>et de</u> ${}_{\varepsilon}O(A_t)$, <u>ces deux groupes étant inclus dans</u> ${}_{\varepsilon}O(A_{z,t})$ <u>de manière évidente</u>.

<u>DEMONSTRATION</u> Posons $u = tz$ et $v = z$. Alors $\alpha(z,t)$ peut s'écrire $\gamma(u,v) \in {}_{\varepsilon}O(A_{u,v})$. De même, $\alpha(t,z) = \gamma(u, uv^{-1})$. De manière explicite, on peut écrire $\gamma(u,v) = \Sigma\gamma_{pq} u^p v^q$ et $\gamma(u,uv^{-1}) = \Sigma\gamma_{pq} u^p u^q v^{-q}$. D'autre part, $\alpha(z,t).\alpha(t,z)$ est homotope à la matrice

$$\begin{pmatrix} \alpha(z,t) & 0 \\ 0 & \alpha(t,z) \end{pmatrix}$$

soit à $\Sigma\alpha_q \theta^q$ où

$$\theta = \begin{pmatrix} v & 0 \\ 0 & uv^{-1} \end{pmatrix} \quad \text{et où} \quad \alpha_q = \begin{pmatrix} \Sigma\alpha_{p,q}u^p & 0 \\ 0 & \Sigma\alpha_{pq}u^p \end{pmatrix}$$

appartient à ${}_{\varepsilon}O(A_u)$ et commute à θ . D'autre part, puisque α_q se présente sous forme de matrice (en bloc) diagonale, elle commute à une homotopie dans ${}_{\varepsilon}O(Z_{u,v})$ de θ en la matrice

$$\begin{pmatrix} 1 & 0 \\ 0 & u \end{pmatrix}$$

[homotopie qui existe d'après les calculs faits en [18] à condition d'ajouter $\Sigma\alpha_{p,q}u^p$ en diagonale un certain nombre de fois]. Grâce à ces calculs, on voit ainsi que $\alpha(z,t)$. $\alpha(t,z)$ est homotope stablement à une matrice de la forme $\Sigma\delta_r u^r = \Sigma\delta_r z^r t^r$ elle-même homotope d'après le même argument à la matrice

$$\begin{pmatrix} \Sigma\,\delta_r z^r & 0 \\ 0 & \Sigma\,\delta_r t^r \end{pmatrix}$$

Le raisonnement que nous venons de faire s'applique aussi aux foncteurs ${}_{\varepsilon}\overline{W}^n(A)$, A non-nécessairement K-régulier. On obtient alors le théorème suivant :

THEOREME On a une décomposition en somme directe

$${}_{\varepsilon}\overline{W}^n(A_z) \approx {}_{\varepsilon}\overline{W}^n(A) \oplus {}_{\varepsilon}\overline{W}^{n+1}(A)$$

pour tout anneau hermitien A satisfaisant aux hypothèses de l'introduction.

Un théorème de ce genre a été démontré la première fois par Novikov [22] pour $n = 0,1$. En ce qui concerne la L-théorie style Quillen, des résultats partiels (pour W_0 , W_1 , L'_0 et L'_1) ont été trouvés par Ranicki (voir la liste des problèmes en L-théorie).

REFERENCES

[1] D.W. ANDERSON — Simplicial K-theory and generalized homology theories I,II (à paraître).

[2] M.F. ATIYAH — K-theory and reality. The quaterly J. of Math. (2) Vol.17 (1966) p.367-386.

[3] M.F. ATIYAH — Bott periodicity and the index of elliptic operators. The Quat J. of Math. Vol.19 (1968) p.113.

[4] H. BASS — Algebraic K-theory. Benjamin. New-York (1967).

[5] H. BASS — Lectures on topics in algebraic K-theory. Tata Institute. Bombay (1967).

[6] H. BASS — K_2 and symbols. Springer Lecture Notes n°108

[7] A. BOREL — C.R. Acad. Sc. Paris,274,sér.A,1972,p. 1700

[8] R. BOTT — The stable homotopy of the classical groups. Annals of Math. Series 2. t.70(1959), p.313-337.

[9] J. DIEUDONNE — La géométrie des groupes classiques. Springer. Berlin 1963.

[10] J. DIXMIER — Les C^*algèbres et leurs représentations. Gauthier-Villars (1964).

[11] S. GERSTEN — On the functor K_2.J.Algebra, 17 (1971), p.212-237.

[12] S. GERSTEN — On the spectrum of algebraic K-theory (à paraître).

[13] M. KAROUBI — Algèbres de Clifford et K-théorie. Ann. Sc. Ec. Norm. Sup. 4^esérie, t.1 (1968), p.161-270.

[14] M. KAROUBI — Séminaire Heidelberg-Saarbrücken-Strasbourg 1967-68. Exposés III, IV et V. Springer Lecture Notes n°136.

[15] M. KAROUBI — Espaces classifiants en K-théorie. Trans.Amer.Math.Soc. Vol.147, N°1 (1970), p.75-115.

[16] M. KAROUBI — La périodicité de Bott en K-théorie générale p.63-95 Ann.Scient.Ec.Norm.Sup. 4^{e}série t.4 (1971), p.

[17] M. KAROUBI et O. VILLAMAYOR — K-théorie algébrique et K-théorie topologique I. Math. Scand. 28 (1971) ,265-307

[18] M. KAROUBI et O. VILLAMAYOR — K-théorie algébrique et K-théorie topologique.II. (à paraître dans Math. Scand.)

[19] J.L. LODAY — Applications algébriques du tore dans la sphère. C.R. Acad.Sc.Paris t.272, série A (1971) p.578-581.

[20] J.L. LODAY — Structures multiplicatives en K-théorie (à paraître)

[21] J. MILNOR — Algebraic K-theory and quadratic forms. Inventiones Math.9 (1970), p.318-344.

[21'] J.MILNOR — Introduction to Algebraic K-theory

[22] S.P. NOVIKOV

[23] O.T. O'MEARA — Introduction to quadratic forms. Springer. Berlin (1963).

[24] D. QUILLEN Conférence au Congrès International des Mathématiciens . Nice (1970) .

[25] G. SEGAL Homotopy everything H-space (à paraître)

[26] J.P. SERRE Cours d'Arithmétique . Presses Universitaires de France . Collection Sup. Paris (1970) .

[27] R.G. SWAN Non-abelian homological algebra and K-theory . Appl. of cat. alg. Proc. of Symp. in Pure math XVII . Amer. Math. Soc. p. 88-123 (1970) .

[28] J.B. WAGONER Delooping classifying spaces in algebraic K-theory (à paraître dans Topology) .

[29] C.T.C. WALL Surgery on compact manifolds . Academic Press . New York (1971) .

Pour un résumé de cet article voir

[30] M. KAROUBI C.R. Acad. Sc. Paris ,273,sér.A,1971,p.1030

ALGEBRAIC L-THEORY

III. TWISTED LAURENT EXTENSIONS

by A.A.Ranicki

Introduction

The algebraic definition of the surgery obstruction groups

$$\left\{\begin{array}{l} L_n^p(\pi) \\ L_n^h(\pi) \\ L_n^s(\pi) \end{array}\right. \text{, for surgery on } \left\{\begin{array}{l} \text{open} \\ \text{compact} \\ \text{compact} \end{array}\right. \text{ manifolds, over } \left\{\begin{array}{l} - \\ \text{finite} \\ \text{simple} \end{array}\right. \text{Poincaré}$$

complexes up to $\left\{\begin{array}{l} \text{proper} \\ - \\ \text{simple} \end{array}\right.$ homotopy, depends on n(mod 4) and a group ring $Z[\pi]$, together with the involution

$$\bar{\ } : Z[\pi] \longrightarrow Z[\pi] \ ; \ \sum_{g\in\pi} n_g g \longmapsto \sum_{g\in\pi} w(g) n_g g^{-1} \qquad (n_g \in Z)$$

given by a group morphism

$$w : \pi \longrightarrow Z_2 = \{1,-1\}$$

(cf.[10]). For finitely presented groups π it is possible to obtain geometrically direct sum decompositions

$$L_n^h(\pi \times Z) = L_n^h(\pi) \oplus L_{n-1}^p(\pi) \qquad ([3])$$

$$L_n^s(\pi \times Z) = L_n^s(\pi) \oplus L_{n-1}^h(\pi) \qquad ([6])$$

The hamiltonian formalism of [4] allowed a unified approach to the three L-theories, and a purely algebraic description of these decompositions. This was done in parts I. and II. of this paper ([5]), which will be denoted I.,II. . In I. there were defined

abelian groups $\left\{\begin{array}{l} U_n(A) \\ V_n(A) \\ W_n(A) \end{array}\right.$, using quadratic forms on $\left\{\begin{array}{l} \text{f.g.projective} \\ \text{f.g.free} \\ \text{based} \end{array}\right.$

A-modules, for any associative ring A with 1 and involution and n(mod 4). It was then shown in II. that there are direct sum decompositions

$$V_n(A_z) = V_n(A) \oplus U_{n-1}(A)$$
$$\widetilde{W}_n(A_z) = W_n(A) \oplus V_{n-1}(A)$$

where $A_z = A[z,z^{-1}]$ is the Laurent extension of A, with involution by $z \mapsto z^{-1}$, and $\widetilde{W}_n(A_z)$ differs from $W_n(A_z)$ in at most one element, of order 2.

Here, we shall generalize I. by considering the intermediate L-theories $\begin{cases} U_n^T(A) \\ V_n^R(A) \end{cases}$, defined using quadratic forms on $\begin{cases} \text{f.g.projective} \\ \text{based} \end{cases}$ A-modules such that all the $\begin{cases} \text{projective classes} \\ \text{Whitehead torsions} \end{cases}$ lie in a prescribed subgroup $\begin{cases} T \subseteq \widetilde{K}_0(A) \\ R \subseteq \hat{K}_1(A) \end{cases}$. The direct sum decompositions of II. generalize to exact sequences

$$\ldots \to U_n^T(A) \xrightarrow{\bar{\varepsilon}} U_n^{\bar{\varepsilon}T}(A_\alpha) \xrightarrow{B} U_{n-1}^{(1-\alpha)^{-1}T}(A) \xrightarrow{C} U_{n-1}^T(A) \to \ldots \quad \text{(Theorem 5.1)}$$

$$\ldots \to V_n^R(A) \xrightarrow{\bar{\varepsilon}} V_n^{\bar{\varepsilon}R}(A_\alpha) \xrightarrow{B} V_{n-1}^{(1-\alpha)^{-1}R}(A) \xrightarrow{C} V_{n-1}^R(A) \to \ldots \quad \text{(Theorem 5.2)}$$

$$\ldots \to V_n^R(A) \xrightarrow{\bar{\varepsilon}} V_n^{\tilde{S}}(A_\alpha) \xrightarrow{B} U_{n-1}^T(A) \xrightarrow{C} V_{n-1}^R(A) \to \ldots \quad \text{(Theorem 5.3)}$$

where A_α is the α-twisted Laurent extension of A (assumed to be such that f.g.free A_α-modules have a well-defined rank) for some automorphism α of A, $\bar{\varepsilon}$ is the inclusion of A in A_α, and C is induced by $1-\alpha$.

For $A = Z[\pi]$ it is possible to identify

$$L_n^p(\pi) = U_n(Z[\pi]) = U_n^{\widetilde{K}_0(Z[\pi])}(A)$$
$$L_n^h(\pi) = V_n(Z[\pi]) = V_n^{\widetilde{K}_1(Z[\pi])}(A)$$
$$L_n^s(\pi) = V_n^{\{\pi\}}(Z[\pi]) \ (= W_n(Z[\pi]), \text{ up to 2-torsion }).$$

The special case $R = \{\pi\}$ of Theorem 5.2, with α given by an automorphism $\alpha : \pi \to \pi$ such that $w\alpha = w : \pi \to Z_2$, is the exact sequence

$$\ldots \to L_n^s(\pi) \to L_n^s(\pi \times_\alpha Z) \to L_{n-1}'(\pi) \to L_{n-1}^s(\pi) \to \ldots$$

of the case H = H' = K of Theorem 10 of [1], where a geometric derivation is announced, following on from some earlier work of F.T.Farrell and W.C.Hsiang. The groups $L'_n(\pi)$ are defined as $L^s_n(\pi)$, except that torsions are measured in $Wh\pi/\ker(1-\alpha: Wh\pi \longrightarrow Wh\pi)$ rather than in the Whitehead group $Wh\pi = \widetilde{K}_1(Z[\pi])/\{\pi\}$. (Thus, if $\alpha = 1$ torsions are not measured at all, and $L'_n(\pi) = L^h_n(\pi)$). It is Cappell (in [1]) who first used the intermediate L-theories.

I am grateful to Professor C.T.C.Wall for sending a preprint to [10] (which contains an earlier account of the intermediate L-theories), and for suggesting that I generalize II. to the twisted case.

I wish to thank the Århus Mathematical Institute and the Battelle Seattle Research Center for their hospitality, and also Trinity College, Cambridge, for partial support of my stays there.

This part of the paper is divided as follows:

This part can be read independently of the previous parts, taking for granted the proofs of the results quoted from I. and II. .

§1. L-theory

The purpose of this section is to introduce some notation, and to recall those definitions and results from I. which will be needed in this part.

Let A be an associative ring with 1, and with an involution, that is a function

$$\bar{\ }: A \longrightarrow A \; ; \; a \longmapsto \bar{a}$$

such that

i) $\overline{(a+b)} = \bar{a} + \bar{b}$

ii) $\overline{(ab)} = \bar{b}.\bar{a}$

iii) $\bar{\bar{a}} = a$

iv) $\bar{1} = 1$

for all $a,b \in A$.

Let $\mathcal{P}(A)$ be the category of finitely generated (f.g.) projective left A-modules. Denote the class of objects of $\mathcal{P}(A)$ by $|\mathcal{P}(A)|$ by $|\mathcal{P}(A)|$. Given $P,Q \in |\mathcal{P}(A)|$, write $\mathrm{Hom}_A(P,Q)$ for the additive group of morphisms $(f:P \to Q) \in \mathcal{P}(A)$.

There is defined a contravariant duality functor, by

$$* : \mathcal{P}(A) \longrightarrow \mathcal{P}(A) \; ; \; \begin{cases} Q \in |\mathcal{P}(A)| \longmapsto \begin{cases} Q^* = \mathrm{Hom}_A(Q,A), \text{ left A-action by} \\ A \times Q^* \longrightarrow Q^*; (a,f) \longmapsto (x \longmapsto f(x).\bar{a}) \end{cases} \\ f \in \mathrm{Hom}_A(P,Q) \longmapsto (f^*: Q^* \longrightarrow P^*; g \longmapsto (x \longmapsto gf(x))). \end{cases}$$

The natural A-module isomorphisms

$$Q \longrightarrow Q^{**} \; ; \; x \longrightarrow (f \longrightarrow \overline{f(x)}) \qquad (Q \in |\mathcal{P}(A)|)$$

allow an identification

$$** = 1 : \mathcal{P}(A) \longrightarrow \mathcal{P}(A) \; .$$

Let

$$f : A \longrightarrow A'$$

be a morphism of rings with involution (such that $f(1) = 1 \in A'$). Give A' an (A',A)-bimodule structure by

$$A' \times A' \times A \longrightarrow A' \; ; \; (a',x,a) \longmapsto a'.x.f(a) \quad .$$

The induced functor

$$f \; : \mathcal{P}(A) \longrightarrow \mathcal{P}(A') \; ; \; \begin{cases} P \longmapsto fP = A' \otimes_A P \\ g \in \mathrm{Hom}_A(P,Q) \longmapsto 1 \otimes g \in \mathrm{Hom}_{A'}(fP,fQ) \end{cases}$$

is such that

$$f(A) = A' \in |\mathcal{P}(A')|$$

and

$$*f = f* \; : \mathcal{P}(A) \longrightarrow \mathcal{P}(A')$$

(up to natural equivalence).

Given $Q \in |\mathcal{P}(A)|$, and $\theta \in \mathrm{Hom}_A(Q,Q^*)$ such that

$$\theta^* = \pm\theta \in \mathrm{Hom}_A(Q,Q^*)$$

(for one of the signs indicated), there is defined a $\pm$hermitian sesquilinear product

$$< > : Q \times Q \longrightarrow A \; ; \; (x,y) \longmapsto <x,y> \equiv \theta(x)(y)$$

with

$$\overline{<x,y>} = \pm <y,x> \in A \qquad (x,y \in Q) \quad .$$

A $\underline{\pm\text{form}}$ (over A) is a pair

$$(Q \in |\mathcal{P}(A)|, \; \varphi \in \mathrm{Hom}_A(Q,Q^*)).$$

We shall be interested only in the $\pm$hermitian products

$$\theta = \varphi \pm \varphi^* : Q \longrightarrow Q^*$$

associated with $\pm$forms (Q,φ).

An $\underline{\text{equivalence of } \pm\text{forms}}$

$$f : (Q,\varphi) \longrightarrow (Q',\varphi')$$

(over the same ground ring A) is an isomorphism $f \in \mathrm{Hom}_A(Q,Q')$

such than

$$f^*\varphi' f - \varphi = \gamma \mp \gamma^* \in \mathrm{Hom}_A(Q,Q^*)$$

for some $\mp$form (Q,γ). Then

$$f^*(\varphi' \pm \varphi'^*)f = \varphi \pm \varphi^* \in \mathrm{Hom}_A(Q,Q^*) \ ,$$

so that equivalences preserve the $\pm$hermitian products associated with $\pm$forms.

The direct sum $\oplus$ in $\mathcal{P}(A)$ generalizes to a sum operation on $\pm$forms: the <u>sum</u> of $\pm$forms is defined by

$$(Q,\varphi) \oplus (Q',\varphi') \ = \ (Q \oplus Q',\varphi \oplus \varphi').$$

A $\pm$form is <u>trivial</u> if it is equivalent to the <u>hamiltonian $\pm$form</u>

$$H_{\pm}(P) = (P \oplus P^*, \begin{pmatrix} 0 & 1 \\ 0 & 0 \end{pmatrix} : P \oplus P^* \longrightarrow P^* \oplus P = (P \oplus P^*)^*;$$
$$(x,f) \longmapsto ((x',f') \longmapsto f(x')) \ )$$

on some $P \in |\mathcal{P}(A)|$.

L-theory considers $\pm$forms up to equivalence because that is how they arise in even-dimensional surgery obstruction theory. Surgery corresponds to the addition of a trivial $\pm$form (or the inverse operation).

A <u>sublagrangian</u> L of a $\pm$form (Q,φ) is a direct summand L of Q such that

i) $j^*(\varphi \pm \varphi^*) \in \mathrm{Hom}_A(Q,L^*)$ is onto,

ii) $j^*\varphi j = \delta \mp \delta^* \in \mathrm{Hom}_A(L,L^*)$ for some $\mp$form (L,δ),

writing $j \in \mathrm{Hom}_A(L,Q)$ for the inclusion. The <u>annihilator</u> of L in (Q,φ),

$$L^{\perp} \ = \ \ker(\ j^*(\varphi \pm \varphi^*) \ : \ Q \longrightarrow L^* \ )$$

is then a direct summand of Q (by i)) containing L as a direct summand (by ii)). Restriction of $\varphi \in \mathrm{Hom}_A(Q,Q^*)$ to a direct complement to L in $L^{\perp}$ defines a $\pm$form $(L^{\perp}/L \ ,\hat{\varphi}\ )$ uniquely up to equivalence.

For example, $L \in |\mathcal{P}(A)|$ is a sublagrangian of

$$(Q,\varphi) = H_{\pm}(L) \oplus (P,\theta)$$

for any $\pm$form (P,θ), with

$$(L^{\perp}/L,\hat{\varphi}) = (P,\theta).$$

The converse holds up to equivalence, by the following version of Witt's theorem in the classical theory of quadratic forms.

Theorem 1.1 _Let_ L _be a sublagrangian of the_ $\pm$_form_ (Q,φ). _The inclusion_

$$j : L \oplus (L^{\perp}/L) \longrightarrow Q$$

extends to an equivalence of $\pm$_forms_

$$f : H_{\pm}(L) \oplus (L^{\perp}/L,\hat{\varphi}) \longrightarrow (Q,\varphi)$$

uniquely up to composition with the self-equivalences

$$\begin{pmatrix} 1 & \theta\mp\theta^* \\ 0 & 1 \end{pmatrix} \oplus 1 \ : H_{\pm}(L) \oplus (L^{\perp}/L,\hat{\varphi}) \longrightarrow H_{\pm}(L) \oplus (L^{\perp}/L,\hat{\varphi})$$

given by $\mp$_forms_ (L^*,θ).

[ ]

A sublagrangian L of a $\pm$form (Q,φ) such that

$$L^{\perp} = L$$

is a _lagrangian_ of (Q,φ).

Corollary 1.2 _A_ $\pm$ _form is trivial if and only if it admits a lagrangian._

[ ]

A _$\pm$formation (over A)_ , $(Q,\varphi;F,G)$, is a $\pm$form (Q,φ) over A, together with a lagrangian F and a sublagrangian G. An _equivalence of $\pm$formations_

$$f : (Q,\varphi;F,G) \longrightarrow (Q',\varphi';F',G')$$

is an equivalence of $\pm$forms

$$f : (Q,\varphi) \longrightarrow (Q',\varphi')$$

such that $f(F) = F'$, $f(G) = G'$.

The sum of ±formations is defined by

$(Q,\varphi;F,G)\oplus(Q',\varphi';F',G') = (Q\oplus Q',\varphi\oplus\varphi';F\oplus F',G\oplus G')$.

A stable equivalence of ±formations

$$[f] : (Q,\varphi;F,G) \longrightarrow (Q',\varphi';F',G')$$

is an equivalence of ±formations

$$f : (Q,\varphi;F,G)\oplus(H_{\pm}(P);P,P^*) \longrightarrow (Q',\varphi';F',G')\oplus(H_{\pm}(P');P',P'^*)$$

defined for some $P,P' \in |\mathcal{P}(A)|$.

A ±formation is elementary if it is equivalent to

$$(H_{\pm}(P);P,\Gamma_{(P,\theta)})$$

for some ∓form (P,θ), where

$$\Gamma_{(P,\theta)} = \{(x,(\theta\mp\theta^*)x)\in P\oplus P^*|x\in P\}$$

is the graph of (P,θ).

L-theory considers ±formations up to stable equivalence because that is how they arise in odd-dimensional surgery obstruction theory. Surgery corresponds to the addition of an elementary ±formation (or the inverse operation).

A hamiltonian complement to a lagrangian L in a ±form (Q,φ) is a lagrangian L' which is a direct complement to L on Q. It follows from Theorem 1.1 that every lagrangian has hamiltonian complements, and that the hamiltonian complements to P^* in $H_{\pm}(P)$ are just the graphs $\Gamma_{(P,\theta)}$ of ∓forms (P,θ), for any $P \in |\mathcal{P}(A)|$.

Corollary 1.3 A ±formation $(Q,\varphi;F,G)$ is elementary if and only if G is a lagrangian sharing a hamiltonian complement with F.

[ ]

Given a lagrangian L in a ±form (Q,φ), and a hamiltonian complement L', the A-module isomorphism

$$L' \longrightarrow L^* \ ; \ x \mapsto (y \mapsto (\varphi\pm\varphi^*)(x)(y))$$

will be used to identify L' with L* (in general). This is an abuse of language, as hamiltonian complements are not unique.

§2. Intermediate U-theories

Let I be an abelian monoid. Given a submonoid J of I, define an equivalence relation $\sim_J$ on I by :

$i \sim_J i'$ if there exist $j, j' \in J$ such that $i \oplus j = i' \oplus j' \in I$.

Denote the quotient monoid $I/\sim_J$ by $I/\bar{J}$, because it depends only on the stabilization of J in I, the submonoid

$$\bar{J} = \{i \in I \mid i \sim_J 0\}$$

Note that $I/\bar{J}$ is an abelian group if and only if for every $i \in I$ there exists $i' \in I$ such that $i \oplus i' \in J$.

Define the abelian group

$$K_0(A) = K(\mathcal{P}(A))$$

as usual. The reduced group

$$\tilde{K}_0(A) = \operatorname{coker}(K_0(Z) \longrightarrow K_0(A))$$

can be regarded as the quotient monoid

$$\{\text{isomorphism classes in } \mathcal{P}(A)\} \Big/ \{\text{isomorphism classes of f.g. free } A\text{-modules}\}.$$

Duality in $\mathcal{P}(A)$ defines an involution of $K_0(A)$

$$* : K_0(A) \longrightarrow K_0(A);\ [P] \longmapsto [P^*]$$

and similarly for $\tilde{K}_0(A)$.

Theorem 3.2 of I. (the case $T = \tilde{K}_0(A)$) generalizes to

Theorem 2.1 *For* n(mod 4) *let* $X_n(A)$ *be the abelian monoid of* $\left\{\begin{matrix}\text{equivalence} \\ \text{stable equivalence}\end{matrix}\right.$ *classes of* $\left\{\begin{matrix}\pm\text{forms} \\ \pm\text{formations}\end{matrix}\right.$ *over* A, *if* $n = \left\{\begin{matrix}2i \\ 2i+1\end{matrix}\right.$ *with* $\pm = (-)^i$.

The monoid morphisms

$$\partial : X_n(A) \longrightarrow X_{n-1}(A);\ \left\{\begin{matrix}(Q,\varphi) \longmapsto (H_{\mp}(Q); Q, \Gamma_{(Q,\varphi)}) \\ (Q,\varphi;F,G) \longmapsto (G^{\perp}/G, \hat{\varphi})\end{matrix}\right. \quad n = \left\{\begin{matrix}2i \\ 2i+1\end{matrix}\right.$$

are such that $\partial^2 = 0$.

The monoid morphisms

$$\sigma : X_n(A) \longrightarrow \tilde{K}_0(A) \; ; \begin{cases} (Q,\varphi) \longmapsto [Q] \\ (Q,\varphi;F,G) \longmapsto [G]-[F^*] \end{cases} \qquad n = \begin{cases} 2i \\ 2i+1 \end{cases}$$

define a chain map

$$\sigma : (X_n(A),\partial) \longrightarrow (\tilde{K}_0(A), 1 + (-)^{n+1} *)$$

of chain complexes of abelian monoids.

Given a $*$-invariant subgroup $T \subseteq \tilde{K}_0(A)$ (that is, $*(T) = T$) define a chain complex of abelian monoids

$$(X^T_n(A),\partial^T) = \sigma^{-1}(T, 1 + (-)^{n+1} *) \qquad (n(\text{mod } 4)) .$$

The subquotient monoids

$$U^T_n(A) = \ker(\partial^T : X^T_n(A) \longrightarrow X^T_{n-1}(A)) / \operatorname{im}(\partial^T : X^T_{n+1}(A) \longrightarrow X^T_n(A))$$

are abelian groups.

A 1-preserving morphism of rings with involution

$$f : A \longrightarrow A'$$

induces morphisms of abelian groups

$$f : U^T_n(A) \longrightarrow U^{T'}_n(A') ; \begin{cases} (Q,\varphi) \longmapsto (A'\otimes_A Q, 1\otimes \varphi) \\ (Q,\varphi;F,G) \longmapsto (A'\otimes_A Q, 1\otimes\varphi; A'\otimes F, A'\otimes G) \end{cases} n= \begin{cases} 2i \\ 2i+1 \end{cases}$$

for any $*$-invariant subgroups $T \subseteq \tilde{K}_0(A)$, $T' \subseteq \tilde{K}_0(A')$ such that $f(T) \subseteq T'$.

[ ]

Following I., II. the groups $U_n^{\tilde{K}_0(A)}(A)$ will be denoted by $U_n(A)$.

A $\begin{cases} \pm \text{ form } (Q,\varphi) \\ \pm \text{formation } (Q,\varphi;F,G) \end{cases}$ is non-singular if $\begin{cases} \varphi \pm \varphi^* \in \operatorname{Hom}_A(Q,Q^*) \text{ is an isomorphism} \\ G \text{ is a lagrangian of } (Q,\varphi) \end{cases}$. Then

$$U^T_n(A) = \begin{cases} \{\text{non-singular } \pm\text{forms} \in X^T_{2i}(A)\} / \overline{\{H_\pm(L) \mid [L] \in T\}} \\ \{\text{non-singular } \pm\text{formations} \in X^T_{2i+1}(A)\} / \overline{\{(H_\pm(P);P,\Gamma_{(P,\theta)}) \mid [P] \in T\}} . \end{cases}$$

Inverses are given by

$$-(Q,\varphi) = (Q,-\varphi) \in U^T_{2i}(A)$$
$$-(Q,\varphi;F,G) = (Q,-\varphi;F^*,G^*) \in U^T_{2i+1}(A) \ .$$

This is clear on noting that the <u>diagonal</u> of a $\pm$form (Q,φ),

$$\Delta_{(Q,\varphi)} = \{ (x,x) \in Q\oplus Q \mid x \in Q \} \ ,$$

is a $\begin{cases}\text{lagrangian} \\ \text{hamiltonian complement to } L\oplus L^*\end{cases}$ in $(Q\oplus Q,\varphi\oplus -\varphi)$, if (Q,φ) is $\begin{cases}\text{non-singular} \\ \text{trivial, with } L,L^* \text{ any hamiltonian complements in } (Q,\varphi)\end{cases}$.

The sum formula of Lemma 3.3 in I. generalizes to

<u>Lemma 2.2</u> $(Q,\varphi;F,G)\oplus(Q,\varphi;G,H) = (Q,\varphi;F,H) \in U^T_{2i+1}(A)$ if $[F],[G],[H] \in T$.

<u>Proof</u>: The identity

$$\begin{aligned}&(Q,\varphi;F,G)\oplus(Q,\varphi;G,H)\oplus[(Q,-\varphi;G^*,G^*)]\\
&\quad\oplus[(Q\oplus Q,\varphi\oplus-\varphi;F\oplus F^*,H\oplus G^*)\oplus(Q\oplus Q,-\varphi\oplus\varphi;\Delta_{(Q,\varphi)},H^*\oplus G)]\\
&= (Q,\varphi;F,H)\oplus[(Q\oplus Q,\varphi\oplus-\varphi;F\oplus F^*,G\oplus G^*)]\\
&\quad\oplus[(Q\oplus Q,\varphi\oplus-\varphi;G\oplus G^*,H\oplus G^*)\oplus(Q\oplus Q,-\varphi\oplus\varphi;\Delta_{(Q,\varphi)},H^*\oplus G)]\end{aligned}$$

is such that each of the $\pm$formations in square brackets is elementary.

[ ]

Let G be an abelian group with involution

$$* : G \longrightarrow G \ ; \ g \longmapsto g^* \ .$$

The Tate cohomology of this Z_2-action is given by groups

$$H^n(G) = \{ x \in G \mid x^* = (-)^n x \}/\{ y + (-)^n y^* \mid y \in G \}$$

defined for n(mod 2), which are abelian of exponent 2.

The exact sequence of Theorem 4.3 in I. (the case $T = \{0\}$, $T' = \tilde{K}_0(A)$) generalizes to

<u>Theorem 2.3</u> <u>Given *-invariant subgroups</u> $T\subseteq T'\subseteq \tilde{K}_0(A)$, <u>there is defined an exact sequence of abelian groups</u>

$$\cdots\longrightarrow H^{n+1}(T'/T) \longrightarrow U^T_n(A) \xrightarrow{1} U^{T'}_n(A) \xrightarrow{\sigma} H^n(T'/T)\longrightarrow \cdots$$

<u>where</u> $H^{n+1}(T'/T) \longrightarrow U^T_n(A);[P]\longmapsto \begin{cases}H\pm(P) \\ (H\pm(P);P,P)\end{cases}$ <u>if</u> $n = \begin{cases}2i \\ 2i+1\end{cases}$.

[ ]

§3. Intermediate V-theories

A based A-module, $\underset{\sim}{Q}$, is a f.g.free A-module Q together with a base $\underset{\sim}{q} = (q_1, \ldots, q_n)$, and n is the rank of $\underset{\sim}{q}$. The dual based A-module $\underset{\sim}{Q}^*$ is Q^* with the base $\underset{\sim}{q}^* = (q_1^*, \ldots, q_n^*)$ given by

$$q_i^*(q_j) = \begin{cases} 1 & \text{if } i=j \\ 0 & \text{otherwise} \end{cases} .$$

Identify $\underset{\sim}{Q}^{**}$ with $\underset{\sim}{Q}$.

Define the abelian groups

$$K_1(A) = GL(A)/E(A) \quad , \quad \widetilde{K}_1(A) = \operatorname{coker}(K_1(Z) \longrightarrow K_1(A))$$

as usual, regarding their elements as the torsions $\tau(f:P \to P)$ of automorphisms $(f:P \to P) \in \mathcal{P}(A)$. There is defined a duality involution

$$* : K_1(A) \longrightarrow K_1(A) \; ; \quad \tau(f:P \to P) \mapsto \tau(f^*:P^* \to P^*) .$$

In dealing with ±forms and ±formations on based A-modules it is more natural to measure torsions not in $\widetilde{K}_1(A)$, but in the slightly larger group $K'(A)$ defined below, which coincides with $\widetilde{K}_1(A)$ if A is such that f.g.free A-modules have a well-defined rank (e.g. $A = Z[\pi]$).

Let $I(A)$ be the abelian monoid of isomorphism classes of triples $(Q, \underset{\sim}{f}, \underset{\sim}{g})$, with Q a f.g.free A-module and $\underset{\sim}{f}, \underset{\sim}{g}$ two bases of Q (not necessarily of the same rank), under the sum operation

$$(Q, \underset{\sim}{f}, \underset{\sim}{g}) \oplus (Q', \underset{\sim}{f}', \underset{\sim}{g}') = (Q \oplus Q', \underset{\sim}{f} \oplus \underset{\sim}{f}', \underset{\sim}{g} \oplus \underset{\sim}{g}') .$$

Let $J(A)$ be the submonoid of $I(A)$ generated by the triples of type

i) $(Q, (f_1, \ldots, f_n), (f_1, \ldots, f_{i-1}, \delta f_i + a f_j, f_{i+1}, \ldots, f_n))$
$(\delta = \pm 1,\ a \in A,\ i \neq j)$

ii) $(Q, \underset{\sim}{f}, \underset{\sim}{g}) \oplus (Q, \underset{\sim}{g}, \underset{\sim}{h}) \oplus (Q, \underset{\sim}{h}, \underset{\sim}{f})$.

The quotient monoid

$$K'(A) = I(A)/\overline{J(A)}$$

is an abelian group in which there is a sum formula

$$(Q, \underset{\sim}{f}, \underset{\sim}{g}) \oplus (Q, \underset{\sim}{g}, \underset{\sim}{h}) = (Q, \underset{\sim}{f}, \underset{\sim}{h}) \in K'(A) .$$

It is therefore possible to regard the elements of $K'(A)$ as the torsions

$$\tau(f:\underset{\sim}{P} \longrightarrow \underset{\sim}{Q}) = (Q,\underset{\sim}{q},f(\underset{\sim}{p})) \in K'(A)$$

of isomorphisms $f \in \mathrm{Hom}_A(P,Q)$ of based A-modules $\underset{\sim}{P},\underset{\sim}{Q}$.

By the Whitehead lemma, the function

$$\widetilde{K}_1(A) \longrightarrow K'(A) ; \tau(f:P \rightarrow P) \longmapsto (P\oplus -P,\underset{\sim}{b},(f\oplus 1)\underset{\sim}{b})$$

is a group morphism, where $-P$ is any projective inverse to P, and $\underset{\sim}{b}$ is any base of $P\oplus -P$. In fact, there is a short exact sequence of abelian groups

$$0 \rightarrow \widetilde{K}_1(A) \rightarrow K'(A) \rightarrow \ker(K_0(Z) \rightarrow K_0(A)) \rightarrow 0$$

where

$$K'(A) \rightarrow \ker(K_0(Z) \longrightarrow K_0(A)) ; (Q,\underset{\sim}{f},\underset{\sim}{g}) \longmapsto [mZ] - [nZ]$$

if $\underset{\sim}{f} = (f_1,\ldots,f_m)$, $\underset{\sim}{g} = (g_1,\ldots,g_n)$. The duality involution

$$* : K'(A) \longrightarrow K'(A) \ ; \ (Q,\underset{\sim}{f},\underset{\sim}{g}) \longmapsto (Q^*,\underset{\sim}{g}^*,\underset{\sim}{f}^*)$$

agrees with that previously defined on $\widetilde{K}_1(A)$, but there is a change of sign in passing to $\ker(K_0(Z) \longrightarrow K_0(A))$.

A <u>based ±form</u> (over A), $(\underset{\sim}{Q},\varphi)$, is a $\pm$form (Q,φ) defined on a based A-module $\underset{\sim}{Q}$. The <u>torsion</u> of $(\underset{\sim}{Q},\varphi)$ is

$$\tau(\underset{\sim}{Q},\varphi) = \begin{cases} \tau(\varphi \pm \varphi^* : \underset{\sim}{Q} \rightarrow \underset{\sim}{Q}^*) & \text{if } (Q,\varphi) \text{ is non-singular} \\ 0 & \text{otherwise} \end{cases} \in \widetilde{K}_1(A) \quad .$$

Let $S \subseteq K'(A)$ be a $*$-invariant subgroup.

An <u>S-equivalence</u> of based $\pm$forms

$$f : (\underset{\sim}{Q},\varphi) \longrightarrow (\underset{\sim}{Q}',\varphi')$$

is an equivalence of $\pm$forms such that

$$\tau(f:\underset{\sim}{Q} \longrightarrow \underset{\sim}{Q}') \in S \ .$$

Now $f^*(\varphi' \pm \varphi'^*)f = (\varphi \pm \varphi^*) \in \mathrm{Hom}_A(Q,Q^*)$, so that

$$\tau(\underset{\sim}{Q},\varphi) - \tau(\underset{\sim}{Q}',\varphi') = \begin{cases} \tau + \tau^* & \text{if } (Q,\varphi) \text{ is non-singular} \\ 0 & \text{otherwise} \end{cases} \in S \subseteq K'(A)$$

where $\tau = \tau(f:\underset{\sim}{Q} \longrightarrow \underset{\sim}{Q}') \in S$.

Given a free sublagrangian L of a $\pm$form (Q,φ) such that $L^{\perp}/L$ is free, it is possible to extend a base $\underset{\sim}{L}\oplus\underset{\sim}{L^{\perp}/L}$ to one of Q uniquely up to simple changes, using any of the equivalences

$$f : H_{\pm}(L)\oplus(L^{\perp}/L,\hat{\varphi}) \longrightarrow (Q,\varphi)$$

given by Theorem 1.1. Call such a base

$$\underset{\sim}{Q} = f(\underset{\sim}{L}\oplus\underset{\sim}{L}^*\oplus\underset{\sim}{L^{\perp}/L})$$

a <u>subhamiltonian base</u> for (Q,φ), and a <u>hamiltonian base</u> if L is a lagrangian.

A <u>based $\pm$formation</u> $(Q,\varphi;\underset{\sim}{F},\underset{\sim}{G})$ is a $\pm$formation $(Q,\varphi;F,G)$ together with bases $\underset{\sim}{f},\underset{\sim}{g},\underset{\sim}{h}$ for $F,G,G^{\perp}/G$ respectively. The <u>torsion</u> of $(Q,\varphi;\underset{\sim}{F},\underset{\sim}{G})$ is

$$\tau(Q,\varphi;\underset{\sim}{F},\underset{\sim}{G}) = (Q,\underset{\sim}{f}\oplus\underset{\sim}{f}^*,\underset{\sim}{g}\oplus\underset{\sim}{g}^*\oplus\underset{\sim}{h}) \in K'(A)$$

with $\underset{\sim}{f}\oplus\underset{\sim}{f}^*$ any hamiltonian base extending $\underset{\sim}{f}$, and $\underset{\sim}{g}\oplus\underset{\sim}{g}^*\oplus\underset{\sim}{h}$ any subhamiltonian base extending $\underset{\sim}{g}\oplus\underset{\sim}{h}$. As shown above, this definition does not depend on the choice of $\underset{\sim}{f}^*,\underset{\sim}{g}^*$.

As before, let $S\subseteq K'(A)$ be a $*$-invariant subgroup.

An <u>S-equivalence</u> of based $\pm$formations

$$f:(Q,\varphi;\underset{\sim}{F},\underset{\sim}{G}) \longrightarrow (Q',\varphi';\underset{\sim}{F}',\underset{\sim}{G}')$$

is an equivalence of $\pm$formations such that

$$\tau(\underset{\sim}{F}\rightarrow\underset{\sim}{F}'),\quad \tau(\underset{\sim}{G}\rightarrow\underset{\sim}{G}'),\quad \tau(\underset{\sim}{G^{\perp}/G}\rightarrow\underset{\sim}{G'^{\perp}/G'}) \in S .$$

Then

$$\tau(Q',\varphi';\underset{\sim}{F}',\underset{\sim}{G}') - \tau(Q,\varphi;\underset{\sim}{F},\underset{\sim}{G}) = \tau-\tau^* \in S\subseteq K'(A)$$

where $\tau = (\tau(\underset{\sim}{F}\rightarrow\underset{\sim}{F}')-\tau(\underset{\sim}{G}\rightarrow\underset{\sim}{G}')-\tau(\underset{\sim}{G^{\perp}/G}\rightarrow\underset{\sim}{G'^{\perp}/G'})) \in S$.

A <u>stable S-equivalence</u> of based $\pm$formations

$$[f] : (Q,\varphi;\underset{\sim}{F},\underset{\sim}{G}) \longrightarrow (Q',\varphi';\underset{\sim}{F}',\underset{\sim}{G}')$$

is an S-equivalence of based $\pm$formations

$$f:(Q,\varphi;\underset{\sim}{F},\underset{\sim}{G})\oplus(H_{\pm}(P);\underset{\sim}{P},\underset{\sim}{P}^*) \longrightarrow (Q',\varphi';\underset{\sim}{F}',\underset{\sim}{G}')\oplus(H_{\pm}(P');\underset{\sim}{P}',\underset{\sim}{P}'^*)$$

defined for some based A-modules $\underset{\sim}{P},\underset{\sim}{P}'$.

Theorem 2.1 has a based analogue:

Theorem 3.1 For n(mod 4) and a $*$-invariant subgroup $S \subseteq K'(A)$ define the abelian monoid $Y^S_n(A)$ of $\begin{cases} \text{S-equivalence classes} \\ \text{stable S-equivalence classes} \end{cases}$ of $\begin{cases} \text{based } \pm\text{forms} \\ \text{based } \pm\text{formations} \end{cases}$ with torsion in S, with $\pm = (-)^i$ if $n = \begin{cases} 2i \\ 2i+1 \end{cases}$.

The monoid morphisms

$$\partial^S : Y^S_n(A) \longrightarrow Y^S_{n-1}(A) ; \begin{cases} (\underset{\sim}{Q},\varphi) \longmapsto (H_{\mp}(Q);\underset{\sim}{Q}, \Gamma_{(\underset{\sim}{Q},\varphi)}) \\ (Q,\varphi;\underset{\sim}{F},\underset{\sim}{G}) \longmapsto (\underset{\sim}{G^{\perp}/G},\hat{\varphi}) \end{cases}$$

are such that $(\partial^S)^2 = 0$. The subquotient monoids

$$V^S_n(A) = \ker(\partial^S : Y^S_n(A) \longrightarrow Y^S_{n-1}(A)) / \overline{\operatorname{im}(\partial^S : Y^S_{n+1}(A) \longrightarrow Y^S_n(A))}$$

are abelian groups.

A 1-preserving morphism of rings with involution

$$f : A \longrightarrow A'$$

induces morphisms of abelian groups

$$f : V^S_n(A) \longrightarrow V^{S'}_n(A') ; \begin{cases} (\underset{\sim}{Q},\varphi) \longmapsto (A'\otimes_A \underset{\sim}{Q}, 1\otimes\varphi) \\ (Q,\varphi;\underset{\sim}{F},\underset{\sim}{G}) \longmapsto (A'\otimes_A Q, 1\otimes\varphi; A'\otimes \underset{\sim}{F}, A'\otimes \underset{\sim}{G}) \end{cases}$$

for any $*$-invariant subgroups $S \subseteq K'(A)$, $S' \subseteq K'(A')$ such that $f(S) \subseteq S'$.

[ ]

Note that

$$V^S_{2i}(A) = \{\text{non-singular based } \pm\text{forms} \in Y^S_{2i}(A)\} / \overline{\{ H_{\pm}(m\underset{\sim}{A}) \mid m > 0 \}}$$

$$V^S_{2i+1}(A) = \{\text{non-singular based } \pm\text{formations} \in Y^S_{2i+1}(A)\} / \overline{\{ (H_{\pm}(P);P, \Gamma_{(\underset{\sim}{P},\theta)}) \mid \tau(\underset{\sim}{P},\theta) \in S \}}$$

Inverses are given by

$$-(\underset{\sim}{Q},\varphi) = (\underset{\sim}{Q},-\varphi) \in V^S_{2i}(A)$$

$$-(Q,\varphi;\underset{\sim}{F},\underset{\sim}{G}) = (Q,-\varphi;\underset{\sim}{F}^*,\underset{\sim}{G}^*) \in V^S_{2i+1}(A) \quad .$$

The sum formula of Lemma 2.2 has a based analogue

Lemma 3.2 $(Q,\varphi;\underset{\sim}{F},\underset{\sim}{G}) \oplus (Q,\varphi;\underset{\sim}{G},\underset{\sim}{H}) = (Q,\varphi;\underset{\sim}{F},\underset{\sim}{H}) \in V^S_{2i+1}(A)$

[ ]

For $S \subseteq \widetilde{K}_1(A)$, this allows the identification of $V^S_{2i+1}(A)$ with the stable unitary group of S-equivalences

$$H_{\pm}(m\underset{\sim}{A}) \longrightarrow H_{\pm}(m\underset{\sim}{A}) \qquad (m > 0)$$

modulo the subgroup generated by those of the type

i) $\begin{pmatrix} f & 0 \\ 0 & f^{*-1} \end{pmatrix}$ where $\tau(f:mA \longrightarrow mA) \in S$

ii) $\begin{pmatrix} 1 & \theta \mp \theta^* \\ 0 & 1 \end{pmatrix}$ for any $\mp$form (mA^*, θ)

iii) $\sigma \oplus \sigma \oplus \ldots \oplus \sigma$ with m copies of

$$\sigma = \begin{pmatrix} 0 & \pm\gamma^{-1} \\ \gamma & 0 \end{pmatrix} : A \oplus A^* \longrightarrow A \oplus A^*$$

where $\gamma : A \longrightarrow A^* ; a \longmapsto (b \longmapsto b\bar{a})$.

This is the kind of definition adopted for the odd-dimensional L-groups in [9] and [10].

The exact sequence of Theorem 2.3 has a based analogue

<u>Theorem 3.3 Given *-invariant subgroups</u> $S \subseteq S' \subseteq K'(A)$, <u>there is defined an exact sequence of abelian groups</u>

$$\ldots \longrightarrow H^{n+1}(S'/S) \rightarrow V^S_n(A) \xrightarrow{1} V^{S'}_n(A) \xrightarrow{\tau} H^n(S'/S) \longrightarrow \ldots$$

<u>with</u>

$$H^{n+1}(S'/S) \longrightarrow V^S_n(A) ; (Q, \underset{\sim}{f}, \underset{\sim}{g}) \longmapsto \begin{cases} (\underset{\sim}{Q} \oplus \underset{\approx}{Q}^*, \begin{pmatrix} 0 & 1 \\ 0 & 0 \end{pmatrix}) \\ (H_{\pm}(Q); \underset{\sim}{Q}, \underset{\approx}{Q}) \end{cases} \quad \text{if } n = \begin{cases} 2i \\ 2i+1 \end{cases}$$

<u>where</u> $\underset{\sim}{Q}$ <u>is</u> Q <u>with base</u> $\underset{\sim}{f}$, <u>and</u> $\underset{\approx}{Q}$ <u>is</u> Q <u>with base</u> $\underset{\sim}{g}$.

[ ]

This is the exact sequence of Theorem 3 of [10].

Following I.,II. denote the groups $\begin{cases} V^{\widetilde{K}_1(A)}_n(A) \\ V^{\{0\}}_n(A) \end{cases}$ by $\begin{cases} V_n(A) \\ W_n(A) \end{cases}$.

It is possible to identify

$$V^{K'(A)}_n(A) = U^{\{0\}}_n(A)$$

Thus if f.g.free A-modules have a well-defined rank (that is,

$\ker(K_0(Z) \longrightarrow K_0(A)) = \{0\}$), then

$$U_n^{\{0\}}(A) = V_n(A) \quad .$$

Otherwise, Theorem 3.3 gives exact sequences

$$0 \longrightarrow V_{2i+1}(A) \longrightarrow U_{2i+1}^{\{0\}}(A) \longrightarrow Z_2 \longrightarrow V_{2i}(A) \longrightarrow U_{2i}^{\{0\}}(A) \longrightarrow 0$$

for i(mod 2).

§4. K-theory of twisted Laurent extensions

The purpose of this section is to recall those K-theoretic definitions and results from [2], [7] and II. which will be needed in this part.

The Laurent extension of A, A_z, is the ring of polynomials $\sum_{j=-\infty}^{\infty} z^j a_j$ in an indeterminate z and its inverse z^{-1}, with coefficients $a_j \in A$ and $\{ j \in Z \mid a_j \neq 0 \}$ finite. Addition is by

$$(\sum_{j=-\infty}^{\infty} z^j a_j) + (\sum_{k=-\infty}^{\infty} z^k b_k) = (\sum_{l=-\infty}^{\infty} z^l (a_l + b_l)) \in A_z$$

and multiplication by

$$(\sum_{j=-\infty}^{\infty} z^j a_j)(\sum_{k=-\infty}^{\infty} z^k b_k) = \sum_{j=-\infty}^{\infty} \sum_{k=-\infty}^{\infty} z^{j+k} a_j b_k \in A_z \quad .$$

There is defined an involution on A_z, by

$$\overline{(\sum_{j=-\infty}^{\infty} z^j a_j)} = \sum_{j=-\infty}^{\infty} z^j \bar{a}_{-j} \in A_z \ .$$

Then A_z is an associative ring with 1 and involution, thus satisfying the conditions imposed on A in §1 above.

The functions

$$\bar{\varepsilon} : A \longrightarrow A_z \ ; \ a \longmapsto a$$

$$\varepsilon : A_z \longrightarrow A \ ; \ \sum_{j=-\infty}^{\infty} z^j a_j \longmapsto \sum_{j=-\infty}^{\infty} a_j$$

are 1-preserving morphisms of rings with involution, such that ε splits $\bar{\varepsilon}$,

$$\varepsilon\bar{\varepsilon} = 1 : A \longrightarrow A \ .$$

Given an automorphism

$$\alpha : A \longrightarrow A$$

(preserving 1 and the involution), define the <u>α-twisted Laurent extension of A</u>, A_α , to be the associative ring with the elements and additive structure of A_Z , but multiplication by

$$z^{-1}az = \alpha(a) \in A_\alpha \qquad (a \in A) \ .$$

The involution defined above for A_Z is also an involution of A_α . Thus A_α satisfies the conditions imposed on A in §1 . Note that A_Z is the special case $A_{1:A\to A}$.

The inclusion

$$\bar{\varepsilon} : A \longrightarrow A_\alpha ; a \longmapsto a$$

is a morphism of rings with involution, though not in general split.

Given $Q \in |\mathcal{P}(A)|$ define $zQ \in |\mathcal{P}(A)|$ by writing z in front of each element of Q, defining addition by

$$zx + zy = z(x+y) \in zQ \qquad (x,y \in Q)$$

and an A-action by

$$A \times zQ \longrightarrow zQ ; \ (a,zx) \longmapsto z\alpha(a)x \ .$$

Then

$$\alpha : K_0(A) \longrightarrow K_0(A) \ ; \ [Q] \longmapsto [zQ] \ .$$

Given $f \in \mathrm{Hom}_A(P,Q)$, define $zf \in \mathrm{Hom}_A(zP,zQ)$ by

$$zf : zP \longrightarrow zQ \ ; \ zx \longmapsto zf(x) \ .$$

Then

$$\alpha : K_1'(A) \longrightarrow K_1'(A) \ ; \ \tau(f:\underset{\sim}{P} \longrightarrow \underset{\sim}{Q}) \longmapsto \tau(zf:z\underset{\sim}{P} \longrightarrow z\underset{\sim}{Q}) \ .$$

Given $Q \in |\mathcal{P}(A)|$ define $Q_\alpha \in |\mathcal{P}(A_\alpha)|$ by extending the action of A on the abelian group

$$Q_\alpha = \sum_{j=-\infty}^{\infty} z^j Q$$

to one of A_α by

$$(z^k a)(z^j x) = z^{j+k}\alpha^j(a)x \in Q_\alpha \qquad (a \in A, x \in Q, \ j,k \in Z) \ .$$

Then

$$\bar{\varepsilon} : K_0(A) \longrightarrow K_0(A_\alpha) ; \ [Q] \longmapsto [Q_\alpha] \ .$$

Given $f \in \mathrm{Hom}_A(P,Q)$ define $f_\alpha \in \mathrm{Hom}_{A_\alpha}(P_\alpha,Q_\alpha)$ by

$$f_\alpha : P_\alpha \longrightarrow Q_\alpha \; ; \; \sum_{j=-\infty}^{\infty} z^j x_j \longmapsto \sum_{j=-\infty}^{\infty} z^j f(x_j) \; .$$

Then

$$\bar{\varepsilon} : K'(A) \longrightarrow K'(A_\alpha) ; \tau(f:\underset{\sim}{P} \to \underset{\sim}{Q}) \mapsto \tau(f_\alpha : \underset{\sim}{P}_\alpha \to \underset{\sim}{Q}_\alpha) .$$

A <u>modular A-base</u> of an A_α-module Q is an A-submodule Q_o of Q such that every $x \in Q$ has a unique expression as

$$x = \sum_{j=-\infty}^{\infty} z^j x_j \qquad (x_j \in Q_o) \quad .$$

If $Q \in |\mathcal{P}(A_\alpha)|$ has a modular A-base Q_o , then $Q_o \in |\mathcal{P}(A)|$, and it is possible to identify

$$Q = (Q_o)_\alpha \; .$$

Given $Q_o \in |\mathcal{P}(A)|$ define complementary A-submodules

$$Q_o^+ = \sum_{j=0}^{\infty} z^j Q_o \qquad Q_o^- = \sum_{j=-\infty}^{-1} z^j Q_o$$

in $Q = (Q_o)_\alpha$. If F,G are modular A-bases of Q then

$$z^N F^+ \subseteq G^+$$

for sufficiently large integers $N \geq 0$ .For such N define the A-module

$$B_N(F,G) = z^N F^- \cap G^+ ,$$

and observe that there is a sum formula

$$B_{M+N}(F,H) = z^M B_N(F,G) \oplus B_M(G,H) \; .$$

This shows that each $B_N(F,G)$ is a f.g.projective A-module, with

$$B_N(F,G) \oplus z^{-N_1} B_{N_1}(G,F) = \sum_{j=-N_1}^{N-1} z^j F \quad ,$$

and also that

$$B : K_1(A_\alpha) \longrightarrow K_0(A) ; \tau(f:G_\alpha \to G_\alpha) \mapsto [B_N(F,G)] - [\sum_{j=0}^{N-1} z^j F]$$

is a well-defined morphism, where $F = f(G)$.

Recall from §8 of [7] the definition of the group $K(A,\alpha)$. Consider pairs

$$(P \in |\mathcal{P}(A)|, \; f \in \mathrm{Hom}_A(P,zP) \text{ isomorphism})$$

under the equivalence relation

$(P,f) \sim (P',f')$ if there exists an isomorphism $g \in \mathrm{Hom}_A(P,P')$ such that $\tau(g^{-1}f'^{-1}(zg)f : P \to P) = 0 \in K_1(A)$.

Then $K(A,\alpha)$ is the abelian group with one generator $[P,f]$ for each equivalence class of pairs (P,f), under the relations

$$[P,f]\oplus[P',f'] = [P\oplus P',f\oplus f'] \ .$$

Given a based A-module $\underset{\sim}{Q}$, define $[Q,\xi] \in K(A,\alpha)$ by

$$\xi:Q \longrightarrow zQ;\ \sum_{i=1}^{n} a_i q_i \longmapsto \sum_{i=1}^{n} z\alpha(a_i)q_i \qquad (a_i \in A)$$

with $\underset{\sim}{q} = (q_1,\ldots,q_n)$ the given base of Q.

The exact sequence of Theorem 9.2 of [7] can be extended to the right by one term, to give

<u>Lemma 4.1</u> <u>The sequence of abelian groups</u>

$$K_1(A) \xrightarrow{1-\alpha} K_1(A) \xrightarrow{j} K(A,\alpha) \xrightarrow{p} K_0(A) \xrightarrow{1-\alpha} K_0(A) \xrightarrow{\bar{\varepsilon}} K_0(A_\alpha)$$

<u>is exact, where</u>

$$j:K_1(A) \longrightarrow K(A,\alpha);\ \tau(f:\underset{\sim}{G} \rightarrow \underset{\sim}{G}) \longmapsto [G,\xi f] - [G,\xi]$$

$$p:K(A,\alpha) \longrightarrow K_0(A);\ [P,f] \longmapsto [P]$$

<u>Proof</u>: Use the A_α-module isomorphisms

$$Q_\alpha \longrightarrow (zQ)_\alpha \ ; \quad \sum_{j=-\infty}^{\infty} z^j x_j \longmapsto \sum_{j=-\infty}^{\infty} z^{j-1}(zx_j)$$

to identify

$$Q_\alpha = (zQ)_\alpha \in |\mathcal{P}(A_\alpha)| \qquad (Q \in |\mathcal{P}(A)|) \ .$$

It follows that the composite

$$K_0(A) \xrightarrow{1-\alpha} K_0(A) \xrightarrow{\bar{\varepsilon}} K_0(A_\alpha)$$

is zero.

Given $[G] - [F] \in \ker(\bar{\varepsilon}:K_0(A) \rightarrow K_0(A_\alpha))$, stabilize F and G until there is defined an isomorphism

$$(F_\alpha \rightarrow G_\alpha) \in \mathcal{P}(A_\alpha) \ .$$

The identity

$$B_{N+1}(F,G) = z^N F \oplus B_N(F,G) = zB_N(F,G) \oplus G$$

shows that

$$[G] - [F] = (1-\alpha)([B_N(F,G)] - [\sum_{j=0}^{N-1} z^j F\ ]) \in \operatorname{im}(1-\alpha:K_0(A) \rightarrow K_0(A)).$$

[ ]

Defining a duality involution

$$*: K(A,\alpha) \longrightarrow K(A,\alpha) ; [P,f] \longmapsto -[P^*, f^{*-1}] ,$$

note that

$$j* = *j : K_1(A) \longrightarrow K(A,\alpha)$$

$$p* = -*p : K(A,\alpha) \longrightarrow K_0(A) \quad .$$

As in §12 of [7], it is possible to combine the results of [2] and [7] to obtain

<u>Theorem 4.2 There is a natural direct sum decomposition</u>

$$K_1(A_\alpha) = K(A,\alpha) \oplus Nil_+(A,\alpha) \oplus Nil_-(A,\alpha)$$

<u>where</u> $Nil_+(A,\alpha) = \{\tau(1+z^{+1}\nu : P_\alpha \to P_\alpha) \mid \nu \in Hom_Z(P,P) \text{ nilpotent}, z\nu \in Hom_A(P,zP)\}$. <u>The inclusion</u>

$$i : K(A,\alpha) \longrightarrow K_1(A_\alpha) ; [P,f] \longrightarrow \tau(f_\alpha : P_\alpha \to (zP)_\alpha = P_\alpha)$$

<u>is split by</u>

$$q : K_1(A_\alpha) \longrightarrow K(A,\alpha) ;$$

$$\tau(f : \underline{G}_\alpha \to \underline{G}_\alpha) \longmapsto [B_{N+1}(F,G), t] - [\sum_{k=0}^{N} z^k F, \xi]$$

<u>where</u> $\underset{\sim}{F} = f(\underset{\sim}{G})$ <u>and</u>

$$t = 1 \oplus \xi^{N+1} f : B_{N+1}(F,G) = zB_N(F,G) \oplus G \longrightarrow zB_N(F,G) \oplus z^{N+1}F = zB_{N+1}(F,G).$$

<u>The duality involution</u>

$$* : K_1(A_\alpha) \longrightarrow K_1(A_\alpha)$$

<u>is such that</u>

$$i* = *i : K(A,\alpha) \longrightarrow K_1(A_\alpha)$$

$$q* = *q : K_1(A_\alpha) \longrightarrow K(A,\alpha) \quad ,$$

<u>and interchanges</u> $Nil_+(A,\alpha)$, $Nil_-(A,\alpha)$.

<u>In the untwisted case</u>, $\alpha = 1 : A \longrightarrow A$, <u>there are defined morphisms</u>

$$\bar{p} : K_0(A) \longrightarrow K(A,1) ; [P] \longmapsto [P,z]$$

$$\bar{j} : K(A,1) \longrightarrow K_1(A) ; [P,f] \longmapsto \tau(z^{-1}f : P \to P)$$

<u>such that</u>

$$K_1(A) \underset{\bar{j}}{\overset{j}{\rightleftarrows}} K(A,1) \underset{\bar{p}}{\overset{p}{\rightleftarrows}} K_0(A)$$

<u>is a direct sum system.</u>

[ ]

Note that

$$ij = \bar{\varepsilon} : K_1(A) \longrightarrow K_1(A_\alpha)$$

$$pq = B : K_1(A_\alpha) \longrightarrow K_0(A)$$

with j,p as in Lemma 4.1, and that in the untwisted case

$$i\bar{p} = B : K_0(A) \longrightarrow K_1(A_z) ; \ [P] \longmapsto \tau(z : P_z \to P_z)$$

$$\bar{j}q = (\varepsilon 000) : K_1(A_z) = \bar{\varepsilon}K_1(A) \oplus \bar{B}K_0(A) \oplus Nil_+(A,1) \oplus Nil_-(A,1) \longrightarrow K_1(A)$$

in the untwisted case.

The relation

$$B^* = -{}^*B : K_1(A_\alpha) \longrightarrow K_0(A)$$

can be obtained directly, from the A-module isomorphism

$$B_N(F^*,G^*) \longrightarrow B_N(F,G)^* ; \ f \mapsto (x \mapsto [f(x)]_0) \quad ,$$

where $[a]_0 = a_0 \in A$ if $a = \sum_{j=-\infty}^{\infty} z^j a_j \in A_\alpha$.

Giving Z the identity involution, define a morphism of rings with involution

$$Z_z \longrightarrow A_\alpha ; \sum_{j=-\infty}^{\infty} z^j n_j \longmapsto \sum_{j=-\infty}^{\infty} z^j n_j . 1 \quad ,$$

and define reduced groups

$$\tilde{K}(A,\alpha) = \mathrm{coker}(K(Z,1) \longrightarrow K(A,\alpha))$$

$$\tilde{K}_1(A_\alpha) = \mathrm{coker}(K_1(Z_z) \longrightarrow K_1(A_\alpha)) \quad .$$

<u>From now on we shall assume that A_α is such that f.g.free A_α-modules have a well-defined rank</u> .

It follows that A also has this property. Lemma 4.1 gives an exact sequence

$$\tilde{K}_1(A) \xrightarrow{1-\alpha} \tilde{K}_1(A) \xrightarrow{j} \tilde{K}(A,\alpha) \xrightarrow{p} \tilde{K}_0(A) \xrightarrow{1-\alpha} \tilde{K}_0(A) \xrightarrow{\bar{\varepsilon}} \tilde{K}_0(A_\alpha)$$

in the reduced groups. Theorem 4.2 gives a direct sum decomposition

$$\tilde{\tilde{K}}_1(A_\alpha) = \tilde{K}(A,\alpha)\oplus Nil_+(A,\alpha)\oplus Nil_-(A,\alpha) \quad .$$

<u>Convention: Given a $*$-invariant subgroup</u> $S\subseteq \tilde{K}(A,\alpha)$ <u>let</u>

$$R = j^{-1}(S)\subseteq \tilde{K}_1(A) \quad , \quad T = p(S)\subseteq \tilde{K}_0(A) \quad .$$

Then $R\subseteq \tilde{K}_1(A)$, $T\subseteq \tilde{K}_0(A)$ are $*$-invariant subgroups.

<u>Theorem 4.3 Given $*$-invariant subgroups</u> $S\subseteq S'\subseteq \tilde{K}(A,\alpha)$, <u>there is defined an exact sequence of Tate cohomology groups</u>

$$\ldots \longrightarrow H^n(R'/R)\xrightarrow{\bar{\varepsilon}} H^n(S'/S)\xrightarrow{B} H^{n-1}(T'/T)\xrightarrow{C} H^{n-1}(R'/R)\rightarrow \ldots$$

<u>with</u> $\bar{\varepsilon}$, B <u>induced by</u> j, p <u>respectively and</u> C <u>the connecting morphism</u>,

$$C : H^n(T'/T)\longrightarrow H^n(R'/R) \ ; \ [x]\longmapsto [j^{-1}(y+(-)^n y^*)]$$

<u>for any</u> $y \in S'/S$ <u>such that</u> $p(y) = x \in T'/T$, <u>associated with the short exact sequence</u>

$$0 \longrightarrow R'/R \xrightarrow{j} S'/S \xrightarrow{p} T'/T \longrightarrow 0 \quad .$$

<u>In the untwisted case</u> $\alpha = 1:A\longrightarrow A$, <u>with</u>

$$S = j(R)\oplus \bar{p}(T) \quad , \quad S' = j(R')\oplus \bar{p}(T')\subseteq \tilde{K}(A,1) = j\tilde{K}_1(A)\oplus \bar{p}\tilde{K}_0(A),$$

<u>there is defined a direct sum system</u>

$$H^n(R'/R)\underset{\varepsilon}{\overset{\bar{\varepsilon}}{\rightleftarrows}} H^n(S'/S)\underset{\bar{B}}{\overset{B}{\rightleftarrows}} H^{n-1}(T'/T) \quad .$$

[ ]

§5. <u>L-theory of twisted Laurent extensions</u>

<u>Theorem 5.1 Given a $*$-invariant subgroup</u> $T\subseteq \tilde{K}_0(A)$, <u>there is defined an exact sequence of abelian groups</u>

$$\ldots \longrightarrow U_n^T(A)\xrightarrow{\bar{\varepsilon}} U_n^{\bar{\varepsilon}T}(A_\alpha)\xrightarrow{B} U_{n-1}^{(1-\alpha)^{-1}T}(A)\xrightarrow{C} U_{n-1}^T(A)\longrightarrow \ldots$$

<u>in a natural way.</u>

<u>The exact sequences associated with $*$-invariant subgroups</u> $T\subseteq T'\subseteq \tilde{K}_0(A)$ <u>combine with the exact sequence of Theorem 2.3 and the Tate cohomology of the short exact sequence</u>

$$0 \rightarrow (1-\alpha)^{-1}T'/(1-\alpha)^{-1}T \xrightarrow{1-\alpha} T'/T \xrightarrow{\bar{\varepsilon}} \bar{\varepsilon}T'/\bar{\varepsilon}T \rightarrow 0 \quad ,$$

<u>to define a commutative diagram</u>

$$\begin{array}{ccccccccc}
 & \vdots & & \vdots & & \vdots & & \vdots & \\
 & \downarrow & & \downarrow & & \downarrow & & \downarrow & \\
\cdots \to & H^{n+1}(T'/T) & \to & H^{n+1}(\bar{\varepsilon}T'/\bar{\varepsilon}T) & \to & H^{n}((1-\alpha)^{-1}T'/(1-\alpha)^{-1}T) & \to & H^{n}(T'/T) & \to \cdots \\
 & \downarrow & & \downarrow & & \downarrow & & \downarrow & \\
\cdots \to & U_n^{T}(A) & \xrightarrow{\bar{\varepsilon}} & U_n^{\bar{\varepsilon}T}(A_\alpha) & \xrightarrow{B} & U_{n-1}^{(1-\alpha)^{-1}T}(A) & \xrightarrow{C} & U_{n-1}^{T}(A) & \to \cdots \\
 & \downarrow & & \downarrow & & \downarrow & & \downarrow & \\
\cdots \to & U_n^{T'}(A) & \xrightarrow{\bar{\varepsilon}} & U_n^{\bar{\varepsilon}T'}(A_\alpha) & \xrightarrow{B} & U_{n-1}^{(1-\alpha)^{-1}T'}(A) & \xrightarrow{C} & U_{n-1}^{T'}(A) & \to \cdots \\
 & \downarrow & & \downarrow & & \downarrow & & \downarrow & \\
\cdots \to & H^{n}(T'/T) & \to & H^{n}(\bar{\varepsilon}T'/\bar{\varepsilon}T) & \to & H^{n-1}((1-\alpha)^{-1}T'/(1-\alpha)^{-1}T) & \to & H^{n-1}(T'/T) & \to \cdots \\
 & \downarrow & & \downarrow & & \downarrow & & \downarrow & \\
 & \vdots & & \vdots & & \vdots & & \vdots &
\end{array}$$

<u>with exact rows and columns</u>.

<u>If</u> $\bar{\varepsilon}T = \bar{\varepsilon}T' \subseteq \widetilde{K}_0(A_\alpha)$, <u>the sequences interlock in a commutative exact braid</u>

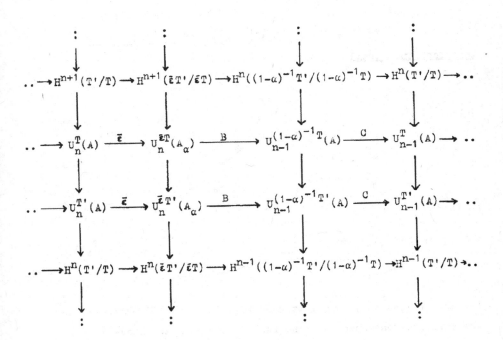

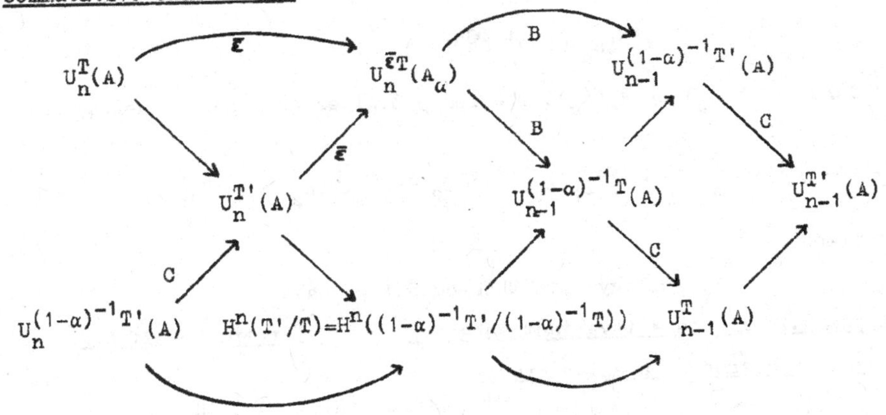

If $(1-\alpha)^{-1}T = (1-\alpha)^{-1}T' \subsetneq \tilde{K}_0(A)$, *the sequences interlock in a braid*

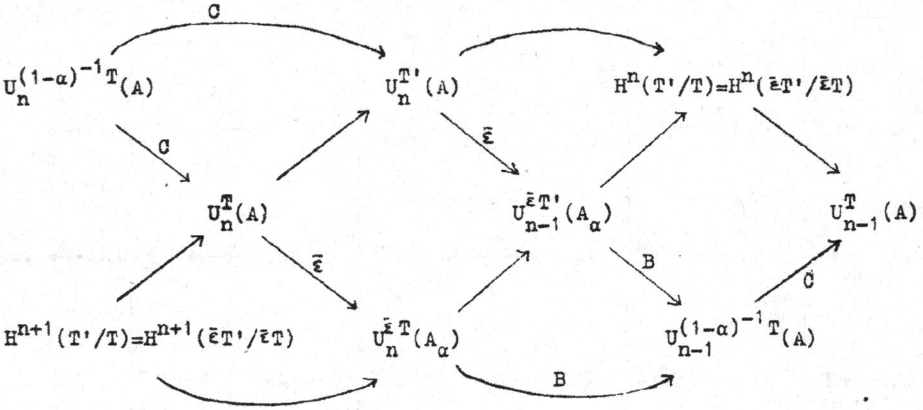

[ ]

(As Wall points out, in a letter of 19th January 1973, these braids are a formal consequence of the larger diagram drawn above.)

Let S_0 be the infinite cyclic subgroup of $\tilde{K}_1(A_\alpha)$ generated by $\tau(\xi: A_\alpha \longrightarrow A_\alpha)$.

Given a $*$-invariant subgroup $R \subseteq \tilde{K}_1(A)$ let

$$\tilde{V}_n^{\bar{\varepsilon}R}(A_\alpha) = V_n^{\bar{\varepsilon}R \oplus S_0}(A_\alpha) \quad ,$$

and denote $V_n^{S_0}(A_\alpha)$ by $\tilde{W}_n(A_\alpha)$. Theorem 3.3 gives an exact sequence

$$0 \longrightarrow V_{2i+1}^{\bar{\varepsilon}R}(A_\alpha) \longrightarrow \tilde{V}_{2i+1}^{\bar{\varepsilon}R}(A_\alpha) \longrightarrow Z_2 \longrightarrow V_{2i}^{\bar{\varepsilon}R}(A_\alpha) \longrightarrow \tilde{V}_{2i}^{\bar{\varepsilon}R}(A_\alpha) \longrightarrow 0$$

for $i \pmod 2$.

By analogy with Theorem 5.1 we have:

Theorem 5.2 Given á $*$-invariant subgroup $R \subseteq \hat{K}_1(A)$ *there is defined an exact sequence of abelian groups*

$$\ldots \longrightarrow V_n^R(A) \xrightarrow{\bar{\varepsilon}} \tilde{V}_n^{\bar{\varepsilon}R}(A_\alpha) \xrightarrow{B} V_{n-1}^{(1-\alpha)^{-1}R}(A) \xrightarrow{C} V_{n-1}^R(A) \longrightarrow \ldots$$

[ ]

with similar naturality and exactness properties.

Given a $*$-invariant subgroup $S \subseteq \widetilde{K}(A,\alpha)$, let

$$\widetilde{V}_n^S(A_\alpha) = V_n^{\widetilde{S}}(A_\alpha)$$

where

$$\widetilde{S} = q^{-1}(S) \subseteq \widetilde{K}_1(A_\alpha) \quad ,$$

with the projection

$$q : \widetilde{K}_1(A_\alpha) \longrightarrow \widetilde{K}(A,\alpha)$$

defined as in Theorem 4.2.

The exact sequence of Theorem 3.3 for $\widetilde{S} \subseteq \widetilde{S}' \subseteq \widetilde{K}_1(A_\alpha)$ can be written as

$$\ldots \longrightarrow H^{n+1}(S'/S) \longrightarrow \widetilde{V}_n^S(A_\alpha) \longrightarrow \widetilde{V}_n^{S'}(A_\alpha) \longrightarrow H^n(S'/S) \longrightarrow \ldots \quad ,$$

using the isomorphism

$$q : \widetilde{S}'/\widetilde{S} \longrightarrow S'/S$$

to identify

$$H^n(\widetilde{S}'/\widetilde{S}) = H^n(S'/S) \quad .$$

In particular,

$$\widetilde{V}_n^S(A_\alpha) = \begin{cases} V_n(A_\alpha) \\ \widetilde{V}_n^{\bar{\varepsilon}R}(A_\alpha) \end{cases} \quad \text{if} \quad S = \begin{cases} \widetilde{K}(A,\alpha) \\ j(R) \quad (R \subseteq \widetilde{K}_1(A)). \end{cases}$$

<u>Theorem 5.3 Given a $*$-invariant subgroup</u> $S \subseteq \widetilde{K}(A,\alpha)$ <u>there is defined an exact sequence of abelian groups</u>

$$\ldots \longrightarrow V_n^R(A) \xrightarrow{\bar{\varepsilon}} \widetilde{V}_n^S(A_\alpha) \xrightarrow{B} U_{n-1}^T(A) \xrightarrow{C} V_{n-1}^R(A) \longrightarrow \ldots$$

<u>in a natural way, with</u> $R = j^{-1}(S) \subseteq \widetilde{K}_1(A)$, $T = p(S) \subseteq \widetilde{K}_0(A)$.

<u>The exact sequences associated with $*$-invariant subgroups</u> $S \subseteq S' \subseteq \widetilde{K}(A,\alpha)$ <u>and the exact sequences of Theorems 2.3,3.3,4.3 combine, to give a commutative diagram</u>

$$
\begin{array}{ccccccccccc}
 & & \vdots & & \vdots & & \vdots & & \vdots & & \\
 & & \downarrow & & \downarrow & & \downarrow & & \downarrow & & \\
\cdots & \longrightarrow & H^{n+1}(R'/R) & \xrightarrow{\bar{\varepsilon}} & H^{n+1}(S'/S) & \xrightarrow{B} & H^{n}(T'/T) & \xrightarrow{C} & H^{n}(R'/R) & \longrightarrow & \cdots \\
 & & \downarrow & & \downarrow & & \downarrow & & \downarrow & & \\
\cdots & \longrightarrow & V_n^R(A) & \xrightarrow{\bar{\varepsilon}} & \tilde{V}_n^S(A_\alpha) & \xrightarrow{B} & U_{n-1}^T(A) & \xrightarrow{C} & V_{n-1}^R(A) & \longrightarrow & \cdots \\
 & & \downarrow & & \downarrow & & \downarrow & & \downarrow & & \\
\cdots & \longrightarrow & V_n^{R'}(A) & \xrightarrow{\bar{\varepsilon}} & \tilde{V}_n^{S'}(A_\alpha) & \xrightarrow{B} & U_{n-1}^{T'}(A) & \xrightarrow{C} & V_{n-1}^{R'}(A) & \longrightarrow & \cdots \\
 & & \downarrow & & \downarrow & & \downarrow & & \downarrow & & \\
\cdots & \longrightarrow & H^{n}(R'/R) & \xrightarrow{\bar{\varepsilon}} & H^{n}(S'/S) & \xrightarrow{B} & H^{n-1}(T'/T) & \xrightarrow{C} & H^{n-1}(R'/R) & \longrightarrow & \cdots \\
 & & \downarrow & & \downarrow & & \downarrow & & \downarrow & & \\
 & & \vdots & & \vdots & & \vdots & & \vdots & &
\end{array}
$$

with exact rows and columns.

If $R = R' \subseteq \tilde{K}_1(A)$, *the sequences interlock in a commutative exact braid*

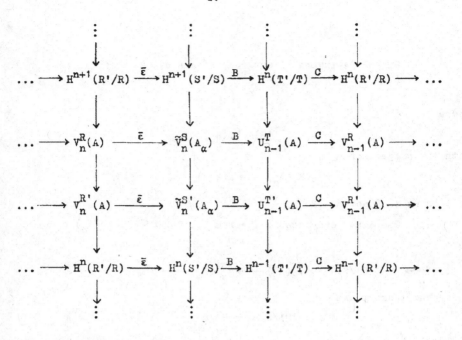

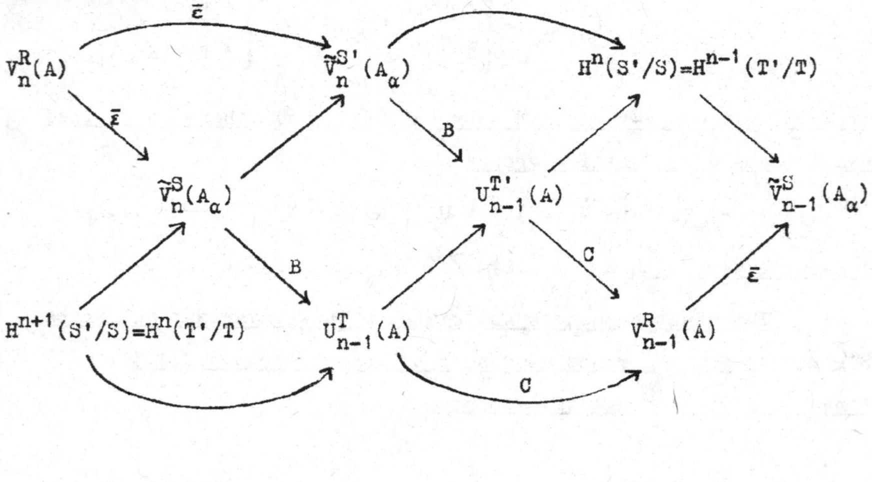

If $T = T' \subseteq \tilde{K}_0(A)$ *, the sequences interlock in a commutative exact braid*

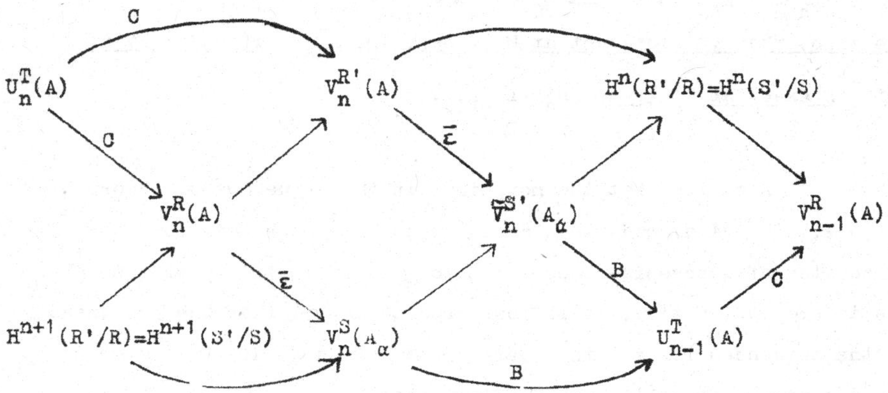

[ ]

In proving the exactness of the sequences of Theorems 5.1,5.2,5.3 (in §6, below) we shall make much use of the following version of Theorem 1 of [8].

Lemma 5.4 Suppose given a commutative diagram of abelian groups and morphisms

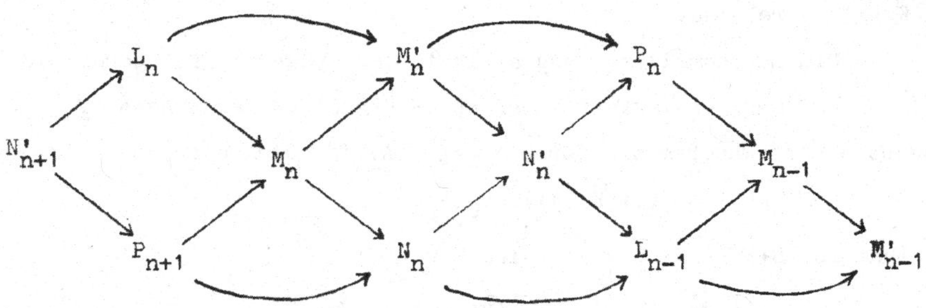

such that the sequences

$$P_{n+1} \longrightarrow M_n \longrightarrow M'_n \longrightarrow P_n \longrightarrow M_{n-1} \longrightarrow M'_{n-1}$$

$$N'_{n+1} \longrightarrow P_{n+1} \longrightarrow N_n \longrightarrow N'_n \longrightarrow P_n$$

are exact.

<u>If the composites of successive morphisms in the sequences</u>

$$L_n \longrightarrow M_n \longrightarrow N_n \longrightarrow L_{n-1} \longrightarrow M_{n-1} \qquad (*)$$

$$L_n \longrightarrow M'_n \longrightarrow N'_n \longrightarrow L_{n-1} \longrightarrow M'_{n-1} \qquad (**)$$

<u>are zero, then (*) is exact at</u> M_n (<u>resp</u>. N_n, L_{n-1}) <u>if and only if</u> <u>(**) is exact at</u> M'_n (<u>resp</u>. N'_n, L_{n-1}) .

[ ]

Assuming that the morphisms in the sequences of Theorems 5.1,5.2,5.3 have already been defined, and are such that the composites of successive ones are zero, and that all the braids are indeed commutative, it follows from Lemma 5.4 that the exactness of the sequences for all the coefficient groups T, R, S(but keeping A and α fixed) is related as for (*), (**).

To see this, note first that for any $*$-invariant subgroup $T \subseteq \widetilde{K}_0(A)$ the exactness of the sequences of Theorem 5.1 for T,and $T \cap (1-\alpha)\widetilde{K}_0(A)$ is related (since

$$(1-\alpha)^{-1}T = (1-\alpha)^{-1}(T \cap (1-\alpha)\widetilde{K}_0(A)) \subseteq \widetilde{K}_0(A) \quad) ,$$

as is that for $T \cap (1-\alpha)\widetilde{K}_0(A)$, $\{0\}$ (since

$$\varepsilon(T \cap (1-\alpha)\widetilde{K}_0(A)) = \{0\} \subseteq \widetilde{K}_0(A_\alpha) \ ).$$

Hence the exactness of the sequences for any two $*$-invariant subgroups $T,T' \subseteq \widetilde{K}_0(A)$ is related.

Similar considerations apply to the sequence of Theorem 5.2.

For any $*$-invariant subgroup $S \subseteq \widetilde{K}(A,\alpha)$ the exactness of the sequences of Theorem 5.3 for S , $S+j\widetilde{K}_1(A)$ is related (since

$$p(S) = p(S+j\widetilde{K}_1(A)) \subseteq \widetilde{K}_0(A) \quad) ,$$

as is that for $S+j\widetilde{K}_1(A)$, $\widetilde{K}(A,\alpha)$ (since

$$j^{-1}(S+j\widetilde{K}_1(A)) = j^{-1}(\widetilde{K}(A,\alpha)) = \widetilde{K}_1(A) \quad).$$

Hence the exactness of the sequences for any two $*$-invariant subgroups $S,S' \subseteq \widetilde{K}(A,\alpha)$ is related.

The sequence of Theorem 5.1 for $T = \{0\} \subseteq \tilde{K}_0(A)$

$$\cdots \longrightarrow V_n(A) \xrightarrow{\bar{\varepsilon}} V_n(A_\alpha) \xrightarrow{B} U_{n-1}^{\tilde{K}_0(A)^\alpha}(A) \xrightarrow{C} V_{n-1}(A) \longrightarrow \cdots$$

coincides with that of Theorem 5.3 for $S = \tilde{K}(A,\alpha)$ (or will be seen to do so, once both are defined).

The sequence of Theorem 5.2 for $R = \tilde{K}_1(A)$

$$\cdots \longrightarrow V_n(A) \xrightarrow{\bar{\varepsilon}} \tilde{V}_n^{\bar{\varepsilon}\tilde{K}_1(A)}(A_\alpha) \xrightarrow{B} V_{n-1}(A) \xrightarrow{C} V_{n-1}(A) \longrightarrow \cdots$$

coincides with that for Theorem 5.3 for $S = j\tilde{K}_1(A) \subseteq \tilde{K}(A,\alpha)$.

Hence the exactness of all the sequences is related.

In proving Theorems 5.1,5.2,5.3 (in §6, below) it will be left to the reader to verify that the definitions of the morphisms B, C are sufficiently natural for the commutativity of the diagrams drawn above (implicitly so for 5.2).

§6. Proof of theorems in §5.

Given a $*$-invariant subgroup $T \subseteq \tilde{K}_0(A)$, define

$$B: U_{2i+1}^{\bar{\varepsilon}T}(A_\alpha) \longrightarrow U_{2i}^{(1-\alpha)^{-1}T}(A); (Q,\varphi;F,G) \longmapsto (P,\theta)$$

where

$$(P,\theta) = (B_N(F_0 \oplus F_0^*, G_0 \oplus G_0^*), [\varphi]_0) \oplus H_\pm\Big(\sum_{j=0}^{N-1} z^j(-F_0)\Big)$$

for any modular A-bases F_0, G_0 of F,G such that

$$[G_0] - [F_0^*] \in T \subseteq \tilde{K}_0(A) \quad ,$$

with $-F_0$ any projective inverse for F_0 , and F_0^* , G_0^* the dual modular A-bases to F_0, G_0 in any hamiltonian complements F^*, G^* to F,G in (Q,φ) , with

$$[\varphi]_0 : Q \longrightarrow \mathrm{Hom}_A(Q,A) \ ; \ x \longmapsto (y \longmapsto [\varphi(x)(y)]_0) \ ,$$

writing $[a]_0$ for $a_0 \in A$ if $a = \sum_{j=-\infty}^{\infty} z^j a_j \in A_\alpha$.

The identity

$$(B_N(F_0 \oplus F_0^*, G_0 \oplus G_0^*), [\varphi]_0) \oplus (z^{-N_1} B_{N_1}(G_0 \oplus G_0^*, F_0 \oplus F_0^*), [\varphi]_0)$$
$$= H\pm\left(\sum_{j=-N_1}^{N-1} z^j F_0 \right) \quad \text{(up to equivalence of } \pm\text{forms over A)}$$

shows that (P,θ) is a non-singular $\pm$form. The identity

$$B_{N+1}(F_0 \oplus F_0^*, G_0 \oplus G_0^*) = z^N(F_0 \oplus F_0^*) \oplus B_N(F_0 \oplus F_0^*, G_0 \oplus G_0^*)$$
$$= (G_0 \oplus G_0^*) \oplus z B_N(F_0 \oplus F_0^*, G_0 \oplus G_0^*)$$

shows that

$$(1-\alpha)[P] = [G_0 \oplus G_0^*] - [z^N(F_0 \oplus F_0^*)] + (1-\alpha)\left[\sum_{j=0}^{N-1} z^j(-F_0 \oplus -F_0^*)\right]$$
$$= ([G_0] - [F_0^*]) + ([G_0^*] - [F_0]) \in T \subseteq \widetilde{K}_0(A) \quad .$$

Hence $(P,\theta) \in U_{2i}^{(1-\alpha)^{-1}T}(A)$.

For $N \geq 0$ so large that

$$z^N F_0^+ \subseteq (G_0 \oplus G_0^*)^+$$

define a $\pm$form over A

$$(P',\theta') = (E_N(F_0, G_0 \oplus G_0^*)/z^N F_0^+ \ , \ [\varphi]_0) \oplus H\pm\left(\sum_{j=0}^{N-1} z^j(-F_0)\right)$$

where

$$E_N(F_0, G_0 \oplus G_0^*) = \{x \in (G_0 \oplus G_0^*)^+ \,|\, [\varphi \pm \varphi^*]_0(x)(z^N F_0^+) = \{0\} \subseteq A \} \quad .$$

Increasing N by 1 adds on

$$H\pm(z^N(F_0 \oplus -F_0^*)) = 0 \in U_{2i}^{(1-\alpha)^{-1}T}(A)$$

to (P',θ'), and for N so large that

$$z^N(F_0 \oplus F_0^*)^+ \subseteq (G_0 \oplus G_0^*)^+$$

the $\pm$forms (P,θ) , (P',θ') coincide, as then

$$E_N(F_0, G_0 \oplus G_0^*) = (F \oplus z^N F_0^{*-}) \cap (G_0 \oplus G_0^*)^+ = z^N F_0^+ \oplus P \quad .$$

Hence $(P,\theta) \in U_{2i}^{(1-\alpha)^{-1}T}(A)$ does not depend on the choice of N or of the hamiltonian complement F^* . The choice of G^* can be dealt with similarly.

If $(Q,\varphi;F,G) = 0 \in U_{2i+1}^{\bar\varepsilon T}(A_\alpha)$, it may be assumed that

$$(Q,\varphi;F,G) = (H_{\pm}(L);L,\Gamma_{(L,\lambda)}) \oplus (H_{\pm}(M);M,M^*)$$

with $[L],[M] \in \bar\varepsilon T$. Choosing

$$F_0 = L_0 \oplus M_0 \qquad (\text{with } [L]_0,[M]_0 \in T) \qquad F_0^* = L_0^* \oplus M_0^*$$

$$G_0^* = L_0^* \oplus M_0 \qquad (G_0^*)^* = L_0 \oplus M_0^* \quad (\text{in } Q) \quad ,$$

note that by symmetry of the definition of B with respect to the lagrangians and their hamiltonian complements

$$B(Q,\varphi;F,G) = B(Q,\varphi;F,G^*)$$

$$= (B_0(F_0 \oplus F_0^*, G_0 \oplus G_0^*),[\varphi]_0) = 0 \in U_{2i}^{(1-\alpha)^{-1}T}(A) \quad .$$

It now only remains to verify that the choice of modular A-bases F_0,G_0 for F,G is immaterial to $(P,\theta) \in U_{2i}^{(1-\alpha)^{-1}T}(A)$.

Let F_0',G_0' be some other modular A-bases of F,G such that

$$[G_0'] - [F_0'^*] \in T \ .$$

Choose $N',N'' \geq 0$ so large that

$$z^{N'}(F_0' \oplus F_0'^*)^+ \subseteq (F_0 \oplus F_0^*)^+ \quad , \quad z^{N''}(G_0 \oplus G_0^*)^+ \subseteq (G_0' \oplus G_0'^*)^+ \quad ,$$

and let $M = N + N' + N''$. Then up to equivalence

$$(B_M(F_0' \oplus F_0'^*, G_0' \oplus G_0'^*),[\varphi]_0)$$

$$= H_{\pm}(z^{N+N''}B_{N'}(F_0'^*,F_0^*)) \oplus (z^{N''}B_N(F_0 \oplus F_0^*, G_0 \oplus G_0^*),[\varphi]_0) \oplus H_{\pm}(B_{N''}(G_0,G_0')) \ ,$$

$$(z^{N''}B_N(F_0 \oplus F_0^*, G_0 \oplus G_0^*),[\varphi]_0) \oplus H_{\pm}(\sum_{j=0}^{N''-1} z^j G_0)$$

$$= (B_N(F_0 \oplus F_0^*, G_0 \oplus G_0^*),[\varphi]_0) \oplus H_{\pm}(\sum_{j=0}^{N''-1} z^{j+N} F_0^*) \quad .$$

Now

$$(1-\alpha)([z^{N+N''}B_{N'}(F_0'^*,F_0^*) \oplus (\sum_{j=0}^{N''-1} z^j(z^N F_0^* \oplus -G_0)) \oplus B_{N''}(G_0,G_0')])$$

$$- (1-\alpha)([\sum_{j=0}^{M-1} z^j F_0'^*] - [\sum_{j=0}^{N-1} z^j F_0^*]) = ([G_0'] - [F_0'^*]) - ([G_0] - [F_0^*]) \in T \subseteq \tilde K_0(A)$$

and so

$$(P',\theta') = (P,\theta) \in U_{2i}^{(1-\alpha)^{-1}T}(A) \quad ,$$

where

$$(P',\theta') = (B_M(F_0' \oplus F_0'^*, G_0' \oplus G_0'^*),[\varphi]_0) \oplus H_{\pm}(\sum_{j=0}^{M-1} z^j(-F_0'))$$

is defined as (P,θ) but with F_0',G_0',M replacing F_0,G_0,N respectively.

Hence

$$B:U^{\bar{\varepsilon}T}_{2i+1}(A_\alpha) \longrightarrow U^{(1-\alpha)^{-1}T}_{2i}(A);(Q,\varphi;F,G) \longmapsto (P,\theta)$$

is a well-defined morphism.

The composite

$$U^T_{2i+1}(A) \xrightarrow{\bar{\varepsilon}} U^{\bar{\varepsilon}T}_{2i+1}(A_\alpha) \xrightarrow{B} U^{(1-\alpha)^{-1}T}_{2i}(A)$$

is zero, sending $(Q,\varphi;F,G) \in U^T_{2i+1}(A)$ to

$$B\bar{\varepsilon}(Q,\varphi;F,G) = (B_0(F_0 \oplus F_0^*, G_0 \oplus G_0^*),[\varphi]_0) = 0 \in U^{(1-\alpha)^{-1}T}_{2i}(A) .$$

Define

$$C:U^{(1-\alpha)^{-1}T}_{2i}(A) \longrightarrow U^T_{2i}(A);(Q,\varphi) \longmapsto (Q,\varphi) \oplus \alpha(Q,-\varphi) \oplus H_\pm(-Q) .$$

This is well-defined because

$$CH_\pm(L) = H_\pm(L \oplus zL \oplus -L \oplus -L) = 0 \in U^T_{2i}(A) \text{ if } [L] \in (1-\alpha)^{-1}T.$$

The composite

$$U^{(1-\alpha)^{-1}T}_{2i}(A) \xrightarrow{C} U^T_{2i}(A) \xrightarrow{\bar{\varepsilon}} U^{\bar{\varepsilon}T}_{2i}(A_\alpha)$$

is zero, sending $(Q,\varphi) \in U^{(1-\alpha)^{-1}T}_{2i}(A)$ to

$$(Q_\alpha,\varphi_\alpha) \oplus (Q_\alpha,-\varphi_\alpha) \oplus H_\pm(-Q_\alpha) = H_\pm((Q \oplus -Q)_\alpha) = 0 \in U^{\bar{\varepsilon}T}_{2i}(A_\alpha)$$

The composite

$$U^{\bar{\varepsilon}T}_{2i+1}(A_\alpha) \xrightarrow{B} U^{(1-\alpha)^{-1}T}_{2i}(A) \xrightarrow{C} U^T_{2i}(A)$$

is zero, as is clear from the identity (valid up to equivalence)

$$\begin{aligned}(B_{N+1}(F_0 \oplus F_0^*, G_0 \oplus G_0^*),[\varphi]_0) &= (B_N(F_0 \oplus F_0^*, G_0 \oplus G_0^*),[\varphi]_0) \oplus H_\pm(z^N F_0) \\ &= \alpha(B_N(F_0 \oplus F_0^*, G_0 \oplus G_0^*),[\varphi]_0) \oplus H_\pm(G_0).\end{aligned}$$

<u>Lemma 6.1</u> <u>The sequence</u>

$$U^T_{2i+1}(A) \xrightarrow{\bar{\varepsilon}} U^{\bar{\varepsilon}T}_{2i+1}(A_\alpha) \xrightarrow{B} U^{(1-\alpha)^{-1}T}_{2i}(A) \xrightarrow{C} U^T_{2i}(A) \xrightarrow{\bar{\varepsilon}} U^{\bar{\varepsilon}T}_{2i}(A_\alpha)$$

<u>is exact for all $*$-invariant subgroups</u> $T \subseteq \vec{K}_0(A)$.

<u>Proof</u>: It has already been verified that the composite of successive morphisms in the sequence is zero. As explained in §5, it is therefore sufficient to consider exactness in the special case $T = \{0\} \subseteq \widetilde{K}_0(A)$,

$$V_{2i+1}(A) \xrightarrow{\bar{\varepsilon}} V_{2i+1}(A_\alpha) \xrightarrow{B} U_{2i}^{\tilde{K}_0(A)^\alpha}(A) \xrightarrow{C} V_{2i}(A) \xrightarrow{\bar{\varepsilon}} V_{2i}(A_\alpha)$$

where $\tilde{K}_0(A)^\alpha = \ker(1-\alpha:\tilde{K}_0(A)\longrightarrow\tilde{K}_0(A))$. (This use of Lemma 5.4 anticipates the definition of

$$C:U_{2i+1}^{(1-\alpha)^{-1}T}(A) \longrightarrow U_{2i+1}^{T}(A) \qquad B:U_{2i}^{\bar{\varepsilon}T}(A_\alpha) \longrightarrow U_{2i-1}^{(1-\alpha)^{-1}T}(A)$$

but no extra exactness properties).

Given $(Q,\varphi) \in \ker(\bar{\varepsilon}:V_{2i}(A)\longrightarrow V_{2i}(A_\alpha))$, it may be assumed that

$$\bar{\varepsilon}(Q,\varphi) = \bar{\varepsilon}H_{\pm}(L)$$

for some f.g.free A-module L. Then

$$(P,\theta) = (B_N(L\oplus L^*,Q),[\varphi]_0)$$

is a non-singular $\pm$form over A such that (up to equivalence)

$$(B_{N+1}(L\oplus L^*,Q),[\varphi]_0) = (Q,\varphi)\oplus\alpha(P,\theta) = (P,\theta)\oplus H_{\pm}(z^N L) .$$

Hence

$$(Q,\varphi) = C(P,\theta) \in \operatorname{im}(C:U_{2i}^{\tilde{K}_0(A)^\alpha}(A) \longrightarrow V_{2i}(A)) ,$$

and the sequence is exact at $V_{2i}(A)$.

Given $(Q,\varphi) \in \ker(C:U_{2i}^{\tilde{K}_0(A)^\alpha}(A)\longrightarrow V_{2i}(A))$, it may be assumed that

$$(Q,\varphi)\oplus\alpha(Q,-\varphi)\oplus\alpha H_{\pm}(-Q) = H_{\pm}(L)$$

for some f.g.free A-module L. Then

$$\begin{aligned}(Q,\varphi) &= (B_1(Q\oplus Q\oplus -Q\oplus -Q^*,L\oplus L^*),[\varphi_\alpha]_0)\\ &= B((Q_\alpha\oplus Q_\alpha,\varphi_\alpha\oplus -\varphi_\alpha)\oplus H_{\pm}(-Q_\alpha);\Delta_{(Q_\alpha,\varphi_\alpha)}\oplus -Q_\alpha\ ,\ L_\alpha\ )\\ &\in \operatorname{im}(B:V_{2i+1}(A_\alpha) \longrightarrow U_{2i}^{\tilde{K}_0(A)^\alpha}(A)) ,\end{aligned}$$

verifying exactness at $U_{2i}^{\tilde{K}_0(A)^\alpha}(A)$.

Given $(Q,\varphi;F,G) \in \ker(B:V_{2i+1}(A_\alpha) \longrightarrow U_{2i}^{\tilde{K}_0(A)^\alpha}(A))$, it may be assumed that

$$(B_N(F_0\oplus F_0^*,G_0\oplus G_0^*),[\varphi]_0) = H_{\pm}(L) .$$

with $[L] \in \tilde{K}_0(A)^\alpha$.

Let

$$P_0 = L\oplus L^* = B_N(F_0\oplus F_0^*, G_0\oplus G_0^*) \quad \in \quad |\mathcal{P}(A)|$$

and define an A_α-module morphism

$$f: P = (P_0)_\alpha \longrightarrow Q$$

by extending the inclusion of P_0 in Q . Let

$$(P, \psi) = \bar{\varepsilon} H_{\pm}(L)$$

and let

$$\theta : P \longrightarrow P^*$$

be the unique A_α-module morphism such that

$$f^*(\varphi\pm\varphi^*)f = \theta\pm\theta^* \in \mathrm{Hom}_{A_\alpha}(P,P^*) \qquad (\theta-\psi)(P_0) \subseteq \sum_{j=1}^{\infty} z^j P_0^* \quad .$$

Define A_α-module morphisms

$$\eta = \begin{pmatrix} 0 & \mp t_\alpha \\ t_\alpha^* & 0 \end{pmatrix} : P^* = L_\alpha^*\oplus L_\alpha \longrightarrow L_\alpha\oplus L_\alpha^* = P$$

$$\xi = 1\oplus t_\alpha^{*-1} : P = L_\alpha\oplus L_\alpha^* \longrightarrow L_\alpha\oplus L_\alpha^*$$

for some isomorphism $t \in \mathrm{Hom}_A(L,zL)$.

Then

$$h_1 = 1\oplus\begin{pmatrix} 1 & \eta \\ 0 & 1 \end{pmatrix} : (Q,\varphi)\oplus H_{\pm}(P) \longrightarrow (Q,\varphi)\oplus H_{\pm}(P)$$

$$h_2 = \begin{pmatrix} 1 & -f\xi & 0 \\ 0 & 1 & 0 \\ \xi^* f^*(\varphi\pm\varphi^*) & -\xi^*\theta\xi & 1 \end{pmatrix} : (Q,\varphi)\oplus H_{\pm}(P) \longrightarrow (Q,\varphi)\oplus H_{\pm}(P)$$

are self-equivalences (over A_α) such that h_1 preserves the lagrangian $F\oplus P$ of $(Q,\varphi)\oplus H_{\pm}(P)$ and h_2 preserves the lagrangian $F\oplus P^*$. It now follows from the sum formula of Lemma 3.2 that

$$(Q',\varphi';F',G') = ((Q,\varphi)\oplus H_{\pm}(P); h(F\oplus P), G\oplus P) \quad (h=h_1h_2)$$

is a $\pm$formation over A_α such that

$$\begin{aligned}(Q',\varphi';F',G') &= ((Q,\varphi)\oplus H_{\pm}(P); F\oplus P, G\oplus P) \\ &= (Q,\varphi;F,G) \in V_{2i+1}(A_\alpha) \ .\end{aligned}$$

Define a modular A-base

$$G_0' = G_0 \oplus P_0$$

for G', giving the hamiltonian complement $G'^* = G^* \oplus P^*$ to G' in (Q',φ') the dual modular A-base

$$G_0'^* = G_0^* \oplus P_0^* \quad .$$

Let

$$Q_0' = G_0' \oplus G_0'^*$$

be the corresponding modular A-base for Q' .

The A-module morphism

$$\nu : Q' \longrightarrow Q' ; \quad \sum_{j=-\infty}^{\infty} z^j x_j \longmapsto \sum_{j=0}^{\infty} z^j x_j \qquad (x_j \in Q_0')$$

is such that

$$\nu(F') \subseteq F'$$

because

$$\nu h(x,y) = \begin{cases} h(x,y) & \text{if } (x,y) \in z^N F_0^+ \oplus P_0^+ \\ h(0,\xi^{-1}\beta(f\xi(y)-x)) & \phantom{\text{if }} (x,y) \in z^N F_0^- \oplus P_0^- \end{cases}$$

where β is the projection

$$\beta = (\ 1 \quad 0\ ) : Q = P_0 \oplus (z^N(F_0 \oplus F_0^*)^+ \oplus (G_0 \oplus G_0^*)^-) \longrightarrow P_0 \quad .$$

It follows that each $x \in F'$ has a unique expression as

$$x = \sum_{j=-\infty}^{\infty} z^j x_j$$

with

$$x_j = z(1-\nu)z^{-1}\nu z^{-j} x \in F' \cap Q_0' \quad ,$$

and so

$$F_0' = F' \cap Q_0'$$

is a modular A-base for F'. (This is precisely the same argument as was used in the untwisted case, in § 2 of II.). Now

$$[F_0'] - [F_0 \oplus P_0] \in \ker(\bar{\varepsilon} : \widetilde{K}_0(A) \to \widetilde{K}_0(A_\alpha)) = \operatorname{im}(1-\alpha : \widetilde{K}_0(A) \to \widetilde{K}_0(A))$$

(by the reduced version of Lemma 4.1), so that

$$(Q,\varphi;F,G) = (Q',\varphi';F',G')$$
$$= \bar{\varepsilon}(H_{\pm}(G_0');F_0',G_0') \in \mathrm{im}(\bar{\varepsilon}:U_{2i+1}^{(1-\alpha)\widetilde{K}_0(A)}(A) \longrightarrow V_{2i+1}(A_\alpha)) \ .$$

We have shown that

$$\ker(B:V_{2i+1}(A_\alpha) \longrightarrow U_{2i}^{\widetilde{K}_0(A)^\alpha}(A)) \subseteq \mathrm{im}(\bar{\varepsilon}:U_{2i+1}^{(1-\alpha)\widetilde{K}_0(A)}(A) \to V_{2i+1}(A_\alpha)) \ .$$

Chasing round the diagram

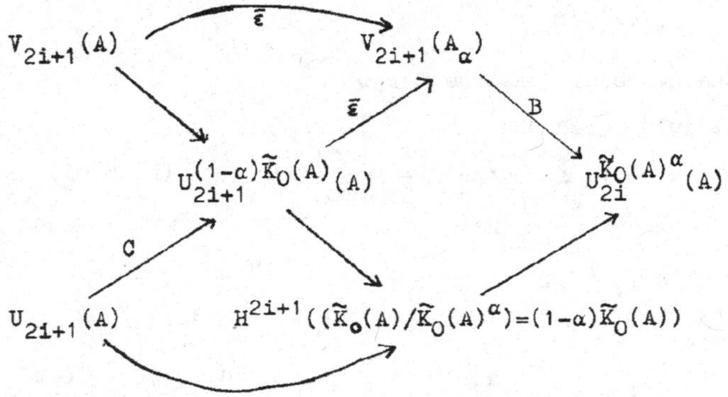

(which is part of a braid, and anticipates the definition of

$$C = 1-\alpha : U_{2i+1}(A) \longrightarrow U_{2i+1}^{(1-\alpha)\widetilde{K}_0(A)}(A) \quad)$$

the exactness of

$$V_{2i+1}(A) \xrightarrow{\bar{\varepsilon}} V_{2i+1}(A_\alpha) \xrightarrow{B} U_{2i}^{\widetilde{K}_0(A)^\alpha}(A)$$

follows.

[ ]

In the untwisted case, $\alpha=1:A \longrightarrow A$, Lemma 6.1 gives a short exact sequence

$$0 \longrightarrow U_{2i+1}^T(A) \xrightarrow{\bar{\varepsilon}} U_{2i+1}^{\bar{\varepsilon}T}(A_z) \xrightarrow{B} V_{2i}(A) \longrightarrow 0$$

which splits, with B split by

$$\bar{B} : V_{2i}(A) \longrightarrow U_{2i+1}^{\bar{\varepsilon}T}(A_z) \ ;$$
$$(Q,\varphi) \longmapsto (Q_z \oplus Q_z, \varphi_z \oplus -\varphi_z; \Delta_{(Q_z,\varphi_z)}, (z\oplus 1)\Delta_{(Q_z,\varphi_z)}) \ .$$

Given a $*$-invariant subgroup $R \subseteq \widetilde{K}_1(A)$ define

$$B : \widetilde{V}^{\bar{\varepsilon}R}_{2i+1}(A_\alpha) \longrightarrow V^{(1-\alpha)^{-1}R}_{2i}(A) ; (Q,\varphi;\underset{\sim}{F},\underset{\sim}{G}) \longmapsto (\underset{\sim}{P},\theta)$$

as follows. Let

$$(P,\theta) = (B_N(F_0 \oplus F_0^*, G_0 \oplus G_0^*), [\varphi]_0)$$

with F_0, G_0 the modular A-bases of F, G generated by the given A_α-bases. Let $\tau_0 \in R$ be such that

$$\tau(Q,\varphi;\underset{\sim}{F},\underset{\sim}{G}) = \bar{\varepsilon}\tau_0 \in \bar{\varepsilon}\, R \subseteq \widetilde{\widetilde{K}}_1(A_\alpha) \quad (=\text{coker}(K_1(Z_z) \to K_1(A_\alpha)))$$

(so that by the reduced version of the exact sequence of Lemma 4.1 τ_0 is unique up to torsions in $R \cap (1-\alpha)\widetilde{K}_1(A)$). Now

$$[P] = B\bar{\varepsilon}(\tau_0) = 0 \in \widetilde{K}_0(A) \;,$$

so that for sufficiently large $N \geq 0$ P is a free f.g. A-module. Applying Theorem 4.2, note that

$$\begin{aligned} q\tau(Q,\varphi;\underset{\sim}{F},\underset{\sim}{G}) = [&B_{N+1}(F_0 \oplus F_0^*, G_0 \oplus G_0^*), \\ &1 \oplus \xi^{N+1} f : zP \oplus (G_0 \oplus G_0^*) \to zP \oplus z^{N+1}(F_0 \oplus F_0^*)] \\ = {}& j\tau_0 \in \widetilde{K}(A,\alpha) \end{aligned}$$

with f defined by

$$f(\underset{\sim}{G} \oplus \underset{\sim}{G}^*) = \underset{\sim}{F} \oplus \underset{\sim}{F}^* \;.$$

Choosing any A-base for P, it follows that

$$j\tau(1 : z\underset{\sim}{P} \oplus (\underset{\sim}{G}_0 \oplus \underset{\sim}{G}_0^*) \longrightarrow \underset{\sim}{P} \oplus z^N(\underset{\sim}{F}_0 \oplus \underset{\sim}{F}_0^*)) = j\tau_0 \in \widetilde{K}(A,\alpha) \;,$$

and so (by Lemma 4.1)

$$\tau(1 : z\underset{\sim}{P} \oplus (\underset{\sim}{G}_0 \oplus \underset{\sim}{G}_0^*) \to \underset{\sim}{P} \oplus z^N(\underset{\sim}{F}_0 \oplus \underset{\sim}{F}_0^*)) - \tau_0 = (1-\alpha)\tau_1 \in \widetilde{K}_1(A)$$

for some $\tau_1 \in \widetilde{K}_1(A)$ which is unique up to torsions in $(1-\alpha)^{-1}R$ (allowing τ_0 to vary). Changing the base of P by τ_1, we can ensure that

$$\tau(1 : z\underset{\sim}{P} \oplus (\underset{\sim}{G}_0 \oplus \underset{\sim}{G}_0^*) \longrightarrow \underset{\sim}{P} \oplus z^N(\underset{\sim}{F}_0 \oplus \underset{\sim}{F}_0^*)) = \tau_0 \in R \subseteq \widetilde{K}_1(A) \quad (*) \;.$$

Let

$$B(Q,\varphi;\underset{\sim}{F},\underset{\sim}{G}) = (\underset{\sim}{P},\theta)$$

with $\underset{\sim}{P}$ in the preferred class of bases of P (unique up to changes in $(1-\alpha)^{-1}R$) satisfying the condition $(*)$. Then

$$(1-\alpha)\tau(\underset{\sim}{P},\theta) = -(\tau_0+\tau_0^*) \in R ,$$

and so we do have an element

$$(\underset{\sim}{P},\theta) \in V_{2i}^{(1-\alpha)^{-1}R}(A)$$

which does not depend on the choice of τ_0 or $\underset{\sim}{P}$. The verification that this does define a morphism

$$B : \tilde{V}_{2i+1}^{\varepsilon R}(A_\alpha) \longrightarrow V_{2i}^{(1-\alpha)^{-1}R}(A)$$

is by analogy with that for

$$B : U_{2i+1}^{\varepsilon T}(A_\alpha) \longrightarrow U_{2i}^{(1-\alpha)^{-1}T}(A)$$

carried out above, taking into account torsions rather than projective classes.

Define also

$$C : V_{2i}^{(1-\alpha)^{-1}R}(A) \longrightarrow V_{2i}^{R}(A) \ ; \ (\underset{\sim}{Q},\varphi) \longmapsto (\underset{\sim}{Q},\varphi) \oplus \alpha(\underset{\sim}{Q}',-\varphi)$$

where $\underset{\sim}{Q}'$ is Q with the base defined by

$$\underset{\sim}{Q}' = (\varphi\pm\varphi^*)^{-1}(\underset{\sim}{Q}^*) ,$$

so that

$$\tau((\underset{\sim}{Q},\varphi)\oplus\alpha(\underset{\sim}{Q}',-\varphi)) = (1-\alpha)\tau(\underset{\sim}{Q},\varphi) \in R \subseteq \tilde{K}_1(A) .$$

Given a $*$-invariant subgroup $S \subseteq \tilde{K}(A,\alpha)$, define

$$B : \tilde{V}_{2i+1}^{S}(A_\alpha) \longrightarrow U_{2i}^{T}(A) \ ; \ (Q,\varphi;\underset{\sim}{F},\underset{\sim}{G}) \longmapsto (B_N(F_0\oplus F_0^*,G_0\oplus G_0^*),[\varphi]_0)$$

with F_0,G_0 the modular A-bases of F,G generated by the given A_α-bases, so that

$$[B_N(F_0\oplus F_0^*,G_0\oplus G_0^*)] = B\tau(Q,\varphi;\underset{\sim}{F},\underset{\sim}{G}) \in T = p(S) \subseteq \tilde{K}_0(A) .$$

Define also

$$C : U_{2i}^{T}(A) \longrightarrow V_{2i}^{R}(A) \ ; \ (Q,\varphi) \longmapsto (\underset{\sim}{Q}',\varphi')$$

with

$$(Q',\varphi') = (Q,\varphi)\oplus\alpha(Q,-\varphi)\oplus H_{\pm}(-Q)$$

$$\underset{\sim}{Q}' = (\underset{\smile}{Q\oplus -Q})\oplus(t(\varphi\pm\varphi^*)^{-1}\oplus 1)(\underset{\smile}{Q\oplus -Q})^*$$

for any projective inverse $-Q$ to Q, and any A-base $(\underset{\smile}{Q\oplus -Q})$, where

$t \in \mathrm{Hom}_A(Q,zQ)$ is any isomorphism such that $[Q,t] \in S$ (and is thus unique up to composition with automorphisms of Q with torsion in $j^{-1}(S) = R \subseteq \widetilde{K}_1(A)$). Now

$$\underset{\approx}{Q}'_\alpha = \underset{\sim}{(Q \oplus -Q)}_\alpha \oplus ((\varphi_\alpha \pm \varphi^*_\alpha)^{-1} \oplus 1)\underset{\sim}{(Q \oplus -Q)}^*_\alpha$$

is a hamiltonian A_α-base for $\overline{\varepsilon}(Q',\varphi')$ such that

$$\tau(1 : \underset{\approx}{Q}'_\alpha \longrightarrow \underset{\sim}{Q}'_\alpha) = i[Q,t] \in \widetilde{\widetilde{K}}_1(A_\alpha) \quad .$$

Applying Theorem 4.2,

$$q\tau(1 : \underset{\approx}{Q}'_\alpha \longrightarrow \underset{\sim}{Q}'_\alpha) = [Q,t] \in S \subseteq \widetilde{K}(A,\alpha) \ ,$$

so that

$$\begin{aligned} j\tau(\underset{\sim}{Q}',\varphi') &= q\overline{\varepsilon}\tau(\underset{\sim}{Q}',\varphi') \\ &= -([Q,t] + [Q,t]^*) \in S \subseteq \widetilde{K}(A,\alpha) \end{aligned}$$

and

$$\tau(\underset{\sim}{Q}',\varphi') \in j^{-1}(S) = R \subseteq \widetilde{K}_1(A) \ .$$

Thus we do have an element

$$(Q',\varphi') \in V^R_{2i}(A)$$

which does not depend on the choice of $\underset{\sim}{(Q \oplus -Q)}$ or t.

The verification that all the morphisms B, C appearing in the sequences

$$V^R_{2i+1}(A) \xrightarrow{\overline{\varepsilon}} \widetilde{V}^{\overline{\varepsilon}R}_{2i+1}(A_\alpha) \xrightarrow{B} V^{(1-\alpha)^{-1}R}_{2i}(A) \xrightarrow{C} V^R_{2i}(A) \xrightarrow{\overline{\varepsilon}} \widetilde{V}^{\overline{\varepsilon}R}_{2i}(A_\alpha)$$

$$V^R_{2i+1}(A) \xrightarrow{\overline{\varepsilon}} \widetilde{V}^S_{2i+1}(A_\alpha) \xrightarrow{B} U^T_{2i}(A) \xrightarrow{C} V^R_{2i}(A) \xrightarrow{\overline{\varepsilon}} \widetilde{V}^S_{2i}(A_\alpha)$$

are well-defined, and that the composite of successive morphisms is zero, is by analogy with that for the sequence of Lemma 6.1. Exactness follows, by the argument of §5.

In particular, in the untwisted case $\alpha = 1 : A \to A$, with

$$S = j(R) \oplus \overline{p}(T) \subseteq \widetilde{K}(A,1) = j\widehat{K}_1(A) \oplus \overline{p}\widetilde{K}_0(A)$$

there is defined a split short exact sequence

$$0 \longrightarrow V^R_{2i+1}(A) \xrightarrow{\overline{\varepsilon}} \widetilde{V}^S_{2i+1}(A_z) \xrightarrow{B} U^T_{2i}(A) \longrightarrow 0 \ ,$$

with splitting morphisms

$$\varepsilon : \tilde{V}^S_{2i+1}(A_z) \longrightarrow V^R_{2i+1}(A)$$

$$\overline{B} : U^T_{2i}(A) \longrightarrow \tilde{V}^S_{2i+1}(A_z) ;$$

$$(Q,\varphi) \longrightarrow ((Q_z \oplus Q_z, \varphi_z \oplus -\varphi_z) \oplus H_{\pm}(-Q_z); \underset{\sim}{L}_z, \S \underset{\sim}{L}_z)$$

where

$$\S = z \oplus 1 : Q_z \oplus (Q_z \oplus -Q_z \oplus -Q^*_z) \longrightarrow Q_z \oplus (Q_z \oplus -Q_z \oplus -Q^*_z)$$

$$\underset{\sim}{L} = \{(x,x,y,0) \in Q_z \oplus Q_z \oplus -Q_z \oplus -Q^*_z \mid (x,y) \in (\underset{\sim}{Q \oplus -Q})\}$$

for any projective inverse $-Q$ to Q, and any A-base $(\underset{\sim}{Q \oplus -Q})$.

Given a $*$-invariant subgroup $T \subseteq \tilde{K}_0(A)$, define

$$B : U^{\bar{\varepsilon}T}_{2i}(A_\alpha) \longrightarrow U^{(1-\alpha)^{-1}T}_{2i-1}(A) ; (Q,\varphi) \longmapsto (H_{\mp}(P_N); P_N, B_N(Q_0,\varphi))$$

as follows, where

$$P_N = \sum_{j=0}^{N-1} z^j Q_0 \quad .$$

Choose a modular A-base Q_0 of Q such that

$$[Q_0] \in T \subseteq \tilde{K}_0(A) \quad ,$$

let

$$\nu : Q \oplus Q^* \longrightarrow (Q_0 \oplus Q^*_0)^+ ; \sum_{j=-\infty}^{\infty} z^j x_j \longmapsto \sum_{j=0}^{\infty} z^j x_j \quad (x_j \in (Q_0 \oplus Q^*_0)),$$

and define

$$B_N(Q_0,\varphi) = \{(z^N(1-\nu)z^{-N}x, \nu(\varphi \pm \varphi^*)x) \in P_N \oplus P^*_N \mid x \in B_N((\varphi \pm \varphi^*)^{-1}Q^*_0, Q_0)\}.$$

Then $B_N(Q_0,\varphi)$ is a lagrangian of $H_{\mp}(P_N)$, with hamiltonian complement

$$B^*_N(Q_0,\varphi) = \{(-\nu y, \nu(\varphi \pm \varphi^*)(1-\nu)y) \in P_N \oplus P^*_N \mid y \in B_N(Q_0, (\varphi \pm \varphi^*)^{-1}Q^*_0)\}.$$

The associated $\mp$hermitian product of $H_{\mp}(P_N)$

$$\begin{pmatrix} 0 & 1 \\ \mp 1 & 0 \end{pmatrix} : P_N \oplus P^*_N \longrightarrow P^*_N \oplus P_N = (P_N \oplus P^*_N)^*$$

restricts to the A-module isomorphism

$$B^*_N(Q_0,\varphi) \longrightarrow B_N(Q_0,\varphi)^* ;$$

$$(-\nu y, \nu(\varphi \pm \varphi^*)(1-\nu)y) \longmapsto ((z^N(1-\nu)z^{-N}x, \nu(\varphi \pm \varphi^*)x) \mapsto [(\varphi \pm \varphi^*)(y)(x)]_0).$$

Hence

$$[B_N(Q_0,\varphi)] = [B_N((\varphi\pm\varphi^*)^{-1}Q_0^*,Q_0)] \in \tilde{K}_0(A)$$

and

$$\begin{aligned}(1-\alpha)([B_N(Q_0,\varphi)]-[P_N^*]) &= ([Q_0]-[z^N Q_0^*])+([z^N Q_0^*]-[Q_0^*]) \\ &= [Q_0]-[Q_0^*] \in T \subseteq \tilde{K}_0(A) \quad ,\end{aligned}$$

so that we do have an element

$$B(Q,\varphi) = (H_{\mp}(P_N);P_N,B_N(Q_0,\varphi)) \in U_{2i-1}^{(1-\alpha)^{-1}T}(A) \ .$$

Increasing N by 1, note that

$$B_{N+1}(Q_0,\varphi) = B_N(Q_0,\varphi)\oplus\{(z^{N+1}(1-\nu)z^{-(N+1)}x,(\varphi\pm\varphi^*)x| \\ x\in(\varphi\pm\varphi^*)^{-1}(z^N Q_0^*)\} \ .$$

Now $B_N^*(Q_0,\varphi)\oplus z^N Q_0$ is a hamiltonian complement in $H_{\mp}(P_{N+1})$ to both $B_{N+1}(Q_0,\varphi)$ and $B_N(Q_0,\varphi)\oplus z^N Q_0^*$. Applying the sum formula of Lemma 2.2,

$$\begin{aligned}(H_{\mp}(P_N);P_N,B_N(Q_0,\varphi)) &= (H_{\mp}(P_{N+1});P_{N+1},B_N(Q_0,\varphi)\oplus z^N Q_0^*) \\ &= (H_{\mp}(P_{N+1});P_{N+1},B_N^*(Q_0,\varphi)\oplus z^N Q_0) \\ &= (H_{\mp}(P_{N+1});P_{N+1},B_{N+1}(Q_0,\varphi)) \\ &\in U_{2i-1}^{(1-\alpha)^{-1}T}(A) \ .\end{aligned}$$

Hence the choice of N is immaterial to $B(Q,\varphi) \in U_{2i-1}^{(1-\alpha)^{-1}T}(A)$.

Let Q_0' be another modular A-base of Q such that

$$[Q_0'] \in T \ ,$$

write

$$P_{N'}' = \sum_{j=0}^{N'-1} z^j Q_0' \quad ,$$

and define

$$\nu' : Q\oplus Q^* \to (Q_0'\oplus Q_0'^*)^+; \ \sum_{j=-\infty}^{\infty} z^j x_j \longmapsto \sum_{j=0}^{\infty} z^j x_j \\ (x_j \in (Q_0'\oplus Q_0'^*)) \ .$$

Let $M \geq 0$ be so large that

$$Q_0' \subseteq \sum_{j=-M}^{M} z^j Q_0 \qquad Q_0 \subseteq \sum_{j=-M}^{M} z^j Q_0'$$

Then $N' = N + 2M$ is sufficiently large for $B_{N'}(Q_0',\varphi)$ to be defined, with

$$B_{N'}((\varphi\pm\varphi^*)^{-1}Q_0'^*,Q_0') = (\varphi\pm\varphi^*)^{-1}(z^{M+N}B_M(Q_0'^*,Q_0^*))$$
$$\oplus z^M B_N((\varphi\pm\varphi^*)^{-1}Q_0^*,Q_0)\oplus B_M(Q_0,Q_0')$$

and

$$B_{N'}(Q_0',\varphi) = \{(z^{N'}(1-\nu')z^{-N'}x,(\varphi\pm\varphi^*)x)\,|\,x\in(\varphi\pm\varphi^*)^{-1}(z^{M+N}B_N(Q_0'^*,Q_0^*))\}$$
$$\oplus\{(x,(\varphi\pm\varphi^*)x)\,|\,x\in z^M B_N((\varphi\pm\varphi^*)^{-1}Q_0^*,Q_0)\}$$
$$\oplus\{(x,\nu'(\varphi\pm\varphi^*)x)\,|\,x\in B_M(Q_0,Q_0')\}\subseteq P_{N'}'\oplus P_{N'}'^* .$$

Now

$$P_{N'}' = z^{M+N}B_M(Q_0',Q_0)\oplus z^M P_N\oplus B_M(Q_0,Q_0')$$

and

$$z^{M+N}B_M(Q_0',Q_0)\oplus z^M B_N^*(Q_0,\varphi)\oplus B_M(Q_0^*,Q_0'^*)$$

is a hamiltonian complement in $H_\mp(P_{N'}')$ to both $B_{N'}(Q_0',\varphi)$ and $z^{M+N}B_M(Q_0'^*,Q_0^*)\oplus z^M B_N(Q_0,\varphi)\oplus B_M(Q_0,Q_0')$. Applying the sum formula of Lemma 2.2,

$$(H_\mp(P_{N'}');P_{N'}',B_{N'}(Q_0',\varphi))$$
$$= (H_\mp(P_{N'}');P_{N'}',z^{M+N}B_M(Q_0'^*,Q_0^*)\oplus z^M B_N(Q_0,\varphi)\oplus B_M(Q_0,Q_0'))$$
$$= (H_\mp(z^{M+N}B_M(Q_0',Q_0));z^{M+N}B_M(Q_0',Q_0),z^{M+N}B_M(Q_0'^*,Q_0^*))$$
$$\oplus\alpha^M(H_\mp(P_N);P_N,B_N(Q_0,\varphi))$$
$$\oplus(H_\mp(B_M(Q_0,Q_0'));B_M(Q_0,Q_0'),B_M(Q_0^*,Q_0'^*))$$
$$= \alpha^M(H_\mp(P_N);P_N,B_N(Q_0,\varphi)) \in U_{2i-1}^{(1-\alpha)^{-1}T}(A) .$$

But $zB_N^*(Q_0,\varphi)\oplus Q_0$ is a hamiltonian complement to $B_{N+1}(Q_0,\varphi)$ in

$H_{\mp}(P_{N+1})$, so that

$$(H_{\mp}(P_N);P_N,B_N(Q_0,\varphi)) = (H_{\mp}(P_{N+1});P_{N+1},B_{N+1}(Q_0,\varphi))$$
$$= \alpha(H_{\mp}(P_N);P_N,B_N(Q_0,\varphi)) \oplus (H_{\mp}(Q_0);Q_0,Q_0^*)$$
$$= \alpha(H_{\mp}(P_N);P_N,B_N(Q_0,\varphi)) \in U_{2i-1}^{(1-\alpha)^{-1}T}(A) \ .$$

Hence

$$B(Q,\varphi) = (H_{\mp}(P_N);P_N,B_N(Q_0,\varphi)) \in U_{2i-1}^{(1-\alpha)^{-1}T}(A)$$

does not depend on the choice of modular A-base Q_0 .

Finally, suppose

$$(Q,\varphi) = \bar{\varepsilon}(Q_0,\varphi_0)$$

for some $(Q_0,\varphi_0) \in U_{2i}^T(A)$. Then

$$B(Q,\varphi) = (H_{\mp}(0);0,B_0(Q_0,\varphi)) = 0 \in U_{2i-1}^{(1-\alpha)^{-1}T}(A) \ .$$

Hence

$$B:U_{2i}^{\bar{\varepsilon}T}(A_\alpha) \longrightarrow U_{2i-1}^{(1-\alpha)^{-1}T}(A);(Q,\varphi) \longmapsto (H_{\mp}(P_N);P_N,B_N(Q_0,\varphi))$$

is well-defined, and such that the composite

$$U_{2i}^T(A) \xrightarrow{\bar{\varepsilon}} U_{2i}^{\bar{\varepsilon}T}(A_\alpha) \xrightarrow{B} U_{2i-1}^{(1-\alpha)^{-1}T}(A)$$

is zero.

The morphism

$$C = 1-\alpha : U_{2i-1}^{(1-\alpha)^{-1}T}(A) \longrightarrow U_{2i-1}^T(A) \ ;$$
$$(Q,\varphi;F,G) \longmapsto (Q,\varphi;F,G) \oplus \alpha(Q,-\varphi;F^*,G^*)$$

is clearly well-defined, and such that the composites of successive morphisms in

$$U_{2i}^{\bar{\varepsilon}T}(A_\alpha) \xrightarrow{B} U_{2i-1}^{(1-\alpha)^{-1}T}(A) \xrightarrow{C} U_{2i-1}^T(A) \xrightarrow{\bar{\varepsilon}} U_{2i-1}^{\bar{\varepsilon}T}(A_\alpha)$$

is zero ($CB = 0$ follows from the relation

$$\alpha B(Q,\varphi) = B(Q,\varphi) \in U_{2i-1}^{(1-\alpha)^{-1}T}(A) \quad ((Q,\varphi) \in U_{2i}^{\bar{\varepsilon}T}(A_\alpha))$$

proved above).

Given a $*$-invariant subgroup $R \subseteq \widetilde{K}_1(A)$, define

$$B : \widetilde{V}_{2i}^{\bar{\varepsilon}R}(A_\alpha) \longrightarrow V_{2i-1}^{(1-\alpha)^{-1}R}(A) \; ; \; (\underset{\sim}{Q},\varphi) \longmapsto (H_{\mp}(P_N);\underset{\sim}{P}_N, \underset{\sim}{B_N(Q_0,\varphi)})$$

as follows. Let $\underset{\sim}{Q}_0$ be the modular A-base of Q generated by the given A_α-base, with the corresponding A-base. Let $N \geq 0$ be so large that $B_N((\varphi\pm\varphi^*)^{-1}Q_0^*,Q_0)$ is a free A-module. Let $\tau_0 \in R$ be such that

$$\tau(\underset{\sim}{Q},\varphi) = \bar{\varepsilon}\tau_0 \in \bar{\varepsilon}R \subseteq \widetilde{\widetilde{K}}_1(A_\alpha) \; .$$

Then, working as in the definition of $B : \widetilde{V}_{2i+1}^{\bar{\varepsilon}R}(A_\alpha) \longrightarrow V_{2i}^{(1-\alpha)^{-1}R}(A)$, there is a preferred class of A-bases $\underline{B_N((\varphi\pm\varphi^*)^{-1}Q_0^*,Q_0)}$, unique up to changes in $(1-\alpha)^{-1}R$ for varying τ_0, such that

$$\tau(1:z\underline{B_N((\varphi\pm\varphi^*)^{-1}Q_0^*,Q_0)} \oplus \underset{\sim}{Q}_0 \longrightarrow \underline{B_N((\varphi\pm\varphi^*)^{-1}Q_0^*,Q_0)} \oplus (\varphi\pm\varphi^*)^{-1}(z^N\underset{\sim}{Q}_0^*))$$
$$= \tau_0 \in R \subseteq K_1(A) \; .$$

Give $B_N(Q_0,\varphi)$ an A-base by choosing one of these, and setting

$$\underline{B_N(Q_0,\varphi)} = \{(z^N(1-\nu)z^{-N}x, \nu(\varphi\pm\varphi^*)x) \in P_N \oplus P_N^* \,|\, x \in \underline{B_N((\varphi\pm\varphi^*)^{-1}Q_0^*,Q_0)}\} \; .$$

Let $\left\{\begin{matrix}\underline{B_{N+1}(Q_0,\varphi)} \\ \underline{\underline{B_{N+1}(Q_0,\varphi)}}\end{matrix}\right.$ stand for $B_{N+1}(Q_0,\varphi)$ with the base

$$\left\{\begin{aligned}\underline{B_{N+1}((\varphi\pm\varphi^*)^{-1}Q_0^*,Q_0)} &= z\underline{B_N((\varphi\pm\varphi^*)^{-1}Q_0^*,Q_0)} \oplus \underset{\sim}{Q}_0 \\ \underline{\underline{B_{N+1}((\varphi\pm\varphi^*)^{-1}Q_0^*,Q_0)}} &= \underline{B_N((\varphi\pm\varphi^*)^{-1}Q_0^*,Q_0)} \oplus (\varphi\pm\varphi^*)^{-1}(z^N\underset{\sim}{Q}_0^*)\end{aligned}\right. .$$

Using the hamiltonian complements given above (in the definition of $B : U_{2i}^{\bar{\varepsilon}T}(A_\alpha) \longrightarrow U_{2i-1}^{(1-\alpha)^{-1}T}(A)$) it can be shown that

$$\begin{aligned}&(H_{\mp}(P_{N+1});\underset{\sim}{P}_{N+1}, \underline{B_{N+1}(Q_0,\varphi)}) \\ &\quad = (H_{\mp}(P_{N+1});\underset{\sim}{P}_{N+1}, z\underline{B_N(Q_0,\varphi)} \oplus z^N\underset{\sim}{Q}_0^*) \\ &\quad = \alpha(H_{\mp}(P_N);\underset{\sim}{P}_N, \underline{B_N(Q_0,\varphi)}) \in V_{2i-1}^{(1-\alpha)^{-1}R}(A)\end{aligned}$$

and similarly

$$(H_{\mp}(P_{N+1});\underset{\sim}{P}_{N+1},\underset{\approx}{B_{N+1}(Q_0,\varphi)})$$
$$= (H_{\mp}(P_{N+1});\underset{\sim}{P}_{N+1},\underset{\sim}{B_N(Q_0,\varphi)} \oplus z^N\underset{\sim}{Q}_0^*)$$
$$= (H_{\mp}(P_N);\underset{\sim}{P}_N,\underset{\sim}{B_N(Q_0,\varphi)}) \in V_{2i-1}^{(1-\alpha)^{-1}R}(A) \quad .$$

Hence

$$(1-\alpha)\tau(H_{\mp}(P_N);\underset{\sim}{P}_N,\underset{\sim}{B_N(Q_0,\varphi)}) = (\tau_0-\tau_0^*) \in R \ ,$$

and we do have an element

$$B(\underset{\sim}{Q},\varphi) = (H_{\mp}(P_N);\underset{\sim}{P}_N,\underset{\sim}{B_N(Q_0,\varphi)}) \in V_{2i-1}^{(1-\alpha)^{-1}R}(A) \ .$$

Define also

$$C = 1-\alpha \ : \ V_{2i-1}^{(1-\alpha)^{-1}R}(A) \longrightarrow V_{2i-1}^{R}(A);$$
$$(Q,\varphi;\underset{\sim}{F},\underset{\sim}{G}) \longmapsto (Q,\varphi;\underset{\sim}{F},\underset{\sim}{G}) \oplus \alpha(Q,-\varphi;\underset{\sim}{F}^*,\underset{\sim}{G}^*) \ .$$

Given a $*$-invariant subgroup $S \subseteq \widetilde{K}(A,\alpha)$ define

$$B:\widetilde{V}_{2i}^{S}(A_\alpha) \longrightarrow U_{2i-1}^{T}(A);(\underset{\sim}{Q},\varphi) \longmapsto (H_{\mp}(P_N);P_N,B_N(Q_0,\varphi))$$

with Q_0 the modular A-base of Q generated by the given A_α-base, so that

$$[B_N(Q_0,\varphi)] = B\tau(\underset{\sim}{Q},\varphi) \in T = p(S) \subseteq \widetilde{K}_0(A) \ .$$

Define also

$$C:U_{2i-1}^{T}(A) \to V_{2i-1}^{R}(A);(Q,\varphi;F,G) \longmapsto (Q',\varphi';\underset{\sim}{F}',\underset{\sim}{G}')$$

as follows. It may be assumed that F is free and that there is defined an isomorphism $t \in \mathrm{Hom}_A(G,zG)$ such that $[G,t] \in S$ .Let

$$(Q',\varphi';F',G') = (Q,\varphi;F,G) \oplus \alpha(Q,-\varphi;F^*,G^*)$$

for any hamiltonian complements F^*,G^* to F,G . Choosing any base for F, let

$$\underset{\sim}{F}' = \underset{\sim}{F} \oplus z\underset{\sim}{F}^* \qquad \underset{\sim}{G}' = (1\oplus t^{*-1})(\underset{\sim}{G\oplus G^*}) \qquad (\underset{\sim}{G\oplus G^*}) = \underset{\sim}{F}\oplus\underset{\sim}{F}^* \ .$$

Now

$$\bar{\varepsilon}\tau(Q',\varphi';\underset{\sim}{F}',\underset{\sim}{G}') = \tau(Q_\alpha\oplus Q_\alpha,\varphi_\alpha\oplus-\varphi_\alpha;(1\oplus\xi_\alpha)(\underset{\sim}{F}\oplus\underset{\sim}{F}^*)_\alpha,(1\oplus t_\alpha^{*-1})(\underset{\sim}{G\oplus G^*})_\alpha)$$
$$= i(*-1)([G,t]-[F^*,\xi]) \in i(S) \subseteq \widetilde{\widetilde{K}}_1(A_\alpha)$$

(i as in Theorem 4.2).

Hence

$$\tau(Q',\varphi';\underset{\sim}{F}',\underset{\sim}{G}') \in j^{-1}(S) = R \subseteq \widetilde{K}_1(A) ,$$

and we do have an element

$$C(Q,\varphi;F,G) = (Q',\varphi';\underset{\sim}{F}',\underset{\sim}{G}') \in V^R_{2i-1}(A) .$$

The verification that the morphisms B, C appearing in the sequences

$$V^R_{2i}(A) \xrightarrow{\bar{\varepsilon}} \widetilde{V}^{\bar{\varepsilon}R}_{2i}(A_\alpha) \xrightarrow{B} V^{(1-\alpha)^{-1}R}_{2i-1}(A) \xrightarrow{C} V^R_{2i-1}(A) \xrightarrow{\bar{\varepsilon}} \widetilde{V}^{\bar{\varepsilon}R}_{2i-1}(A_\alpha)$$

$$V^R_{2i}(A) \xrightarrow{\bar{\varepsilon}} \widetilde{V}^S_{2i}(A_\alpha) \xrightarrow{B} U^T_{2i-1}(A) \xrightarrow{C} V^R_{2i-1}(A) \xrightarrow{\bar{\varepsilon}} \widetilde{V}^S_{2i-1}(A_\alpha)$$

are well-defined, and that the composite of successive morphisms is zero, is by analogy with that for the sequence

$$U^T_{2i}(A) \xrightarrow{\bar{\varepsilon}} U^{\bar{\varepsilon}T}_{2i}(A_\alpha) \xrightarrow{B} U^{(1-\alpha)^{-1}T}_{2i-1}(A) \xrightarrow{C} U^T_{2i-1}(A) \xrightarrow{\bar{\varepsilon}} U^{\bar{\varepsilon}T}_{2i-1}(A_\alpha)$$

which was dealt with above.

We can now apply the trick (first used in [4]) of introducing a new Laurent variable to deduce the exactness of these sequences from that of Lemma 6.1.

Note first that for $*$-invariant subgroups

$$S = j(R) \oplus \bar{p}(T) \subseteq \widetilde{K}(A,1) = j\widetilde{K}_1(A) \oplus \bar{p}\widetilde{K}_0(A)$$

there is defined a morphism

$$\bar{B} : U^T_{2i-1}(A) \longrightarrow \widetilde{V}^S_{2i}(A_z) ; (Q,\varphi;F,G) \longmapsto (\underset{\sim}{G}_z \oplus \underset{\sim}{G}^*_z, \begin{pmatrix} \lambda & -z\gamma \\ \delta & (1-z)(\lambda_1 \pm \lambda_1^*) \end{pmatrix})$$

with $\underset{\sim}{G}$ any base for G (which may be assumed to be free), and

$$\begin{pmatrix} \lambda \pm \lambda^* & \gamma \\ \delta & \lambda_1 \pm \lambda_1^* \end{pmatrix} : G \oplus G^* \longrightarrow G^* \oplus G$$

an expression for

$$\begin{pmatrix} 0 & 1 \\ 0 & 0 \end{pmatrix} : F \oplus F^* \longrightarrow F^* \oplus F ,$$

for any hamiltonian complements F^*, G^* to F, G in (Q,φ).

It was shown in §3 of II. that this does define a morphism $\overline{B}$, and that

$$V^R_{2i}(A) \underset{\varepsilon}{\overset{\overline{\varepsilon}}{\rightleftarrows}} \widetilde{V}^S_{2i}(A_z) \underset{\overline{B}}{\overset{B}{\rightleftarrows}} U^T_{2i-1}(A)$$

is a direct sum system, if $S = \{0\}$ or $\widetilde{K}(A,1)$. The proof generalizes immediately to any S of type $j(R)\oplus\overline{p}(T)$.

Let z' be an invertible indeterminate over A_α .

Identify $(A_\alpha)_{z'}$ with $(A_{z'})_{\alpha'}$, where

$$\alpha' : A_{z'} \longrightarrow A_{z'} \; ; \; \sum_{j=-\infty}^{\infty} z'^j a_j \longmapsto \sum_{j=-\infty}^{\infty} z'^j \alpha(a_j) \quad ,$$

and write $A_{\alpha,z'}$ for this double Laurent extension of A.

Let $\left\{\begin{matrix} S_0 \\ S'_0 \end{matrix}\right.$ be the infinite cyclic subgroup of $\left\{\begin{matrix} \widetilde{K}_1(A_\alpha) \\ \widetilde{K}_1(A_z) \end{matrix}\right.$ generated by

$\left\{\begin{matrix} \tau(\xi : A_\alpha \longrightarrow A_\alpha) \\ \tau(\xi' : A_{z'} \longrightarrow A_{z'}) \end{matrix}\right.$, where $\left\{\begin{matrix} \xi \in \mathrm{Hom}_{A_\alpha}(A_\alpha, A_\alpha) \\ \xi' \in \mathrm{Hom}_{A_{z'}}(A_{z'}, A_{z'}) \end{matrix}\right.$ is multiplication on the right by $\left\{\begin{matrix} z \\ z' \end{matrix}\right.$. Define

$$\widetilde{\widetilde{W}}_n(A_{\alpha,z'}) = V_n^{\overline{\varepsilon}(z')S_0 \oplus \overline{\varepsilon}(\alpha)S'_0}(A_{\alpha,z'}) \qquad (n \pmod 4)$$

where $\left\{\begin{matrix} \overline{\varepsilon}(z') : A_\alpha \longrightarrow A_{\alpha,z'} \\ \overline{\varepsilon}(\alpha) : A_{z'} \longrightarrow A_{\alpha,z'} \end{matrix}\right.$ is the inclusion. The preimage of

$$\widetilde{K}(A,1)^{\alpha'} = j\widetilde{K}_1(A)^\alpha \oplus \overline{p}\widetilde{K}_0(A)^\alpha \subseteq \widetilde{K}(A,1)$$

under the projection

$$q : \widetilde{K}_1(A_{z'}) = \overline{\varepsilon}\widetilde{K}_1(A) \oplus \overline{B}K_0(A) \oplus \mathrm{Nil}_+(A,1) \oplus \mathrm{Nil}_-(A,1)$$
$$\longrightarrow \widetilde{K}(A,1) = j\widetilde{K}_1(A) \oplus \overline{p}\widetilde{K}_0(A)$$

(as defined in Theorem 4.2) is

$$\widetilde{\widetilde{K}(A,1)^{\alpha'}} = \overline{\varepsilon}\widetilde{K}_1(A)^\alpha \oplus \overline{B}(T_0) \oplus \mathrm{Nil}_+(A,1) \oplus \mathrm{Nil}_-(A,1) \subseteq \widetilde{K}_1(A_{z'}) \quad ,$$

where

$$T_0 = (1-\alpha)^{-1}(\mathrm{im}(K_0(Z) \longrightarrow K_0(A)) \subseteq K_0(A) \; .$$

Further,

$$(1-\alpha')^{-1}(S'_0) = \overline{\varepsilon}\widetilde{K}_1(A)^\alpha \oplus \overline{B}(T_0) \oplus \mathrm{Nil}_+(A,1)^{\alpha'} \oplus \mathrm{Nil}_-(A,1)^{\alpha'} \subseteq \widetilde{K}_1(A_{z'}),$$

where

$$\mathrm{Nil}_\pm(A,1)^{\alpha'} = \{ \tau \in K_1(A_{z'}) \mid \nu \in \mathrm{Hom}_A(P,P) \text{ nilpotent },$$
$$\tau = \tau(1+\nu z'^{\pm 1} : P_{z'} \to P_{z'}) = \tau(1+(z\nu)z'^{\pm 1} : (zP)_{z'} \to (zP)_{z'}) \in \widetilde{K}_1(A_{z'})\}.$$

Hence

$$\tilde{V}_n^{\tilde{K}(A,1)^{\alpha'}}(A_{z'}) = V_n^{\widetilde{\tilde{K}(A,1)^{\alpha'}}}(A_{z'}) \qquad \text{(by definition)}$$
$$= V_n^{(1-\alpha')^{-1}(S_0')}(A_{z'}) \quad (= V_n(A_{z'}) \text{ if } \alpha = 1)$$

by the exact sequence of Theorem 3.3 .

All the squares of shape $\begin{array}{c}\longrightarrow\\ \downarrow\quad\downarrow\\ \longrightarrow\end{array}$, $\begin{array}{c}\longrightarrow\\ \uparrow\quad\uparrow\\ \longrightarrow\end{array}$ in the diagram

$$\begin{array}{ccccccccc}
V_{2i}(A) & \xrightarrow{\bar{\varepsilon}(\alpha)} & V_{2i}(A_\alpha) & \xrightarrow{B(\alpha)} & U_{2i-1}^{\tilde{K}_0(A)^\alpha}(A) & \xrightarrow{C(\alpha)} & V_{2i-1}(A) & \xrightarrow{\bar{\varepsilon}(\alpha)} & V_{2i-1}(A_\alpha) \\
\bar{B}(z')\downarrow\uparrow B(z') & & \bar{B}(z')\downarrow\uparrow B(z') & & \bar{B}(z')\downarrow\uparrow B(z') & & \bar{B}(z')\downarrow\uparrow B(z') & & \bar{B}(z')\downarrow\uparrow B(z') \\
\tilde{W}_{2i+1}(A_{z'}) & \xrightarrow{\bar{\varepsilon}(\alpha')} & \tilde{\tilde{W}}_{2i+1}(A_{\alpha,z'}) & \xrightarrow{B(\alpha')} & \tilde{V}_{2i}^{\tilde{K}(A,1)^{\alpha'}}(A_{z'}) & \xrightarrow{C(\alpha')} & \tilde{W}_{2i}(A_{z'}) & \xrightarrow{\bar{\varepsilon}(\alpha')} & \tilde{\tilde{W}}_{2i}(A_{\alpha,z'}) \\
\varepsilon(z')\downarrow\uparrow\bar{\varepsilon}(z') & & \varepsilon(z')\downarrow\uparrow\bar{\varepsilon}(z') & & \varepsilon(z')\downarrow\uparrow\bar{\varepsilon}(z') & & \varepsilon(z')\downarrow\uparrow\bar{\varepsilon}(z') & & \varepsilon(z')\downarrow\uparrow\bar{\varepsilon}(z') \\
W_{2i+1}(A) & \xrightarrow{\bar{\varepsilon}(\alpha)} & \tilde{W}_{2i+1}(A_\alpha) & \xrightarrow{B(\alpha)} & V_{2i}^{\tilde{K}_1(A)^\alpha}(A) & \xrightarrow{C(\alpha)} & W_{2i}(A) & \xrightarrow{\bar{\varepsilon}(\alpha)} & \tilde{W}_{2i}(A_\alpha)
\end{array}$$

commute, except for those round the shaded area, the columns are direct sum systems, and the rows through $\tilde{W}_{2i+1}(A_{z'}), W_{2i+1}(A)$ are exact (being the special cases $S_0' \subseteq \tilde{K}_1(A_{z'})$, $\{0\} \subseteq \tilde{K}_1(A)$ of the sequence of Theorem 5.2 in the range of dimensions considered in §6). It was shown in Lemma 3.4 of II. that the square

$$\begin{array}{ccc}
V_{2i}(A_\alpha) & \xrightarrow{B(\alpha)} & U_{2i-1}^{\tilde{K}_0(A)^\alpha}(A) \\
\bar{B}(z')\downarrow & & \downarrow\bar{B}(z') \\
\tilde{\tilde{W}}_{2i+1}(A_{\alpha,z'}) & \xrightarrow{B(\alpha')} & \tilde{V}_{2i}^{\tilde{K}(A,1)^{\alpha'}}(A_{z'})
\end{array}$$

skew-commutes for $\alpha = 1$. The proof generalizes immediately to the twisted case (for any α). It follows that both the squares round the shaded area (in the large diagram above) skew-commute, and that the row through $V_{2i}(A)$ is exact as well. But this is the special case $T = \{0\}$ of the sequence of Theorem 5.1 in the range of dimensions not already covered in §6. As explained in §5, this suffices to complete the proof of Theorems 5.1,5.2,5.3.

§7. Lower L-theories

Bass has defined lower K-groups $K_p(A)$ for $p < 0$, with natural split injections

$$\overline{B} : K_p(A) \longrightarrow K_{p+1}(A_z) ,$$

such that

$$K_{p+1}(A_z) = \overline{\varepsilon} K_{p+1}(A) \oplus \overline{B} K_p(A) \oplus Nil_+^{(p)}(A) \oplus Nil_-^{(p)}(A) .$$

There is defined a duality involution

$$* : K_p(A) \longrightarrow K_p(A)$$

for all $p < 0$, with

$$\overline{B}* = -*\overline{B} : K_p(A) \longrightarrow K_{p+1}(A_z)$$

$$*(Nil_\pm^{(p)}(A)) = Nil_\mp^{(p)}(A) .$$

In II. there were defined "lower L-theories" $L_n^{(p)}(A)$, for $p < 0$ and $n(\mathrm{mod}\ 4)$, by

$$L_n^{(p)}(A) = \ker(\varepsilon : L_{n+1}^{(p+1)}(A_z) \longrightarrow L_{n+1}^{(p+1)}(A))$$

with $L_n^{(0)}(A) = U_n(A)$.

Given a $*$-invariant subgroup $Q \subseteq K_0(A)$ let $\widetilde{Q} \subseteq \widetilde{K}_0(A)$ be the subgroup to which the natural projection $K_0(A) \to \widetilde{K}_0(A)$ sends Q, and define

$$L_n^Q(A) = U_n^{\widetilde{Q}}(A) \quad (n(\mathrm{mod}\ 4)).$$

Assuming inductively that $L_n^{Q'}(A_z)$ has already been defined for all $*$-invariant subgroups $Q' \subseteq K_{p+1}(A_z)$, define

$$L_n^Q(A) = \ker(\varepsilon : L_{n+1}^{\overline{\varepsilon} K_{p+1}(A) \oplus \overline{B} Q}(A_z) \longrightarrow L_{n+1}^{K_{p+1}(A)}(A))$$

for $*$-invariant subgroups $Q \subseteq K_p(A)$, $p < 0$.

Theorem 2.3 gives

<u>Theorem 7.1</u> <u>There is defined an exact sequence of abelian groups</u>

$$\ldots \longrightarrow H^{n+1}(Q'/Q) \longrightarrow L_n^Q(A) \longrightarrow L_n^{Q'}(A) \longrightarrow H^n(Q'/Q) \longrightarrow \ldots$$

<u>for $*$-invariant subgroups</u> $Q \subseteq Q' \subseteq K_p(A)$, $p < 0$.

[ ]

In particular, it follows that

$$L_n^Q(A) = \begin{cases} L_n^{(p+1)}(A) & \text{if } Q = \begin{cases} \{0\} \subseteq K_p(A) \\ K_p(A) \end{cases} \\ L_n^{(p)}(A) & \end{cases} .$$

Theorem 5.1 gives

<u>Theorem 7.2 There is defined an exact sequence of abelian groups</u>

$$\ldots \longrightarrow L_n^Q(A) \xrightarrow{\bar{\varepsilon}} L_n^{\bar{\varepsilon}Q}(A_\alpha) \xrightarrow{B} L_{n-1}^{(1-\alpha)^{-1}Q}(A) \xrightarrow{C} L_{n-1}^Q(A) \longrightarrow \ldots$$

<u>in a natural way, for *-invariant subgroups</u> $Q \subseteq K_p(A)$, $p < 0$.

[ ]

A lower L-theoretic analogue of Theorem 5.3 requires a lower K-theoretic analogue of Theorem 4.2. So far, this is only available in the untwisted case:

<u>Theorem 7.3 Let</u> $Q = \bar{\varepsilon}(R) \oplus \bar{B}(S) \subseteq K_{p+1}(A_z)$, <u>for some *-invariant subgroups</u> $R \subseteq K_{p+1}(A)$, $S \subseteq K_p(A)$ $(p < 0)$. <u>Then there is defined a direct sum system</u>

$$L_n^R(A) \underset{\varepsilon}{\overset{\bar{\varepsilon}}{\rightleftarrows}} L_n^Q(A_z) \underset{\bar{B}}{\overset{B}{\rightleftarrows}} L_{n-1}^S(A)$$

<u>in a natural way</u>.

[ ]

References

[1] S.Cappell — A splitting theorem for manifolds and surgery groups. Bull. Amer. Math. Soc. 77(1971),281-286.

[2] F.T.Farrell and W.C.Hsiang — A formula for $K_1R_\alpha[T]$. In Proc. Symp. in Pure Math. 17 (Categorical Algebra). Amer. Math. Soc. (1970), 192 - 218.

[3] S.Maumary — Proper surgery groups. (These proceedings).

[4] S.P.Novikov — The algebraic construction and properties of Hermitian analogues of K-theory for rings with

involution, from the point of view of Hamiltonian formalism. Some applications to differential topology and the theory of characteristic classes. I. Izv. Akad. Nauk S.S.S.R. ser. mat. 34(1970) 253-288 II. ibid., 34(1970) 475 - 500.

[5] A.A.Ranicki *Algebraic L-theory: I.Foundations. II.Laurent extensions.* to appear in Proc. Lond. Math. Soc.

[6] J.L.Shaneson *Wall's surgery obstruction groups for* $G\times Z$. Ann. of Maths. 90(1969) 296-334.

[7] L.C.Siebenmann *A total Whitehead torsion obstruction to fibering over the circle*. Comm. Math. Helv. 45(1970) 1-48.

[8] C.T.C.Wall *On the exactness of interlocking sequences*. l'Ens. Math. 12(1966) 95-100.

[9] C.T.C.Wall *Surgery on compact manifolds*. Academic Press (1970).

[10] C.T.C.Wall *Foundations of algebraic L-theory*. (These proceedings).

Trinity College,
Cambridge.

Surgery and Unitary K_2

by

Richard Sharpe

§1. Introduction. In this note we give an intuitive exposition of the main ideas in the author's papers [1] and [2]. The point of view we adopt is that of representation theory, in which one attempts to represent an algebraic object (in our case the unitary Steinberg group) on a geometric object (in our case, a surgery problem). By a thorough understanding of the geometric situation, one can deduce consequences for the algebraic object. In sections 2 and 3, we describe the geometric situation; section 4 gives the first hint of a connection between KU_2 and surgery problems; section 5 shows how the geometry predicts some properties of the unitary Steinberg group; section 6 contains a precisely stated theorem, and some computations.

§2. Surgery on Odd Dimensional Manifolds. (The references for this section are [1], [4], or see Shaneson's article in these proceedings.) Let $f: M^{2k+1} \to X^{2k+1}$ be a normal map. Then f determines an element $\theta \in L^s_{2k+1}(\pi) = SU(Z[\pi])/RU(Z[\pi])$, where $\pi = \pi_1(X)$, where SU is a stable unitary group, and RU is essentially the commutator subgroup. The invariant θ vanishes iff f is normally cobordant to a simple homotopy equivalence. We describe briefly how one obtains θ: First assume f is highly connected, so that $f_*: H_*(M) \to H_*(X)$ is an isomorphism for $* \neq k, k+1$. Then choose imbedded framed spheres $(S^k \times D^{k+1})_i$ $(i = 1,2,\ldots \ell)$ in M representing generators for $K_k(M) = \mathrm{Ker}\{H_k(M) \to H_k(X)\}$. Let U be the boundary connected sum of these spheres, and set $M_0 = M - \text{int } U$. Then $K_{k+1}(M_0, \partial U) \subset K_k(\partial U)$, and the submodule is actually a totally isotropic direct summand of half the rank of $K_k(\partial U)$. Moreover $K_k(U) \subset K_k(\partial U)$ (the map induced by the framing of the spheres) is also a totally isotropic direct summand of half the rank. Hence there is an automorphism $A \in SU(Z[\pi])$ of $K_k(\partial U)$ mapping $K_k(U)$ to $K_{k+1}(M_0, \partial U)$. In fact there are many such A's, but they all differ by right multiplication by elements of the form $\begin{pmatrix} a & 0 \\ 0 & a^{*-1} \end{pmatrix}\begin{pmatrix} 1 & b \\ 0 & 1 \end{pmatrix}$, where a is simple, and b has the form $x - (-1)^k x^*$.

Note that if $A = I$, then the composite $K_{k+1}(M_0, \partial U) \subset K_k(\partial U) \to K_k(U)$ is an isomorphism, so

$$
\begin{array}{ccccccccc}
0 \to K_{k+1}(M) \to & K_{k+1}(M, U) & \to & K_k(U) & \to K_k(M) \to 0 \\
 & \wr\wr & \nearrow\cong & & \\
 & K_{k+1}(M_0, \partial U) & & &
\end{array}
$$

shows $K_*(M) = 0$ for $* = k$ and $k+1$, so $f: M \to X$ is a homotopy equivalence.

§3. Relative Surgery. (The references here are [1], [4]). Consider a relative even dimensional $(n = 2k+2)$ normal map $f: (N, M) \to (Y, X)$. The picture of this is:

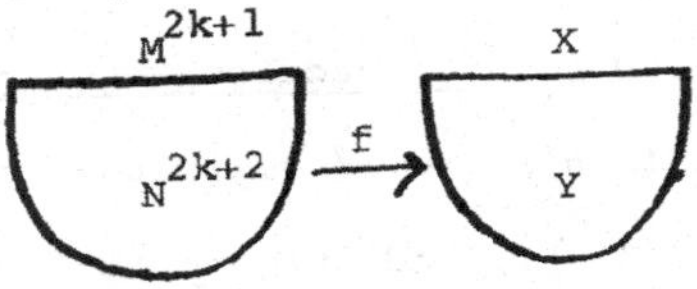

Again we assume f is highly connected so that the only remaining K-groups are those occurring in the exact sequence $0 \to K_{k+1}(M) \to K_{k+1}(N) \xrightarrow{i_*} K_{k+1}(N, M) \to K_k(M) \to 0$, where i_* is dual to the intersection form on $K_{k+1}(N)$. Now we choose some imbedded framed spheres in M representing generators of $K_k(M)$. Some persistence allows us to get $K_{k+1}(N, M)$ free and, moreover, the imbedded framed spheres $(S^k \times D^{k+1})_i$ in M extend to immersed framed discs $(D^{k+1} \times D^{k+1})_i$ in N, such that these discs give a basis for $K_{k+1}(N,M)$. These discs intersect each other in points, and give rise to an intersection matrix P. Also, using the dual basis for $K_{k+1}(N)$ gives a matrix Q for the intersection form on $K_{k+1}(N)$. The picture for this situation is:

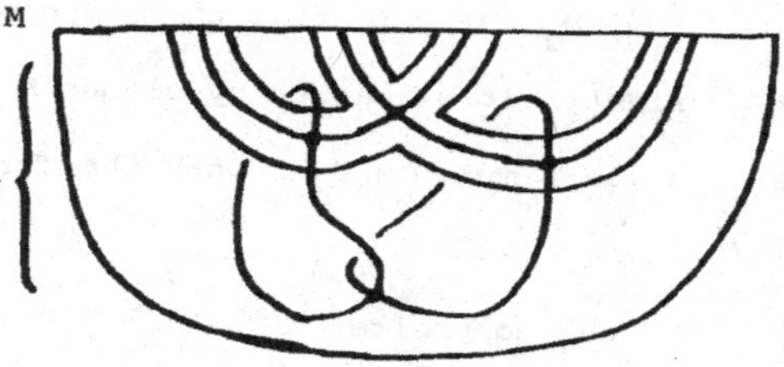

Further persistence yields $A = \begin{pmatrix} 1 & 0 \\ -P & 1 \end{pmatrix}\begin{pmatrix} 0 & 1 \\ (-1)^k & 0 \end{pmatrix}\begin{pmatrix} 1 & 0 \\ -Q & 1 \end{pmatrix}$ up to right multiplication by the indeterminacy of A (cf. §2)(all matrices interpreted over the ring $Z[\pi_1(Y)]$).

From this point one can go on to give a "Grothendieck group" description of the relative even dimensional surgery obstruction group (cf. [3] or [4] appendix 17G).

§4. <u>Connection Between</u> KU_2 <u>and Surgery Problems</u>. Now $RU(Z[\pi])$ is approximately the (perfect) commutator subgroup of $SU(Z[\pi])$, and is generated by "elementary unitary matrices." Moreover, these elementary unitary matrices correspond to elementary geometric operations on a surgery problem. On the other hand $StU(Z[\pi])$, the unitary Steinberg group, which is approximately the universal central extension of $RU(Z[\pi])$, is also generated by "elementary unitary matrices." Thus one might expect an action of

$StU(Z[\pi])$ on surgery problems in the following way: represent an element $y \in StU(Z[\pi])$ as a product of "elementary matrices" $y = y_n y_{n-1} \cdots y_1$. Then if $f_0: M^{2k+1} \to X$ is a surgery problem, one can construct cobordisms of it using y_1, y_2 etc., interpreted as elementary geometric operations. The picture is:

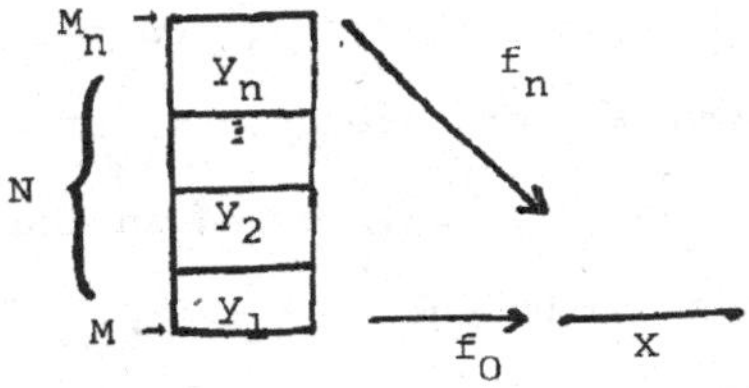

The odd dimensional invariant $A' \in SU(Z[\pi])$ at the top end of this cobordism is just $y_n y_{n-1} \cdots y_1 A$, where $A \in SU(Z[\pi])$ is the invariant at the bottom end, and the y's are interpreted as ordinary matrices. In particular, if $M = X$, f_0 is the identity, and $y \in KU_2(Z[\pi])$, then $f_n: M_n \to X$ at the top end is a simple homotopy equivalence. Thus in this case $y \in KU_2(Z[\pi])$ seems to determine an absolute even dimensional (= 2k+2) surgery problem, which yields an invariant on $L_{2k+2}(\pi)$. Roughly one might expect an epimorphism $KU_2(Z[\pi]) \to L_{2k+2}(\pi)$!

§5. Geometrical Prediction of a Bruhat-Like Decomposition for $StU(\Lambda)$.

Let $x \in RU(Z[\pi])$. By a theorem of Wall, ([4], p. 66) one can construct a normal map $f: M^{2k+1} \to X$ whose invariant A (cf. §2) is exactly x. But now $[x] = 0$ in $L_{2k+1}(\pi)$, so there is a normal cobordism N of $f: M^{2k+1} \to X$ to a simple homotopy equivalence. Thus we get in this way a relative even dimensional surgery problem

$(N,M) \to (X\times I, X\times 0)$, (which is a simple homotopy equivalence on the <u>other</u> boundary of N to $X \times 1$). Thus as in §3 we can write $x = \begin{pmatrix} I & \\ -PI & \end{pmatrix}\begin{pmatrix} 0 & I \\ (-1)^k & 0\end{pmatrix}\begin{pmatrix} I & \\ -Q & I\end{pmatrix}\begin{pmatrix} a & \\ & a^{*-1}\end{pmatrix}\begin{pmatrix} 1 & b \\ 0 & 1\end{pmatrix}$. Here $\pi_1(X\times I) \approx \pi_1(X\times 0)$ so this is a decomposition of arbitrary elements of $RU(Z[\pi])$ into five factors of very simple type. In fact, one can obtain this kind of decomposition for elements of $StU(Z[\pi])$. On the other hand we can ask whether these decompositions are unique. It is reasonable to expect that they are not, and that they depend on the cobordism N. Even if we fix N, they are not unique. In fact, in the construction of §3, at one point we chose immersed framed discs $(D^{k+1} \times D^{k+1})_i$ in N extending the framed imbedded spheres $(S^k \times D^{k+1})_i \subset M$, such that these discs represented a basis of $K_k(N,M)$. Changing the choice of the extensions will change the P and Q we get. In fact one can show that any new P' and Q' one can get in this way is related to the old ones by the equations
$$P' = P+X+(-1)^{k+1}X^*+X^*QX$$
$$Q' = B^{-1}QB^{-1*}$$
where $B = I + (-1)^{k+1}QX$ is a simple matrix (which is the matrix representing the basis change on $K_k(N,M)$). This gives a precise description of the non uniqueness of the Bruhat-like decomposition on $RU(Z[\pi])$ arising from different choices with respect to a fixed N. Following the scheme of §4, different choices of N should correspond to different liftings of x to $StU(Z[\pi])$. This suggests that the relations among Bruhat decompositions coming from a <u>fixed</u> choice of N should be <u>all the relations among Bruhat-like decompositions in $StU(Z[\pi])$</u>. This is in fact true, and is the main result of [2].

§6. The Periodicity Sequence and Some Computations. Although the previous sections used as base ring only the integral group rings of finitely presented groups, it turns out that all the conclusions about $StU(\Lambda)$ hold for an arbitrary ring with involution Λ. Now we shall describe precisely the groups we use:

$$SU_n^{\pm}(\Lambda) = \{A = \begin{pmatrix} \alpha & \beta \\ \gamma & \delta \end{pmatrix} \in E_{2n}(\Lambda) \mid A^*\begin{pmatrix} 0 & 1 \\ \pm 1 & 0 \end{pmatrix}A = \begin{pmatrix} 0 & 1 \\ \pm 1 & 0 \end{pmatrix} \text{ and } \alpha^*\gamma, \delta^*\beta \text{ have the form } x \mp x^*\}.$$

$$SU^{\pm}(\Lambda) = \varinjlim SU_n^{\pm}(\Lambda)$$

$$EU^{\pm}(\Lambda) = [SU^{\pm}(\Lambda), SU^{\pm}(\Lambda)]$$

$KU_0^{\pm}(\Lambda)$ = the Grothendieck group of stable, simple cogredience classes of simple matrices of the form $X \pm X^*$

$$KU_1^{\pm}(\Lambda) = SU^{\pm}(\Lambda)/EU^{\pm}(\Lambda)$$

$$KU_2^{\pm}(\Lambda) = H_2(EU^{\pm}(\Lambda)).$$

These groups are related by the following exact Periodicity Sequence:

$$KU_1^{\mp}(\Lambda) \to K_2(\Lambda)/\{c+c^*/c \in K_2(\Lambda)\} \to KU_2^{\pm}(\Lambda) \to KU_0^{\mp}(\Lambda) \to Z/dZ$$

(where $d = 2$ or 4 depending on whether the sequence contains $KU_0^-(\Lambda)$ or $KU_0^+(\Lambda)$). (Remark: A similar sequence may be obtained from Karoubi's work [5].) This periodicity sequence can be used to show:

1) Let k be a field with trivial involution of characteristic $\neq 2$. Then

i) $k^*/k^{*2} \to H_2(SL(k)) \to H_2(Sp(k)) \to I_2(k) \to 0$ is exact

$a \longmapsto \{a,-1\}$; the map $H_2(SL(k)) \to H_2(Sp(k))$ is the hyperbolic map

where $I_2(k)$ is the subgroup of the Wittgroup of k on

which the rank and determinant vanish.

ii) $H_2(SO(k)) \approx H_2(SL(k)) \oplus Z/2Z$

These two results have been obtained independently by A. Bak using his unitary symbols.

2) Let Z be the integers with the trivial involution. Then

i) $H_2(Sp^{quad}(Z)) = Z$

ii) $H_2(SO(Z)) = Z_2 \oplus Z_2$

Here, by $Sp^{quad}(Z)$ we mean that not only is a symplectic form preserved, but also a quadratic form as well (precisely, $Sp^{quad}(Z) = SU^{-}(Z)$).

Columbia University,
NEW YORK, N.Y. 10027

Bibliography

[1] Sharpe, R.W. "Surgery on compact manifolds: the bounded even dimensional case," to appear in the Annals of Math.

[2] Sharpe, R.W. "On the structure of the unitary Steinberg group," to appear in the Annals of Math.

[3] Sharpe, R.W. "Surgery of compact manifolds: the bounded even dimensional case," Yale thesis, 1970.

[4] Wall, C.T.C. Surgery on Compact Manifolds, Academic Press, 1970.

[5] Karoubi, M. "Modules quadratiques et periodicite en K-theorie" (Preprint).

SURGERY GROUPS AND INNER AUTOMORPHISMS

Lawrence R. Taylor

The object of this paper is to investigate the map induced on surgery groups by an inner automorphism of the group. To describe our results, let $\omega: G \longrightarrow Z_2$ be a homomorphism. $L_n(G,\omega)$ is the nth Wall group. This notation is ambiguous as there are Wall groups L_n^s for simple homotopy equivalence, L_n^h for homotopy equivalence, and there are other possibilities (see section 1 for a thorough discussion). L_n denotes any of these groups.

$Aut(G,\omega)$ denotes the group of automorphisms of G which preserve ω. There is a homomorphism $G \longrightarrow Aut(G,\omega)$ which takes an element in G to its inner automorphism. Wall groups are functors, so there is a homomorphism $Aut(G,\omega) \longrightarrow Aut(L_n(G,\omega))$. Finally there is a standard homomorphism from Z_2 to the automorphism group of any abelian group. If the group is written additively, $-1 \in Z_2$ just goes to multiplication by -1.

Theorem 1:

$$\begin{array}{ccc} G & \longrightarrow & Aut(G,\omega) \\ \Big\downarrow{\scriptstyle \omega} & & \Big\downarrow \\ Z_2 & \longrightarrow & Aut(L_n(G,\omega)) \end{array}$$

commutes.

Corollary 1.1: If $G \xrightarrow{\omega} Z_2$ is trivial, then any inner automorphism induces the identity on $L_n(G,\omega)$.

Corollary 1.2: If $(\text{Center } G) \xrightarrow{\omega} Z_2$ is onto, then $L_n(G,\omega)$ is a Z_2-vector space, and any inner automorphism induces the identity.

Here are examples of non-trivial actions on $L_n(G,\omega)$. Let $\alpha: Z \longrightarrow Aut(Z) \cong Z_2$ be onto, and let K be the semi-direct product of Z and Z by α. There is a homomorphism $\omega: K = Z \times_\alpha Z \longrightarrow Z \longrightarrow Z_2$ which is onto. By Wall [5], page 171, $L_1(K,\omega) \cong Z$ and $L_2(K,\omega) \cong Z \oplus Z_2$. Any element $k \in K$ such that $\omega(k) = -1$ gives an inner automorphism of K

which does not induce the identity on L_1 and L_2. $K \times Z \times Z$ with the obvious ω has inner automorphisms which are the identity on neither L_1, L_2, L_3, nor L_4.

Cappell and Shaneson [1] have defined surgery groups $\Gamma_n(\mathcal{F})$, where $\mathcal{F}: Z[G] \longrightarrow \Lambda'$ is an epimorphism of rings with involution. Any $g \in G$ induces an automorphism of $\mathcal{F}$ by acting on $Z[G]$ via conjugation by g and on Λ' via conjugation by the image of g in Λ'. This gives a homomorphism $G \longrightarrow \mathrm{Aut}(\mathcal{F})$. The Γ_n are functors, so there is a homomorphism $\mathrm{Aut}(\mathcal{F}) \longrightarrow \mathrm{Aut}(\Gamma_n(\mathcal{F}))$. As with Wall groups, Γ_n denotes any of the possible torsions or projective classes which can be used to manufacture Cappell-Shaneson groups.

Theorem 2:

$$\begin{array}{ccc} G & \longrightarrow & \mathrm{Aut}(\mathcal{F}) \\ \Big\downarrow{\scriptstyle \omega} & & \Big\downarrow \\ Z_2 & \longrightarrow & \mathrm{Aut}(\Gamma_n(\mathcal{F})) \end{array}$$

commutes.

There are also relative Wall and Γ groups associated to any groupoid of finite type. If $\mathcal{G}$ is the groupoid, there is given a homomorphism of groupoids $\omega: \mathcal{G} \longrightarrow Z_2$. $\mathrm{Aut}(\mathcal{G}, \omega)$ will denote the automorphism group of the groupoid whose elements preserve ω. Each element of $\prod_{G \in \mathcal{G}} G$ gives a collection of automorphisms, f_G, where f_G is just the inner automorphism on G given by the component of G in the product. $(\mathcal{G}) \subseteq \prod_{G \in \mathcal{G}} G$ is the subgroup such that $\{f_G\} \in \mathrm{Aut}(\mathcal{G})$. $(\mathcal{G})_\omega$ is the subgroup such that $\omega(g_{G_i}) = \omega(g_{G_j})$ where G_i, $G_j \in \mathcal{G}$ are arbitrary and g_G is the component of G in the product.

Conjecture:

$$\begin{array}{ccc} (\mathcal{G})_\omega & \longrightarrow & \mathrm{Aut}(\mathcal{G}, \omega) \\ \Big\downarrow{\scriptstyle \omega} & & \Big\downarrow \\ Z_2 & \longrightarrow & \mathrm{Aut}(L_n(\mathcal{G}, \omega)) \end{array}$$

commutes.

There is a similar conjecture for Γ groups.

Almost all that I can prove is

Theorem 3: The conjecture is true for the Wall groups L^s_{2k}, L^h_{2k}, and L^h_{2k+1} if $\mathcal{G}$ is a pair. I know nothing about relative Γ groups.

Remarks: The obvious versions of Corollaries 1.1 and 1.2 are valid for Theorems 2 and 3, and all are proved by simple diagram chasing.

Notice that the groups are not assumed to be finitely generated or finitely presented.

Section 1: Preliminaries.

Wall and Γ groups can actually be defined for any rings with involution, not just integral group rings. Even Wall and Γ groups are obtained by a Grothendieck construction from modules with quadratic form plus additional structure. If Λ is a ring with involution, and, if $A \subset \tilde{K}_0(\Lambda)$ is a subgroup invariant under the induced involution, then we can define Wall groups $L^A_{2k}(\Lambda,-)$ by insisting that our module be a projective module whose image in $\tilde{K}_0$ is contained in A. If $B \subset \tilde{K}_1(\Lambda)$ is invariant under the induced involution, and, if we insist that our modules be free and based, then we can define Wall groups $L^B_{2k}(\Lambda,-)$ by insisting that the torsion of the adjoint map lie in B.

Even Γ groups can also be defined. Given $\mathcal{F}:\Lambda \longrightarrow \Lambda'$ a local epimorphism (see [1]) we have quadratic modules over Λ with nice properties when tensored over Λ'. If $A \subset \tilde{K}_0(\Lambda')$ and $B \subset \tilde{K}_1(\Lambda')$ are invariant subgroups, we can define $\Gamma^A_{2k}(\mathcal{F})$ and $\Gamma^B_{2k}(\mathcal{F})$ by insisting in the first case that our modules when tensored are projective with class in A and in the second case that our modules are free and based over Λ and the adjoint map when tensored over Λ' has torsion in B (this uses lemma 1.2 of [1]).

Hopefully the above motivates considering $Q_k(\Lambda,-)$, the category whose objects are $\xi \in [P, \lambda, \mu]$ where P is a right Λ-module; $\lambda: P \times P \longrightarrow \Lambda$ is a map; $\mu: P \longrightarrow \Lambda/\{x+(-1)^k\bar{x}\}$ is a map; and the five relations of Wall [5] page 45 are satisfied (with $\chi_N(x) = 0$). Morphisms are just Λ-module homomorphisms which preserve this extra structure. $Q^t_k(\Lambda,-)$ is a similar category: the only difference is that now we require P to be free and based. The first two paragraphs of this section just say that Wall and Γ groups, with any torsions,etc. are constructed naturally from various subcategories of Q_k and Q_k^t. We remark that any map of rings with involution induces a functor on Q_k and on Q_k^t.

Odd Wall and Γ groups are more tricky. L_{2k+1}^B can be defined as a quotient of $U(\Lambda;B)$, where $U(\Lambda;B) \subseteq U(\Lambda)$ is the subgroup of the infinite unitary group over Λ whose elements are those matrices in $U(\Lambda)$ with torsions in B. L_{2k+1}^A can be defined directly (Novikov [4]) or by a theorem of Farrell-Wagoner which equates L_{2k+1} to L_{2k} of some other ring. We will use this latter method.

Γ_{2k+1}^B can be defined as those elements of $L_{2k+1}^B(\Lambda')$ which come from elements in $U(\Lambda)$ with a special property. One might define $\Gamma_{2k+1}^A = \{ x \in L_{2k+1}^A(\Lambda') \mid 2x$ is in the image of $\Gamma_{2k+1}^{\tilde{K}_1(\Lambda')}(\mathcal{F}) \longrightarrow L_{2k+1}^{K_1(\Lambda')}(\Lambda') \longrightarrow L_{2k+1}^A(\Lambda') \}$. I know of no use for groups like Γ^A, so the definition is not very important.

As is usual with surgery groups, the proof divides into an even and an odd case.

Section 2: The Even Case.

Let Λ be a ring with involution $^-$, and suppose u is a unit of Λ such that $u\bar{u} = \pm 1$. It follows that $\bar{u}u = u\bar{u}$. There is an isomorphism of rings with involution $r_u : \Lambda \longrightarrow \Lambda$ given by $r_u(x) = u^{-1}xu$. Any such map induces functors $u_* : Q_k(\Lambda,-) \longrightarrow Q_k(\Lambda,-)$ and $u_* : Q_k^t(\Lambda,-) \longrightarrow Q_k^t(\Lambda,-)$.

On the category Q_k (or Q_k^t) there is a functor c such that $c([P, \lambda, \mu]) = [P, -\lambda, -\mu]$. Note $c \circ c$ is the identity.

Theorem: Let $\xi \in Q_k(\Lambda,-)$. Then $u_*(\xi)$ is naturally isomorphic to $\begin{cases} \xi & \text{if } u\bar{u} = 1 \\ c(\xi) & \text{if } u\bar{u} = -1 \end{cases}$. If $\xi \in Q_k^t(\Lambda,-)$ then the torsion of the equivalence is in $\{u\}$, where $\{u\}$ is the subgroup of $\tilde{K}_1(\Lambda)$ generated by the unit u.

proof: Under u_*, the module P goes to $P \otimes_\Lambda \Lambda = P^u$. where Λ is made into a Λ-module using r_u. Let $i : P \longrightarrow P^u$ be the natural map. Then i is r_u-linear, that is, $x \times a = i(i^{-1}(x) \cdot uau^{-1})$ where $\times$ is the product in P^u and $\cdot$ is the product in P. $u_*\lambda(x,y) = r_u(\lambda(i^{-1}(x), i^{-1}(y)))$ and

$u_*\mu(x) = r_u^k\mu(i^{-1}(x))$ where r_u^k is the map induced on $\Lambda/\{x+(-1)^k\bar{x}\}$ by r_u. If P is based, use i to base P^u.

Consider $\xi^u = [P^u, \lambda_u, \mu_u]$ where $\lambda_u(x,y) = \bar{u}\,\lambda(i^{-1}(x), i^{-1}(y))u$ and $\mu_u(x) = \bar{u}\,\mu(i^{-1}(x)u$. Clearly

$$\xi^u = \begin{cases} u_*(\xi) & \text{if } u\bar{u} = 1 \\ c(u_*(\xi)) & \text{if } u\bar{u} = -1 \end{cases}.$$ Hence $\xi^u \in Q_k$ or Q_k^t.

Define $j: P \longrightarrow P^u$ by $j(x) = i(x) \times u^{-1} = i(x \cdot u^{-1})$. j is a Λ-module isomorphism and one easily checks

$$\begin{array}{ccc} P \times P & \xrightarrow{\lambda} & \Lambda \\ \downarrow{\scriptstyle j\times j} & \nearrow{\scriptstyle \lambda_u} & \\ P^u \times P^u & & \end{array} \qquad \text{and} \qquad \begin{array}{ccc} P & \xrightarrow{\mu} & \Lambda/\{x+(-1)^k\bar{x}\} \\ \downarrow{\scriptstyle j} & \nearrow{\scriptstyle \mu_u} & \\ P^u & & \end{array}$$

commute. Hence, in $Q_k(\Lambda,-)$, ξ and ξ^u are isomorphic.

If $\xi \in Q_k^t(\Lambda,-)$ then the torsion of the isomorphism clearly lies in $\{u\}$.

Since $c \circ c = id$, $u_*(\xi) = \begin{cases} \xi & \text{if } u\bar{u} = 1 \\ c(\xi) & \text{if } u\bar{u} = -1 \end{cases}$, where here equal means isomorphic. It is easy to check that j gives a natural isomorphism. Q.E.D.

The theorem plus the discussion in section 1 almost proves Theorems 1 and 2. We need only further note that, if ξ goes to an element of L_{2k} (or Γ_{2k}), then $c(\xi)$ goes to the inverse element.

For Theorem 3, we have the following situation. We have rings with involution Λ and $\Lambda_1, \ldots, \Lambda_n$; maps $h_i: \Lambda_i \longrightarrow \Lambda$ of rings with involution; and units $u \in \Lambda$, $u_i \in \Lambda_i$ such that $u\bar{u} = u_i\bar{u}_i = \pm 1$ and each

$$\begin{array}{ccc} \Lambda_i & \xrightarrow{h_i} & \Lambda \\ \downarrow{\scriptstyle r_{u_i}} & & \downarrow{\scriptstyle r_u} \\ \Lambda_i & \xrightarrow{h_i} & \Lambda \end{array}$$ commutes.

Wall's description ([5] page 72) of the groups $L_{2k}^s(\Lambda;\Lambda_1,\ldots,\Lambda_n)$ (dropping mention of torsions gives L_{2k}^h) and our theorem easily show r_u induces $\begin{cases} id & \text{if } u\bar{u} = 1 \\ -id & \text{if } u\bar{u} = -1 \end{cases}$ on these relative Wall groups.

Section 3: The Odd Case

Again Λ is a ring with involution - and u is a unit such that $u\bar{u} = \pm 1$. r_u induces a map on $U(\Lambda)$ which covers the map induced on Wall groups. $U(\Lambda)$ is the limit of the finite unitary groups $U_{2n}(\Lambda)$. r_u induces maps $u_*: U_{2n}(\Lambda) \longrightarrow U_{2n}(\Lambda)$ which takes the matrix (a_{ij}) to $(u^{-1}a_{ij}u)$. Let $c_n(u)$ be the diagonal matrix

$$\left(\begin{array}{ccc|ccc} u^{-1} & & 0 & & & \\ & \ddots & & & 0 & \\ 0 & & u^{-1} & & & \\ \hline & & & \bar{u} & & 0 \\ & 0 & & & \ddots & \\ & & & 0 & & \bar{u} \end{array}\right)_{2n \times 2n}$$

with u^{-1} in the first n positions and $\bar{u}$ in the last n. Given $M \in U_{2n}(\Lambda)$, M' denotes the matrix with basis e_i, f_i changed to e_i, $-f_i$ (see Wall [5] page 62).

Theorem: $c_n(u) \in U_{2n}(\Lambda)$ and if $M \in U_{2n}(\Lambda)$,

$$u_*(M) = \begin{cases} c_n(u)\cdot M\cdot c_n(u)^{-1} & \text{if } u\bar{u} = 1 \\ c_n(u)\cdot M'\cdot c_n(u)^{-1} & \text{if } u\bar{u} = -1 \end{cases}.$$

proof: That $c_n(u) \in U_{2n}(\Lambda)$ is just a formal check. Our equation for $u_*(M)$ is obvious. Q.E.D.

Since Wall groups are abelian, and since M' is the inverse in the Wall group, we have proved Theorem 1 for $\{u\} \subseteq B \subseteq \tilde{K}_1(\Lambda)$.

Now Farrell-Wagoner [3] showed $L^A_{2k-1}(\Lambda,-)$ and $L^A_{2k}(\lambda\Lambda,-)$ are naturally isomorphic. $\lambda\Lambda = \ell\Lambda/m\Lambda$, where $\ell\Lambda$ is the ring of locally-finite matrices over Λ and $m\Lambda$ is the ideal in Λ of matrices with only finitely many non-zero entries. Note $A \subseteq K_0(\Lambda) = \tilde{K}_1(\lambda\Lambda)$ by Farrell-Wagoner [2].

Naturality shows

$$\begin{array}{ccc} L^A_{2k-1}(\Lambda) & \xrightarrow{r_u} & L^A_{2k-1}(\Lambda) \\ \downarrow & & \downarrow \\ L^A_{2k}(\lambda\Lambda) & \xrightarrow{r_\mu} & L^A_{2k}(\lambda\Lambda) \end{array}$$

commutes, where μ is the infinite diagonal matrix with all u's. Our result for the even case now carries over to this case.

The groups Γ_{2k-1} are subgroups of odd Wall groups so we are done in this case also.

For $L^h_{2k-1}(\Lambda;\Lambda_1,\ldots,\Lambda_n)$ we use a relative version of the above Farrell-Wagoner theorem to place ourselves in the case $L^s_{2k}(\lambda\Lambda;\lambda\Lambda_1,\ldots,\lambda\Lambda_n)$, which we know.

The reader may wonder why we do not just use the well known Shaneson-Wall splitting theorem instead of this strange theorem of Farrell-Wagoner. The point is that the Shaneson-Wall formula is known only for finitely generated, finitely presented integral group rings and rings with 1/2, whereas the Farrell-Wagoner formula is valid for an arbitrary ring with involution. Non-finitely generated groups play a role in surgery on paracompact manifolds, so we wish to avoid any finiteness assumptions on G.

1. S. Cappell and J. Shaneson, The codimension two placement problem and homology equivalent manifolds, preprint, 1972.
2. T. Farrell and J. Wagoner, Infinite matrices in algebraic K-theory and topology, preprint, U.C. Berkeley, 1971.
3. T. Farrell and J. Wagoner, private communication.
4. S. Novikov, Algebraic construction and properties of Hermitian analogs of K-theory over rings with involution from the viewpoint of Hamiltonian formalism. Applications to differential topology and the theory of characteristic classes. I., IZV. Akad. Nauk SSSR Ser. Mat. 34(1970) = Math. USSR IZV. 4(1970), 257-292.
5. C.T.C. Wall, Surgery on compact manifolds, Academic Press, 1971.

MAYER-VIETORIS SEQUENCES IN HERMITIAN K-THEORY

Sylvain E. Cappell*

PREFACE

This lecture will describe the consequences for unitary K-theory of infinite groups of my geometric splitting theorems. The geometric splitting theorems for manifolds (Section 1) lead to a formula for the surgery groups of free products, Theorem 5, and to Mayer-Vietoris type exact sequences for amalgamated free products and related constructions, especially Theorems 8 and 9 (Section 2). From these results, the surgery groups of many infinite groups can be computed. In particular, the surgery groups of some geometrically important infinite groups, free groups in Theorem 16, fundamental groups of surfaces, etc. are explicitly given (Section 3).

Some earlier results on surgery groups of infinite groups are described below in Section 2 and also in J. Shaneson's lecture in these Proceedings. The appendix of the present paper describes a problem in unitary algebraic K-theory whose solution would lead to further extensions of the geometric and hence of the algebraic results of this lecture.

* The author is a Sloan Foundation fellow and was partially supported by an N.S.F. grant.

1. The h and s splitting problems

We shall be concerned with the following situation. Let Y be a connected closed $(n+1)$ dimensional manifold (or Poincaré complex) and X an n-dimensional connected closed submanifold (or sub-Poincaré complex), $j:X \subset Y$, $n \geq 5$,* with $\pi_1(X) \to \pi_1(Y)$ injective. Moreover, we assume X satisfies any one of the following three equivalent conditions, which are trivially satisfied if both X and Y are orientable:

(1) X cuts some neighborhood of itself in Y into two components.

(2) The normal bundle of X in Y is trivial.

(3) $j^*\omega_1(Y) = \omega_1(X)$. Here, ω_1 denotes the first Stiefel-Whitney class.

Now, let W be a differentiable, piecewise-linear (P.ℓ.), or topological manifold and $f:W \to Y$ a homotopy equivalence. The map f is said to be "splittable", or more precisely "splittable along X", if it is homotopic to a map, which we continue to call f, transverse regular to X (whence $f^{-1}(X)$ is an n-dimensional submanifold of Y), with the restrictions of f to $f^{-1}(X) \to X$ and to $f^{-1}(Y-X) \to (Y-X)$ being homotopy equivalences. If $(Y-X)$ has 2 components this means that f restricts to a homotopy equivalence to each. In the present paper, for a differentiable (respectively; P.ℓ., topological) manifold, "submanifold" means a differentiable (resp; P.ℓ. locally-flat, topological locally-flat) submanifold.

---

* A special result for $n=4$ is discussed in Theorem 3. In [CS1], results are proved on the stable splitting problem for $n=5$ by using special low-dimensional arguments and the results of the present paper.

<u>h-Splitting Problem:</u> When is W h-cobordant to a manifold W' with the induced homotopy equivalence $f':W' \to Y$ splittable along X ?

<u>s-Splitting Problem:</u> When is $f:W \to Y$ splittable along X ?

Corresponding to the number of components of $Y-X$, these problems have two cases. Let $G = \pi_1(Y)$ and $H = \pi_1(X)$: as we assumed $\pi_1(X) \to \pi_1(Y)$ injective, we have $H \subset G$. This discussion and the methods of the present paper, also apply equally well to relative splitting problems.

<u>Case A :</u> $Y-X$ has two components. In this case, let Y_1 and Y_2 denote the closures in Y of the two components of $Y-X$, so that $Y = Y_1 \cup_X Y_2$. Set $G_i = \pi_1(Y_i)$, $i = 1,2$, ; the inclusion $X \subset Y_i$, induces $\xi_i:H \to G_i$ with , as $H \subset G$, ξ_i an inclusion, $i = 1,2$. By Van Kampen's theorem, G is the free product with amalgamation $G = G_1 *_H G_2$.

<u>Case B :</u> $Y-X$ has one component. In this case, let Y' denote the manifold with boundary obtained by cutting Y' along X ; that is, the boundary of Y' is $X_1 \cup X_2$, $X_1 \cong X_2 \cong X$ and Y is obtained from Y' by identifying X_1 with X_2 . Set $J = \pi_1(Y)$. Corresponding to the inclusions $X_i \subset Y'$, there are two maps $\xi_i:H \to J$ which are injective as $\pi_1(X) \subset \pi_1(Y)$. (To be more precise about basepoints, choose $p \in X$ and correspondingly $p_i \in X_i$, $i = 1,2$, Let γ be a path in Y' from p_1 to p_2 . There is the obvious inclusion $H = \pi_1(X_1,p_1) \to \pi_1(Y',p_1) = J$. Using the identification $[\gamma]:\pi_1(Y',p_2) \to \pi_1(Y',p_1)$ induced from γ , we get another inclusion $H = \pi_1(X_2,p_2) \to \pi_1(Y',p_2) \xrightarrow{[\gamma]} \pi_1(Y',p_1) = J$. The loop γ in Y , represents $t \in G$. Then from two applications

of Van Kampen's theorem we get $G \cong J *_H \{t\}$ where we have:

Definition: For $\xi_i : H \to J$, $i = 1,2$, two injective group homomorphisms, let

$$J *_H \{t\} = Z * J / \langle \{t^{-1}\xi_1(u)t\xi_2(u)^{-1} \mid u \in H\} \rangle$$

where Z is an infinite cyclic group generated by t. As usual $\langle \{P\} \rangle$ denotes the smallest normal subgroup containing $\{P\}$.

This $J *_H \{t\}$ notation [W1] is concise. Note, however, that the group obtained depends on the inclusions ξ_i.

The most general previous result in case B is the splitting theorem of Farrell and Hsiang for $G = Z \times_\alpha H$ [FH1]. In the notation introduced above, this corresponds to both ξ_1 and ξ_2 surjective, and $G = Z \times H$ corresponds to $\xi_1 = \xi_2$ being surjective. Earlier results on Case B were obtained by Browder and Levine [BL], [B2], in the setting of the fibration problem, the determination of which high-dimensional manifolds fibre over a circle, for $G = Z$. The fibration problem is related to the problem of deciding when an open high dimensional manifold is the interior of some closed manifold. This was solved in the simply connected case by Browder, Levine and Livesay [BLL] and in general by Siebenmann [S1]. Siebenmann's result implies a splitting theorem for certain open manifolds. The high-dimensional fibration problem was solved by Farrell [F].

Case A was solved by Browder [B1] for Y_1, Y_2 and X all simply connected. As a consequence of the development of relative non-simply connected surgery theory [W2], Wall showed that the problem could always be solved for* $G_1 = G_2 = G = H$. R. Lee [L1] made an important advance when he solved the problem for n even, $H = 0$, and G having no 2-torsion.

All the above splitting theorems for compact manifolds are special cases of Theorems 1 and 2 of the present paper. We treat Cases A and B by the same geometric methods. From the geometric results on Case A (respectively, Case B) we derive the Mayer-Vietoris sequence for surgery groups of Theorem 8 (resp; Theorem 9). The extension of the geometric results from the differentiable and P.ℓ. cases to topological manifolds makes use of topological transversality [KS] and surgery [L2] [KS]. We require some algebraic notation.

Definition: A subgroup H of a group G is said to be square-root closed in G if, for all $g \in G$, $g^2 \in H$ implies $g \in H$.

In [C1] such subgroups were characterized by an equivalent condition called "two-sided subgroup". The formulation of the condition in terms of square-root closed subgroups was suggested to me by C. Miller.

Examples:

(1) If H is normal in G , $H \triangleleft G$, then H is square-root closed in G if and only if G/H has no elements of order 2. In particular, the trivial subgroup is square-root closed in G if and only if G has no elements of order 2.

*Wall's result is more general than the quoted statement and requires only $H = G_1$ and hence $G_2 = G$. However, we are here concerned only with $\pi_1(X) \to \pi_1(Y)$ injective.

(2) Any subgroup of a finite group of odd order is square-root closed. In general, a subgroup H of a finite group G is square-root closed if and only if H contains all elements of G of 2-primary order.

(3) H is square-root closed in $G_1 *_H G_2$ if and only if H is square-root closed in both G_1 and G_2.

(4) Given inclusions $\xi_i : H \to J$, $i = 1,2$, then H is square-root closed in $J *_H \{t\}$ if and only if both $\xi_1(H)$ and $\xi_2(H)$ are square-root closed in J. In particular, (or from (1) above):

(5) H is square-root closed in $Z \times_\alpha H$.

(6) Let G be a free group and H a subgroup generated by a non-square element of G. Then H is square-root closed in G. (This is used in computing below the surgery groups of all two-manifolds.)

(7) If H is square-root closed in G, $Z \times H$ is square-root closed in $Z \times G$.

Note that for X an n-dimensional closed submanifold with trivial normal bundle of the $(n+1)$ dimensional manifold Y with $Y-X$ having two components, $Y = Y_1 \cup_X Y_2$ (respectively; one component with $Y = Y'/$(identifying two copies of X) and with $\xi_i : \pi_1(X) \to \pi_1(Y')$ the induced maps), $\pi_1(X) \to \pi_1(Y)$ is injective with square-root closed image if and only if this is true for each of the induced maps $\pi_1(X) \to \pi_1(Y_i)$ (resp; $\xi_i : \pi_1(X) \to \pi_1(Y')$, $i = 1,2$). This follows from example 3 (resp; 4) of square-root closed subgroups above.

We describe first, in Theorem 1, the fundamental groups for which we show that the h or s splitting problems can, for any homotopy equivalence, be solved. For a group G, $Wh(G)$ denotes the Whitehead group of G, and $\tilde{K}_o(G)$ denotes the reduced projective class group of the ring $Z[G]$.

Theorem 1 : Let Y be a closed manifold* of dimension $n+1$, $n \geq 5$ with $\pi_1(Y) = G$ and X a closed submanifold* of dimension n of Y with trivial normal bundle and $\pi_1(X) = H \subset G$ a square-root closed subgroup. Assume $Y-X$ has two components (respectively; one component) with fundamental groups G_i and $\xi_i : H \to G_i$ (resp; group J and $\xi_i : H \to J$), $i = 1,2$ the induced maps.

(i) If $\tilde{K}_o(H) = 0$, or $\xi_{1*} - \xi_{2*} : \tilde{K}_o(H) \to \tilde{K}_o(G_1) \oplus \tilde{K}_o(G_2)$ (resp; $\xi_{1*} - \xi_{2*} : \tilde{K}_o(H) \to \tilde{K}_o(J)$) is injective or even just $H^{n+1}(Z_2; \mathrm{Ker}(\tilde{K}_o(H) \to \tilde{K}_o(G_1) \oplus \tilde{K}_o(G_2))) = 0$ (resp; $H^{n+1}(Z_2; \mathrm{Ker}(\tilde{K}_o(H) \xrightarrow{\xi_{1*} - \xi_{2*}} \tilde{K}_o(J))) = 0$) then for any homotopy equivalence $f: W \to Y$, W a closed manifold, W is h-cobordant to some manifold W' with the induced homotopy equivalence $f': W' \to Y$ splittable.

(ii) If $Wh(G_1) \oplus Wh(G_2) \to Wh(G)$ (resp; $Wh(J) \to Wh(G)$) is surjective, then every homotopy equivalence $f: W \to Y$, W a closed manifold, is splittable.

Note that the hypothesis of (i) is always satisfied if one of the inclusions $H \to G_i$ has a retraction. The hypothesis of both

---

* Or Poincaré-complex.

(i) and (ii) are satisfied for H a member of a large class of groups constructed by Waldhausen [W1].

The ring $Z[H]$ acquires, as usual, a conjugation obtained from the homomorphism defined by the first Stiefel-Whitney class of X, $\omega_1 : \pi_1(X) = H \to Z_2 = \{\pm 1\}$ by the formula $\bar{g} = \omega_1(g)g^{-1}$, $g \in H \subset Z[H]$. Of course, if X is orientable $\omega_1(g) = 1$ for all $g \in H$. This involution of $Z[H]$ induces the Z_2 action on $\tilde{K}_o(H)$ referred to in (i) of Theorem 1. There is a relative form of Theorem 1 in which we begin with a splitting of $\partial W \to \partial Y$ along ∂X and consider the problem of extending this to a splitting of the homotopy equivalence $W \to Y$ along X.

A note to the reader: The reader unfamiliar with the algebraic k-theory referred to below may prefer to concentrate on Theorem 1, and ignore the refinements necessary for the more general Theorem 2. The basic geometric ideas are much the same and the algebraic K-theory needed to prove Theorem 1 is almost trivial. Theorem 1 suffices to prove the results of Section 2 when the relevant Whitehead groups or projective class groups are zero.

A study of the obstructions to the solution of the general h and s splitting problems, as in Theorem 2, requires the notion of Whitehead torsion [M] [Wh] and some algebraic K-theory. Following Waldhausen we call the groups $G_1 *_H G_2$ or $J *_H \{t\}$, produced from the inclusions $\xi_i : H \to G_i$ or $\xi_i : H \to J$, $i = 1,2$, , generalized free products. In [W1], a number of general results on Whitehead groups of generalized free products are proved. Stallings had previously given a decomposition formula for $Wh(G_1 * G_2)$ and showed that the Whitehead group of a free group is zero [St].

Higman had earlier shown that $Wh(Z) = 0$. Bass, Heller and Swann proved a decomposition formula for $Wh(Z \times H)$ and showed that the Whitehead groups of free abelian groups were zero [BHS]. For a general group H, their formula for $Wh(Z \times H)$ involves two copies of a certain group of $Z[H]$-linear nilpotent maps. These results were generalized by Farrell and Hsiang to obtain a similar description of $Wh(Z \times_\alpha H)$ [FH2]. Waldhausen's results on the Whitehead groups of generalized free products [W1] extends these results. He proved the exactness of the sequences

$$Wh(H) \xrightarrow{\xi_{1*}-\xi_{2*}} Wh(G_1) \oplus Wh(G_2) \xrightarrow{\beta} Wh(G_1 *_H G_2)$$

$$Wh(H) \xrightarrow{\xi_{1*}-\xi_{2*}} Wh(J) \xrightarrow{\beta} Wh(J *_H \{t\})$$

where the map β is induced from the obvious inclusions. He showed that $Ker(K_o(H) \xrightarrow{\xi_{1*}-\xi_{2*}} K_o(G_1) \oplus K_o(G_2))$ (respectively; $Ker(K_o(H) \xrightarrow{\xi_{1*}-\xi_{2*}} K_o(J)))$ is a direct summand of $Coker(\beta)$; in fact, the inclusion of the summand $\delta : Ker(\xi_{1*}-\xi_{2*} : \tilde{K}_o(H)) \to Coker(\beta)$ is easily described directly by an extension of the map of [BHS] for $G = Z \times H$ from $\tilde{K}_o(H)$ to $Wh(Z \times H)/Wh(H)$. He further obtained a description of the other direct summand of CoKer β as a certain group of $Z[H]$ linear nilpotent maps. This generalizes the Bass-Heller-Swann Nil group. Note, however, that as $Wh(Z \times H) = Wh(H) \oplus \tilde{K}_o(H) \oplus \widetilde{Nil}(H) \oplus \widetilde{Nil}(H)$, in this case Waldhausen's group of nilpotent maps reduces to two copies of the reduced Nil groups. Furthermore, Waldhausen shows that the second summand of coker β, his group of nilpotent maps, often vanishes. We write

$$\phi:Wh(G_1 *_H G_2) \to Ker(\tilde{K}_o(H) \xrightarrow{\xi_{1_*}-\xi_{2_*}} \tilde{K}_o(G_1) \oplus \tilde{K}_o(G_2))$$

$$(resp; \quad \phi:Wh(J *_H \{t\}) \to Ker(\tilde{K}_o(H) \xrightarrow{\xi_{1_*}-\xi_{2_*}} \tilde{K}_o(J)))$$

for the maps defined by Waldhausen; we also write for the quotient map to the second summand of Coker β

$$\eta:Wh(G_1 *_H G_2) \to Wh(G_1 *_H G_2)/Wh(G_1)+Wh(G_2)+Ker(\tilde{K}_o(H)\to\tilde{K}_o(G_1)\oplus\tilde{K}_o(G_2))$$

$$(resp; \eta:Wh(J*_H\{t\})\to Wh(J*_H\{t\})/Wh(J)+Ker(\tilde{K}_o(H) \xrightarrow{\xi_{1_*}-\xi_{2_*}} \tilde{K}_o(J))) .$$

Directly, η can be described as the projection of $Wh(G_1 *_H G_2)$ or $Wh(J *_H \{t\})$ to $CoKer(\delta)$; the notation does not presume that $Ker(\xi_{1_*}-\xi_{2_*}:\tilde{K}_o(H))$ splits from $Wh(G_1 *_H G_2)$ or $Wh(J *_H \{t\})$ but only from the quotient Coker (β) .

Given as above inclusions $\xi_i:H \to G_i$ (resp; $\xi_i:H \to J$) and given homomorphisms $\omega_i:G_i \to Z_2 = \{\pm1\}$, $i = 1,2,$ (resp; $\omega_J:J \to \{\pm1\}$, $\omega':Z \to \{\pm1\}$ with Z generated by t) with $\omega_1\xi_1 = \omega_2\xi_2$ (resp; $\omega_J\xi_1 = \omega_J\xi_2$) , there is a unique extension of ω_1 and ω_2 to a homomorphism $\omega:G_1 *_H G_2 \to \{\pm1\}$ (of ω_J and ω' to a homomorphism $\omega:J *_H \{t\} \to \{\pm1\}$). The map ω and its restrictions to H and to G_1 and G_2 (resp; J) are used, as usual, to define anti-involutions $x \to \overline{x}$ of the rings $Z[H]$ and $Z[G_1]$, $Z[G_2]$, $Z[G_1 *_H G_2]$ (resp; $Z[J]$, $Z[J *_H \{t\}]$) by the formula $\overline{g} = \omega(g)g^{-1}$ $g \in G_1 *_H G_2 \subset Z[G_1 *_H G_2]$ (resp; $g \in J *_H \{t\} \subset Z[J *_H \{t\}]$). Using these anti-involutions, which are compatible with all the obvious ring inclusions, Z_2 actions $x \to x^*$ are induced in the usual manner on the projective class groups and Whitehead groups [M] of these rings. These Z_2 actions are compatible with ξ_{i_*} and η and $\phi(x^*) = -\phi(x)^*$. This generalizes the known results for $G = Z \times_\alpha H$. In that case, the Z_2 action switches both copies of the reduced groups of nilpotent maps [FH2].

For an abelian group C equipped with a Z_2 action, $x \to x^*$, $x \in C$, we make the usual identification $H^k(Z_2;C) \cong \{x\varepsilon C | x = (-1)^k x^*\}/\{(x+(-1)^k x^* | x\varepsilon C\}$. If $x \in Wh(G_1 *_H G_2)$ (resp; $x \in Wh(J *_H \{t\})$ with $x-(-1)^k x^* \in$ Image β we write $\overline{\Phi}(x)$ for the element in $H^{k+1}(Z_2;Ker(\tilde{K}_o(H) \to \tilde{K}_o(G_1) \oplus \tilde{K}_o(G_2)))$

(resp; $H^{k+1}(Z_2;Ker(\tilde{K}_o(H) \xrightarrow{\xi_{1_*}-\xi_{2_*}} \tilde{K}_o(J))))$ represented by $\phi(x)$.
Similarly, we write $\overline{\eta}(x)$ for the element in

$H^k(Z_2;Wh(G_1 *_H G_2)/Wh(G_1)+Wh(G_2)+Ker(\tilde{K}_o(H) \to \tilde{K}_o(G_1)\oplus\tilde{K}_o(G_2)))$

(resp; $H^k(Z_2;Wh(J *_H \{t\})/Wh(J)+Ker(\tilde{K}_o(H) \xrightarrow{\xi_{1_*}-\xi_{2_*}} \tilde{K}_o(J))))$

represented by $\eta(x)$.

Theorem 2 : Let Y be a closed manifold or Poincaré-complex of dimension $n+1$, $n \geq 5$ with $\pi_1(Y) = G$, and X a closed submanifold or sub-Poincaré-complex of dimension n of Y with trivial normal bundle and with $\pi_1(X) = H$, $H \subset G$ a square-root closed subgroup. Assume $Y-X$ has two components (respectively; one component) with fundamental groups G_1 and G_2 (resp; group J with $\xi_i:H \to J$, $i = 1,2$ being the induced maps). Assume given a homotopy equivalence $f:W \to Y$, W a closed manifold; denote its Whitehead torsion by $\tau(f) \in Wh(G)$. Then:

(i) W *is* h-*cobordant to some manifold* W' *with the induced homotopy equivalence* $f':W' \to Y$ *splittable along* X *if and only if* $\bar{\phi}(\tau(f)) \in H^{n+1}(Z_2; \mathrm{Ker}(\tilde{K}_o(H) \to \tilde{K}_o(G_1) \oplus \tilde{K}_o(G_2)))$

(resp; $\bar{\phi}(\tau(f)) \in H^{n+1}(Z_2; \mathrm{Ker}(\tilde{K}_o(H) \xrightarrow{\xi_{1*}-\xi_{2*}} \tilde{K}_o(J)))$)

is zero.

(ii) *The map* f *is splittable along* X *if and only if the image of* $\tau(f)$ *in* $Wh(G)/Wh(G_1)+Wh(G_2)$ (resp; $Wh(G)/Wh(J)$) *is zero and further, for* n *odd,* $\theta(f) \in H^{n+1}(Z_2; Wh(G)/Wh(G_1)+Wh(G_2) + (\mathrm{Ker}(\tilde{K}_o(H) \to \tilde{K}_o(G_1) \oplus \tilde{K}_o(G_2)))$ (resp;

$H^{n+1}(Z_2; Wh(G)/Wh(J)+\mathrm{Ker}(\tilde{K}_o(H) \xrightarrow{\xi_{1*}-\xi_{2*}} \tilde{K}_o(J))))$ is zero.

Here, for n odd and $\tau(f) \in$ Image β , we define $\theta(f) = \bar{\eta}(x)$, where x is the Whitehead torsion of any h-cobordism, which must from (i) above exist, of W to a split manifold. We show that $\theta(f)$ is a well-defined invariant of the homotopy class of f and assumes, for the relative splitting problem, all values in the given cohomology group. The group in whose cohomology it takes its value is isomorphic to the group of nilpotent maps of Waldhausen. He has shown that this group is zero for H a member of a large class of groups including free groups, free abelian groups twisted products of Z , classical know groups, etc., and G_1 , G_2 , J any groups [W1]. Waldhausen has further extended the set of groups for which this nil-group is known to be zero. However, his original conjecture on the vanishing of this nil-group for H a member of a very large class of groups he constructed is not, apparently, known at this time. However, Theorem 2 can be used to show that for H square-root closed in G the (2k+1) cohomology of

Z_2 with coefficients in this nilgroup is zero.

The homomorphism of $G \to \{\pm 1\}$, used to produce the anti-involutions of $Z[G]$, $Z[H]$, $Z[G_1]$ and $Z[G_2]$ (resp; $Z[J]$), is again defined by using the first Stiefel-Whitney class of Y. As in Theorem 1, this produces the Z_2 action on the Whitehead groups and projective class groups. There is also a relative form of Theorem 2 in which we begin with W a compact manifold with boundary ∂W, Y a Poincaré-complex with boundary ∂Y, $(X,\partial X)$ a proper co-dimension 1 sub-Poincaré-complex of $(Y,\partial Y)$ and the homotopy equivalence of pairs $f:(W,\partial W) \to (Y,\partial Y)$ already split along $\partial X \subset \partial Y$. The obstructions to producing an h-cobordism, fixed on ∂W of W to a manifold split along X, or of extending the splitting of $\partial W \to \partial Y$ along ∂X to a splitting of W along X are the same as in the absolute case.

In an important special case, we weaken both the dimension and the square-root closed restraints. Denote the connected sum of manifolds or Poincaré-complexes P and Q of the same dimension by $P \# Q$. To define the connected sum of Poincaré-complexes, use for example, the fact that an n-dimensional Poincaré-complex, $n > 4$, is homotopy equivalent to a complex with one cell in dimension n [W2].

Theorem 3 : *Let $f:W \to Y$ be a (respectively; simple) homotopy equivalence with W an n-dimensional P.ℓ. manifold and Y an n-dimensional manifold or Poincaré-complex, $n > 4$. Assume that either (i) $\pi_1(Y)$ has no elements of order 2, or (ii) $n \equiv 3 \pmod 4$, and Y is orientable.*

Then, if Y is a connected sum, $Y = P \# Q$, then

$W = P' \# Q'$, P' and Q' P.ℓ. manifolds with $f_1 : P' \to P$, $f_2 : Q' \to Q$ (resp; simple) homotopy equivalences and f homotopic to $f_1 \# f_2$.

For n an odd number greater than 5 and $\pi_1(Y)$ having no elements of order 2, Theorem 3 was first proved by R. Lee [L].* Reformulating Theorem 3, we have:

Connected-Sum Homotopy Criteria: A P.ℓ. manifold W of dimension greater than 4 without elements of order 2 in its fundamental group, or orientable of dimension $4k+3$, is a non-trivial connected sum of P.ℓ. manifolds if and only if W is homotopy equivalent to a connected sum of Poincaré-complexes P and Q , where P and Q are not homotopy equivalent to spheres.

An important special case of Theorem 2, is the Farrel-Hsiang splitting theorem [FH1]. As is well-known, the Farrell fibering theorem [F] could be derived from it. Here for a group C equipped with an automorphism α , we write C^{α} for $\{x \in C | \alpha(x) = x\}$.

Corollary 4 : (Farrell-Hsiang Splitting Theorem)
Let Y be a closed manifold or Poincaré complex of dimension $n+1$, $n \geq 5$ with $\pi_1(Y) = Z \times_{\alpha} H$ and with X a submanifold or sub-Poincaré complex of dimension n of Y with $\pi_1(X) = H$. Then if $f : W \to Y$ is a homotopy equivalence of closed manifolds with Whitehead torsion $\tau(f) \in Wh(Z \times_{\alpha} G)$, then

(i) W is h-cobordant to a manifold W' with the induced homotopy equivalence $f' : W' \to Y$ splittable along X if and only if $\bar{\Phi}(\tau(f) \in H^{n+1} (Z_2; \tilde{K}_o(H)^{\alpha *})$ is 0 .

---

* For $n > 5$ and $\pi_1(Y) = 0$ Theorem 3 was first proved by Browder [B1].

(ii) $f:W \to Y$ *is splittable along* X *if and only if image* $(\tau(f))$ *in* $Wh(Z \times_\alpha H)/Wh(H)$ *is* 0 .

For the general splitting problem, in which H is not assumed square-root closed in G , we construct, if $\overline{\phi}(\tau(f)) = 0$ in both Case A and B, a normal cobordism of W to split homotopy equivalence. Moreover, the surgery obstruction of this normal cobordism goes to 0 in the "surgery obstruction group" of the ring $Z[\frac{1}{2}][\pi_1(Y)]$. Hence for Wall groups over group rings $R[G]$, $Z[\frac{1}{2}] \subset R \subset Q$, we can demonstrate the existence of Mayer-Vietoris sequences of the type constructed in §2 without the square-root closed hypothesis. In another paper, we will construct the obstruction groups to the general splitting problem and relate these groups of unitary nil-potent maps to the computation of surgery groups. See in this connection the appendix of this paper. This detailed analyses is not needed in the present paper, because of the square-root closed hypothesis. Algebraically, H square-root in G is used in the proof of Theorems 1 and 2 to observe that as a $Z[H]$ bi-module, $Z[G] \cong Z[H] \oplus C \oplus D$ where the involution on $Z[G]$ switches C and D . For a discussion of further possible extensions of the present results, see the appendix.

We briefly outline the proof of Theorems 1 and 2. First we try to produce a submanifold $f^{-1}(X)$ in W^{n+1} homotopy equivalent to X^n by working inside W . While this works well for fixing up the low dimensions of $f^{-1}(X)$, in the middle dimension range this becomes geometrically difficult. So for $n = 2k$, we go as far as we can up to the middle dimension of $f^{-1}(X)$, and measure the remaining difficulty in terms of certain $Z[H]$-linear nilpotent maps of projective $Z[H]$-modules. Then working <u>outside</u> of W , we construct a normal cobordism of W to a homotopy equivalent split manifold. Of course, we want to replace this cobordism by an s-cobordism or h-cobordism. We compute the obstruction to doing just that, as a surgery obstruction, in terms of the $Z[H]$-linear nilpotent maps. If the subgroup $\pi_1(X)$ is square-root closed in $\pi_1(Y)$, we show algebraically that any surgery obstruction constructed in this way from a nilpotent map essentially vanishes. (More precisely, we decompose the module of the Hermitian form into two summands on which, separately, the intersection form (λ of [W2]) vanishes. Moreover, as $\pi_1(X)$ is square-root closed in $\pi_1(Y)$, it contains all the elements of order 2. Hence, on the given subspaces, the self-intersection form (μ of [W2]) takes values in $Z[\pi_1(X)]$. Thus, the surgery obstruction is in the image $L_{n+2}(\pi_1(X)) + L_{n+2}(\pi_1(Y-X)) + L_{n+2}(\pi_1(Y))$. It can therefore be changed to zero by a further normal cobordism without affecting the splitting.)

For $n = 2k-1$, we could get a weak form of the result of the present paper by crossing with a circle to get into a dimension in which the splitting problem has already been solved. Of course, this employs the observation that if H is square-root closed in G, $Z \times H$ is square-root closed in $Z \times G$. We can then try to split along $S^1 \times X$ and use the Farrell-Hsiang theorem to remove the extraneous circle. However, the obstructions arising in the use of the Farrell-Hsiang theorem are difficult to relate to our initial data. Hence this only suffices to prove Theorem 1 and not Theorem 2. (Of course, Theorem 2 implies Theorem 1, but not conversely.)

Instead of this, we use for $n = 2k-1$, a direct geometric construction. After working in W to improve $f^{-1}(X)$ below the pair of middle dimensions, we construct an <u>explicit</u> geometric splitting of $S^1 \times W$ using the construction already developed for $n = 2k$. We then measure explicitly the obstruction to removing the circle factor and show it vanishes under the hypothesis of Theorem 2.

A general splitting result, related to the results announced in [Cl] was also announced by W.C. Hsiang and G. Swarup.

2. Mayer-Vietoris Sequences for Surgery Groups

We use the splitting theorems of §1 together with the geometric interpretation of Wall's surgery groups [W2] to produce a decomposition formula for free products, Theorem 5, and Mayer-Vietoris sequences for surgery groups, especially Theorems 8 and 9. In §3 these general results are applied to explicitly compute the surgery groups of infinite groups. J.L. Shaneson [Sh] and C.T.C. Wall [W2] previously obtained a formula for the surgery groups of $Z \times G$ and computed the surgery groups of the free abelian group on m generators. They used the Farrel-Hsiang splitting theorem [FH1] (for $m=1$, results of [BL] [B]).* Later results of Farrell and Hsiang and of R. Lee on the surgery groups of infinite groups are described below.

Let $L_n^h(G,\omega)$ (respectively; $L_n^s(G,\omega)$) denote Wall's group of surgery obstructions for the problem of obtaining (resp; simple)

* Some classification results for manifolds with fundamental group Z were earlier obtained by Browder [B].

homotopy equivalences for manifolds of dimension n, fundamental group G, and orientation homomorphism $\omega : G \to Z_2 = \{\pm 1\}$. When the homomorphism ω is trivial, which corresponds geometrically to the study of orientable manifolds, we write $L_n^h(G) = L_n^h(G,\omega)$ and $L_n^s(G) = L_n^s(G,\omega)$. For groups G equipped with homomorphisms $\omega : G \to Z_2$, $L_n^h(G,\omega)$ (resp; $L_n^s(G,\omega)$) is a functor to abelian groups with $L_n^h = L_{n+4}^h$ and $L_n^s = L_n^{s+4}$. The group $L_{2k}^h(G,\omega)$ (resp; $L_{2k}^s(G,\omega)$) is defined as the reduced Grothondieck group of (resp; based) $(-1)^k$ Hermitian forms over the ring $Z[G]$. This ring is equipped with the anti-involution $x \to \bar{x}$, $x \in Z[G]$ defined by $\bar{g} = \omega(g)g^{-1}$, $g \in G \subset Z[G]$. The group $L_{2k+1}^h(G,\omega)$ (resp; $L_{2k+1}^s(G,\omega)$) is defined as a quotient of a (resp; special) unitary group of $(-1)^k$ Hermitian forms over $Z[G]$.

The Mayer-Vietoris sequences below will first be stated for the surgery groups of oriented manifolds; that is, for the homomorphism ω being trivial. The modifications of notation for the unoriented case will be described afterwards. We write

$$L_n^h(G) \cong \tilde{L}_n^h(G) \oplus L_n^h(0) \quad \text{and} \quad L_n^s(G) \cong \tilde{L}_n^s(G) \oplus L_n^s(0) ,$$

where $\tilde{L}_n^h$ and $\tilde{L}_n^s$ denote the reduced surgery group. Recall, essentially from [KM] (see [B3] [W2])

$$L_n^s(0) = L_n^h(0) = \begin{cases} Z & n=4k \\ 0 & n=2k+1 \\ Z_2 & n=4k+2 \end{cases}$$

<u>Theorem 5 : Let G_1 and G_2 be finitely presented groups with no elements of order 2 or with $n=4k$. Then</u>

$$\tilde{L}_n^h(G_1 * G_2) = \tilde{L}_n^h(G_1) \oplus \tilde{L}_n^h(G_2)$$

$$\tilde{L}_n^s(G_1 * G_2) = \tilde{L}_n^s(G_1) \oplus \tilde{L}_n^s(G_2)$$

For the case n even and G_1 and G_2 without elements of order 2, Theorem 5 was first proved by R. Lee [L1]. In §3, Theorem 5 is used to compute all the surgery groups of free groups.

In the most general situation considered below, it is not possible to express the precise* Mayer-Vietoris sequences in terms of only the functors L_n^h or L_n^s, or even both of them. Already for $Z \times G$ when the Whitehead group of G is not zero, it was necessary, in describing the results to employ both L^s and L^h [Sh] or surgery groups of pairs [W2]. The statements of the general Mayer-Vietoris sequences of the present paper** employ the notion of surgery groups relative to a self-dual subgroup of the Whitehead group, introduced in [C1] for this purpose. These groups, generalizing L_n^s and L_n^h, are related to each other and to L_n^s and L_n^h by exact sequences, Theorem 6 and Corollary 7, with the "difference terms" consisting of 2-torsion groups explicitly described as subquotients of the Whitehead group [C1]. This generalizes Rothenberg's exact sequence which relates L_n^h to L_n^s [Sh].

Recall, for a group G equipped with a homomorphism $\omega: G \to Z_2 = \{\pm 1\}$, the usual anti-involution $x \to \bar{x}$ of $Z[G]$ determined by $\bar{g} = \omega(g)g^{-1}$ $g \in G \subset Z[G]$ is used to define the

---

* This can be done after tensoring the surgery groups with $Z[\frac{1}{2}]$ but that procedure loses some information.

** Our results can also be expressed in terms of the L^s and L^h surgery groups of pairs and quadrads.

Z_2 action $v \to v^*$ on $\tilde{K}_0(G)$ and $Wh(G)$ [M]. For a subgroup B of $Wh(G)$, write B^* for $\{v^* \in Wh(G) | v \in B\}$. The subgroup B is called self-dual if $B = B^*$. The group $L_n^B(G,\omega)$ is the surgery obstruction group for obtaining homotopy equivalences with Whitehead torsion in B. In particular $L_n^{\{0\}}(G,\omega) = L_n^s(G,\omega)$ and $L_n^{Wh(G)}(G,\omega) = L_n^h(G,\omega)$. The groups $L_{2K}^B(G,\omega)$ consist of $(-1)^k$ Hermitian forms over $Z[G]$ which are based forms, up to changes of basis by automorphisms with torsion in B. Similarly, $L_{2k+1}^B(G,\omega)$ is a quotient of the group of $Z[G]$ $(-1)^k$ unitary matrices with torsion in B.

Theorem 6 : [Cl] Let $A \subset B$ be self-dual subgroups of $Wh(G)$, G a finitely presented group, with the Z_2 action on $Wh(G)$ the usual one induced from a homomorphism $\omega : G \to Z_2 = \{\pm 1\}$. Then there is a long exact sequence,

$$\ldots \to H^{n+1}(Z_2;B/A) \to L_n^A(G,\omega) \to L_n^B(G,\omega) \to H^n(Z_2;B/A) \to \ldots$$

For $A=0$ and $B = Wh(G)$, the above is the Rothenberg exact sequence [Sh].

Corollary 7 : Let G be a finitely presented group with $\omega : G \to Z_2$ inducing, as usual, a Z_2 action on $Wh(G)$. Then if A is a subgroup of $Wh(G)$ with $A = A^*$, there are long exact sequences

$$\ldots \to H^{n+1}(Z_2;Wh(G)/A) \to L_n^A(G,\omega) \to L_n^h(G,\omega) \to H^n(Z_2;Wh(G)/A) \to \ldots$$

and

$$\ldots \to H^{n+1}(Z_2;A) \to L_n^s(G,\omega) \to L_n^A(G,\omega) \to H^n(Z_2;A) \to \ldots$$

Theorem 8 : (A Mayer-Vietoris type sequence for surgery groups; Case A). Let H , G_1 , G_2 be finitely presented groups with $\xi_i : H \to G_i$ an inclusion of H as a square-root closed subgroup of G_i , $i = 1,2$. Let $A = \mathrm{Ker}(\phi : Wh(G_1 *_H G_2) \to \tilde{K}_o(H))$. Then, there is a long exact sequence

$$\ldots \to L_n^h(H) \xrightarrow{\xi_{1*} - \xi_{2*}} L_n^h(G_1) \oplus L_n^h(G_2) \xrightarrow{\gamma_{1*} + \gamma_{2*}} L_n^A(G_1 *_H G_2) \xrightarrow{\partial} L_{n-1}^h(H) \to \ldots$$

where $\gamma_i : G_i \to G_1 *_H G_2$, $i = 1,2$ are the obvious inclusions. Here L_n^A is related to L_n^h by the long exact sequence, for $C = \mathrm{Ker}(\tilde{K}_o(H) \to \tilde{K}_o(G_1) \oplus \tilde{K}_o(G_2))$

$$\ldots \to H^n(Z_2;C) \to L_n^A(G_1 *_H G_2) \to L_n^h(G_1 *_H G_2) \to H^{n-1}(Z_2;C) \to \ldots$$

Theorem 9 : (A Mayer-Vietoris type sequence for surgery groups; Case B). Let H , J be finitely presented groups with $\xi_i : H \to J$, $i = 1,2$ two inclusions of H as a square-root closed subgroup of J . Let $A = \mathrm{Ker}(\phi : Wh(J *_H \{t\} \to \tilde{K}_o(H))$. Then there is a long exact sequence

$$\ldots \to L_n^h(H) \xrightarrow{j} L_n^h(J) \xrightarrow{\gamma *} L_n^A(J *_H \{t\}) \xrightarrow{\partial} L_{n-1}^h(H) \to \ldots$$

where $\gamma : J \to J *_H \{t\}$ is the obvious inclusion and $j = \xi_{1*} - \xi_{2*}$. Here L_n^A is related to L_n^h by the long exact sequence, for $C = \mathrm{Ker}(\xi_{1*} - \xi_{2*} : \tilde{K}_o(H) \to \tilde{K}_o(J)))$

$$\ldots \to H^n(Z_2;C) \to L_n^A(J) \to L_n^h(J) \to H^{n-1}(Z_2;C) \to \ldots$$

These Mayer-Vietoris type sequences have many variant forms and corollaries obtained by working relative to various subgroups of the Whitehead group or by considering important special cases; some of these are presented below. Using Corollary 7, the groups

L_n^A above could also be related to L_n^s . Theorems 8 and 9 follow most directly from the following variant form of Theorem 2.

Theorem 10 : Hypothesis as in Theorem 2. Then there is an h-cobordism $(V;W,W')$ with Whitehead torsion x satisfying $\phi(x) = 0$, $\phi(x) \in \tilde{K}_o(H)$, and with the induced homotopy equivalence $f':W' \to Y$ splittable along X , if and only if $\phi(\tau(f)) = 0$.

Note that part (i) of the following variant form of Theorem 8 is always satisfied if one of the inclusions $\xi_i : H \to G_i$ has a retraction. Part (iv) of Corollaries 11 and 12 can be used to compute surgery groups relative to various self-dual subgroups of the Whitehead group.

Corollary 11 : Let H , G_1 , G_2 be finitely presented groups with H a square-root closed subgroup of both G_1 and G_2 . Then if either

(i) $H^i(Z_2;\mathrm{Ker}(\tilde{K}_o(H) \to \tilde{K}_o(G_1) \oplus \tilde{K}_o(G_2)) = 0$, $i > 1$

setting $\alpha_1 = h$, $\alpha_2 = h$, $\alpha_3 = h$, $\alpha_4 = h$

or (ii) if $H^i(Z_2;\mathrm{Ker}(Wh(H) \to Wh(G_1) \oplus Wh(G_2)) = 0$, $i > 1$

and $H^k(Z_2;Wh(G_1 *_H G_2)/Wh(G_1)+Wh(G_2)+\mathrm{Ker}(\tilde{K}_o(H) \to \tilde{K}_o(G_1)+\tilde{K}_o(G_2))) = 0$, $k > 1$ *

setting $\alpha_1 = s$, $\alpha_2 = s$, $\alpha_3 = s$, $\alpha_4 = s$

or (iii) if $H^i(Z_2;\mathrm{Ker}\phi) = 0$, $i > 1$

setting $\alpha_1 = h$, $\alpha_2 = h$, $\alpha_3 = h$, $\alpha_4 = s$

---

* Using Theorem 2, we can show that this group is always zero for H square-root closed in G and $k \equiv 0 \pmod 2$.

or (iv) <u>if</u> B_i <u>is a self-dual subgroup of</u> $Wh(G_i)$, $i = 1,2$ <u>and</u> $\xi_i : H \to G_i$, $\nu_i : G_i \to G_1 *_H G_2$ <u>denote the given inclusions,</u> $i = 1,2$, <u>setting</u>

$$\alpha_1 = (\xi_{1_*} - \xi_{2_*})^{-1}(B_1 \oplus B_2) \subset Wh(H) ,$$

$$\alpha_2 = B_1 \quad , \quad \alpha_3 = B_2 ,$$

$$\alpha_4 = (\nu_{1_*} \oplus \nu_{2_*})(B_1 \oplus B_2) \oplus (Wh(G_1 *_H G_2)/Wh(G_1) + Wh(G_2) + \mathrm{Ker}(\tilde{K}_o(H) \to \tilde{K}_o(G_1) \oplus \tilde{K}_o(G_2))) \subset Wh(G_1 *_H G_2) \ ^*$$

<u>then there is a long exact sequence</u>

$$\to L_n^{\alpha_1}(H) \to L_n^{\alpha_2}(G_1) \oplus L_n^{\alpha_3}(G_2) \to L_n^{\alpha_4}(G_1 *_H G_2) \xrightarrow{\partial} L_{n-1}^{\alpha_1}(H) \to \ldots$$

<u>Corollary 12 :</u> <u>Let</u> H , J <u>be finitely presented groups with</u> $\xi_i : H \to J$, $i = 1,2$ <u>inclusions of</u> H <u>as a square-root closed subgroup of</u> J . <u>Then if either</u>

(i) $H^i(Z_2; \mathrm{Ker}(\tilde{K}_o(H) \xrightarrow{\xi_{1_*} - \xi_{2_*}} \tilde{K}_o(J)) = 0 \quad , \quad i > 1$

<u>setting</u> $\alpha_1 = h$, $\alpha_2 = h$, $\alpha_3 = h$, $j = \xi_{1_*} - \xi_{2_*}$

---

* This case (iv) employs Waldhausen's result that $Wh(G_1 *_H G_2)/Wh(G_1) + Wh(G_2) + \mathrm{Ker}(K_o(H) \quad K_o(G_1) + K_o(G_2))$, isomorphic to his <u>reduced</u> group of nilpotent maps, is a direct summand of $Wh(G_1 *_H G_2)$.

or (ii) <u>if</u> $H^i(Z_2;\mathrm{Ker}(Wh(H) \xrightarrow{\xi_{1_*}-\xi_{2_*}} Wh(J)) = 0$, $i > 1$

<u>and</u> $H^k(Z_2;Wh(J*_H\{t\})/Wh(J)+\mathrm{Ker}(\tilde{K}_o(H) \xrightarrow{\xi_{1_*}-\xi_{2_*}} \tilde{K}_o(J))) = 0$

$k > 1$ *

<u>setting</u> $\alpha_1 = s$, $\alpha_2 = s$, $\alpha_3 = s$, $j = \xi_{1_*}-\xi_{2_*}$

or (iii) <u>if</u> $H^i(Z_2;\mathrm{Ker}\phi) = 0$, $i > 1$

<u>setting</u> $\alpha_1 = h$, $\alpha_2 = h$, $\alpha_3 = s$, $j = \xi_{1_*}-\xi_{2_*}$

or (iv) <u>if</u> B <u>is a self dual-subgroup of</u> Wh(J) <u>and</u> $\nu:J \to J *_H \{t\}$ <u>denotes the given inclusion, setting</u>

$\alpha_1 = (\xi_{1_*}-\xi_{2_*})^{-1}(B) \subset Wh(H)$

$\alpha_2 = B \subset Wh(J)$

<u>and</u> $\alpha_3 = \nu_* B \oplus Wh(J*_H\{t\})/Wh(J)+\mathrm{Ker}(\tilde{K}_o(H) \xrightarrow{\xi_{1_*}-\xi_{2_*}} \tilde{K}_o(J)) \subset$

$Wh(J*_H\{t\})$ **

<u>then there is a long exact sequence</u>

$$\to L_n^{\alpha_1}(H) \xrightarrow{j} L_n^{\alpha_2}(J) \to L_n^{\alpha_3}(J*_H\{t\}) \xrightarrow{\partial} L_{n-1}^{\alpha_1}(H) \to \cdots$$

Taking ξ_1 and ξ_2 to be surjective, Theorem 9 and Corollary 12 can be used to compute surgery groups of $Z \times_\alpha H$. A result on surgery groups of $Z \times_\alpha H$, extending the results of

---

* Using Theorem 2 we can show that this group is always zero for $k \equiv 0 \pmod 2$; it is also always zero for $J*_H\{t\} = Z \times_\alpha H$.

** This case (iv) employs Waldhausen's result that $Wh(J*_H\{t\})/Wh(J)\oplus\mathrm{Ker}(\tilde{K}_o(H) \to \tilde{K}_o(J))$, isomorphic to his <u>reduced</u> group of nilpotent maps is a direct summand of $Wh(J*_H\{t\})$.

Shaneson [Sh] and Wall [W2] on $Z \times H$, was first given in notes of Farrell and Hsiang. In the present paper, to express our description of the surgery groups of $Z \times_\alpha H$ [C1], we again employ surgery groups relative to subgroups of the Whitehead group. (On surgery groups of $Z \times_\alpha H$ see also [W2].) Recall that for a group C equipped with an automorphism β, we write $C^\beta = \{x \in C | \beta(x) = x\}$.

Corollary 13 : Let H be a finitely presented group and $\alpha : H \to H$ an isomorphism. Then if

(i) letting $\alpha_1 = Wh(H)^{\alpha_*}$, $\alpha_2 = s$, $\alpha_3 = s$

or (ii) letting $\alpha_1 = h = \alpha_2$, $\alpha_3 = \mathrm{Image}(Wh(H) \to Wh(Z \times_\alpha H)) \cong$

$$Wh(H)/(1-\alpha_*)(Wh(H))\ ^*, \quad j = 1-\alpha_*$$

or (iii) more generally, for B a self dual subgroup of $Wh(H)$ letting $\alpha_1 = (1-\alpha_*)^{-1}(Wh(H)) \subset Wh(H)$

$$\alpha_2 = B \quad \text{and} \quad \alpha_3 = \mathrm{Image}(B \to Wh(Z \times_\alpha H))$$

there is a long exact sequence**

$$\cdots \to L_n^{\alpha_1}(H) \xrightarrow{j} L_n^{\alpha_2}(H) \to L_n^{\alpha_3}(Z \times_\alpha H) \xrightarrow{\partial} L_{n-1}^{\alpha_1}(H) \to \cdots$$

Corollary 14 : Let H be a finitely presented group with $\alpha : H \to H$ an isomorphism. Then if

---

* This last isomorphism of Whitehead groups is due to Farrell and Hsiang [FH2] and to Siebenmann [S2].

** In cases (i) and (iii), because of the choice of subgroups of the Whitehead groups, it may be impossible to induce 1_* or α_* from $L_n^{\alpha_1}$ to $L_n^{\alpha_2}$ and hence to write j as a difference.

(i) $H^i(Z_2;\tilde{K}_o(H)^{\alpha_*}) = 0$, $i > 1$

setting $\alpha_1 = h$, $\alpha_2 = h$, $\alpha_3 = h$

or (ii) *if* $H^i(Z_2;Wh(H)^{\alpha_*}) = 0$, $i > 1$ *setting* $\alpha_1 = s$, $\alpha_2 = s$, $\alpha_3 = s$

or (iii) *if* $H^i(Z_2;Wh(H)/(1-\alpha_*)(Wh(H))) = 0$, $i > 1$

setting $\alpha_1 = h$, $\alpha_2 = h$, $\alpha_3 = s$

or (iv) *if* $H^i(Z_2;(1-\alpha_*)Wh(H)) = 0$, $i > 1$

setting $\alpha_1 = h$, $\alpha_2 = s$, $\alpha_3 = s$

there is a long exact sequence, with $j = 1-\alpha_*$

$$\to L_n^{\alpha_1}(H) \xrightarrow{j} L_n^{\alpha_2}(H) \to L_n^{\alpha_3}(Z \times_\alpha H) \xrightarrow{\partial} L_{n-1}^{\alpha}(H) \to \ldots$$

Taking $\alpha=1$ in the above, we obtain the results of Shaneson [Sh] and Wall [W2] on the surgery groups of $Z \times H$ in several forms:

Corollary 15 : Let H *be a finitely presented group. Then*

(i) $L_{n+1}^s(Z \times H) = L_{n+1}^s(H) \oplus L_n^h(H)$

(ii) *if* $H^i(Z_2;Wh(H)) = 0$, $i > 1$, *then* $L_{n+1}^s(Z\times H) = L_{n+1}^s(H) \oplus L_n^s(H)$

(iii) *if* $H^i(Z_2;\tilde{K}_o(H)) = 0$, *then* $L_{n+1}^h(Z \times H) = L_{n+1}^h(H) \oplus L_n^h(H)$

The splitting results of Theorem 2 have consequences, not used in the present paper, for the study of the Z_2 action on the Whitehead groups. These can be used to further study the relations between L_n^s and L_n^h and between surgery groups taken relative to various subgroups of the Whitehead group.

In another paper, we shall discuss the computation of surgery groups when H is not square-root closed in $G_1 *_H G_2$ or in $J *_H \{t\}$. The methods of the present paper show that surgery groups, in that case, of $G_1 *_H G_2$ and $J *_H \{t\}$ contain as a direct summand a certain group of unitary nil-potent maps over $Z[H]$. This group is related to Waldhausen's group of nilpotent maps [W1]. It is zero if H is a square-root closed subgroup; we do not know if it is ever non-zero. However, for "surgery groups" over the group-rings $R[H]$, $R[G_1]$, $R[G_2]$, $R[G_1 *_H G_2]$, $R[J]$, $R[J *_H\{t\}]$ for $Z[\frac{1}{2}] \subset R \subset Q$, we can show (see §1) that Mayer-Vietoris sequences are always valid, without employing the square-root closed condition.

<u>Non-oriented case:</u> The Mayer-Vietoris sequences and their consequences, Theorems 8 and 9, Corollaries 11, 12, 13, 14, 15, carry over, with slight changes of notation to the unoriented case in which the groups may have non-trivial homomorphisms to $Z_2 = \{\pm 1\}$. We indicate these changes without restating the results. The square-root closed condition is unchanged.

In Theorem 8 and Corollary 11 (resp; Theorem 9 and Corollary 12; Corollary 13 and 14; Corollary 15) we now assume that the groups G_1, G_2, H (resp; J, H, Z generated by t ; Z and H ; H) are equipped with homomorphisms ω_{G_1}, ω_{G_2}, ω_H (resp;

ω_J, ω_H , ω_Z ; ω_Z , ω_H ; ω_H) to $Z_2 = \{\pm 1\}$ which coincide on H (resp; with $\omega_J \xi_1 = \omega_J \xi_2$; with $\xi_H = \xi_H \alpha$; -). These homomorphisms extend to a unique homomorphism of $G_1 *_H G_2$ (resp; $J *_H \{t\}$; $Z \times_\alpha H$; $Z \times H$, trivial on Z) to $Z_2 = \{\pm 1\}$. With these maps to Z_2 , all the exact sequences are again as stated; in Theorem 9 and Corollary 12 (i), (ii) and (iii) (resp; Corollary 13 (ii) and Corollary (14) the definition of j should be changed, in this general situation, to $\xi_{1_*} - \omega_Z(t)\xi_{2_*}$ (resp; $1-\omega_Z(t)\alpha_*$).

3. Computations of Surgery Groups of Infinite Groups

We give some examples of the application of the results of §2 to the computation of the surgery groups of some geometrically important infinite groups. Let Z^m (resp; Z_2^m) denote the free abelian group (resp; the vector space over the integers modulo 2) on m generators.

As a consequence of Theorem 5, we get:

<u>Theorem 16 :</u> <u>Let</u> F^m <u>denote the free group on</u> m <u>generators</u>. <u>Then</u>,

$$L_n^h(F^m) = L_n^s(F^m) = \begin{cases} Z & n \equiv 0 \pmod 4 \\ Z^m & n \equiv 1 \pmod 4 \\ Z_2 & n \equiv 2 \pmod 4 \\ Z_2^m & n \equiv 3 \pmod 4 \end{cases}$$

For n even, the above result was first proved by R. Lee [L1].

Corollary 17 : [C2] *Let π be a classical knot group* with finitely generated commutator subgroup. Then the map given by abelianization, $\pi \to Z$, induces for all n an isomorphism $L_n^s(\pi) \to L_n^s(Z)$.*

F. Quinn has extended Corollary 17 to other classical knot groups. He proves that for a large class of groups, the Mayer-Vietoris sequences of §2 imply that the L_n functors behave as a kind of homology theory with the homotopy groups of G/Top as coefficient groups [Q1].

However, the conjecture that the conclusion of Corollary 17 should hold for all classical knot groups is not apparently known. Such a result certainly does not extend to knots $S^2 \subset S^4$ [C2].

Theorem 16 and the results of §2 lead to a computation of the surgery groups of the fundamental groups of all closed 2-dimensional manifolds.

Theorem 18: *Let M be a closed two-dimensional manifold. Then the following is a table of values of $L_n^s(\pi_1(M),\omega) = L_n^h(\pi_1(M),\omega)$ where $\omega:\pi_1(M) \to Z_2 = \{\pm 1\}$ is the homomorphism given by the first Stiefel-Whitney class of M.*

---

* That is, for some knot $S^1 \subset S^3$, $\pi = \pi_1(S^3-S^1)$. The commutator subgroup of such a group is finitely generated if and only if the knot complement (S^3- neighborhood of S^1) fibres over a circle. [N].

Table of Values of $L_n(\pi_1(M),\omega)$

	n=	4k	4k+1	4k+2	4k+3
oriented M =	S^2	Z	0	Z_2	0
	T_g	$Z \oplus Z_2$	Z^{2g}	$Z \oplus Z_2$	Z_2^{2g}
unoriented M =	RP^2	Z_2	0	Z_2	0
	$K \# T_g$	$Z_2 \oplus Z_2$	Z^{2g+1}	$Z \oplus Z_2$	Z_2^{2g+2}

Here, T_g denotes the oriented surface of genus g (for $g = 1$ computed in [Sh] and [W2]) RP^2 denotes the projective plane (computed in [W2]), K denotes the Klein bottle, and # denotes connected sum.

APPENDIX

The Category of Unitary Nilpotent Maps

A key algebraic step in the proof of the geometric splitting theorem of §1, from which the Mayer-Vietoris sequences for surgery groups of §2 are derived, is the solution of the algebraic problem about unitary nilpotent maps described below. It is in this algebraic step that we employ the square-root closed condition. So, we will pose the problem of eliminating entirely (which seems unlikely) or weakening this restriction on the groups.*

We recall from [W1] the notion of a nilpotent object and introduce the category of a unitary nilpotent objects. Assume given finitely presented groups H, G_1, G_2 with inclusions $H \to G_i$, $i = 1,2$, . Let $\widetilde{Z[G_i]}$ denote the $Z[H]$ bi-submodule of $Z[G_i]$ generated by $g \in \{G_i - H\}$. By a nilpotent object over $(H;G_1,G_2)$, that is an element of the set $\widetilde{Nil}(H;G_1,G_2)$, we mean a quadruple (P,Q,ρ_1,ρ_2) where

(i) P and Q are free right $Z[H]$ modules

(ii) $\rho_1 : P \to Q \otimes_{Z[H]} \widetilde{Z[G_1]}$, $\rho_2 : Q \to P \otimes_{Z[H]} \widetilde{Z[G_2]}$

are $Z[H]$ linear maps which are nilpotent in the sense that there exist some finite filtrations of $Z[H]$ modules

$$P = P_0 \supset P_1 \supset \dots \supset P_n = 0$$

$$Q = Q_0 \supset Q_1 \supset \dots \supset Q_n = 0$$

---

* We describe the problem for surgery groups of $G = G_1 *_H G_2$ (case A). There is also an entirely parallel problem for $G = J *_H \{t\}$ (case B).

so that the maps ρ_i raise the degree of the filtration - i.e. that $\rho_1(P_i) \subset Q_{i+1} \otimes_{Z[H]} \widetilde{Z[G_1]}$, $\rho_2(Q_i) \subset P_{i+1} \otimes_{Z[H]} \widetilde{Z[G_2]}$. Equivalently, one can define "composites" of ρ_1 and ρ_2 to suitable long tensor products. Thus
$\rho_2\rho_1 : P \to P \otimes_{Z[H]} \widetilde{Z[G_2]} \otimes_{Z[H]} \widetilde{Z[G_1]}$ etc. The nilpotency condition is equivalent to $(\rho_2\rho_1)^m = 0$ for m sufficiently large.

Waldhausen [W1] showed that a group formed by equivalence classes of objects of this set $\widetilde{Nil}(H;G_1,G_2)$ is a direct summand of the Whitehead group of $G_1 *_H G_2$.

Now we will define the category of a unitary nilpotent objects over $(H;G_1,G_2)$. To simplify the notation, we assume that the orientation homomorphisms to $Z_2 = \{\pm 1\}$ of these groups, determining as usual the involutions of the group rings, are trivial and that H contains all the elements of order 2 in G_1 and G_2 . The usual involution on group rings is denoted by $x \to \bar{x}$.

An object of $U\,\widetilde{Nil}_k(H;G_1,G_2)$ is a triple (P,ρ_1,ρ_2) , where P^* denotes the dual of the $Z[H]$ module P , and with:

(i) (P,P^*,ρ_1,ρ_2) is an object of $\widetilde{Nil}(H;G_1,G_2)$

(ii) Letting $< , >$ denote the usual evaluation map $P \times P^* \to Z[H]$ and the induced map $(P \otimes_{ZH} Z[G_i]) \times (P^* \otimes_{ZH} Z[G_i]) \to Z[G_i]$, then

$$< x, \rho_1(y) > = (-1)^k \; \overline{< y, \rho_1(x) >} \qquad x,y \in P$$

$$< \rho_2(x), y > = (-1)^k \; \overline{< \rho_2(y), x >} \qquad x,y \in P^*$$

$$\forall x \in P, \; \exists v \in \widetilde{Z[G_1]} \qquad < x, \rho_1(x) > = v + (-1)^k \, \bar{v}$$

$$\forall x \in P^*, \; \exists v \in \widetilde{Z[G_2]} \qquad < \rho_2(x), x > = v + (-1)^k \, \bar{v} \quad .$$

Now define a map $\phi: U\, \widetilde{Nil}_k(H;G_1,G_2) \to L^h_{2k}(G_1 *_H G_2)$ by letting, for $(P,\rho_1,\rho_2) \in U\, \widetilde{Nil}_k(H;G_1,G_2)$ $\phi(P,\rho_1,\rho_2)$ be represented by the Hermitian form (M,λ,μ) over $Z[G]$, $G = G_1 *_H G_2$ defined by

(i) $M = (P \oplus P^*) \otimes_{Z[H]} Z[G]$

(ii) $$(x,y) = \begin{cases} \langle x,y \rangle & x \in P,\ y \in P^* \\ \langle x,\rho_1(y) \rangle & x,y \in P \\ \langle \rho_2(x),y \rangle & x,y \in P^* \end{cases}$$

(iii) $\mu(x) \in \widetilde{Z[G_1]}/\{v-(-1)^k \bar{v} | v \in Z[G_1]\}$, $x \in P$

$\mu(x) \in \widetilde{Z[G_2]}/\{v-(-1)^k \bar{v} | v \in Z[G_2]\}$, $x \in P^*$

U Nil Problem: Is the map $\bar{\Phi}: U\, \widetilde{Nil}_k(H;G_1,G_2) \to L^h_{2k}(G_1 *_H G_2)/L^h_{2k}(G_1) + L^h_{2k}(G_2)$, induced by ϕ , trivial?

For given $(H;G_1,G_2)$ a geometric splitting theorem for fundamental groups $G_1 *_H G_2$ is implied by the triviality of $\bar{\Phi}$ (in conjunction with a similar result on L^s_{2k}). Hence if $\bar{\Phi}$ is trivial, there is a Mayer-Vietoris type sequence for the Wall surgery groups of $G_1 *_H G_2$.

Theorem: If H is square-root closed in G_1 and G_2 , or equivalently H square-root closed in $G_1 *_H G_2$, then the map $\bar{\Phi}$ is trivial.*

It is easy to verify that, even if H is not square-root closed in $G_1 *_H G_2$, the map $\bar{\Phi}$ is trivial if we introduce $\frac{1}{2}$ into the rings and consider Hermitian forms over $R[G]$, $Z[\frac{1}{2}] \subset R \subset Q$.

---

* We also have a similar result, for H square-root closed in $G_1 *_H G_2$ on the groups L^s_{2k} .

Added in proof : Since completing the present paper the author has produced examples of unsplittable homotopy equivalences, described general codimension one splitting obstruction groups, and extended the results of the present paper to get a general description of the Wall groups of $G_1 *_H G_2$ and $J *_H \{t\}$. See in this connection the following papers of the author : "Splitting obstructions for Hermitian forms and manifolds with $Z_2 \subset \pi_1$ ", to appear in Bull. Amer. Math. Soc., 1973; "On connected sums of manifolds", "Unitary nilpotent groups and Hermitian K-theory, I and II","Manifolds with infinite fundamental group, I and II","On the homotopy invariance of higher signatures", to appear.

BIBLIOGRAPHY

[BHS] Bass, H., A. Heller and R.G. Swan, "The Whitehead group of a polynomial extension", Publ. Math. I.H.E.S. 22 (1964), pp. 61-79.

[B1] Browder, W., "Embedding 1-connected manifolds", 72 (1966) pp. 225-231.

[B2] Browder, W., "Manifolds with $\pi_1 = Z$ ", Bull. Amer. Math. Soc. 72 (1966), pp. 238-244.

[B3] Browder, W., "Surgery on simply-connected manifolds", Springer-Verlag, to appear.

[BL] Browder, W., and J. Levine, "Fibering manifolds over a circle", Comm. Math. Helv., 40 (1966), pp. 153-160.

[BLL] Browder, W., J. Levine and G.R. Livesay, "Finding a boundary for an open manifold", Amer. J. Math., 87 (1965), 1017-1028.

[C1] Cappell, S.E., "A splitting theorem for manifolds and surgery groups", Bull. Amer. Math. Soc. 77 (1971), 281-286.

[C2] Cappell, S.E., "Superspinning and knot complements", Topology of Manifolds (Proceedings of the 1969 Georgia Conference), Markham Press, 1971, 358-383.

[F] Farrell, F.T., "The obstruction to fibering a manifold over a circle", Indiana Univ. Math. Jour. 21 (1971), 315-346.

[FH1] Farrell, F.T., and W.C. Hsiang, "Manifolds with $\pi_1 = G \times_\alpha T$", Amer. J. Math., to appear.

[FH2] Farrell, F.T., and W.C. Hsiang, "A formula for $K_1 R_\alpha[T]$", Proceedings of Symposia in Pure Mathematics, XVII (1970), 172-218.

[KM] Kervaire, M.A., and J.W. Milnor, "Groups of homotopy spheres I" Ann. of Math. 77 (1963), pp. 504-537.

[KS] Kirby, R.C., and L.C. Siebenmann, "Foundations of Topology", Not. Amer. Math. Soc. 16 (1969), p.848.

[L1] Lee, R., "Splitting a manifold into two parts", Mimeographed notes, Inst. Advanced Study, 1969.

[L2] Lees, J., "Immersions and surgeries of topological manifolds", Bull. Amer. Math. Soc. 75, (1969), pp. 529-534.

[M] Milnor, J.W., "Whitehead torsion", Bull. Amer. Math. Soc. 72, (1966), pp. 358-426.

[N] Neuwirth, L., "Knot groups", Ann. of Math. Studies, 56.

[Q1] Quinn, F., "B Top_n and the surgery obstruction", Bull. Amer. Math. Soc. 77, (1971), pp. 596-600.

[S1] Siebenmann, L.C., "The obstruction to finding the boundary of an open manifold", Princeton Univ. Thesis.

[S2] Siebenmann, L.C., "A total Whitehead torsion obstruction to fibering over the circle", Comm. Math. Helv. 45 (1970), pp.1-48.

[Sh] Shaneson, J.L., "Wall's surgery obstruction groups for $Z \times G$", Ann. of Math. 90 (1969), pp. 296-334.

[St] Stallings, J., "Whitehead torsion of free products", Ann. of Math. 82 (1965), pp. 354-363.

[W1] Waldhausen, F., "Whitehead groups of generalized free products", To appear.

[W2] Wall, C.T.C., "Surgery on compact manifolds", Academic Press, 1970.

[Wh] Whitehead, J.H.C., "Simple homotopy types", Amer. J. Math. 72 (1950), pp. 1-51.

Groups of Singular Hermitian Forms

Sylvain E. Cappell*

In this lecture, we will describe some algebraic results and problems on the groups of "singular", or rather "very weakly non-singular" Hermitian forms $\Gamma_{2k}(\mathcal{F})$. These groups were introduced and studied jointly by J. Shaneson and myself in the course of developing general methods of classifying embeddings of manifolds in codimension two. We showed, using a classification theory for homology equivalent manifolds, that the solutions of these geometric problems depend largely on the relevant fundamental groups. Geometric aspects of our results are discussed elsewhere. (See [CS2] [CS3] [CS5]; detailed proofs [CS6]; applications to non-locally flat embeddings [J] [CS4].) For $\mathcal{F}$ an isomorphism, these groups reduce to Wall's surgery obstruction groups of non-singular Hermitian forms [W 1]. However, unlike the situation with Wall's groups or relative Wall groups, the groups $\Gamma_{2k}(\mathcal{F})$, in most of our geometrically arising situations, are not finitley generated.

* The author is a Sloan Foundation fellow and was partially supported by an N.S.F. grant.

Let π be a finitely presented group equipped with a homomorphism $\omega:\pi \to Z_2 = \{\pm 1\}$. The integral group ring $Z\pi$ has the conjugation determined by $\bar{g} = \omega(g)\, g^{-1}$ for $g \in \pi \subset Z\pi$. Let $\mathfrak{F}: Z\pi \to \Lambda$ denote a homomorphism of rings (with 1 and) with involution. In our geometric applications, the most important case is when $\Lambda = Z\pi'$ and $\mathfrak{F}$ is induced from an epimorphism of groups $\pi \to \pi'$. Hence, we will assume below that $\mathfrak{F}$ is surjective. However, much of the following discussion applies equally well if $\mathfrak{F}$ is <u>locally epic</u>; i.e. $\forall\, \lambda_1, \lambda_2, \ldots, \lambda_k \in \Lambda$, $\exists$ a unit u of Λ with $\lambda_1 u, \ldots, \lambda_k u$ all in $\mathfrak{F}(Z\pi)$.

Define $Wh(\mathfrak{F}) = K_1(\Lambda)/\mathfrak{F}(\pm\pi)$. Of course, if $\mathfrak{F}$ is induced from an epimorphism $\pi \to \pi'$, then $Wh(\mathfrak{F}) = Wh(\pi')$, the Whitehead group of π'. An isomorphism of <u>stably</u> based Λ-modules will be called <u>$\mathfrak{F}$-simple</u>, or just <u>simple</u> if there is no danger of confusion, iff it represents the zero element of $Wh(\mathfrak{F})$. A stable basis of a Λ-module will be said to be in the ($\mathfrak{F}$)-<u>preferred class</u>, with respect to a given stable basis, if and only if the stable automorphsim given by change of basis is simple with respect to the given basis.

Let $\eta = \pm 1$. Let $I_\eta = \{\lambda - \eta\bar{\lambda} \mid \lambda \in \mathbb{Z}\pi\}$. By a <u>special</u> η-<u>form</u> over $\mathfrak{F}$ is meant a triple (H,ϕ,μ), H a finitely-generated (right) $\mathbb{Z}\pi$-module, $\phi: H \times H \to \mathbb{Z}\pi$ a $\mathbb{Z}$-bilinear map, $\mu: H \to \mathbb{Z}\pi/I_\eta$, satisfying the following properties:

(Q1) $\phi(x,y\lambda) = \phi(x,y)\lambda \quad \forall\, x,y \in H$ and $\forall\, \lambda \in \mathbb{Z}\pi$;

(Q2) $\phi(x,y) = \eta\overline{\phi(y,x)} \quad \forall\, x,y \in H$;

(Q3) $\phi(x,x) = \mu(x) + \eta\mu(x) \quad \forall\, x \in H$;

(Q4) $\mu(x+y) - \mu(x) - \mu(y) \equiv \phi(x,y) \bmod I_\eta$,

$\forall\, x,y \in H$;

(Q5) $\mu(x\lambda) = \bar{\lambda}\mu(x)\lambda$, $\forall x \in H$;

(Q6) $H_\Lambda = H \otimes_{\mathbb{Z}\pi} \Lambda$ is stably based, and the map

$A\phi_\Lambda : H_\Lambda \to \mathrm{Hom}_\Lambda(H_\Lambda, \Lambda)$ given by $A\phi_\Lambda(x)(y) = \phi_\Lambda(x,y)$, ϕ_Λ induced by ϕ is a simple isomorphism with respect to a preferred class of stable bases and its dual.

Note that by (Q6) , $(H_\Lambda, \phi_\Lambda, \mu_\Lambda)$ is a special η-Hermitian form over Λ , in the sense of Wall [W1].

The special η-forms form a semi-group under orthogonal direct sum, denoted $\perp$.

We say the η-form α is strongly equivalent to zero (write $\alpha \overset{s}{\approx} 0$) if $\exists$ a submodule $K \subset H$ with the following properties:

(PS1) $\phi(x,y) = 0$ and $\mu(x) = 0$ $\forall x,y \in K$; and

(PS2) The image of K_Λ in H_Λ is a <u>subkernel</u> in the sense of Wall [W1, Lemma 5.3].

The submodule K will be called a <u>pre-subkernel.</u>

If $\alpha = (H,\phi,\mu)$ is an η-form, we define $-\alpha = (H,-\phi,-\mu)$.

<u>Lemma</u> . $\alpha \perp (-\alpha) \perp \kappa \overset{s}{\approx} 0$, <u>a kernel over</u> $\mathbb{Z}\pi$.

<u>Pf:</u> Let $\alpha = (H,\phi,\mu)$. Adding a <u>kernel</u> over $\mathbb{Z}\pi$[W1, p.47], we may assume that H_Λ is free. Let $K \subset H \oplus H$ be the diagonal submodule; i.e., $K = \{(x,x) \mid x \in H\}$ .Clearly $\phi \perp (-\phi)$ and $\mu \perp (-\mu)$ vanish on K . The same argument as in [W1], Lemma 5.4] shows that K satisfies PS2).

Now say $\alpha \sim \beta$ if and only if $\alpha \perp (-\beta) \approx 0$. Let $\Gamma_\eta(\mathfrak{F})$ be the set of equivalences classes of η-forms under the equivalence relation generated by $\sim$; $\perp$ induces the structure of an abelian group on $\Gamma_\eta(\mathfrak{F})$. We also write $\Gamma_\eta(\mathfrak{F}) = \Gamma_{2k}(\mathfrak{F})$ for $\eta = (-1)^k$. Note that the η-form α represents zero in $\Gamma_\eta(\mathfrak{F})$ if and only if $\exists$ an η-form β with $\beta \approx 0$ and $\alpha \perp \beta \approx 0$. Clearly $\Gamma_\eta(\mathfrak{F})$ depends functorially on $\mathfrak{F}$.

In the geometric applications, it is important to not restrict ourselves to η-froms (H,ϕ,μ) with H a free module. However, we have the following:

Lemma. Each η-form $\alpha = (H,\phi,\mu)$ is equivalent to a form $\alpha_0 = (H_0 \phi_0 \mu_0)$ with H_0 free, with a basis whose image in $H_0 \otimes \Lambda$ is in the preferred class.

Proof: After adding a kernel over $\mathbb{Z}\pi$, if necessary, let $y_1,\ldots,y_m$ be a basis of H_Λ in the preferred class. Since $\mathfrak{F}$ is an epimorphism, we may write $y_i = x_i \otimes 1$, $x_i \in \mathbb{Z}\pi$. Let H_0 be the free $\mathbb{Z}\pi$-module with basis $x_1,\ldots,x_m$ and let $p: H_0 \to H$ be the homomorphism determined by $p(x_i) = x_i$. Let $\phi_0(x,y) = \phi(px,py)$, $\mu_0(x) = \mu(px)$. The base $x_1,\ldots,x_m$ provides $(H_0)_\Lambda$ with a basis also, and it is clear that $\alpha_0 = (H_0, \phi_0, \mu_0)$ is an η-form, with $x_1 \otimes 1,\ldots,x_m \otimes 1$ in the preferred class. Let $K = \{(px,x) \mid x \in H_0\}$; then (PS1) for K is clear and (PS2) follows by [W1, 5.4]. Hence $K \subset H \oplus H_0$ is a pre-subkernel for $\alpha \perp (-\alpha_0)$; i.e., $\alpha \sim \alpha_0$.

Lemma. The η-form α represents zero in $\Gamma_\eta(\mathfrak{F})$ if and only if $\exists$ a kernel κ over $\mathbb{Z}\pi$ with $\alpha \perp \kappa \approx 0$.

Proof. Suppose α represents zero. Then $\beta \approx 0$ with $\alpha \perp \beta \approx 0$. Write $\beta = (H,\phi,\mu)$. Then, by (PS1) and (PS2), $\exists$ elements $x_1,\ldots,x_r$ in H such that ϕ and μ vanish on the submodule spanned by these elements and such that $\exists\ y_1,\ldots,y_r$ with $x_1 \otimes 1,\ldots,x_r \otimes 1, y_1 \otimes 1,\ldots,y_r \otimes 1$ a preferred basis of H_Λ, with ϕ_Λ and μ_Λ trivial on the submodule spanned by $y_1 \otimes 1,\ldots,y_j \otimes 1$, and with $\phi_\Lambda(x_i \otimes 1, yj \otimes 1) = \delta_{ij}$. The elements $y_i,\ldots,y_r$ are found by lifting to H a suitable basis of the dual subkernel to the image in H_Λ of a presubkernel; the $x_1,\ldots,x_r$ arise by lifting a suitable basis of this image to the presubkernel itself; again this uses the fact that $\mathfrak{F}$ is onto.

Let H_0 be the free $\mathbb{Z}\,\pi$-module on $\{x_1,\ldots,x_r, y_1,\ldots,y_r\}$. Let $p{:}H_0 \to H$ be the homomorphism with $p(x_i) = x_i$ and $p(y_i) = y_i$. Define $\phi_0(x,y) = \phi(px,py)$ and $\mu_0(x,y) = \mu(px)$. Let $(H_0)_\Lambda$ have the stable basis $x_1 \otimes 1,\ldots,y_r \otimes 1$ (there is a certain abuse of notation here). It is easy to see that if K is a pre-subkernel for $\alpha \perp \beta$, then $((\mathrm{id}_\alpha) \perp p)^{-1}K$ is a pre-subkernel for $\alpha \perp \beta_0$, where id_α denotes the identity automorphism of α. In particular $\alpha \perp \beta_0 \approx 0$.

Now let κ be the kernel over $\mathbb{Z}\,\pi$ of dimension $2r$, with standard basis $e_1,\ldots,e_1$, $f_i,\ldots,f_r$; i.e., if $\kappa = (K, \rho, \nu)$, $\rho(e_i,f_j) = \delta_{ij}$, $\rho(e_i,e_j) = \rho(f_i,f_j) = \nu(e_i) = \nu(f_j) = 0$, and $e_1,\ldots,e_r$, $f_1,\ldots,f_r$ is in the preferred class. We define a homomorphism $h{:}\ \beta_0 \to \kappa$ by setting

$$h(x_i) = \sum_{j=1}^{r} e_j(\eta\ \phi_0(y_j,x_i)) \quad \text{and}$$

$$h(y_k) = f_k + \gamma_k e_k + \sum_{j>k} e_j(\eta\ \phi_0\ (y_j,y_k)) .$$

where $\gamma_k \equiv \mu_0(y_k) \bmod I_\eta$. It is easy to verify that h preserves forms by checking it on basis elements. For example,
$\rho(h(x_i), h(y_k)) = \eta\overline{\phi_0(y_k,x_i)} = \eta^2\phi_0(x_i,y_k) = \phi_0(x_i,y_k)$.
Further $(id_\alpha \perp h) \otimes id_\Lambda = id_{\alpha_\Lambda} \perp (h \otimes id_\Lambda)$ is an isomorphism of the special η-Hermitian forms $(\alpha \perp \beta_0)$ and $(\alpha \perp \kappa)_\Lambda$. It now follows that if L is a pre-subkernel for $\alpha \perp \beta_0$, then $(id_\alpha \perp h)(N)$ is a pre-subkernel for $\alpha \perp \kappa$; i.e., $\alpha \perp \kappa \approx 0$.

If we omit the words "simple" and "preferred class of basis" from the above discussion, we obtain groups $\Gamma^h_\eta(\mathfrak{F}) = \Gamma^h_{2k}(\mathfrak{F})$, $\eta = (-1)^k$. The lemmas remain valid, by easier versions of the same proofs. When we wish to emphasize the distinction between the types of groups, we write $\Gamma_\eta = \Gamma^s_\eta$. There are natural homomorphisms $\Gamma^s_\eta \to \Gamma^h_\eta$.

For $e = s,h$ there are natural homomorphisms $L^e_{2k}(\pi,w) \to \Gamma^e_{2k}(\mathfrak{F})$ and $\Gamma^e_{2k}(\mathfrak{F}) \to L^e_{2k}(\Lambda,\pm\mathfrak{F}\pi)$; the second homomorphism will be generically called j_*, and is clearly an epimorphism. By $L^e_{2k}(\Lambda,\pm\mathfrak{F}\pi)$ we mean $L^h_{2k}(\Lambda)$ if $e = h$ and, for $e = s$, the Wall group of special η-Hermitian forms defined using vanishing in $Wh(\mathfrak{F})$ as the criterion for "simplicity". If $\Lambda = \mathbb{Z}\pi'$ and $\mathfrak{F}$ is induced by a homomorphism of groups, $L^e_{2k}(\Pi,\pm\mathfrak{F}) = L^e_{2k}(\pi',w')$.

For $\Gamma^h_\eta(\mathfrak{F})$, one actually needs only that $\mathfrak{F}$ be locally epic.

Note that for $\Lambda = \mathbb{Z}\pi$ and $\mathfrak{F}$ = identity, $\Gamma^e_{2k}(\mathfrak{F}) = L^e_{2k}(\pi,w)$. As another essentially known example, let R be an extension of $\mathbb{Z}$ contained in the rationals. Let $\mathfrak{F}$ be the inclusion of $\mathbb{Z}\pi$ in $R\pi$. Then $\mathfrak{F}$ is locally epic. Using the lemmas above it is not hard to show that $\Gamma^h_{2k}(\mathfrak{F}) = L^h_{2k}(R\pi)$.

In [CS6, Chapter I, § 2] the groups Γ_n^h are also defined algebraically for n odd as a certain subgroup of the Wall group of the ring Λ, $L_{2k-1}^h(\Lambda)$, $n=2k-1$. We also define there $\Gamma_{2k-1}^s(\Lambda)$ algebraically. These definitions are determined by the geometric problems and condsiderations. Precisely, define $\Gamma_{2k-1}^h(\mathfrak{F})$ to be the subgroup of elements of $L_{2k-1}^h(\Lambda)$ that have representatives $\alpha \in U(\Lambda)$ with the property that $\exists$ a presubkernel of the standard kernel over $Z\pi$ whose image in the standard kernel Λ, after tensoring with Λ, is precisely the image under α of the standard subkernel. One sees that this is a subgroup using the orthogonal sum representation of addition in $L_{2k}^h(\Lambda)$. Write $j_*: \Gamma_{2k-1}^h(\mathfrak{F}) \to L_{2k-1}^h(\Lambda)$ for the inclusion. Using geometric methods, we show that j_* is bijective if $\mathfrak{F}$ is induced from an epimorphism of groups $\pi \to \pi'$ with compatible homomorphisms to $Z_2 = \{\pm 1\}$ determining the conjugations on $Z\pi$ and $\Lambda = Z\pi'$; it appears likely that j_* is always bijective. However, this would probably require a direct algebraic argument.

We also define geometrically relative Γ-groups, and these fit into the usual long exact sequence with the absolute groups. The Γ functors are periodic with period 4; i.e., $\Gamma_n(\mathfrak{F}) = \Gamma_{n+4}(\mathfrak{F})$.

By contrast with the situation for n odd, the group $\Gamma_n(\mathfrak{F})$ for n even are, even in elementary examples, usually not finitely generated. This difference reflects itself in many qualitative differences in the results on codimension two embeddings between odd and even dimensions.

Before discussing computations of the groups $\Gamma_{2k}(\mathfrak{F})$, an example of an application of the methods of [CS6] to an embedding classification problem will be outlined. A key point in the example is the dependence of the classification results only on the relevant fundamental groups. The embedding problem is solved by studying an associated problem in the classification of homology equivalent manifolds. For manifolds with fundamental group π , $\Gamma_n(\mathcal{G})$ is the surgery obstruction group to obtaining homology equivalence with local coefficients in Λ .

Our example is the solution of the problem of giving a geometric description of the algebraic periodicity observed by Levine in the high-dimensional knot cobordism groups [K1] [K2] [L1] [L2] [L3]. (For topological knot periodicity [CS 1]). Using an algebraic description of the odd-dimensional knot cobordism groups in terms of groups of Seifert matrices[1] associated to Seifert surfaces, he showed that these groups were periodic with period 4. In place of Levine's groups, we use a description of the cobordism group of high-dimensional knots with zero Arf invariant or index in terms of the group of Hermitian forms $\Gamma_{2k}(Z[Z] \to Z)$. Kervaire had previously shown that the knot cobordism groups were zero in even dimensions[2] and infinitely generated in odd dimensions. A knot cobordism group was first introduced by Fox and Milnor, in the low dimensional case [FM].

By comparing both sets with certain Γ-groups, a one to one correspondence is set up between the concordance classes of embeddings in the usual homotopy class, for M^n a simply connected manifold, of $S^k \times M$ in $S^{k+2} \times M$ with the concordance classes of embeddings of S^{n+k} in S^{n+k+2} . Thus, every embedding of $S^k \times M$

---

1 The use of groups of Seifert matrices (or more precisely, the associated Seifert forms) has been extended to more general situations by Matsumoto [M].

2 We reproved this with our methods and extended it to the equivariant case.

in $S^{k+2} \times M$ is seen to be concordant to one which is "unknotted" outside a small ball. Taking $M = \mathbb{C}P^2$, this gives a geometric procedure for beginning with an embedding $S^k \subset S^{k+2}$, taking its product with $\mathbb{C}P^2$ to get $S^k \times CP^2 \subset S^{k+2} \times \mathbb{C}P^2$, and concording that embedding to one knotted only in a ball. Looking at the knot in the ball, we thus obtain an embedding of $S^{k+4} \subset S^{k+6}$. This procedure is a geometric analogue of the isomorphism $\Gamma_n(\mathcal{F}) = \Gamma_{n+4}(\mathcal{F})$ and induces our <u>geometric knot cobordism periodicity isomorphism.</u> (See problem 12 of [S1] in these proceedings on related algebraic problem.)

These methods are also applicable to the study of invariant and fixed submanifolds of manifolds equipped with group actions and to equivariant knot cobordism [CS3] [CS4][CS5][CS6]. (For related algebraic problems, see problems 12 and 14 of [S1]). In particular, a positive solution to problems 13 of [S1], or even something a little weaker, would lead to a characterization of an equivariant knot cobordism class by the knot cobordism class and the given actions on the spheres.

The groups $\Gamma_n(\mathcal{F})$ are also obstruction groups for codimension two splitting problems. It would in this connection, be useful when studying closed manifolds to know when the map $j_*: L^e_{2k}(\pi,\omega) \to \Gamma^e{}_{2k}(\mathcal{F})$ defined above is zero? When is it injective?

These methods also apply to the problem of determining when, in codimension two, sufficiently close embeddings are concordant. This is solved by introducing, for ξ a 2-plane bundle over a manifold M, the set $C(M,\xi)$ of semi-local knots of a manifold M; that is, $C(M,\xi)$ consists of the concordance classes of embeddings, homotopic to the o-section, of M in the total space of ξ.

We show that $C(M,\xi)$ is in a natural way, for dimension M not small, an abelian group and compute this group in terms of some $\Gamma_{2k}(\mathcal{J})$ groups for dimension M odd and in terms of Wall surgery groups for dimension M even. It would be useful, in this connection, to have results on problem 14 of [S1] and to have computations of $\Gamma_{2k}(Z[Z_p] \to Z)$ when the map of rings is induced from the trivial map $Z_p \to e$. For $p = 2$ we have (see an appendix of [CS6]):

Theorem: There is an exact sequence

$$0 \to \Gamma_{2k}(Z[Z_2] \to Z) \to L_{2k}(e) \oplus L_{2k}(e;Z_{(2)}) \xrightarrow{\beta} L_{2k}(e;\dot{Z}_2) \to 0$$

Here $L_{2k}(e;Z_{(2)})$ (respectively; $L_{2k}(e;Z_2)$) denotes the Wall group of the ring of integers localized at 2, $Z_{(2)}$ (resp; integers modulo 2, Z_2). The map β is given by Arf-invariants. A complete set of invariants for $L_{2k}(e;Z_{(2)})$ is essentially well known. It is not a finitely presented group. The formula of the theorem suggests several generalizations. For example, there should perhaps be a formula for $\Gamma_{2k}(Z[Z_p] \to Z)$ in terms of $L_{2k}(e;Z_{(p)}(\xi))$, ξ a p-th root of unity.

For ξ a trivial 2 plane-bundle over M, we write $C(M,\xi) = C(M)$. The computation of $C(T^n)$, T^n the n-torus $S^1 \times \ldots \times S^1 = (S^1)^n$ employs the formula for Γ-groups of polynomial extension rings described below. For piecewise-linear embeddings we have:

Theorem [CS6]: $C(T^n) = C(S^n) \oplus [\Sigma(T^n\text{-point}) ; G/PL]$, $C(S^n)$ is isomoprhic to the knot cobordism group of knots S^n in S^{n+2}.
This leads to a description of a concordance class of embeddings of T^n in $T^n \times D^2$ in terms of an n dimensional knot cobordism class and just the arf invariants and indices of lower dimensional knots introduced along subtori.

For $\mathfrak{g} : Z\pi \to \Lambda$, we write $\mathfrak{g} \times Z$ (the motivation for the notation is geometric) for the induced map of rings $Z\pi[t,t^{-1}] \to \Lambda[t,t^{-1}]$. Here the involutions on $Z\pi$ and Λ are extended by the formula $\overline{ut^i} = \bar{u}t^{-i}$ for u in Λ or $Z\pi$. The following formula for $\Gamma_n(\mathfrak{g} \times Z)$ is analogous[1] to the results of [S2] and [W] on Wall groups of polynomial rings. It is proved using a geometric splitting theorem (15.1 of [CS6]), valid only in <u>odd</u> dimensions, for homology equivalent manifolds. This splitting principle is analogous to the Farrell-Hsiang splitting theorem [FH] for homotopy equivalent manifolds. It is proved by adapting the methods used to prove the general splitting theorems of [C1] [C2]. (See problem 10 of [S1]).

<u>Theorem:</u> (14.1 of [CS6]). <u>Let $\mathfrak{g}$ be induced from a homomorphism of groups</u> $\pi \to \pi'$. <u>Then</u> $\Gamma_n^s(\mathfrak{g} \times Z) = \Gamma_n^s(\mathfrak{g}) \oplus L_{n-1}^h(\pi')$.

The groups Γ_n^s and Γ_n^h are related by an exact sequence (15.2 of [CS6]) analogous to the Rothenberg sequence [S2].

---

1 In fact, it reduces to [S2, 5.1], see also [W1 §13], for n odd. However, this reduction employs the fact that j_* is an isomoprhism in odd dimension and in [CS6], j_* is shown to be surjective using the present result for n even.

BIBLIOGRAPHY

[C1] Cappell, S., A splitting theorem for manifolds and surgery groups. Bull. A.M.S. 77(1971) 281-286.

[C2] Cappell, S., Mayer-Vietoris sequences for surgery groups, (these proceedings).

[CS1] Cappell, S., and Shaneson, J.L., Topological knots and knot cobordism. Topology, to appear

[CS2] Cappell, S., and Shaneson, J.L., Submanifolds, group actions, and knots I. Bull. A.M.S., to appear.

[CS3] Cappell, S., and Shaneson, J.L., Submanifolds, group actions, and knots II. Bull. A.M.S., to appear.

[CS4] Cappell, S., and Shaneson, J.L., Non-locally flat embeddings Bull. A.M.S., to appear.

CS5] Cappell, S., and Shaneson, J.L., Submanifolds of codimension two and homology equivalent manifolds, Ann. Inst. Fourier, to appear.

[CS6] Cappell, S., and Shaneson, J.L., The codimension two placement problem and homology equivalent manifolds, to appear.

[FH] Farrell, F.T., and Hsiang, W.C., Manifolds with $\pi_1 = G \times_\alpha T$, to appear. (See also Bull. A.M.S. 74(1968) 548-553).

[FM] Fox, R.H., and Milnor, J., Singularities of 2-spaces in 4-space and equivalence of knots. Bull. A.M.S. 63(1965) 406.

[J] Jones, L., Three characteristic classes measuring the obstruction to P.L. local unknottedness, to appear.

[K1] Kervaire, M.A., Les noeuds de dimension superieure. Bull. Soc. Math. de France 93 (1965), 225-271.

[K2] Kervaire, M.A., Knot cobordism in codimension two, "Manifolds-Amsterdam 1970", Springer-Verlag 197 (1970).

[L1] Levine, J., Knot cobordism in codimension two. Comment. Math. Helv. 44 (1968), 229-244.

[L2] Levine, J., Invariants of knot cobordism, Inventiones Math. 8 (1969), 98-110.

[L3] Levine, J., Unkotting spheres in codimension two. Topology 4 (1965), 9-16.

[M] Matsumoto, Y., Knot cobordism groups and surgery in codimension two, to appear.

[S1] Shaneson, J.l., Some problems in Hermitian K-theory, these proceedings.

[S2] Shaneson,J.L., Wall's surgery obstruction groups for Z × G . Ann. of Math. 90 (1969), 296-334.

[W1] Wall, C.T.C., "Surgery on compact manifolds". Academic Press, 1970.

Proper surgery groups and Wall-Novikov groups

Serge Maumary *
IAS Princeton
and UC Berkeley

The lifting of a surgery problem of closed manifolds to a covering leads usually to a proper surgery problem on open locally compact manifolds, and this procedure gives by the present work new informations about the original problem. This is the motivation of proper surgery. In [8], proper surgery groups are constructed formally as in [9, §9] and our goal has been to "compute" these groups in terms of Wall-Novikov groups (both [8] and the present work have been done simultaneously and ignoring each other). I am indebted to W.Browder, J.Wagoner, R.Lee, A.Ranicki for useful and friendly conversations, and to L.Taylor who pointed out a gap.

1. Notations and conventions

We consider exclusively locally compact manifolds M and CW-complexes X of finite dimension, and proper maps f between them (i.e. f^{-1}(compact) is compact).
If X is connected, we can choose a fondamental sequence of ngbd of ∞: $X_1 \supset X_2 \supset X_3 \supset \ldots$ formed by subcomplexes X_n with only non compact components (in finite number). We denote by $\overline{X-X_n}$ any finite subcomplex of X such that $\overline{X-X_n} \cup X_n = X$, and let $\dot{X}_n = \overline{X-X_n} \cap X_n$ which is a finite subcomplex containing the frontier of X_n in X. For any pointed connected CW-complex A, with associated universal covering $\widetilde{A}$, one usually denotes by C(A) the chain complex of cellular chains on $\widetilde{A}$ with integer coefficients (and if $B \subset A$, then C(A) mod C(A,B) is denoted by C(A,B)) We denote by $C(X_n)$ the family $C(X_n^i)$ obtained by choosing implicitely one base point in each connected component X^i of X_n.

* Supported by grant of Fonds National Suisse. SG 58

Similarly, we denote by $Z\pi_1 X_n$ the family of rings $Z\pi_1 X_n^i$ and by a $Z\pi_1 X_n$-module M, we mean a family of $Z\pi_1 X_n^i$ - modules M^i. The homology of $C(X_n)$ is denoted by $H_k(X_n)$, while the homology of its dual $C^*(X_n)$, the family of $\mathrm{Hom}_{\pi_1 X_n^i}(C(X_n^i), Z\pi_1 X_n^i)$, is denoted by $H^k(X_n)$, where $C(X_n^i)$ is given the right structure via the anti-automorphism $\alpha \to w(\alpha)\alpha^{-1}$ of $Z\pi_1 X_n^i$, w being given by some fixed homomorphism $\pi_1 X \to \pm 1$.
The U-groups of [4] or [5] will be denoted by $L_m^p(G)$ while $L_m(G)$ denotes the ordinary Wall groups (or V-groups). As an inner automorphism of G induces $\pm$ identity on $L_m^p(G)$, $L_m^p(\pi_1 X_n^i)$ is well defined and we write $L_m^p(\pi_1 X_n) = \bigoplus_i L_m^p(\pi_1 X_n^i)$. Similarly $X_{n+1} \to X_n$ induces a unique homomorphism on L_m^p. In particular, $\varprojlim_n L_m^p(\pi_1 X_n)$ only depends on X, as well as $\varprojlim_n{}^1 L_m^p(\pi_1 X_n)$. As for the latter it maybe useful to recall Milnor's definition of $\varprojlim^1$ of an inverse system of abelian groups $A_1 \overset{\#}{\leftarrow} A_2 \overset{\#}{\leftarrow} A_3 \overset{\#}{\leftarrow} \ldots$: this is the Coker of $\prod_{n>1} A_n \xrightarrow{1-S} \prod_{n>1} A_n$, where $(1-S)(a_1, a_2, a_3, \ldots) = (a_1 - a_2^{\#}, a_2 - a_3^{\#}, \ldots)$. Observe that a subsequence of $\{A_n\}$ gives the same result : e.g. $\varprojlim^1 A_{2n+1} \approx \varprojlim^1 A_n$ by mapping $(a_1, a_2, a_3, \ldots)$ to $(a_1 + a_2^{\#}, a_3 + a_4^{\#} \ldots)$ in the range of 1-S.

2. Homology and cohomology inverse systems

Having implicitely choosen one base point for each connected component of X_n, we join the base points of X_{n+1} to those of X_n by paths in X_n (in this usy, a tree grows in each connected component of X_1). The latter determine maps $\tilde{X}_{n+1}^j \to \tilde{X}_n^i$ and so pseudo-linear homomorphisms $C(X_{n+1}^j) \to C(X_n^i)$.

This gives rise to an <u>inverse system</u> $\{C(X_n)\}$. Note that $\bigoplus_j \left(Z\pi_1 X_n^i \underset{Z\pi_1 X_{n+1}^j}{\otimes} C(X_{n+1}^j) \right)$ (one summand for each j such that $X_{n+1}^j \subset X_n^i$) is isomorphic to the subcomplex of $C(X_n^i)$ determined by $\tilde{X}_n^i \,|\, X_{n+1}$

Two choices of base points and paths give two inverse systems related by a diagram of subsequences

$$\begin{array}{ccccccc}
C(X_{r_1}) & \longleftarrow & C(X_{r_2}) & \longleftarrow & C(X_{r_3}) & \longleftarrow & \\
\downarrow & \nwarrow & \downarrow & \nwarrow & \downarrow & \nwarrow & \\
C(X_{s_1}) & \longleftarrow & C(X_{s_2}) & \longleftarrow & C(X_{s_3}) & \longleftarrow &
\end{array}$$

which commutes up to the action of $\pi_1 X_n^i$ on itself by inner automorphisms. Such a diagram is called a <u>conjugate equivalence</u>. Similarly, the families of cochain complexes $C_c^*(X_n, \dot{X}_n) \overset{\text{def}}{=} \varinjlim_r C^*(X_n, \dot{X}_n \cup X_r)$ form an inverse system by excision and is also well defined up to conjugate equivalence. Now, any element $[X] \in \varprojlim_r H_m(X, X_r; Z)$ (homology with coefficients extended by w : $Z\pi_1 X \to Z$) gives by cap products (see [1]) a commutative diagram

$$\begin{array}{ccccc}
C_c^*(X_1, \dot{X}_1) & \longleftarrow & C_c^*(X_2, \dot{X}_2) & \longleftarrow & \\
\downarrow \cap[X] & & \downarrow \cap[X] & & \\
C(X_1) & \longleftarrow & C(X_2) & \longleftarrow &
\end{array}$$

i.e. by definition a <u>morphism</u> of inverse systems. The latter is called an <u>equivalence</u> if there is an "inverse" morphism $\{C(X_n)\} \overset{\Psi}{\longrightarrow} \{C_c^*(X_n, \dot{X}_n)\}$, i.e. a commutative diagram of subsequences

$$\begin{array}{ccccccc}
C_c^*(X_{r_1}, \dot{X}_{r_1}) & \longleftarrow & & & C_c^*(X_{r_2}, \dot{X}_{r_2}) & \longleftarrow & \\
\downarrow \cap[X] & & \nwarrow \Psi_1 & & \downarrow \cap[X] & & \nwarrow \Psi_2 \\
C(X_{r_1}) & \leftarrow & C(X_{s_1}) & \leftarrow & C(X_{r_2}) & \leftarrow & C(X_{s_2}) \leftarrow
\end{array}$$

where Ψ_n is pseudo-linear, and $r_1 < s_1 < r_2 < s_2 < \ldots$

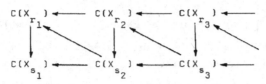

Observe that in this case, we can assume $r_n = s_{n-1} = n$ without loss of generality. When $\cap[X]$ is an equivalence, we say that $[X]$ is a <u>m-fundamental class</u> at ∞, and that X is <u>properly Poincaré</u> at ∞. This turns out to be an invariant of the proper homotopy type of X. Now, a proper map f: $M \to X$ of properly Poincaré complexes is said of <u>degree 1</u> if $f^*\ [M] = [X]$. By confusing X with the mapping cylinder of f, and denoting the k+1-homology of $C(X_n, M_n)$, resp. $C_c^*(X_n, \dot{X}_n \cup M_n)$ by $K_k(M_n)$, resp. $K_c^k(M_n, \dot{M}_n)$, where $M_n = X_n \cap M$, $\dot{M}_n = \dot{X}_n \cap M$ we get again inverse systems $\{K_k(M_n)\}$ and $\{K_c^k(M_n, \dot{M}_n)\}$ well defined up to conjugate equivalence. If $\dot{M}_n = \partial M_n$, then the composition :

$$\Psi : K_{m-k}(M_n) \overset{\partial}{\to} H_{m-k}(M_n) \overset{\text{Poincaré}}{\simeq} H_c^k(M_n, \partial M_n) \to K_c^k(M_n, \dot{M}_n)$$ turns out to be a canonical equivalence of inverse systems with an inverse shifting n by 4 (and so shifting n by 1 on a subsequence). Of course in the above, $H_*(M_n)$ and $H_c^*(M_n, \partial M_n)$ are with $\pi_1 X_n$-coefficients.

3. <u>Homology and cohomology direct systems</u>

For $r<n<s$, let $C^*(X_n, X_s)_r$ be the family $\bigoplus_j \mathrm{Hom}_{Z\pi_1 X_r^i}$

$$\left(Z\pi_1 X_r^i \underset{Z\pi_1 X_n^j}{\otimes} C(X_n^j, X_s),\ Z\pi_1 X_r^i\right)$$ and let $C_c^*(X_n)_r$ be the family $\lim\limits_{\overrightarrow{s}} C^*(X_n, X_s)_r$. For r fixed, the restriction maps $C_c^*(X_n)_r \to C_c^*(X_{n+1})_r$ determine a <u>direct system</u> $\{C_c^*(X_n)_r\}$.

Similarly, if $C(X_n, \dot{X}_n)_r$ denotes the family $\bigoplus_j \left(Z\pi_1 X_r^i \underset{Z\pi_1 X_n^j}{\otimes} C(X_n^j, \dot{X}_n^j)\right)$ (chains of $\tilde{X}_r^i | X_n$ mod $\tilde{X}_r^i | X_r - X_n$), then the quotient maps $C(X_n, \dot{X}_n)_r \to C(X_{n+1}, \dot{X}_{n+1})_r$ form a direct system, for r fixed and $n > r$. Now, given a proper map f: $M \to X$ of degree 1, if we write $K_*(M_n, \dot{M}_n)_r$, resp $K_c^*(M_n)_r$, for the homology of $C(X_n, \dot{X}_n \cup M_n)_r$, resp $C_c^*(X_n, M_n)_r$, we find again an equivalence Ψ :

$$\{K_{m-k}(M_n, \dot{M}_n)_r\} \to \{K_c^k(M_n)_r\}.$$

Of course, these direct systems are well defined only up to conjugate equivalence, the latter notion being the same as for inverse systems.

4. End homology and cohomology

The dual of $C_c^*(X_n,\dot{X}_n)$ is canonically isomorphic to $C'(X_n,\dot{X}_n) \overset{\text{notat.}}{=}$ $= \varprojlim_s C(X_n,\dot{X}_n \cup X_s)$, which is nothing but the chain complex of locally finite chain on X_n mod $\dot{X}_n$, with $Z\pi_1X_n$-coefficients. The quotient complex $C'(X_n,\dot{X}_n)/C(X_n,\dot{X}_n)$ yields the end homology $H_*^e(X_n)$ by definition. As usually, the cochain complex $\varinjlim_s C^*(X_s)_n$ yields the end cohomology $H_e^*(X_n)$ by definition. Now, one can prove [see 3] that, if [X] is a m-fundamental class at ∞ coming from $C_m'(X;Z)$, then $\cap[X]$ gives rise to an isomorphism $H_e^k(X_n) \approx H_{m-k}^e(X_n)$. All this applies to a proper map $f : M \to X$ of degree 1, to yield an isomorphism $K_e^k(M_n) \approx K_{m-k}^e(M_n)$. Our end homology can be viewed as an ε-construction (see [8] or [2]) with π_1X_n-coefficients as follows : consider the diagram of families of pointed subcomplexes

$$\begin{array}{cccc}
(X_n,x_n) & (X_n,x_{n+1}) & (X_n,x_{n+2}) & \dots \\
 & \cup & \cup & \\
 & (X_{n+1},x_{n+1}) & (X_{n+1},x_{n+2}) & \dots \\
 & & \cup & \\
 & & (X_{n+2},x_{n+2}) & \dots \\
 & & & \dots
\end{array}$$

Then let $\mu C(X_s)_n$ be the quotient complex $\prod_{r>s} C(X_s,x_r)_n / \bigoplus_{r>s} C(X_s,x_r)$ and $\varepsilon C(X_n) = \varprojlim_{s>n} \mu C(X_s)_n$. An isomorphism $C^e(X_n) \approx \varepsilon C(X_n)$ arises by decomposing $z \in C'(X_n)$ into $z_n \oplus z'_{n+1} \in C(X_n,X_{n+1}) \oplus C'(X_{n+1})_n$, then z'_{n+1} into $z_{n+1} \oplus z'_{n+2}$, and so forth.

5. The category of (inverse or direct) systems

If one considers systems of families of modules $\{A_n\}$ over $\{Z\pi_1 X_n\}$ as abstract objects and takes their equivalence classes by the relation of (conjugate) equivalence, and if one does the same thing for the morphisms $\{A_n\} \to \{B_n\}$, then it is routine to verify that one gets an abelian category (see [3], compare [7]). A more specific result is the following.

Proposition (see [3]) let $\{C(n)\}$ be a system of chain complexes, each of the form $0 \to C_\ell(n) \xrightarrow{\partial} \dots \to C_1(n) \to C_0(n) \to 0$ where $\ell > 0$ is fixed independant of n and $C_k(n)$ is free of countable rank. Suppose that the associated homology systems $\{H_k(n)\}$ are equivalent to 0 for all $k < \ell$. Then there is an equivalence $\{H_\ell(n)\} \to \{P_n\}$, where each P_n is a projective countably generated module and each homomorphism $H_\ell(n) \to P_n$ is injective. Moreover, in the system $\{P_n\}$, one can assume that the image of $P_{n+1} \to P_n$ a direct summand, in particular also projective.

These two results essentially allow us to elaborate an algebraic Whitehead torsion for proper homotopy equivalence (compare [8]).

6. Proper surgery

It is well known that any surgery rel, boundary on a compact m-submanifold of M^m extends to M, and similarly for a closed bicollared submanifold V^{n-1}. By definition, a proper surgery on M is the result of a diverging sequence of disjoint such surgeries. We distinguish the following particular case of carving out $\mathbb{R}^q \subset M$. Let $f: M \to X$ be a proper normal map (relative to ξ on X), $\varphi: \mathbb{R}^q \to M$ be a proper embedding, $\Psi: \mathbb{R}^{q+1}_+ \to X$ a proper map such that $\Psi \mid \mathbb{R}^q = f_0 \varphi$ $(\mathbb{R}^q = \partial \mathbb{R}^{q+1}_+)$.
Now the normal bundle of φ is trivial (because $\mathbb{R}^q$ is contractible) and we form W^{m+1} by gluing $M \times I$ and $\mathbb{R}^{q+1}_+ \times D^{m-q}$ along $\mathbb{R}^q \times D^{m-q} \subset M \times 1$.

As $(M\times I)\cup\mathbb{R}^{q+1}_+$ is a proper deformation retract of W, we can extend f to F : $W\to X\times I$ by using ψ . The stable trivialisation of $\tau_M \oplus f^*\xi$ on M extends to a stable trivialisation of $\tau_W \oplus F^*\xi$ on W because W retracts by deformation on M x 0. Now, W is a cobordism between M and $M' = M - \varphi(\mathbb{R}^q)$. The inclusions $M \subset W \supset M' \cup D^{mq}$ are homotopy equivalences

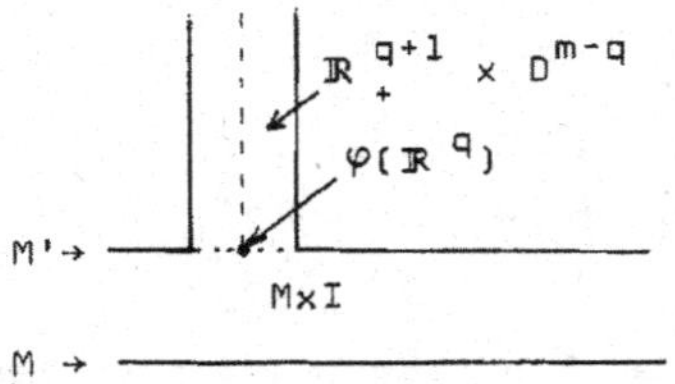

One can observe that M also results form M' by first a (m-q)-surgery and then carving out $\mathbb{R}^{m-q}$. To each cocompact submanifold $M_n \subset M$ corresponds a cocompact submanifold $M'_n \subset M'$ of the following shape : $M'_n = (M_n \cup \text{q-handle}) - \mathbb{R}^q$

7. Preliminary surgeries

Let M be an open m-manifold, X a proper Poincaré complex at ∞ and $f : M\to X$ a proper normal map of degree 1. We assume that X is connected and so we can choose cocompact subcomplexes X_n in X which have only non compact connected components. We can assume that each $\dot{X}_n$ is bicollared, and that f is transversal on each of them (see [1]). Then $f^{-1}(X_n)$ is a cocompact submanifolds $M_n \subset M$, such that $\partial M_n = f^{-1}(\dot{X}_n)$ and $\overline{M_n - M_{n+1}} = f^{-1}(\overline{X_n - X_{n+1}})$. Clearly, if $m=2q$, resp. $2q+1$, $q \geq 3$, we can assume that each map

$\overline{M_n - M_{n+1}} \xrightarrow{f} \overline{X_n - X_{n+1}}$ is q-connected, while $\partial M_n \xrightarrow{f} \dot{X}_n$ is q-1, resp. q-connected. In particular, f is bijective on ends spaces.

When m= 2q+1, we can improve still the connectivity of f as follows. Each module of the family $K_q(M_n, M_{n+1}) \overset{def}{=} H_{q+1}(X_n, M_n \cup X_{n+1})$ is finitely generated, and each generator can be represented by an embedded q-sphere S^q in $\overline{M_n - M_{n+1}}$, provided with a nulhomotopy D^{q+1} in $\overline{X_n - X_{n+1}}$. We pipe S^q to ∞, getting $\mathbb{R}^q \underset{proper}{\subset} M_n$ and extend D^q into $\mathbb{R}^{q+1}_+ \xrightarrow{proper} X_n$:

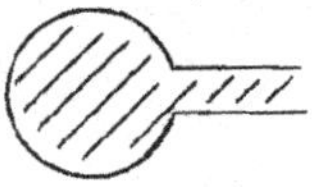

Then the process of carving out $\mathbb{R}^q \subset M_n$ allows to kill each $K_q(M_n, M_{n+1})$. An immediate consequence is $K^q_c(M_n) = 0$, hence the direct system $\{K_{q+1}(M_n, \partial M_n)_r\}$ is equivalent to 0 by duality. A more involved argument (see [3]) shows that $K'_q(M_n) \overset{def}{=} H'_{q+1}(X_n, M_n)$ also vanishes hence the inverse system $\{K^{q+1}(M_n, \partial M_n)\}$ is equivalent to 0 by duality. Moreover, the inverse system $\{K_q(M_n)\}$ and the direct system $\{K_q(M_n, \partial M_n)_r\}$ are both equivalent to systems of projective countably generated modules (ibid).

8. The case m=2q+1, M open

Assuming the preliminininary surgery already done the starting situation is described by a commutative square

$$\begin{array}{ccc} K_c^{q+1}(M_n, \partial M_n)_r & \longrightarrow & K_c^{q+1}(M_n)_r \\ \uparrow \Psi & & \uparrow \overline{\Psi} \\ K_q(M_n)_r & \longrightarrow & K_q(M_n, \partial M_n)_r \end{array} \qquad r < n.$$

where Ψ, resp. $\overline{\Psi}$, are equivalences of inverse, resp. direct, systems (r being fixed, n variable > r), with inverse equivalences shifting n by +1.

The fundamental duality property of this square is the following ± commutative diagrams of exact sequences

$$\begin{array}{ccccccccc} 0 \to & K^q(M_n,\partial M_n)_r & \to & K^q_e(M_n)_r & \to & K^{q+1}_c(M_n,\partial M_n)_r & & \\ & \uparrow \bar{\Psi}^* & & \uparrow \Psi^e & & \uparrow \Psi & & \\ & K'_{q+1}(M_n)_r & \to & K^e_{q+1}(M_n)_r & \to & K_q(M_n)_r & \to & 0 \end{array}$$

$$\begin{array}{ccccccccc} 0 \to & K^q(M_n)_r & \to & K^q_e(M_n)_r & \to & K^{q+1}_c(M_n)_r & & \\ & \uparrow \Psi^* & & \uparrow \Psi^e & & \uparrow \bar{\Psi} & & \\ & K_{q+1}(M_n,\partial M_n)_r & \to & K^e_{q+1}(M_n)_r & \to & K_q(M_n,\partial M_n)_r & \to & 0 \end{array}$$

where Ψ^* is the composition $K'_{q+1}(M_n,\partial M_n)_r \overset{\text{can.}}{\to}$ dual $K^{q+1}_c(M_n,\partial M_n)_r \overset{\text{dual }\Psi}{\to}$ dual $K_q(M_n)_r \simeq K^q(M_n)_r$, and similarly for Ψ^*, and $\Psi^e = \varinjlim_n \Psi^*$ is actually an isomorphism (see [3]). One sees that both Ψ and $\bar{\Psi}$ are induced by Ψ^e. Our aim is to improve the initial arbitrary choice of $X_n, \dot{X}_n$ in the mapping cylinder X of $M \overset{f}{\to} X$ so as to get Ψ bijective. One cannot do this for X itself but one can replace X by any complex simply homotopy equivalent to X rel. M. The first step is the following.

<u>Lemma</u> : Ker Ψ and Ker Ψ are finitely generated. Proof (sketched): using the results of §5, one finds an equivalence $\{K^{q+1}_c(M_n,\partial M_n)_r\} \overset{\text{inj.}}{\to} \{P_n\}$, Where each P_n is projective, the image of $P_{n+2} \to P_n$ being a direct summand P'_n. By composition with Ψ we get an equivalence $\alpha:\{K_q(M_n)\} \to \{P_{n-1}\}$ such that ker α =ker Ψ and im $\alpha = P'_n$, which is projective. Hence ker α is a direct summand. But ker Ψ is contained in the kernel of $K_q(M_{n+1})_r \to K_q(M_n)_r$, which is finitely generated, hence so is ker Ψ, as direct summand. The same argument applies to $\bar{\Psi}$. This shows actually that, for a <u>subsequence</u>, the kernel of $K_q(M_n)_{n-1} \overset{\Psi}{\to} K^{q+1}_c(M_n,\partial M_n)_{n-1}$ is finitely generated, and similarly for $\bar{\Psi}$.

The first improvement is to replace X_n by $X_{n+1} \cup M_n$ and $\dot{X}_n$ by $X_{n+1} \cup \dot{M}_n$ where $\dot{M}_n = \overline{M_n - M_{n+1}}$

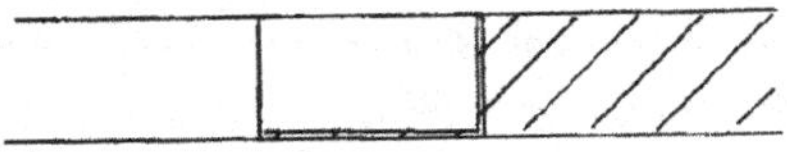

Then, in the square

$$\begin{array}{ccc} K_c^{q+1}(M_n, \dot{M}_n) & \longrightarrow & K_c^{q+1}(M_n) \\ \psi \uparrow & & \uparrow \overline{\psi} \\ K_q(M_n) & \longrightarrow & K_q(M_n, \dot{M}_n) \end{array}$$

ker ψ and ker $\overline{\psi}$ are finitely generated. The second improvement is to enlarge X_n inside $\overline{X_n - X_{n+1}}$ with $\overline{M_n - M_{n+2}} \cup e^{q+1}$, to kill ker $\overline{\psi}$:

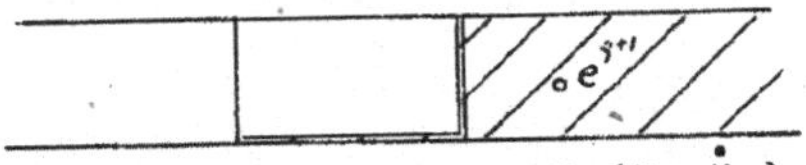

By taking the quotient map, we find $K_q(M_n, \dot{M}_n) \to K_c^{q+1}(M_n)$ injective, and by the fundamental duality property we can restablish ψ and the initial square (see [3]). Assuming ψ injective, we can enlarge both X_n and $\dot{X}_n$ inside $\overline{X_{n-1} - X_n}$ with $\overline{M_{n-1} - M_{n+2}} \cup e^{q+2}$ to kill ker ψ. By taking the quotient map, we find $K_q(M_n) \overset{\psi}{\to} K_c^{q+1}(M_n, \dot{M}_n)$ injective, and we restablish $\overline{\psi}$ and the square by the fundamental duality property again. By using the proof of the above lemma, both $K_q(M_n)$ and $K_q(M_n, \dot{M}_n)$ are seen to be projective (ibid). Then one can still kill the kernel of the map $K_q(\dot{M}_n)^{\#} \to K_q(M_n)$ where $\#$ means with $\pi_1 X_n$-coefficients, and this will make ψ bijective (ibid). Then the fundamental duality property implies that ψ is injective. Now, the commutative diagram of exact sequence

$$\begin{array}{ccccccc} 0 \to & K^q(\dot{M}_n)^{\#} & \to & K_c^{q+1}(M_n, \dot{M}_n) & \to & K_c^{q+1}(M_n) & \\ & \uparrow & & \uparrow & & \uparrow & \\ 0 \to & K_q(\dot{M}_n)^{\#} & \to & K_q(M_n) & \to & K_q(M_n, \dot{M}_n) & \to 0 \end{array}$$

shows that ψ induces an isomorphism $K_q(\dot{M}_n)^{\#} \simeq K^q(\dot{M}_n)^{\#}$, i.e. a non degenerated quadratic projective finitely generated $Z\pi_1 X_n$ module $< K_q(\dot{M}_n) >$

Proposition : the quadratic form on $< K_q(M_n) >$ so obtained satisfies the following properties :

i) it is induced by the (degenerated) intersection form on $K_q(\partial M_r)^{\#}$ for some $r > n$, hence determine an element of $L^p_{2q}(\pi_1 X_n)$.

ii) it is defined stably, and the operation of carving out a trivial proper embedded $\mathbb{R}^q \subset M$ (bounding $\mathbb{R}^{q+1}_+ \underset{proper}{\subset} M$) adds a trivial free hyperbolic module

iii) there is a canonical equivalence between the quadratic $Z\pi_1 X_n$-modules $< K_q(\dot{M}_n)>$ and the $Z\pi_1 X_n$-extension of $< K_q(\dot{M}_{n+1})>$. In other words the sequence $< K_q(\dot{M}_n)>$ is an element of $\varprojlim_n L^p_{2q}(\pi_1 X_n)$

iv) the latter is well defined by the normal map $f: M \to X$, and is a cobordism invariant. For the proof of this proposition, we refer to [3]. As a result, we get a homomorphism σ : $L_m(eX) \to \varprojlim_n L^p_{m-1}(\pi_1 X_n)$ for m odd. Here, $L_m(eX)$ is the group of proper "surgery data over X at ∞", (same definition as in [8], but use only proper h.e. at ∞ in defining 0) and satisfies actually an exact sequence
$L^p_m(\pi_1 X) \xrightarrow{I} L_m(X) \to L_m(eX) \to 0$, where $L_m(X)$ is the proper surgery group (see [8] for its construction).

Proposition : ker σ is isomorphic to $\varprojlim{}^1 L_{2q+1}(\pi_1 X_n)$.
The idea of the proof is to construct a map $\varprojlim{}^1 L_{2q+1}(\pi_1 X_n) \xrightarrow{I}$ Ker σ and an injective left inverse (see [3]).

Theorem (partial exact sequence) : for m odd, one has an exact sequence $\Pi_m \xrightarrow{1-s} L_m(\pi_1 X) \oplus \Pi \xrightarrow{I} L_m(X) \xrightarrow{\sigma} \Pi^p_{m-1} \xrightarrow{1-s} L^p_{m-1}(\pi_1 X) \oplus \Pi^p_{m-1}$

where Π_m is the product $\prod_{n>1} L_m(\pi_1 X_n)$, and S is the <u>shifting map</u>.
More precisely, $(1-S)(a_1, a_2, a_3, \dots) = (-a_1^{\#}, a_1 - a_2^{\#}, a_2 - a_3^{\#}, \dots)$

for $a_n \in L_m(\pi_1 X_n)$, # denoting the homomorphisms

$L_m(\pi_1 X) \leftarrow L_m(\pi_1 X_1) \leftarrow L_m(\pi_1 X_2) \leftarrow \dots$

Proof : observe that ker (1-S) is the subgroup of $\varprojlim_n L^p_{m-1}(\pi_1 X_n)$ vanishing in $L^p_{m-1}(\pi_1 X)$. The range of σ is in ker (1-S) by the proof of iii in prop.above, replacing $\langle K_q(\dot{M}_n)\rangle$ by ϕ and $\langle K_q(\dot{M}_{n+1})\rangle$ by the π_1X-extension of $\langle K_q(\dot{M}_n)\rangle$. The exactness $\mathrm{Im}\sigma = \ker(1-S)$ is seen by constructing a cobordism between $N \xrightarrow{1} N$ and a proper h.e. $N' \to N$, where N is an open 2q-manifold provided with a 1-equivalence $N \to X$. The various map τ are also constructed by cobordism on a 2q-manifold, and $\tau_o(1-S)$ vanishes. Hence we get induced maps τ satisfying the commutative diagram of exact sequences

$$\begin{array}{ccccccccc}
\varprojlim_n L_m(\pi_1 X_n) & \to & L_m(\pi_1 X) & \to & L_m(X) & \to & L_m(eX) & \to & 0 \\
 & & & & \uparrow\bar{\tau} & & \uparrow\bar{\tau}\ \uparrow\sigma & & \\
\varprojlim_n L_m(\pi_1 X_n) & \to & L_m(\pi_1 X) & \to & \mathrm{Coker}(1-S) & \to & \varprojlim L_m(\pi_1 X_n) & \to & 0
\end{array}$$

By the latter proposition, the right $\bar{\tau}$ is injective, hence so is the middle one. This proves the exactness Ker τ = Im (1-S). We also know that $\sigma_o\tau = 0$. The exactness Ker σ = Im τ is a result of the above diagram

9. The case m=2q+2, M open

Assuming the preliminary surgery already done, we are left (as in) the case m odd) with only one inverse system $\{K_{q+1}(M)_n)_r\}$ and one direct system $\{K_{q+1}(M_n, \partial M_n)_r\}$ not equivalent to 0. Following Wall's idea for the compact case, we want to consider the surgery data $M \xrightarrow{f} X$ as the union of two surgery cobordisms $M^o \cup V \to X^o \cup H$ along their common boundary $U \to \partial H$.

Lemma (see [8 chap.II th.3]) : X has the simple homotopy type of a CW-complex $X^o_{\partial H} \cup H$, where H is a locally finite m-handlebody of 0 and 1-handles. Actually, H is a regular ngbd of a tree in R^m, with 1-handles attached.

Proposition : assuming X of the above form, one can find a codimension 0-submanifold V of M such that, if $M^o = \overline{M-V}$, $f(M^o) \subset X^o$ and $f(V) \subset H$ up to a proper homotopy of f. Actually, V is a locally finite handlebody of 1, q and q+1-handles, formed by a regular ngbd of the union of immersed spheres $S^{q+1} \to M$ piped to ∞.
The proof relies on the same geometrical arguments than [6].
We refer to this as a Mayer-Vietoris decomposition of $M \xrightarrow{f} X$.
Actually, the ngbd of ∞ in ∂H, resp ∂V, can be chosen such that their frontier $\partial \dot{H}_n$, resp $\partial \dot{V}_n$, is S^{2q}, resp $S^q \times S^q$, and $f(\partial \dot{V}_n) \subset \partial \dot{H}_n$. This implies that $K_q(\partial \dot{V}_n)$ is a free hyperbolic module (with the intersection form). Then we can modify the choices of the ngbd of ∞: X^o_n in X^o, and the choice of $\dot{X}^o_n$, as in the proof of iv in the first prop. of §8 to get $K_q(\dot{M}^o_n)$ as a projective Lagrangian plane in $K_q(\partial \dot{V}_n)$. This determines an element of $L^p_{2q+1}(\pi_1 X_n)$ and we have results similar to those in §8, with m replaced by m+1.

REFERENCES

1. W.Browder : Surgery on simply connected manifolds. Springer 1971.

2. T.Farrell-J.Wagoner : Algebraic torsion for infinite simple homotopy types. Infinite matrices in algebraic K-theory and topology Comm. Math. Helv. 1972

3. S.Maumary : Proper surgery groups, Berkeley mimeo notes 1972

4. S.P.Novikov : Algebraic construction and properties of hermitian analogs of K-theory ... kv.Akad. Nauk SSR Ser.Mat.Tom 34 1970 Math, USSR kv vol.4,2, 1970

5. Ranicki : Algebraic L-theories, these Proceedings

6. R.Sharpe : thesis Yale 1970

7. L.Siebenmann : Infinite simple homotopy types Indag. Math. 32, 5 1970

8. L. Taylor : thesis Berkeley 1971

9. C.T.C. Wall : Surgery on compact manifolds Acad. Press 1970

Induction in Equivariant K - Theory and Geometric Applications

Ted Petrie

Department of Mathematics

Rutgers University

§0. Introduction

This paper presents some algebraic situations and problems which arise from comparing two R valued bilinear forms $< >_\Lambda$, $< >_\Gamma$ on R orders Λ and Γ. Here R will be a P.I.D. which arises by localizing $R(S^1)$, the complex representation ring of S^1, at a set P of prime ideals in $R(S^1)$. These forms occur geometrically in the following way: Let X and Y be closed smooth S^1 manifolds of even dimension. Under mild assumptions on X and Y, there are non-degenerate $R = R(S^1)_P$ valued symmetric bilinear forms on $\Lambda = K^*_{S^1}(X)_P$ and $\Gamma = K^*_{S^1}(Y)_P$ constructed from the algebraic structure of each and the Atiyah-Singer Index homomorphism [5].

An S^1 map $f: X \longrightarrow Y$ induces an R algebra homomorphism $f^*: \Lambda \longrightarrow \Gamma$. A natural geometric assumption concerning f leads to the situation in which f^* is a monomorphism and $\Lambda, \Gamma, \Theta = \pi_{i=1}^{n} R$ are R orders in the semisimple F algebra $\Theta \otimes_R F$. Using the non-degenerate bilinear forms $< >_\Lambda$ and $< >_\Gamma$, we define an induction homomorphism $f_*: \Gamma \longrightarrow \Lambda$. In particular, $f_*(1) \in \Lambda$ is an interesting algebraic invariant of the situation.

Knowledge of this invariant $f_*(1)$ translates into important geometric information comparing the differential structures of X and Y and the representations of S^1 on the normal bundle to the fixed set $X^{S^1} \subset X$ and on the normal bundle to the fixed set $Y^{S^1} \subset Y$.

The layout of the paper is as follows:

§1. This section sets forth the algebraic situation. We record a number of algebraic results concerning the induction homomorphism f_* which arises from non-degenerate bilinear forms on two algebras, Λ and Γ, and an algebra map f^* between them. We pose a basic and very important question about the invariant $f_*(1)$.

§2. This section sets forth a geometric situation which motivated the above algebra and to which this algebra applies. Roughly the geometric set up is this: We want to compare the differential structures on two smooth S^1 manifolds X and Y when there is an S^1 equivariant map f from X to Y which induces a homotopy equivalence on the underlying smooth manifolds $|X|$ and $|Y|$. The algebraic tool for this study is complex equivariant K theory $K^*_{S^1}(\ )$ and the basic new theorem which allows us to apply §1 is that there are non-degenerate bilinear forms on

$$K^*_{S^1}(X)_P/\text{Torsion} \quad \text{and} \quad K^*_{S^1}(Y)_P/\text{Torsion}$$

defined by the Atiyah-Singer Index homomorphism [5].

§3. This section provides some non-trivial illustrations of the algebra of §1. These are recorded because (as of now) examples are hard to produce and because they illustrate the results of Sections 1 and 2. Moreover, they indicate the possibility of constructing more examples in the geometric situation and point where to look for such examples.

§1. Algebraic Situation

We start with the ring $Z[t,t^{-1}]$. Among other things, this is the complex representation ring of the group S^1 of complex numbers of norm 1. If $\lambda_1, \lambda_2, \cdots \lambda_n$ are integers, then the term

$$t^{\lambda_1} + \cdots + t^{\lambda_n} \in Z[t,t^{-1}]$$

is viewed as the complex n dimensional representation of S^1 in which $t \in S^1$ is represented by the $n\times n$ diagonal matrix with eigenvalues t^{λ_i}

The ring $Z[t,t^{-1}]$ has two important sets of prime ideals. We denote these by P and $P_1 \subset P$. Their definition is this:

(1) P is the set of prime ideals defined by cyclotomic polynomials $\phi_m(t)$ as m ranges over all integers.

(2) P_1 is the set of prime ideals defined by cyclotomic polynomials $\phi_{p^r}(t)$ where p is a prime integer and r is arbitrary.

The basic ring over which all our algebra takes place is the ring R obtained from localizing $Z[t,t^{-1}]$ at the set P. This means that elements of R are fractions a/b where b is prime to all the prime ideals of P. Another way of saying this is that $b = b(t)$ does not vanish at any root of unity. If A is any $Z[t,t^{-1}]$ module set

$$A_P = A \underset{Z[t,t^{-1}]}{\otimes} R$$

and

$$A_{P_1} = A \underset{Z[t,t^{-1}]}{\otimes} R_{P_1}.$$

Here R_{P_1} denotes R localized at P_1. Another description for R_{P_1} is that it consists of fractions

$$^{a}/_{b} \quad a,b \in Z[t,t^{-1}]$$

and b is prime to all the prime ideals of P_1. Observe that R is a principle ideal domain. Denote its field of fractions by F.

The only prime ideals of R are those in P. Each $p \in P$ defines a valuation on R and a norm $\|\ \|_p$ on F written $x \longrightarrow \|x\|_p$ $X \in F$. In order to be explicit, if $^{a}/_{b} \in F$ and $p = (\phi_m(t))$ then $\|{}^{a}/_{b}\|_p = \phi_m(t)^k$ where

$$\phi_m(t)^{\alpha} \| a, \quad \phi_m(t)^{\beta} \| b \quad \text{and} \quad k = \alpha - \beta.$$

For now Θ will denote a commutative R algebra which is free of finite rank over R. Then there are two cannonical homomorphisms from Θ to R

(3) $\det_{\Theta}: \Theta \longrightarrow R$. Each element $\theta \in \Theta$ defines an R linear transformation from Θ to itself, namely left multiplication. Choosing an R base for Θ represents this transformation as a matrix over R. The determinant of this transformation is called $\det_{\Theta}(\theta)$. Thus $\det_{\Theta}$ is a multiplicative homomorphism.

(4) $\mathrm{tr}_{\Theta}: \Theta \longrightarrow R$. As in (3), view an element $\theta \in \Theta$ as an R linear transformation. Its trace is called $\mathrm{tr}_{\Theta}(\theta)$. Thus tr_{Θ} is an R linear homomorphism.

Assumption C There is an element $Id^{\Theta} \in Hom_R(\Theta, R)$ such that the R valued bilinear form on Θ defined by

$$\langle\theta_1,\theta_2\rangle_{\Theta} = Id^{\Theta}(\theta_1\cdot\theta_2)$$

is non-degenerate. If $\Theta = \Pi R$ is the product of copies of R, we require Id^{Θ} to be the trace homomorphism tr_{Θ}.

Remark: The bilinear form $\langle\ \rangle_{\Theta}$ is non-degenerate iff the associated homomorphism $\Phi(\langle\ \rangle_{\Theta}) : \Theta \longrightarrow Hom_R(\Theta, R)$ is an isomorphism of R modules. Here

$$\Phi(\langle\ \rangle_{\Theta})(\theta_1)[\theta_2] = \langle\theta_1,\theta_2\rangle.$$

Remark: More generally we want Θ to be an algebra graded by the integers mods 2. So $\Theta = \Theta_o \oplus \Theta_1$ and $\Theta_i\,\Theta_j \subset \Theta_{i+j}$ where $i+j$ is taken mod 2. Moreover, if $\theta \in \Theta_i$ and $\theta' \in \Theta_j$ $\theta\cdot\theta' = (-1)^{i\cdot j}\theta'\cdot\theta$. We reserve this remark for meditation after the contents of the simpler situation have been digested.

For us an R order in Θ will denote a subalgebra Λ of Θ such that $\Lambda \underset{R}{\otimes} F = \Theta \underset{R}{\otimes} F$. The inclusion of Λ in Θ is denoted by i_Λ.

Let $\Lambda \subset \Theta$ be a fixed order.

Assumption A: There is an element $\alpha_\Lambda \in \Lambda$ such that $i_\Lambda(\alpha_\Lambda)$ is a unit of $\Theta \otimes_R F$ and the F valued homomorphism

$$\lambda \longrightarrow Id^{\Theta} \otimes 1_F \left({}^{i_\Lambda(\lambda)}\!\big/_{i_\Lambda(\alpha_\Lambda)} \right)$$

is actually R valued. Call this homomorphism $Id^\Lambda : \Lambda \longrightarrow R$ and write

$$Id^{\Lambda}(\lambda) = Id^{\Theta}\left({}^{i_{\Lambda}(\lambda)}/_{i_{\Lambda}(\alpha_{\Lambda})}\right)$$

Assumption B: <u>The</u> R <u>valued bilinear form on</u> Λ <u>defined by</u> $\langle\lambda_1,\lambda_2\rangle_{\Lambda} = Id^{\Lambda}(\lambda_1\cdot\lambda_2)$ is <u>non-degenerate</u>.

Now suppose Γ is a second R order in Θ with distinguished element $\alpha_{\Gamma} \in \Gamma$ which satisfies the above two assumptions. In addition, we make the

Assumption D: <u>$\Lambda \subset \Gamma \subset \Theta$</u>.

Let i_{Γ} denote the inclusion of Γ in Θ and f^* the inclusion of Λ in Γ.

Because the bilinear forms on Λ, Γ, and Θ are non-degenerate, we have <u>induction homomorphisms</u>

$$(i_{\Lambda})_*: \Theta \longrightarrow \Lambda \;; \qquad \langle (i_{\Lambda})_*(\theta),\lambda\rangle_{\Lambda} = \langle\theta, i_{\Lambda}(\lambda)\rangle_{\Theta}$$

$$(i_{\Gamma})_* : \Theta \longrightarrow \Gamma \qquad \langle (i_{\Gamma})_*(\theta),\gamma\rangle_{\Gamma} = \langle\theta, i_{\Gamma}(\gamma)\rangle_{\Theta}$$

$$f_* : \Gamma \longrightarrow \Lambda \qquad \langle f_*(\gamma),\lambda\rangle_{\Gamma} = \langle\gamma, f^*(\lambda)\rangle_{\Gamma}$$

These induction homomorphisms satisfy

Proposition 4.

$$(i_{\Lambda})_*(1) = \alpha_{\Lambda}, \quad i_{\Lambda}(i_{\Lambda})_*(\theta) = [i_{\Lambda}(i_{\Lambda})_*(1)]\cdot\theta$$

$$(i_{\Gamma})_*(1) = \alpha_{\Gamma}, \quad i_{\Gamma}(i_{\Gamma})_*(\theta) = [i_{\Gamma}(i_{\Gamma})_*(1)]\cdot\theta$$

$$f^*f_*(\gamma) = [f^*f_*(1)]\cdot\gamma \qquad \gamma \in \Gamma$$

$$f_*(f^*(\lambda)\gamma) = \lambda f_*(\gamma) \qquad \lambda \in \Lambda, \quad \gamma \in \Gamma$$

Proof: We show that $f^*f_*(\gamma) = f^*f_*(1)\cdot\gamma$. Since i_Γ and i_Λ are algebra monomorphisms and since $i_\Gamma f^* = i_\Lambda$, it sufficies to show that

$$i_\Lambda f_*(\gamma) = {i_\Lambda(\alpha_\Lambda)}/{i_\Gamma(\alpha_\Gamma)} \cdot i_\Gamma(\gamma).$$

In particular, this implies $i_\Lambda f_*(1) = i_\Lambda(\alpha_\Lambda)/i_\Gamma(\alpha_\Gamma)$. Let $\lambda \in \Lambda$, $\gamma \in \Gamma$ then

$$<i_\Lambda f_*(\gamma), i_\Lambda(\lambda)/i_\Lambda(\alpha_\Lambda)>_{\Theta} = <f_*(\gamma),\lambda>_\Lambda = <\gamma, f^*(\lambda)>_\Gamma$$

$$= <i_\Gamma(\gamma),\ i_\Gamma f^*(\lambda)/i_\Gamma(\alpha_\Gamma)>_{\Theta} =$$

$$<i_\Lambda(\alpha_\Lambda)/i_\Gamma(\alpha_\Gamma) \cdot i_\Gamma(\gamma),\ i_\Lambda(\lambda)/i_\Lambda(\alpha_\Lambda)>_{\Theta}$$

Since the bilinear form on $\Theta \otimes_R F$ is non-degenerate, $i_\Lambda f_*(\gamma) = i_\Lambda(\alpha_\Lambda)/i_\Gamma(\alpha_\Gamma) \cdot i_\Gamma(\gamma)$ as desired.

Corollary 5. <u>The elements</u> $f_*(1) \in \Lambda$, $\alpha_\Lambda = (i_\Lambda)_*(1) \in \Lambda$ <u>and</u> $\alpha_\Gamma = (i_\Gamma)(1) \in \Gamma$ <u>are related by</u>

$$i_\Lambda f_*(1) = i_\Lambda(\alpha_\Lambda)/i_\Gamma(\alpha_\Gamma) \in \Theta.$$

The particular invariants of this algebraic situation which interest us are $f_*(1)$, $(i_\Lambda)_*(1) = \alpha_\Lambda$ and $(i_\Gamma)_*(1) = \alpha_\Gamma$. What can we say about these elements? Let $U(R)$ denote the multiplicative group of units of R and R^2 the subset of squares of R. Then we have

Proposition 6. <u>Let</u> $\det_\Gamma : \Gamma \longrightarrow R$ <u>and</u> $\det_\Theta : \Theta \longrightarrow R$ <u>be the determinant homomorphism. Then</u>

(i) $\det_\Gamma f^*f_*(1) \in U(R)\cdot R^2$

(ii) $\det_{\Theta} i_\Lambda(\alpha_\Lambda) \in U(R)\cdot R^2$

(iii) $\det_{\Theta} i_\Gamma(\alpha_\Gamma) \in U(R)\cdot R^2$

Proof: Take the case (i), $1 \in \Gamma$. If we take an R base for Λ and for Γ, then f^* is represented by a square matrix over R. The homomorphism f_* is the composition of these homomorphisms:

$$\Gamma \xrightarrow[\cong]{\Phi(\langle\ \rangle_\Gamma)} \mathrm{Hom}_R(\Gamma,R) \xrightarrow{f^{**}} \mathrm{Hom}_R(\Lambda,R) \underset{\cong}{\overset{\Phi(\langle\ \rangle_\Lambda)^{-1}}{\rightleftarrows}} \Lambda$$

where f^{**} is $\mathrm{Hom}_R(f^*,R)$. If we take a dual base for Γ and a dual base for Λ we obtain isomorphisms

$$j_\Gamma : \Gamma \longrightarrow \mathrm{Hom}_R(\Gamma,R) \quad\text{and}\quad j_\Lambda : \Lambda \longrightarrow \mathrm{Hom}_R(\Lambda,R)$$

and the matrix representing $j_\Lambda^{-1}\, f^{**}\, j_\Gamma$ is the transpose of f^*. There are isomorphisms $P_\Gamma : \Gamma \longrightarrow \Gamma$ and $P_\Lambda : \Lambda \longrightarrow \Lambda$ such that

$$j_\Gamma\, P_\Gamma = \Phi(\langle\ \rangle_\Gamma), \qquad j_\Lambda\, P_\Lambda = \Phi(\langle\ \rangle_\Lambda).$$

Thus $f_* = \Phi(\langle\ \rangle_\Lambda)^{-1} f^{**} \Phi(\langle\ \rangle_\Gamma) = P_\Lambda^{-1}\, j_\Lambda^{-1}\, f^{**}\, j_\Gamma\, P_\Gamma$; so

$$\det f_* = (\det P_\Lambda)^{-1}\cdot \det(j_\Lambda^{-1} f^{**} j_\Gamma)\cdot \det(P_\Gamma) = (\det P_\Lambda)^{-1} \det P_\Gamma \det f^*$$

because the determinant of the transpose of a matrix is the determinant of the matrix.

From Proposition 4 we have

$$f^*f_*(\gamma) = f^*f_*(1)\cdot\gamma \quad \text{for } \gamma \in \Gamma.$$

This means that

$$\det f^* \det f_* = \det_\Gamma f^*f_*(1).$$

But $\det f^* \det f_* = (\det P_\Lambda)^{-1} \det P_\Gamma \cdot (\det f^*)^2$ from above. Since P_Λ and P_Γ are isomorphisms $(\det P_\Lambda)^{-1}$ and $\det P_\Gamma$ are in $U(R)$.

Corollary 7. <u>The elements</u> $\det_\Lambda (f_*(1))$, $\det_\Lambda(\alpha_\Lambda)$ <u>and</u> $\det_\Gamma(\alpha_\Gamma)$ <u>are in</u> $U(R)\cdot R^2$.

Proof: This follows from the fact that $\det_\Gamma \circ f^* = \det_\Lambda$, $\det_\Theta \circ i_\Lambda = \det_\Lambda$ and $\det_\Theta \circ i_\Gamma = \det_\Gamma$.

Remark: Actually we have proved a stronger statement which is useful to have. Here is an appealing way to state it: Since R is a principle ideal domain, any finitely generated R module M is a direct sum of cyclic modules, say

$$M = {}^{R}/_{(P_1)} \oplus \cdots \oplus {}^{R}/_{(P_n)}$$

where (P_i) denotes the principle ideal generated by $P_i \in R$.

<u>Definition</u>: The product ideal $(P_1 \cdot P_2 \cdots P_n)$ is called the order of M, written $\mathrm{ord}_R(M)$.

Corollary 7. <u>Let</u> A_1, A_2 <u>be two</u> R <u>algebras which are free</u> R <u>modules of the same rank: Let</u> $i: A_1 \longrightarrow A_2$ <u>be an inclusion of algebras. Let</u> $\mathrm{Id}^{A_2} \in \mathrm{Hom}_R(A_2, R)$, $a_1 \in A_1$, <u>and suppose the bilinear forms</u> $\langle\ \rangle_{A_1}$ <u>and</u> $\langle\ \rangle_{A_2}$ <u>defined by</u>

$$\langle x,y\rangle_{A_1} = \mathrm{Id}^{A_2} \otimes 1_F\left(\frac{i(x\cdot y)}{i(a_1)}\right) \quad , \quad x,y \in A_1$$

$$\langle z,w\rangle_{A_2} = \mathrm{Id}^{A_2}(z\cdot w), \quad z,w \in A_2$$

<u>are</u> R <u>valued and non-degenerate. Let</u> $i_*: A_2 \longrightarrow A_1$ <u>be the induction homomorphism. Then</u> $i_*(1) = a_1 \in A_1$ <u>and the</u> R <u>ideals</u> $(\det_{A_1} i_*(1))$ <u>and</u> $(\mathrm{ord}_R({}^{A_2}/_{A_1}))^2$ <u>are equal</u>.

For reasons which become apparent in the next section, we impose this additional condition on the orders Λ & Γ with $\Lambda \subset \Gamma$:

Assumption E: $\Lambda_{P_1} = \Gamma_{P_1}$ i.e., $f^*_{P_1}$ is an isomorphism.

Then

Corollary 8. $f_*(1)$ <u>is a unit of</u> Λ_{P_1}.

Proof: From the equation

$f^*f_*(\gamma) = f^*f_*(1)\cdot\gamma$ and the fact that $f^*_{P_1}$ is an isomorphism (hence $(f_*)_{P_1}$ is an isomorphism), we conclude that $f^*f_*(1)$ is a unit of Γ_{P_1}. Since f^* is an algebra morphism, $f_*(1)$ is a unit of Λ_{P_1}.

Corollary 9. <u>For any algebra homomorphism</u> $r: \Theta \longrightarrow R$, $\|rf_*(1)\|_p = 1$ <u>for all</u> $p \in P_1$.

Proof: $rf_*(1)$ is a unit of R_p.

An especially interesting case to study for applications occurs when $\Theta = \Pi_{i=1}^n R$ is the product of n copies of R. Here we can even enrich the structure already at hand.

Assumption F: <u>The element</u> $\alpha_\Lambda \in \Lambda$ (also $\alpha_\Gamma \in \Gamma$) <u>has this form: For some integer</u> m, <u>called the dimension of</u> α_Λ, <u>there are</u> n, m <u>dimensional complex representations</u> $\Lambda_1, \Lambda_2, \cdots \Lambda_n$ of S^1 (<u>respectively</u> $\Gamma_1, \Gamma_2, \cdots \Gamma_n$) <u>and</u> n <u>integers</u> μ_i (<u>respectively</u> ν_i) <u>such that</u>

$$i_\Lambda(\alpha_\Lambda)_i = t^{\mu_i}\lambda_{-1}(\Lambda_i) \in R \qquad i = 1,2, \cdots n$$

$$i_\Gamma(\alpha_{\Gamma_i}) = t^{\nu_i}\lambda_{-1}(\Gamma_i) \in R \qquad i = 1,2, \cdots n$$

Here $\lambda_{-1}(M) = \sum(-1)^i\lambda^i(M)$ is the alternating sum of the exterior powers of the complex S^1 representation M. It is an element of $R(S^1) \subset R$.

There is an abundant set of algebra endomorphisms ψ^k $k = 0,1, \cdots n \cdots$ of R. These are the Adams operations. If $^a/_b \in R$ then

$$\psi^k(^a/_b) = {}^{a(t^k)}/_{b(t^k)} \in R$$

for $a = a(t)$, $b = b(t) \in R(S^1) = Z[t,t^{-1}]$.

These operations define Algebra endomorphisms ψ^k of Θ. Precisely if $\theta = (\theta_1, \theta_2, \cdots \theta_n) \in \Theta$ $\theta_i \in R$ $\psi^k(\theta) = (\psi^k(\theta_1), \cdots \psi^k(\theta_n))$.

Assumption H: The orders Λ, Γ, Θ are closed under the Adams operations.

With all these assumptions on the orders Λ, Γ, Θ and $\alpha_\Gamma \in \Gamma$ we have reason to suspect that the answer to the following basic question is yes: Let $\varepsilon_\Lambda : \Lambda \longrightarrow \Lambda \otimes_R Q$ denote the obvious surjection. Here Q is made an R module by $^a/_b \longrightarrow {}^{a(1)}/_{b(1)} \in Q$.

<u>Basic Question</u>: <u>Is $\varepsilon_\Lambda f_*(1) = 1$?</u>

For the importance of this question, see §2 Theorem 2 and the subsequence discussion.

In the last section we exhibit examples where $f_*(1) \neq 1 \in \Lambda$ but $\varepsilon_\Lambda f_*(1) \in \Lambda \otimes_R Q$ is 1.

§2. Geometric Situation

Definition: Let G be a compact Lie group. A smooth G manifold Y consists of a pair $(|Y|, \chi)$ where $|Y|$ is a smooth manifold and χ is a representation of G in the group $\mathrm{Diff}\,(|Y|)$ of diffeomorphisms of $|Y|$ such that the map $G \times |Y| \xrightarrow{\mu} |Y|$ defined by

$$\mu(g,y) = \chi(g)[y] \qquad g \in G, y \in |Y|$$

is smooth. For brevity we write $y \in Y$ and $\chi(g)[y] = gy$.

If X is a G manifold also, a map $f:X \longrightarrow Y$ must satisfy

$$f(gx) = gf(x) \quad \text{for all} \quad x \in X.$$

The question which motivates the geometrical study is this:

Q.1 <u>If</u> M <u>is a smooth manifold on which</u> G <u>acts effectively and if</u> N <u>is homotopy equivalent to</u> M, <u>is there an effective</u> G <u>action on</u> N? We say G acts effectively on M if there is a G manifold Y with $|Y| = M$ and for every $g \in G$, $g \neq 1$, there is a $y \in Y$ with $gy \neq y$.

This question is concerned with the relationship between the set of differential structures on a fixed homotopy type and the diffeomorphism groups of the various differential structures. It reflects the classification scheme of smooth manifolds, namely by fixing a homotopy type and describing all smooth manifolds within the given homotopy type.

A fundamental question which is related to Q.1 and is of interest in its own right is:

Q.2 <u>If</u> G <u>acts effectively on a smooth manifold, what are the relations among the representations of</u> G <u>on the tangent spaces at the points fixed by</u> G <u>and the global invariants of the manifold, eg. its Pontrjagin classes and its cohomology</u>?

I now indicate the program for dealing with Q.1 and the bearing of Q.2.

We introduce the set $S_G(Y)$ attached to the smooth closed G manifold Y. It consists of equivalence classes of pairs (X,f) where X is a smooth closed G manifold and $f:X\longrightarrow Y$ is a G map such that

(1) $|f| : |X|\longrightarrow|Y|$ is a homotopy equivalence. Here $|f|$ means the underlying map to f obtained by neglecting its relation to G.

(2) $|f^G| : |X^G|\longrightarrow|Y^G|$ is a homotopy equivalence between the G fixed point sets.

Two elements (X_i, f_i) $i = 0,1$ are equivalent if there is a G map $\phi : X_0\longrightarrow X_1$ such that

(1') ϕ is a G homotopy equivalence.

(2') $f_1\phi$ is G homotopic to f_0. The equivalence class of (X,f) is denoted by $[X,f]$. The element $[Y, \text{identity}] \in S_G(Y)$ is called the trivial element.

The set $S_G(Y)$ is introduced for the purpose of studying Q.1.

The principle algebraic tool for studying this geometric situation is the cohomology theory on the category of G spaces defined by equivariant complex K theory, $K^*_G(\ )$. If X is a compact G manifold, $K^*_G(X)$ is an $R(G)$ algebra with unit. Here $R(G)$ is the complex representation ring of G. [1]

If η is a complex G line bundle over X, then

$$\psi^k(\eta) = \eta^k = \eta \underset{C}{\otimes} \eta \underset{C}{\otimes} \cdots \underset{C}{\otimes} \eta$$

is another complex G line bundle over X. In fact there are algebra endomorphisms ψ^k of $K^*_G(X)$ for $k = 0,1 \cdots$ which extend the k power operation on complex G line bundles over X. These are the Adams operations. As a special case take $G = S^1$, the group of complex numbers of norm 1 and X a point P with trivial action of S^1.

Then $K^*_{S^1}(P) = R(S^1)$ is the ring $Z[t,t^{-1}]$ and if $a = a(t) \in Z[t,t^{-1}]$, $\psi^k(a) = a(t^k)$.

Henceforth we deal with the case $G=S^1$. Recall that $R = R(S^1)_P$.

If $[X,f] \in S_{S^1}(Y)$, we set

(i) $\Lambda = K^*_{S^1}(Y)_P/_{T_Y}$

(ii) $\Gamma = K^*_{S^1}(X)_P/_{T_X}$

(iii) $\Theta = K^*_{S^1}(X^{S^1})_P \cong K^*_{S^1}(Y^{S^1})_P$

See condition 2 above for the isomorphism (iii). Here T_Y and T_X are the R torsion subgroups of the indicated algebras. There is no R torsion in $K^*_{S^1}(X^{S^1})_P$ because

$$K^*_{S^1}(X^{S^1}) = K^*(X^{S^1}) \otimes_Z R(S^1)_P \quad \text{and} \quad R(S^1)_P$$

contains the rationals.

We have a commutative diagram of R algebra morphisms

$$\begin{array}{ccc} \Lambda & \xrightarrow{f^*} & \Gamma \\ & {\scriptstyle i_\Lambda} \searrow \quad \swarrow {\scriptstyle i_\Gamma} & \\ & \Theta & \end{array}$$

It follows from (1) the Localization and Completion Theorems [2] and [3] of Atiyah-Segal, and some elementary algebra that

(3) i_Λ, i_Γ, f^* <u>are monomorphisms and induce isomorphisms over</u> F. <u>In particular, if</u> X^{S^1} <u>consists of isolated fixed points</u>

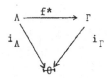

$\Theta = \Pi_{i=1}^{n} R$ is the product of copies of R one for each point of X^{S^1}.
Θ is an R order in $\Theta \otimes_R F$ and $\Lambda \subset \Gamma \subset \Theta$ are R orders in $\Theta \otimes_R F$.

In general, the fixed point set X^{S^1} consists of several components. In each component $X_i^{S^1}$ select a point p_i and let $X_0^{S^1}$ denote the collection of points p_i. In the case X^{S^1} consists of isolated points $X^{S^1} = X_0^{S^1}$. Anyway

$$\Theta_0 = K^*_{S^1}(X_0^{S^1})_P = \prod_{i=1}^{d} R$$

where d is the number of components of X^{S^1}. A point x of Θ_0 has coordinates $x_{p_i} \in R$ labeled by the points $p_i \in X_0^{S^1}$ and $x_{p_i} = x_{p_i}(t)$ is a rational function of t.

The S^1 tangent bundle of the smooth closed S^1 manifold X is denoted by TX. The Atiyah-Singer index homomorphism from $K^*_{S^1}(TX)$ to $R(S^1)$ induces an R homomorphism

$$Id^X: K^*_{S^1}(TX)_P \longrightarrow R$$

When the smooth manifold $|X|$ is a spin^{c}manifold we have a Thom isomorphism

$$\psi^X: K^*_{S^1}(X)_P \longrightarrow K^*_{S^1}(TX)_P \quad [7]$$

and the composition $Id^X \circ \psi^X$ induces an R homomorphism Id^Γ

$$Id^\Gamma: \Gamma \longrightarrow R$$

Moreover, we have this

Theorem 1 [5]: The R valued bilinear form on Γ defined by

$$<x,y>_\Gamma = Id^\Gamma(x \cdot y) \qquad x,y \in \Gamma$$

is non-degenerate. Here

$$\Gamma = K^*_{S^1}(X)_P/\text{Torsion}.$$

In order to proceed in the spirit of the first section we need to show that the homomorphism Id^Γ has the form

$$\mathrm{Id}^\Gamma(\gamma) = \mathrm{Id}^{\Theta}\left[\frac{i_\Gamma(\gamma)}{i_\Gamma(\alpha_\Gamma)}\right], \quad \gamma \in \Gamma,$$

for some $\alpha_\Gamma \in \Gamma$.

The fact that $|X|$ is a spin ^{c} manifold means by definition that its tangent bundle $T|X|$ has a spin ^{c} structure. The normal bundle $N = NX$ is a <u>complex</u> S^1 bundle over X^{S^1}. Its underlying vector bundle $|N|$ obtained by forgetting the S^1 action, is a complex vector bundle, so it has a spin ^{c} structure.

Since $T|X^{S^1}| \oplus |N| = T|X|\,|_{|X^{S^1}|}$ and since $T|X|\,|_{|X^{S^1}|}$ and $|N|$ have spin ^{c} structures, $T|X^{S^1}|$ has a spin ^{c} structure i.e., $|X^{S^1}|$ is a spin ^{c} manifold and we have a Thom isomorphism

$$\psi^{X^{S^1}} : K^*_{S^1}(X^{S^1})_P \longrightarrow K^*_{S^1}(TX^{S^1})_P$$

It follows from [2] that the index homomorphism Id^X is given by

$$\mathrm{Id}^X(x) = \mathrm{Id}^{X^{S^1}}\left((Ti)^*(x)/_{\lambda_{-1}(N\otimes C)}\right)$$

where $Ti\colon TX^{S^1} \longrightarrow TX$ is the inclusion. It follows from [7] that

$$(\psi^{X^{S^1}})^{-1}(Ti)^*\,\psi^X(u) = i^*(\mu)\cdot\delta(N)$$

where $i\colon X^{S^1} \longrightarrow X$ is the inclusion and $\delta(N) = [\Delta_+(N) - \Delta_-(N)]$ are the differences of the two complex vector bundles over X^{S^1} assigned to the spin ^{c} structure on N by the half spin representations

Δ_+ and Δ_- of the group spin^c. In particular if N^* denotes the complex conjugate of the S^1 bundle N, then

$\delta(N)\delta(N^*) = \lambda_{-1}(N \otimes C) \in K^*_{S^1}(X^{S^1})_P = \Theta$. Moreover, there is a unit

$\mu \in \Theta$ such that

$$\delta(N^*) = \mu \cdot \lambda_{-1}(N) \in \Theta.$$

In view of this information, we can write for $\gamma \in \Gamma$

$$\mathrm{Id}^\Gamma(\gamma) = \mathrm{Id}^X \circ \psi^X(\gamma) =$$

$$\mathrm{Id}^{X^{S^1}} \left[\frac{(Ti)^* \psi^X(\gamma)}{\lambda_{-1}(N \otimes C)}\right] =$$

$$\mathrm{Id}^{X^{S^1}} \circ \psi^{X^{S^1}} \circ \{(\psi^{X^{S^1}})^{-1} (Ti)^* \psi^X\}\left[\frac{\gamma}{\lambda_{-1}(N \otimes C)}\right]$$

$$= \mathrm{Id}^{X^{S^1}} \circ \psi^{X^{S^1}} \left[\frac{i^*(\gamma) \cdot \delta(N)}{\lambda_{-1}(N \otimes C)}\right] =$$

$$\mathrm{Id}^{X^{S^1}} \circ \psi^{X^{S^1}} \left[\frac{i^*(\gamma)}{\delta(N^*)}\right] = \mathrm{Id}^{\Theta}\left(\frac{i_\Gamma(\gamma)}{i_\Gamma(\alpha_\Gamma)}\right)$$

where $i_\Gamma = i^*$, $\delta(N^*) = i_\Gamma(\alpha_\Gamma)$, and $\alpha_\Gamma = (i_\Gamma)_*(1) \in \Gamma$. Note that $(i_\Gamma)_*$ is defined because of Theorem 1.

Thus the algebras

$$\Gamma = K^*_{S^1}(X)_P \,/\, \mathrm{Torsion}$$

$\Theta = K^*_{S^1}(X^{S^1})_P$, the element $\alpha_\Gamma \in \Gamma$ which is defined by

$i_\Gamma(\alpha_\Gamma) = \delta(N^*) = \mu \cdot \lambda_{-1}(N) \in \Theta$ where μ is some unit of Θ, the homomorphisms

$$\mathrm{Id}^\Gamma = \mathrm{Id}^X \circ \psi^X \quad \text{and} \quad \mathrm{Id}^\Theta = \mathrm{Id}^{X^{S^1}} \circ \psi^{X^{S^1}}$$

and the bilinear forms $< \;>_\Gamma$ and $< \;>_{\Theta}$ satisfy Assumptions A and B of §1. Because of this, we have the algebraic consequences of §1. I shall point out the geometric implications of these algebraic consequences.

Suppose that $K^*_{S^1}(Y)$ has no $(t-1) = p$ torsion. Then the natural map from $K^*_{S^1}(Y)$ to $K^*(|Y|)$ induces a homomorphism

$$\varepsilon_\Lambda : \Lambda \longrightarrow \Lambda \otimes_R Q = K^*(|Y|) \otimes Q$$

Moreover, composing the Chern character isomorphism

$$ch: K^*(|Y|) \otimes_Z Q \longrightarrow H^*(|Y|), Q)$$

with ε_Λ gives

$$\phi_Y = ch \circ \varepsilon_\Lambda$$

a homomorphism of R modules. Here Q is a ring over R via the augmentation $\varepsilon : R \longrightarrow Q$ defined by

$$\varepsilon \left[{}^{a(t)}/_{b(t)} \right] = \left[{}^{a(1)}/_{b(1)} \right] \in Q.$$

Now suppose that $[X,f] \in S_{S^1}(Y)$, that $|Y|$ and hence $|X|$ are closed spin^c manifolds. Let $A(|X|)$ and $A(|Y|)$ denote the cohomology classes in $H^*(|X|,Q)$ and $H^*(|Y|,Q)$ associated to the tangent bundles of $|X|$ and $|Y|$ by the power series

$({}^{x}/_{2}/_{\sinh}\ {}^{x}/_{2})$ [7]; so $A(|X|)$ is a polynomial in the Pontrjagin classes of $|X|$. Since $|f|: |X| \longrightarrow |Y|$ is a homotopy equivalence, there is a map $g: |Y| \longrightarrow |X|$ which is a homotopy inverse of $|f|$. Since the term of degree zero of the classes $\hat{A}(|X|)$ and $\hat{A}(|Y|)$ is 1, they are units in their respective cohomologies and

$A(|Y|)/_{g^*A(|X|)}$ is a unit of $H^*(|Y|,Q)$.

Because of Theorem 1, there is an induction homomorphism $f_*: \Gamma \longrightarrow \Lambda$ and

Theorem 2. $\phi_Y\ f_*(1) = g^*A(|X|)/_{A(|Y|)} \in H^*(|Y|,Q)$. Thus if $\varepsilon_\Lambda\ f_*(1)=1$, then $\phi_Y\ f_*(1) = 1$ and $|f|^*$ preserves Pontrjagin classes because the class $A(|Y|)$ determines the Pontrjagin classes of $|Y|$.

This theorem illustrates the importance of understanding the algebraic properties of $f_*(1)$, the invariant of the bilinear forms $< >_\Lambda$ on Λ, $< >_\Gamma$ on Γ and the algebra map $f^*: \Lambda \longrightarrow \Gamma$.

Here are other consequences of Theorem 1. The normal bundle of X^{S^1} in X is a complex S^1 bundle $N = NX$ over X^{S^1} so the class

$$\lambda_{-1}(N) = \textstyle\sum_i (-1)^i \lambda^i(N) \in K^*_{S^1}(X^{S^1}))$$

defines an element denoted again by $\lambda_{-1}(N)$ in Θ. As Θ is a free R algebra we have the multiplicative homomorphism $\det_\Theta: \Theta \longrightarrow R$.

Theorem 3. The R ideals $(\det_\Theta \lambda_{-1}(N))$ and $(\mathrm{ord}_R\ {}^{\Theta}/_{\Gamma})^2$ are equal. In particular, $\det_\Theta(\lambda_{-1}(N)) \in U(R)\cdot R^2$.

Proof: There is a unit $\mu \in \Theta$ such that

$$i_\Gamma(i_\Gamma)_*(\theta) = \mu\cdot\lambda_{-1}(N)\cdot\theta \qquad \theta \in \Theta.$$

This follows from the fact already mentioned that $\alpha_\Gamma = (i_\Gamma)_*(1)$, $i_\Gamma(\alpha_\Gamma) = \delta(N^*)$ and $\delta(N^*) = \mu\cdot\lambda_{-1}(N)$ for some $\mu \in U(\Theta)$. From §1 Proposition 6 and Corollaries 7 & 8, $\det_\Theta(i_\Gamma(\alpha_\Gamma)) \in U(R)\cdot R^2$ and $(\det_\Theta(i_\Gamma(\alpha_\Gamma))) = (\mathrm{ord}_R\ {}^{\Theta}/_{\Gamma})^2$. Since $i_\Gamma(\alpha_\Gamma) = \mu\cdot\lambda_{-1}(N)$, $\mu \in U(\Theta)$, the result follows.

To understand this theorem, suppose $X^{S^1} = X_0^{S^1}$. Then for each $p_i \in X_0^{S^1}$, the restriction of N to p_i is just the restriction of TX to p_i, written TX_{p_i} and is a complex S^1 module so

$$TX_{pi} = t^{\lambda_1(p_i)} + \cdots + t^{\lambda_m(p_i)} \qquad 2m = \dim_R |X|.$$

Then $\det_{\Theta} \lambda_{-1}(N) = \prod_{p_i} \prod_{j=1}^{m} (1-t^{\lambda_j(p_i)})$ and for this to be in $U(R)\cdot R^2$, each integer $|\lambda_j(p_i)|$ (the absolute value of $\lambda_j(p_i)$) must occur an even number of times in the collection of $n\cdot m$ integers $\{|\lambda_j(p_i)|\}$ where n = number of $p_i \in X^{S^1}$. Compare [4].

This is a result which holds for any closed smooth S^1 manifold X which has a spin ^{c} structure on its underlying manifold $|X|$.

Returning to the case $[X,f] \in S_{S^1}(Y)$, suppose we wish to compare the S^1 representations NX_{p_i} with the representation $NY_{f(p_i)}$ $p_i \in X_0^{S^1}$. These representations are related to the induction homomorphism through the formula

(4) $i_\Lambda f_*(1) = \mu\cdot\lambda_{-1}(NY)/_{\lambda_{-1}(NX)} \in \Theta$ for some unit $\mu \in \Theta$ and the

p_i th coordinate of $i_\Lambda f_*(1)$ in Θ_0 is

(5) $i_\Lambda f_*(1)_{p_i} = \mu_{p_i}\cdot\lambda_{-1}(NY_{f(p)}/_{\lambda_{-1}(NX_{p_i})} \in R.$

Since $|f|$ is a homotopy equivalence, it follows from the completion theorem [3] and some elementary algebra, that

(6) $f^*_{p_1} : \Lambda_{p_1} \longrightarrow \Gamma_{p_1}$ is an isomorphism.

This implies that $f_*(1) \in \Lambda_{p_1}$ is a unit (Corollary 8, §1) which in turn implies that $i_\Lambda f_*(1)_{p_i}$ is a unit in R_{p_1}. But then

$$|| i_\Lambda f_*(1)_{p_i} ||_p = 1$$

for all $p \in P_1$ (Corollary 9, §1).

Corollary 4. <u>For each</u> $p_i \in X_0^{S^1}$ <u>and each prime ideal</u> p <u>of</u> P_1

$$\left\| \lambda_{-1}(NY_{f(p_i)}) / \lambda_{-1}(NX_{p_i}) \right\|_p = 1$$

Remark 1: If $\left\| \lambda_{-1}(NY_{f(p_i)}) / \lambda_{-1}(NX_{p_i}) \right\|_p = 1$ for all $p \in P$, then the underlying real representations of $NY_{f(p_i)}$ and NX_{p_i} are equivalent.

Remark 2: The content of Corollary 4, is that $\lambda_{-1}(NY_{f(p_i)}) / \lambda_{-1}(NX_{p_i})$ is in R and the only possible cyclotomic polynomials dividing this element are $\phi_m(t)$ where m is a <u>composite</u> integer.

Remark 3: Because of Remark 1, if $f_*(1)$ is a unit of Λ, the representations $NY_{f(p_i)}$ and NX_{p_i} are equivalent for all p_i.

For further remarks concerning Corollary 4, the reader may consult [7], Theorem 2.6, p. 139.

Let $[X,f] \in S_{S^1}(Y)$. We summarize the connections between

(i) the invariant $f_*(1) \in \Lambda = K^*_{S^1}(Y)_P / \text{Torsion}$

(ii) the differential structures of $|X|$ and $|Y|$

(iii) The complex S^1 modules NX_{p_i} and $NY_{f(p_i)}$, $p_i \in X_0^{S^1}$ defined by the action of S^1 on the normal bundle of X^{S^1} in X at p_i and on the normal bundle of Y^{S^1} in Y at $f(p_i)$.

We assume that $K^0_{S^1}(Y)$ has no $(t-1) = p$ torsion (e.g., if $H^{odd}(|Y|,Q) = 0$).

Roughly the connection lies in the fact that the homomorphism

$$\Lambda \xrightarrow{\varepsilon_\Lambda} \Lambda \underset{R}{\otimes} Q = K^*(|Y|)\otimes Q \xrightarrow{ch} H^*(|Y|,Q)$$

gives information about the differential structrues while the homomorphism

$$\Lambda \xrightarrow{i_\Lambda} \Theta$$

gives information about the S^1 modules $NX^{S^1}_{p_i}$ and $NY_{f(p_i)}$. Specifically,

$$(7) \qquad ch\ \varepsilon_\Lambda\ f_*(1) = g^* A(|X|)/A(|X|)$$

where $g: |Y| \longrightarrow |X|$ is a homotopy inverse of $|f|$.

$$(8) \qquad i_\Lambda f_*(1)_{p_i} = \frac{\lambda_{-1}(NY_{f(p_i)})}{\lambda_{-1}(NX_{p_i})} \in R$$

where NX_{p_i} and $NY_{f(p_i)}$ are the complex S^1 modules defined by the representation of S^1 in the normal fibers to the fixed set X^{S^1} at the points p_i and $f(p_i)$.

(9) The representations NX_{p_i} and $NY_{f(p_i)}$ are not arbitrary but satisfy in addition to (8) the condition

$$\| \lambda_{-1}(NY_{f(p_i)})/\lambda_{-1}(NX_{p_i}) \|_p = 1$$

for all $p \in P_1$.

Remark: Condition (7) is a very strong relation between $f_*(1)$ and the differential structure on $|X|$ and $|Y|$ because $A(|X|)$ and $A(|Y|)$ determine the Pontrjagin classes of $|X|$ and $|Y|$.

Let me close this section by describing a geometric example which illustrates both the geometry and the algebra.

Let p and q be relatively prime integers and M and N the following complex 2 dimensional representations of S^1.

$$M = t^1 + t^{pq} \quad \varepsilon \quad R(S^1) \quad = \quad Z[t,t^{-1}]$$

$$N = t^p + t^q \quad \varepsilon \quad R(S^1)$$

Let W be any complex representation of S^1 and set $w = \dim_C W$. Let $\Omega = \lambda(M)\otimes_C W$ [1] be the indicated $4w$ dimensional representation of S^1. This complex $4w$ dimensional representation of S^1 defines an action of S^1 on the space of complex lines in C^{4w} i.e., complex projective space of dimension $4w-1$. The resulting S^1 manifold we call $P(\Omega)$.

Theorem 5 [6]. <u>There is a non trivial element</u> $[X,f] \varepsilon S_{S^1}(P(\Omega))$ <u>with</u> $|X|$ <u>a real algebraic manifold diffeomorphic to</u> $|P(\Omega)|$, <u>complex projective</u> $4n-1$ <u>space, and the action of</u> S^1 <u>on</u> $|X|$ <u>defined by</u> X <u>is real algebraic. If</u>

$$\Lambda = K^*_{S^1}(P(\Omega))_P$$

$$\Gamma = K^*_{S^1}(X)_P$$

<u>Then</u> $f_*: \Gamma \longrightarrow \Lambda$ <u>is defined and</u>

$$f_*(1) = \left(\lambda_{-1}(M) \middle/ \lambda_{-1}(N)\right) \cdot 1_\Lambda = \frac{(1-t)(1-t^{pq})}{(1-t^p)(1-t^q)} \cdot 1_\Lambda$$

<u>where</u> 1_Λ <u>is the identity of</u> Λ. <u>In particular</u>,

$$\varepsilon_Y f_*(1) = 1 \varepsilon K^*(|P(\Omega)|)\otimes Q = \Lambda \otimes_R Q \quad \text{but} \quad f_*(1) \neq 1_\Lambda.$$

<u>If the representation</u> Ω <u>has no multiple eigenvalue</u>,

1. $\lambda(M)$ is the exterior algebra of M. It defines a complex representation of S^1.

$P(\Omega)^{S^1}$ consists of isolated fixed points and X^{S^1} also consists of isolated fixed points. So in this case $NX = TX|X^{S^1}$ and $NY = TY|P(\Omega)^{S^1}$ are the restrictions of the respective tangent bundles to the fixed point sets X^{S^1} and Y^{S^1}. Moreover, for any $p \in X^{S^1}$

$$\lambda_{-1}(NP(\Omega)_{f(p)})/\lambda_{-1}(NX_p) = \lambda_{-1}(M)/\lambda_{-1}(N) \in R$$

In particular

$$\left\| \lambda_{-1}(NP(\Omega)_{f(p)})/\lambda_{-1}(NX_p) \right\|_p = \left\| \lambda_{-1}(M)/\lambda_{-1}(N) \right\|_p = 1$$

for all primes of R not dividing

$$\frac{(1-t)(1-t^{pq})}{(1-t^p)(1-t^q)} .$$

Remark: The algebras Λ and Γ of this geometric example are discussed in the next section.

§3. Algebraic Examples

The reader will soon convince himself that if Λ, Θ are fixed R orders in the sense of §1 and $\alpha_\Lambda \in \Lambda$ is given such that all the assumptions of section 1 are satisfied, it is very difficult to find an R order Γ and element $\alpha_\Gamma \in \Gamma$ such that

$$\Lambda \subset \Gamma \subset \Theta, \qquad \Lambda_{P_1} = \Gamma_{P_1} \text{ (without } \Lambda = \Gamma)$$

and all the assumptions of §1 are satisfied for (Γ, α_Γ). None-the-less, such orders Γ do exist and are important as Theorem 5 of §2 shows.

The first example gives a non trivial case of such a Γ which arose from geometrical considerations. The second example, due to Kervaire, gives a candidate for an example Γ. We use the algebra of section 1 to show that Γ does not have a non-degenerate bilinear form $\langle\ \rangle_\Gamma$ of the type dictated by the assumptions of §1. On-the-other hand it does fulfill the other conditions. In particular, it is invariant under the Adams operations on Θ.

In these examples $\Theta = \Pi R$ will be the product of copies of R; also we can even work over $R(S^1) = Z[t,t^{-1}]$ instead of $R(S^1)_p$; so R will denote either $Z[t,t^{-1}]$ or $Z[t,t^{-1}]_p$ depending on the readers taste.

Example 1. We let $R = Z[t,t^{-1}]$ and $A = R[x,x^{-1}]$. Let M and N be 2 complex representations of S^1 of the same dimension r such that

$$(1) \quad \lambda_{-1}(M)/\lambda_{-1}(N) \in R$$

$$(2) \quad \varepsilon\left[\lambda_{-1}(M)/\lambda_{-1}(N)\right] = 1$$

where $\varepsilon: R \longrightarrow Z$ is the augementation. Let $\lambda(M)$ denote the total exterior algebra of M so

$$\lambda(M) = \lambda_+(M) \oplus \lambda_-(M)$$

$$\lambda_+(M) = \sum_i \lambda^{2i}(M)$$

$$\lambda_-(M) = \sum_i \lambda^{2i-1}(M)$$

Let W be any complex representation of S^1 such that if

$$\lambda(M) \otimes W = \sum_{i=1}^{n} t^{\lambda_i} \in R = Z[t,t^{-1}],$$

then the λ_i's are distinct integers. Set $\Omega_{\pm} = \lambda_{\pm}(M) \otimes W$, $\Omega = \lambda(M) \otimes W$. Then

$$\lambda_{-x}(\Omega) = \sum (-1)^i x^i \lambda^i(\Omega) \in A$$

and we let

$$\Lambda = A/(\lambda_{-x}(\Omega))$$

Let $\Theta(\Omega)$ denote the product of $n = \dim_c \Omega$ copies of R. The coordinates of a point $z \in \Theta = \Theta(\Omega)$ are indexed by the exponents λ_i defining Ω; so we write $z = (z_{\lambda_i})$ where $z_{\lambda_i} \in R$ is the coordinate of cooresponding to λ_i.

I claim that Λ is an R order in Θ. To see this define an R algebra homomorphism $a: \Lambda \longrightarrow \Theta$ by $a(x)_{\lambda_i} = t^{-\lambda_i}$. Observe that since

$$\Omega = \sum_{i=1}^{n} t^{\lambda_i}, \quad \lambda_{-x}(\Omega) = \prod_{i=1}^{n} (1-xt^{\lambda_i})$$

so

$$a(\lambda_{-x}(\Omega)) = 0$$

and a induces a monomorphism $i_\Lambda: \Lambda \longrightarrow \Theta$. We write $\Lambda \subset \Theta$.

The element $b_+ = x^{\frac{-n}{2}} \lambda_{-x}(\Omega_+)$ $(n \equiv 0(2^n)$ in Λ is divisible by

$$d = \lambda_{-1}(M)/_{\lambda_{-1}(N)} \in R.$$

in Θ. Let $\gamma_+ = \frac{1}{d} \cdot b_+ \in \Theta$.

Since $\dim_C \lambda_*(M) = 2^r$, $n = \dim_C \lambda(M) \otimes W = 2^r \cdot w$ where $w = \dim_C W$. It is easy to check that an R basis for Λ is

$$1, x, \cdots x^{2^{r-1}w-1}, b_+, b_+x \cdots b_+x^{2^{r-1}w-1}.$$

Let $\Gamma \subset \Theta$ be the R submodule generated by

$$1, x, \cdots x^{2^{r-1}w-1}, \gamma_+x \cdots \gamma_+x^{2^{r-1}w-1}.$$

Then $\Lambda \subset \Gamma \subset \Theta$ and $\Lambda_p = \Gamma_p$ for all prime ideals of R not dividing d.

The inclusion of Γ in Θ is denoted by i_Γ and the inclusion of Λ in Γ by f^*.

Let $\alpha_\Lambda \in \Lambda$ be defined by

$$i_\Lambda(\alpha_\Lambda)_{\lambda_i} = \prod_{j|j\neq i} (1 - t^{\lambda_j - \lambda_i})$$

and $\alpha_\Gamma \in \Gamma$ be defined by $i_\Gamma \alpha_\Gamma = {}^1/_d \, i_\Lambda(\alpha_\Lambda)$.

Let $Id^\Theta : \Theta \longrightarrow R$ be tr_Θ (See §1). Then the bilinear form $<\ >_\Lambda$ on Λ defined by

$$<x,y>_\Lambda = Id^\Theta\left(\frac{i_\Lambda(x \cdot y)}{i_\Lambda(\alpha_\Lambda)}\right)$$

is non-degenerate. Here x, y $x \cdot y \in \Lambda$.

Let the matrix of this form with respect to the basis

$$1, x \cdots x^{2^{r-1}w-1}, b_+, \; b_+x \cdots b_+x^{2^{r-1}w-1}$$

be

$$\Delta = \begin{pmatrix} A & C \\ C^t & B \end{pmatrix}.$$

I claim the bilinear form on Γ defined by

$$\langle u,v\rangle_\Gamma = \mathrm{Id}^{\Theta}\left(\frac{i_\Gamma(u)\cdot i_\Gamma(v)}{i_\Gamma(\alpha_\Gamma)}\right)$$

is non-singular. To see this, note that the matrix of $\langle\ \rangle_\Gamma$ with respect to the basis

$$1, x, \cdots x^{2^{r-1}w-1}, \gamma_+, \; \gamma_+x, \cdots\cdot \gamma_+ x^{2^{r-1}w-1}$$

is

$$\Delta^1 = \begin{pmatrix} dA & C \\ C^t & \frac{1}{d}B \end{pmatrix}$$

so $\det \Delta^1 = \det \Delta$ is a unit of R, i.e., $\langle\ \rangle_\Gamma$ is non-singular.

In the case $\dim_c M$ is 2, Γ is <u>an R order</u> in Θ. This is probably true for $\dim_c M$ arbitrary. If this is so, this construction provides us with a whole host of examples of <u>R orders</u> $\Lambda \subset \Gamma \subset \Theta$ with $\Gamma \neq \Lambda$ and

(1) All are closed under the Adams operations ψ^k.

(2) $\Lambda_p = \Gamma_p$ for all primes p of R not dividing

$$d = \lambda_{-1}(M) \big/ \lambda_{-1}(N)$$

(3) There are non-singular bilinear forms $\langle\ \rangle_\Lambda$ and $\langle\ \rangle_\Gamma$ associated with their algebras Λ and Γ.

(4) $f_*(1) = d\cdot 1 \in \Lambda$ and $\varepsilon_\Lambda f_*(1) = 1 \in \Lambda \otimes_R Z$.

Here in more detail is a special case of the preceding: The representations N, M and W are respectively

$$N = t^p \oplus t^q$$

$$M = t + t^{pq}$$

W = 1, the trivial 1-dimensional S^1 module. Then

$$d = \lambda_{-1}(M)/\lambda_{-1}(N) = \frac{(1-t)(1-t^{pq})}{(1-t^p)(1-t^q)} \in R$$

$$\Lambda = A/((1-x)(1-xt)(1-xt^{pq})(1-xt^{pq+1}))$$

$$i_\Lambda(x^{-1}) = (t^0,\ t^1,\ t^{pq},\ t^{pq+1}) \in \Theta$$

$$b_+ = (x^{-1}-1)(x^{-1}-t^{pq+1}) \in \Lambda$$

$$i_\Gamma(\gamma_+) = i_\Lambda(b_+)/_d = (0,\ -t(1-t)(1-t^{pq}),\ -t^{pq}(1-t)(1-t^{pq}),0)$$

The equation $\gamma_+^2 = -(1-t^p)(1-t^q)x^{-1}\gamma_+$ is easily checked and shows that the submodule Γ of Θ generated by $1, x^{-1}, \gamma_+, \gamma_+x^{-1}$ is a <u>subalgebra</u> of Θ, i.e., an R order. It is also closed under the Adams operations ψ^k.

The elements $\alpha_\Lambda \in \Lambda$ and $\alpha_\Gamma \in \Gamma$ are defined by

$$i_\Lambda(\alpha_\Lambda) = ((1-t)(1-t^{pq})(1-t^{pq+1}),\ (1-t^{-1})(1-t^{pq-1})(1-t^{pq}),$$
$$(1-t^{-pq})(1-t^{-pq+1})(1-t),\ (1-t^{-pq-1})(1-t^{-pq})(1-t^{-1}))$$

$$i_\Gamma(\alpha_\Gamma) = ((1-t^p)(1-t^q)(1-t^{pq+1}),\ -t^{-1}(1-t^p)(1-t^q)(1-t^{pq-1}),$$
$$-t^{-pq}(1-t^p)(1-t^q)(1-t),\ t^{pq+2}(1-t^p)(1-t^q)(1-t^{-pq-1}).$$

Example 2. This example is due to Kervaire.

Let $\Lambda = A/I$ where I is the ideal $((1-x)(1-xt^2)(1-xt^3)(1-xt^{13}))$ in A. Then Λ is an R order in $\Theta = R\times R\times R\times R$. To see this define an R algebra morphism $i_\Lambda\colon \Lambda \longrightarrow \Theta$ by

$$i_\Lambda(x^{-1}) = (1,\ t^2,\ t^3,\ t^{13}) \in \Theta.$$

In fact there is an R base $e_0, e_1, e_2, e_3 \in \Lambda$ defined by

$$i_\Lambda(e_0) = (1,\ 1,\ 1,\ 1)$$

$$i_\Lambda(e_1) = (0,\ t^2-1,\ t^3-1,\ t^{13}-1)$$

$$i_\Lambda(e_2) = (0,\ 0,\ (t^2-1)(t^3-t^2),\ (t^{13}-1)(t^{13}-t^2))$$

$$i_\Lambda(e_3) = (0,\ 0,\ 0,\ (t^{13}-1)(t^{13}-t^2)(t^{13}-t^3)).$$

Let $\Gamma \subset \Theta$ be the free R submodule of Θ with base f_i $i = 0, 1, 2, 3$ where $f_i = e_i$, $i = 0, 1, 2$, and $f_3 = {}^1/_{\phi_{10}} \cdot e_3$.

$$\phi_{10} = \phi_{10}(t) = \frac{(1-t)(1-t^{10})}{(1-t^2)(1-t^5)} \in Z[t,t^{-1}].$$

Then Γ is an order in Θ closed under the Adams operations ψ^k, $\Lambda \subset \Gamma$ and $\Lambda_{P_1} = \Gamma_{P_1}$.

Define $\alpha_\Lambda \in \Lambda$ by

$$i_\Lambda(\alpha_\Lambda) = \prod_{j}{}_{j\neq i} (1-t^{a_j-a_i}) \in R$$

for $i = 0, 1, 2, 3$ where the integers (a_0, a_1, a_2, a_3) are $(0, 2, 3, 13)$. Let $Id^\Theta\colon \Theta \longrightarrow R$ be the trace homomorphism of §1, i.e., $Id^\Theta = tr_\Theta$. Then the R valued bilinear form $< >_\Lambda$ of Λ defined by

$$<\lambda_1, \lambda_2>_\Lambda = Id^\Theta\left({}^{i_\Lambda(\lambda_1 \lambda_2)}/_{i_\Lambda(\alpha_\Lambda)}\right)$$

is non-degenerate.

Note that the dimension of α_Λ is 3. I claim that there is no $\alpha_\Gamma \in \Gamma$ of dimension 3 such that the bilinear form on Γ defined by α_Γ and the inclusion i_Γ of Γ in $\mathcal{H}$ is non-degenerate. Here is a proof:

The 4 complex 3 dimensional S^1 modules defining the dimension of α_Λ are

$$\Lambda_0 = t^2 + t^3 + t^{13}$$

$$\Lambda_1 = t^{-2} + t^1 + t^{11}$$

$$\Lambda_2 = t^{-3} + t^{-1} + t^{10}$$

$$\Lambda_3 = t^{-13} + t^{-11} + t^{10}$$

The associated integers μ_i are all 1. There must be 4 complex 3 dimensional S^1 modules defining the dimension of α_Γ:

$$\Gamma_0 = t^{x_{01}} + t^{x_{02}} + t^{x_{03}}$$

$$\Gamma_1 = t^{x_{11}} + t^{x_{12}} + t^{x_{13}}$$

$$\Gamma_2 = t^{x_{21}} + t^{x_{22}} + t^{x_{23}}$$

$$\Gamma_3 = t^{x_{31}} + t^{x_{32}} + t^{x_{33}}$$

where x_{ij}'s are integers. Moreover, there should exist integers μ_i $i = 0, 1, 2, 3$ such that

(a) $i_\Gamma(\alpha_\Gamma)_i = t^{\mu_i} \lambda_{-1}(\Gamma_i)$

(b) $i_\Lambda f_*(1)_i = t^{-\mu_i} \lambda_{-1}(\Lambda_i) \Big/ \lambda_{-1}(\Gamma_i) \quad \in R$

(c) $\det_\Lambda f_*(1) = t^{-\sum_{i=0}^{3} \mu_i} \prod_{i=0}^{3} \lambda_{-1}(\Lambda_i) \Big/ \lambda_{-1}(\Gamma_i)$

is a unit $\cdot(\det f^*)^2$; see §1, Proposition 6, Corollary 7 and Corollary 7'.

Clearly, $\det f^* = \phi_{10}$ so (c') $\det_\Lambda f_*(1) = \mu\cdot\phi_{10}^2$ for some unit μ of R. By inspection we see that there are no possible choices for integers x_{ij} which satisfy (a), (b), and (c').

Remark 1. In this example A could be either $R(S^1)[x,x^{-1}]$ or $R[x,x^{-1}]$.

Remark 2. The above argument shows that there is no smooth S^1 action on a smooth manifold M homotopy equivalent to complex projective 3 space defining a smooth S^1 manifold X with

$$K^*_{S^1}(X)_P \Big/ \text{Torsion} = \Gamma.$$

Perhaps there is a topological action with this property?

REFERENCES

[1] Atiyah, M. F., K-Theory, Benjamin 1967.

[2] Atiyah, M. F. and Segal, G., Index of elliptic operators, II, Ann. of Math., 87 (1968) 484-530.

[3] ___________, Equivariant K-theory and completion, J. of Diff. Geometry, 3, No. 1 (1969) 1-18.

[4] Kosniowski, C., Applications of the holomorphic Lefschez formula, Bull. London Math. Soc. (1969).

[5] Petrie, T., Lectures on S^1 manifolds, (to appear).

[6] _________, Real algebraic actions of S^1 on complex projective spaces, (to appear).

[7] _________, Smooth S^1 actions on homotopy complex projective spaces and related topics, Bull A.M.S., 78, No. 2 (1972) 105-153.

Vol. 215: P. Antonelli, D. Burghelea and P. J. Kahn, The Concordance-Homotopy Groups of Geometric Automorphism Groups. X, 140 pages. 1971. DM 16,–

Vol. 216: H. Maaß, Siegel's Modular Forms and Dirichlet Series. VII, 328 pages. 1971. DM 20,–

Vol. 217: T. J. Jech, Lectures in Set Theory with Particular Emphasis on the Method of Forcing. V, 137 pages. 1971. DM 16,–

Vol. 218: C. P. Schnorr, Zufälligkeit und Wahrscheinlichkeit. IV, 212 Seiten. 1971. DM 20,–

Vol. 219: N. L. Alling and N. Greenleaf, Foundations of the Theory of Klein Surfaces. IX, 117 pages. 1971. DM 16,–

Vol. 220: W. A. Coppel, Disconjugacy. V, 148 pages. 1971. DM 16,–

Vol. 221: P. Gabriel und F. Ulmer, Lokal präsentierbare Kategorien. V, 200 Seiten. 1971. DM 18,–

Vol. 222: C. Meghea, Compactification des Espaces Harmoniques. III, 108 pages. 1971. DM 16,–

Vol. 223: U. Felgner, Models of ZF-Set Theory. VI, 173 pages. 1971. DM 16,–

Vol. 224: Revêtements Etales et Groupe Fondamental. (SGA 1). Dirigé par A. Grothendieck XXII, 447 pages. 1971. DM 30,–

Vol. 225: Théorie des Intersections et Théorème de Riemann-Roch. (SGA 6). Dirigé par P. Berthelot, A. Grothendieck et L. Illusie. XII, 700 pages. 1971. DM 40,–

Vol. 226: Seminar on Potential Theory, II. Edited by H. Bauer. IV, 170 pages. 1971. DM 18,–

Vol. 227: H. L. Montgomery, Topics in Multiplicative Number Theory. IX, 178 pages. 1971. DM 18,–

Vol. 228: Conference on Applications of Numerical Analysis. Edited by J. Ll. Morris. X, 358 pages. 1971. DM 26,–

Vol. 229: J. Väisälä, Lectures on n-Dimensional Quasiconformal Mappings. XIV, 144 pages. 1971. DM 16,–

Vol. 230: L. Waelbroeck, Topological Vector Spaces and Algebras. VII, 158 pages. 1971. DM 16,–

Vol. 231: H. Reiter, L^1-Algebras and Segal Algebras. XI, 113 pages. 1971. DM 16,–

Vol. 232: T. H. Ganelius, Tauberian Remainder Theorems. VI, 75 pages. 1971. DM 16,–

Vol. 233: C. P. Tsokos and W. J. Padgett. Random Integral Equations with Applications to stochastic Systems. VII, 174 pages. 1971. DM 18,–

Vol. 234: A. Andreotti and W. Stoll. Analytic and Algebraic Dependence of Meromorphic Functions. III, 390 pages. 1971. DM 26,–

Vol. 235: Global Differentiable Dynamics. Edited by O. Hájek, A. J. Lohwater, and R. McCann. X, 140 pages. 1971. DM 16,–

Vol. 236: M. Barr, P. A. Grillet, and D. H. van Osdol. Exact Categories and Categories of Sheaves. VII, 239 pages. 1971. DM 20,–

Vol. 237: B. Stenström, Rings and Modules of Quotients. VII, 136 pages. 1971. DM 16,–

Vol. 238: Der kanonische Modul eines Cohen-Macaulay-Rings. Herausgegeben von Jürgen Herzog und Ernst Kunz. VI, 103 Seiten. 1971. DM 16,–

Vol. 239: L. Illusie, Complexe Cotangent et Déformations I. XV, 355 pages. 1971. DM 26,–

Vol. 240: A. Kerber, Representations of Permutation Groups I. VII, 192 pages. 1971. DM 18,–

Vol. 241: S. Kaneyuki, Homogeneous Bounded Domains and Siegel Domains. V, 89 pages. 1971. DM 16,–

Vol. 242: R. R. Coifman et G. Weiss, Analyse Harmonique Non-Commutative sur Certains Espaces. V, 160 pages. 1971. DM 16,–

Vol. 243: Japan-United States Seminar on Ordinary Differential and Functional Equations. Edited by M. Urabe. VIII, 332 pages. 1971. DM 26,–

Vol. 244: Séminaire Bourbaki – vol. 1970/71. Exposés 382–399. IV, 356 pages. 1971. DM 26,–

Vol. 245: D. E. Cohen, Groups of Cohomological Dimension One. V, 99 pages. 1972. DM 16,–

Vol. 246: Lectures on Rings and Modules. Tulane University Ring and Operator Theory Year, 1970–1971. Volume I. X, 661 pages. 1972. DM 40,–

Vol. 247: Lectures on Operator Algebras. Tulane University Ring and Operator Theory Year, 1970–1971. Volume II. XI, 786 pages. 1972. DM 40,–

Vol. 248: Lectures on the Applications of Sheaves to Ring Theory. Tulane University Ring and Operator Theory Year, 1970–1971. Volume III. VIII, 315 pages. 1971. DM 26,–

Vol. 249: Symposium on Algebraic Topology. Edited by P. J. Hilton. VII, 111 pages. 1971. DM 16,–

Vol. 250: B. Jónsson, Topics in Universal Algebra. VI, 220 pages. 1972. DM 20,–

Vol. 251: The Theory of Arithmetic Functions. Edited by A. A. Gioia and D. L. Goldsmith VI, 287 pages. 1972. DM 24,–

Vol. 252: D. A. Stone, Stratified Polyhedra. IX, 193 pages. 1972. DM 18,–

Vol. 253: V. Komkov, Optimal Control Theory for the Damping of Vibrations of Simple Elastic Systems. V, 240 pages. 1972. DM 20,–

Vol. 254: C. U. Jensen, Les Foncteurs Dérivés de $\varprojlim$ et leurs Applications en Théorie des Modules. V, 103 pages. 1972. DM 16,–

Vol. 255: Conference in Mathematical Logic – London '70. Edited by W. Hodges. VIII, 351 pages. 1972. DM 26,–

Vol. 256: C. A. Berenstein and M. A. Dostal, Analytically Uniform Spaces and their Applications to Convolution Equations. VII, 130 pages. 1972. DM 16,–

Vol. 257: R. B. Holmes, A Course on Optimization and Best Approximation. VIII, 233 pages. 1972. DM 20,–

Vol. 258: Séminaire de Probabilités VI. Edited by P. A. Meyer. VI, 253 pages. 1972. DM 22,–

Vol. 259: N. Moulis, Structures de Fredholm sur les Variétés Hilbertiennes. V, 123 pages. 1972. DM 16,–

Vol. 260: R. Godement and H. Jacquet, Zeta Functions of Simple Algebras. IX, 188 pages. 1972. DM 18,–

Vol. 261: A. Guichardet, Symmetric Hilbert Spaces and Related Topics. V, 197 pages. 1972. DM 18,–

Vol. 262: H. G. Zimmer, Computational Problems, Methods, and Results in Algebraic Number Theory. V, 103 pages. 1972. DM 16,–

Vol. 263: T. Parthasarathy, Selection Theorems and their Applications. VII, 101 pages. 1972. DM 16,–

Vol. 264: W. Messing, The Crystals Associated to Barsotti-Tate Groups: With Applications to Abelian Schemes. III, 190 pages. 1972. DM 18,–

Vol. 265: N. Saavedra Rivano, Catégories Tannakiennes. II, 418 pages. 1972. DM 26,–

Vol. 266: Conference on Harmonic Analysis. Edited by D. Gulick and R. L. Lipsman. VI, 323 pages. 1972. DM 24,–

Vol. 267: Numerische Lösung nichtlinearer partieller Differential- und Integro-Differentialgleichungen. Herausgegeben von R. Ansorge und W. Törnig. VI, 339 Seiten. 1972. DM 26,–

Vol. 268: C. G. Simader, On Dirichlet's Boundary Value Problem. IV, 238 pages. 1972. DM 20,–

Vol. 269: Théorie des Topos et Cohomologie Etale des Schémas. (SGA 4). Dirigé par M. Artin, A. Grothendieck et J. L. Verdier. XIX, 525 pages. 1972. DM 50,–

Vol. 270: Théorie des Topos et Cohomologie Etale des Schémas. Tome 2. (SGA 4). Dirigé par M. Artin, A. Grothendieck et J. L. Verdier. V, 418 pages. 1972. DM 50,–

Vol. 271: J. P. May, The Geometry of Iterated Loop Spaces. IX, 175 pages. 1972. DM 18,–

Vol. 272: K. R. Parthasarathy and K. Schmidt, Positive Definite Kernels, Continuous Tensor Products, and Central Limit Theorems of Probability Theory. VI, 107 pages. 1972. DM 16,–

Vol. 273: U. Seip, Kompakt erzeugte Vektorräume und Analysis. IX, 119 Seiten. 1972. DM 16,–

Vol. 274: Toposes, Algebraic Geometry and Logic. Edited by. F. W. Lawvere. VI, 189 pages. 1972. DM 18,–

Vol. 275: Séminaire Pierre Lelong (Analyse) Année 1970–1971. VI, 181 pages. 1972. DM 18,–

Vol. 276: A. Borel, Représentations de Groupes Localement Compacts. V, 98 pages. 1972. DM 16,–

Vol. 277: Séminaire Banach. Edité par C. Houzel. VII, 229 pages. 1972. DM 20,–